Second Order Parabolic Differential Equations

Second Order Parabolic Differential Equations

Gary M. Lieberman
Iowa State University

NEW JERSEY • LONDON • SINGAPORE • BEIJING • SHANGHAI • HONG KONG • TAIPEI • CHENNAI

Published by

World Scientific Publishing Co. Pte. Ltd.

5 Toh Tuck Link, Singapore 596224

USA office: 27 Warren Street, Suite 401-402, Hackensack, NJ 07601

UK office: 57 Shelton Street, Covent Garden, London WC2H 9HE

British Library Cataloguing-in-Publication Data
A catalogue record for this book is available from the British Library.

First published 1996
Reprinted 1998, 2005

SECOND ORDER PARABOLIC DIFFERENTIAL EQUATIONS

ISBN 981-02-2883-X

Printed in Singapore.

PREFACE

My goal in writing this book was to create a companion volume to "Elliptic Partial Differential Equations of Second Order" by David Gilbarg and Neil S. Trudinger. Like that book, this one is an essentially self-contained exposition of the theory of second order parabolic (instead of elliptic) partial differential equations of second order, with emphasis on the theory of certain initial-boundary value problems in bounded space-time domains. In addition to the Cauchy-Dirichlet problem, which is the parabolic analog of the Dirichlet problem, I also study oblique derivative problems. Preparatory material on such topics as functional analysis and harmonic analysis is included to make the book accessible to a large audience. (Dave and Neil, I hope I succeeded.)

This book would not be possible withour the help of many people. First and foremost are David Gilbarg, my Ph.D. advisor, who started me on the road to *a priori* estimates; Neil Trudinger, whose continued encouragment, collaboration, and invitations to the Centre for Mathematics and its Applications (in fact, several chapters of this book were written at the C. M. A.) kept me going; and Howard Levine, who showed me the glory of parabolic equations. Next, but just as important, are my family: my parents, Alvin and Tillie Lieberman, without whom I would not have been possible; and my wife, Linda Lewis Lieberman, and my step-children, Ben and Jenny Lewis, who joined me during the writing of this book. Without Linda, Ben, and Jenny, the project may have been completed more quickly, but definitely not more pleasantly. Without Olga Ladyzhenskaya and Nina Ural′tseva, the whole field of *a priori* estimates would be much smaller; I thank them for their pioneering work, continued advances, and constant inspiration. I also thank Cliff Bergman, Jan Nyhus, Ruth deBoer, and Mike Fletcher for their invaluable assistance with and education on $\mathcal{A}\mathcal{M}\mathcal{S}$-TEX and LATEX. Russell Brown and Wei Hu were my collaborators on the material in Section 6.1, and it was their prodding that led to a useful version of the results therein.

I also thank my editors at World Scientific over the years: J. G. Xu, Lam Poh Fong, and Chow Mun Zing.

There were many others who provided encouragement and advice. In particular, Sunčica Čanic, Emmanuele DiBenedetto, Nicola Garofalo, Nina Ivochkina, Nikolai Krylov, Paul Sacks, Mikhail Safonov, and Michael Smiley deserve special mention. My final thanks are to all those people who gave me advice through the years and who shared their work with me.

July, 1996
Ames, Iowa USA
Gary M. Lieberman

PREFACE TO REVISED EDITION

In this edition are several, relatively minor changes. I have tried to correct all the typographical errors from the first edition. (There is an errata sheet at www.public.iastate.edu/~lieb/book/errata.pdf.) I have rewritten Section 2 on the strong maximum principle to show more clearly its connection with the weak Harnack inequality. In addition, Chapter VII has been rewritten to take advantage of the recent work of Lihe Wang (based on ideas of Luis Caffarelli) concerning the proof of the Calderón-Zygmund estimates. Although this method takes a few more steps, it is closer conceptually to the method used here to prove Schauder estimates. In converting to LaTeX, I have adjusted spacing (and sometimes wording) to eliminate the nasty overfull messages. Finally, I have added some references and exercises to keep up with the current state of affairs.

June, 2005
Ames, Iowa USA
Gary M. Lieberman

CONTENTS

CHAPTER I

INTRODUCTION

1. Outline of this book

In Chapter II, we prove maximum principles for general, linear parabolic operators:

$$Pu = -u_t + a^{ij}D_{ij}u + b^iD_iu + cu,$$

where we use the summation convention and notation discussed in Section I.3. Under weak hypotheses on the coefficients of this operator (specifically, that the matrix (a^{ij}) be positive semidefinite and that c be bounded from above), the inequalities $Pu \geq 0$ in Ω and $u \leq 0$ on $\mathscr{P}\Omega$ (the parabolic boundary of Ω) imply that $u \leq 0$ in Ω. Nirenberg's strong maximum principle implies that either $u < 0$ or $u \equiv 0$ on a suitable subset of Ω provided the conditions on the coefficients are strengthened only slightly. Also, maximum estimates for solutions of various initial-boundary value problems are examined.

Chapter III introduces the notion of weak solution for the problem

$$-u_t + D_i(a^{ij}D_ju) = f \text{ in } \Omega, u = \varphi \text{ on } \mathscr{P}\Omega$$

when a^{ij} is a constant, positive definite matrix and f and φ are sufficiently smooth. Unlike most discussions of this problem, we deal with noncylindrical domains directly. In particular, we show that this problem is solvable in sufficiently small space-time balls. General properties of weak solutions and weak derivatives are studied also. From the theory in balls, we derive an existence theorem in a wide class of domains via a version of the Perron process.

The key to our study of linear equations is the Schauder-type estimates in Chapter IV. These estimates relate the norms of solutions of initial-boundary value problems for these equations to the norms of the known quantities in the problems. Specifically, the norms are parabolic Hölder norms, which can be considered as bounds on fractional derivatives of the functions. The main estimates were first proved by Barrar [13, 14] and Friedman [87, 88] using a representation formula for solutions of the inhomogeneous heat equation. Our proof is based on Campanato's approach [41] and does not use any representation formulae, but it does need the existence result of Chapter III. The estimates of Chapter IV are used in Chapter V to prove the existence, uniqueness, and regularity of solutions for various initial-boundary value problems under several different hypotheses on the regularity of the data.

Chapter VI contains a further investigation of weak solutions. The class of equations involved is much larger than that considered in Chapter III and the definition of weak solution must be appropriately expanded. The first part of the chapter is concerned with existence, uniqueness, and regularity questions for weak solutions in suitable spaces. The second part covers the pointwise properties of weak solutions, in particular the (Hölder) continuity of these solutions. These properties are proved via estimates of the solutions in terms of weak information on the data of the problem; such estimates are an important part of our study of nonlinear equations.

Our study of linear equations is completed in Chapter VII, which is devoted to strong solutions of parabolic equations. These are functions having weak derivatives in L^p for some $p > 1$ and satisfying the equation only almost everywhere. We obtain Schauder-type estimates in L^p and pointwise estimates for strong solutions which are analogous to the estimates in Chapter VI for weak solutions. We also prove a boundary Hölder estimate for the normal derivative of a solution of the Cauchy-Dirichlet problem in terms of pointwise estimates of the coefficients of the equation. This estimate, originally proved by Krylov [170], is an important element in the nonlinear theory.

We begin our study of nonlinear equations in Chapter VIII, which introduces two fixed point theorems. The first one, due to Schauder [287], is an infinite dimensional version of the Brouwer fixed point theorem in $\mathbb{R}^n$ and is useful for studying the Cauchy-Dirichlet problem for quasilinear equations. The second method, due to Lieberman [202, 206], is a variant of a fixed point theorem of Kirk and Caristi [158] related to the contraction mapping principle and is useful for studying nonlinear oblique derivative problems with quasilinear equations and fully nonlinear equations (with either Cauchy-Dirichlet data or oblique derivative data). Our point of view on existence is generally to prove local existence first (that is, existence for a small time interval) and then global existence. Such an approach is especially important in blow-up theory. Local existence is often proved using relatively simple *a priori* estimates, but these estimates are the key to the global existence theory. The Cauchy-Dirichlet problem and the oblique derivative problem are considered separately.

The *a priori* estimates for the Cauchy-Dirichlet problem fall naturally into four basic types:

(1) the maximum of the absolute value of the solution,
(2) the maximum of the length of the gradient on the parabolic boundary,
(3) the maximum of the length of the gradient in the domain,
(4) a Hölder norm for the gradient.

These estimates are the topics of Chapters IX, X, XI, and XII, respectively, and each presupposes the preceding ones.

Chapter IX applies the maximum principle of Chapter II and the maximum estimates of Chapter VI to get pointwise bounds on the solution. In this chapter, we also generate some comparison principles which are important for other, later estimates.

Chapter X is devoted to a boundary gradient bound, which turns out to be the key estimate in the existence theory. We develop parabolic analogs of the Serrin curvature conditions for elliptic equations [290], which relate the geometry of the boundary to the structure of the differential equation. As in [290], we also show that these conditions are necessary to solve the problem with arbitrary data. All results are based on the comparison principle.

The gradient estimate is the topic of Chapter XI. The basis for proving such an estimate goes back to Bernstein [17]: show that the square of the length of the gradient of the solution satisfies a suitable differential inequality. Unlike Chapters IX and X, this chapter uses quite detailed structure conditions, and often a change of dependent variable is useful.

Chapter XII applies the Hölder estimates of Chapters VI and VII to the gradient of the solution. Assuming that a bound is known for the gradient of the solution, we prove the Hölder estimates under very weak assumptions on the equation in question. The chapter finishes with some examples which illustrate the various structure conditions.

The oblique derivative problem is examined in Chapter XIII. Since the techniques are so similar to those for the Cauchy-Dirichlet problem, many proofs in this chapter are sketched or omitted. An interesting feature of our study of the oblique derivative problem is that we are not restricted to the conormal problem. For example, with the heat equation $-u_t + \Delta u = f(X)$, we may use the capillarity boundary condition $(1+|Du|^2)^{1/2} Du \cdot \gamma + \psi(X) = 0$, where Du denotes the gradient of u and γ is the interior spatial normal. If f, ψ and the domain are smooth enough, we shall show that this problem has a solution for any smooth enough initial data (satisfying a certain compatibility condition) provided ψ satisfies the necessary condition for a smooth solution: $|\psi| < 1$.

In Chapter XIV, we consider a simple class of fully nonlinear equations, modeled on the parabolic Bellman equation:

$$-u_t + \inf_{v \in \mathscr{V}} \{ \sum_{i,j=1}^{n} a_v^{ij}(X) D_{ij} u \} = 0,$$

where (a_v^{ij}) is a uniformly bounded, uniformly smooth family of uniformly positive definite matrices. Such equations were first studied systematically by Evans and Lenhart [79] and Krylov [169, 170]. The hypotheses of these authors are successfully weakened here by using simplifications due to Safonov [282, 284].

Finally, Chapter XV looks at a class of nonuniformly parabolic fully nonlinear equations, modeled on the parabolic Monge-Ampère equation:

$$-u_t \det D^2 u = \psi(X, u, Du).$$

Elliptic versions of this equation and its generalizations were first studied by Caffarelli, Nirenberg, and Spruck [33–35] and Ivochkina [122–127]. Based on their work, we prove estimates and existence results for the parabolic equations. Our choice of parabolic Monge-Ampère equation was first identified by Krylov [167] as a suitable parabolic version of the equation $\det D^2 u = \psi$, and the corresponding generalizations were recognized by Reye [281] as worthy of study. Other parabolic Monge-Ampère equations and their generalizations are discussed briefly. The most noteworthy is that of Ivochkina and Ladyzhenskaya:

$$-u_t + (\det D^2 u)^{1/n} = \psi.$$

This equation and its variants are studied in [47, 130, 131, 133, 134, 242, 338].

2. Further remarks

Although the notes in this book may seem quite extensive, we have really only scratched the surface of the subject of parabolic equations. An examination of any recent issue of Mathematical Reviews shows several dozen articles each month. Even restricting attention to single, nondegenerate equations of second order (as we have here) leaves a staggering number of articles to read and assimilate. A complete bibliography would be much longer than this book and it would be obsolete before it appeared in print. Alternative sources of material which is directly relevant to the issues raised here are [68, 115, 172, 183, 193, 271, 276, 278].

There are also a number of subjects not covered in this book. Some of them are systems of parabolic equations and equations of higher order (see [4, 183]), degenerate equations ([63]), and probabilistic aspects of parabolic equations ([69, 172]).

In an effort to keep this book at a manageable size, I have given a single point of view for most topics; for example, the Schauder estimates of Chapter IV are proved using the Campanato method, which reappears in the nonlinear theory (although an alternative proof could be given using methods developed in Chapter XIV). Many other proofs of these estimates are known; in fact, an entire book could be written just on the various means of proving them. Similarly, we have emphasized Moser iteration to prove most estimates for weak solutions; deGiorgi iteration is also an appropriate and frequently used technique, but it appears only briefly in Section VI.13. As a result of this single point of view, experts will certainly find a favorite technique missing. For example, I have avoided use of representation of solutions via Green's function and its cousin, potential representations. In addition, the method of viscosity solutions does not appear here.

3. Notation

In this book, we adopt the following conventions. First we use $X = (x,t)$ to denote a point in $\mathbb{R}^{n+1}$ with $n \geq 1$; x will always be a point in $\mathbb{R}^n$. We also write $Y = (y,s)$. Superscripts will be used to denote coordinates, so $x = (x^1, \ldots x^n)$. Norms on $\mathbb{R}^n$ and $\mathbb{R}^{n+1}$ are given by

$$|x| = \left(\sum_{i=1}^{n}(x^i)^2\right)^{1/2} \quad \text{and } |X| = \max\{|x|, |t|^{1/2}\},$$

respectively. A basic set of interest to us is the cylinder

$$Q(X_0,R) = Q(R) = \{X \in \mathbb{R}^{n+1} : |X - X_0| < R, t < t_0\}.$$

We also use the ball

$$B(x_0,R) = \{x \in \mathbb{R}^n : |x - x_0| < R\}.$$

We generally follow tensor notation. As previously indicated, superscripts denote coordinates of points in $\mathbb{R}^n$. Also, subscripts denote differentiation with respect to x. In particular,

$$D_iu = \frac{\partial u}{\partial x^i}, D_{ij}u = \frac{\partial^2 u}{\partial x^i \partial x^j}$$

for u a sufficiently smooth function and i and j integers between 1 and n, inclusive. We also write Du for the vector $(D_1u, \ldots, D_nu)$ and D^2u for the matrix $(D_{ij}u)$. On the other hand, we write u_t (or occasionally D_tu) for $\partial u/\partial t$. In addition, we follow the summation convention that any term with a repeated index i is summed over $i = 1$ to n. For example,

$$b^iD_iu = \sum_{i=1}^{n} b^iD_iu.$$

To be formally correct, we should demand (as was the case in this example) that one occurrence of the index be a superscript and the other be a subscript, but we shall abuse this aspect of the summation convention when convenient.

A word of warning about superscripts and subscripts is also in order. Both will be used as generic indices and to indicate differentiation with respect to other variables. In addition, superscripts will also be used to indicate exponentiation. It should be clear from the context which meaning is intended.

We use Ω to denote a bounded domain in $\mathbb{R}^{n+1}$, that is, Ω is an open connected subset of $\mathbb{R}^{n+1}$, so for any $X_0 \in \Omega$, there is a positive number R such that the space-time ball $\{X : |x - x_0|^2 + (t - t_0)^2 < R^2\}$ is a subset of Ω. For a fixed number t_0, we write $\omega(t_0)$ for the set of all points (x,t_0) in Ω, and we define $\gamma(X_0)$ to be the unit inner normal to $\omega(t_0)$ at X_0 provided $\omega(t_0)$ is not empty. We also write $I(\Omega)$ for the set of all t such that $\omega(t)$ is nonempty. Since Ω is connected, $I(\Omega)$ will be an open interval. As usual, $\partial\Omega$ denotes the topological boundary of Ω. The parabolic boundary $\mathscr{P}\Omega$ will be defined in Chapter II.

We also use $|\Omega|$ to denote the Lebesgue measure of the set Ω, and we define $\operatorname{diam}\Omega = \sup\{|x-y| : X,Y \text{ in } \Omega\}$, so $\operatorname{diam}\Omega$ is not the usual diameter of a set. When Ω is a cylinder $\Omega = \omega \times (a,b)$, then $\operatorname{diam}\Omega$ is the usual diameter of the cross-section ω, but if Ω is noncylindrical, $\operatorname{diam}\Omega$ may be strictly larger than the diameter of any cross-section $\omega(t)$.

CHAPTER II

MAXIMUM PRINCIPLES

Introduction

An important tool in the theory of second order parabolic equations is the maximum principle, which asserts that the maximum of a solution to a homogeneous linear parabolic equation in a domain must occur on the boundary of that domain. In fact, this maximum must occur on a special subset of the boundary, called the parabolic boundary. The strong maximum principle asserts that the solution is constant (at least in a suitable subdomain) if the maximum occurs anywhere other than on the parabolic boundary. The maximum principle is used to prove uniqueness results for various boundary value problems, L^∞ bounds for solutions and their derivatives, and various continuity estimates as well.

1. The weak maximum principle

As noted in Chapter I, we consider linear operators L defined by

$$Lu = a^{ij}(X)D_{ij}u + b^i(X)D_iu + c(X)u - u_t \tag{2.1}$$

in an $(n+1)$-dimensional domain Ω. We assume that L is weakly parabolic. In other words,

$$a^{ij}(X)\xi_i\xi_j \geq 0 \text{ for all } \xi \in \mathbb{R}^n \text{ and all } X \in \Omega. \tag{2.2}$$

We also write $\mathscr{T}$ for $\sum a^{ii}$, the trace of the matrix (a^{ij}).

For a domain $\Omega \subset \mathbb{R}^{n+1}$, we define the parabolic boundary $\mathscr{P}\Omega$ to be the set of all points $X_0 \in \partial\Omega$ such that for any $\varepsilon > 0$, the cylinder $Q(X_0,\varepsilon)$ contains points not in Ω. In the special case that $\Omega = D \times (0,T)$ for some $D \subset \mathbb{R}^n$ and $T > 0$, $\mathscr{P}\Omega$ is the union of the sets $B\Omega = D \times \{0\}$ (which is the *bottom* of Ω), $S\Omega = \partial D \times (0,T)$ (which is the *side* of Ω), and $C\Omega = \partial D \times \{0\}$ (which is the *corner* of Ω). These sets (and their analogs for more general domains) will play an important role in the theory of initial-boundary value problems.

The simplest version of the maximum principle is the following.

LEMMA 2.1. *If $u \in C^{2,1}(\overline{\Omega} \setminus \mathscr{P}\Omega) \cap C^0(\overline{\Omega})$, if $Lu > 0$ in $\overline{\Omega} \setminus \mathscr{P}\Omega$ and if $u < 0$ on $\mathscr{P}\Omega$, then $u < 0$ in $\overline{\Omega}$.*

PROOF. Suppose, to the contrary, that $u \geq 0$ somewhere in $\overline{\Omega}$. Let $t^* = \inf\{t : u(x,t) \geq 0 \text{ for some } X \in \overline{\Omega}\}$. By continuity and the assumption that $u < 0$ on

$\mathscr{P}\Omega$, there is $X^* = (x^*, t^*) \in \overline{\Omega} \setminus \mathscr{P}\Omega$ such that $u(X^*) = 0$. Since $u(\cdot, t^*)$ attains its maximum at x^*, we have $Du(X^*) = 0$, and hence $Lu(X^*) = -u_t(X^*) + a^{ij}(X^*)D_{ij}u(X^*)$. Since we also have $u_t(X^*) \geq 0$ and $D^2u(X^*) \leq 0$, it follows that $Lu(X^*) \leq 0$, contradicting $Lu > 0$. Therefore $u < 0$ in $\overline{\Omega}$. □

Since oblique derivative boundary conditions will play an important role in this book, we formulate a further version of this maximum principle which also pertains to oblique derivative boundary conditions. For $X_0 \in \mathscr{P}\Omega$, we say that an $(n+1)$-vector β *points into* $\overline{\Omega}$ if $\beta^{n+1} \leq 0$ and there is a positive constant ε such that $X_0 + h\beta \in \overline{\Omega}$ if $0 < h < \varepsilon$. For such a β, if the function u is continuous at X_0 and k is a constant, we say that $\beta \cdot \partial u(X_0) \leq k$ if

$$\limsup_{h \to 0^+} \frac{u(X_0 + h\beta) - u(X_0)}{h} \leq k,$$

and $\beta \cdot \partial u(X_0) \geq k$ if

$$\liminf_{h \to 0^+} \frac{u(X_0 + h\beta) - u(X_0)}{h} \geq k,$$

where we write ∂u for the full space-time gradient (Du, u_t) of u. Since these inequalities are just the appropriate inequalities on the directional derivates of u at X_0, it follows that $\beta \cdot \partial u(X_0) = k$ if and only if the corresponding directional derivative $\partial u / \partial \beta$ exists and is equal to k at X_0. Analogous to our operator L is the boundary operator M defined by

$$Mu = \beta \cdot \partial u + \beta^0 u \tag{2.3}$$

for some vector field β such that $\beta(X_0)$ points into $\overline{\Omega}$ at X_0 for each $X_0 \in \mathscr{P}\Omega$ and scalar function β^0. Inequalities on Mu are defined in the obvious way. In particular, we have $u \leq 0$ is equivalent to $Mu \geq 0$ if we choose $\beta \equiv 0$ and $\beta^0 = -1$. With these conventions, we have the following extension of Lemma 2.1.

LEMMA 2.2. *If* $u \in C^{2,1}(\overline{\Omega} \setminus \mathscr{P}\Omega) \cap C^0(\overline{\Omega})$, *if* $Lu > 0$ *in* $\overline{\Omega} \setminus \mathscr{P}\Omega$ *and if* $Mu > 0$ *on* $\mathscr{P}\Omega$, *then* $u < 0$ *in* $\overline{\Omega}$.

PROOF. This time, we note that if $X^* \in \mathscr{P}\Omega$, then $Mu(X^*) \leq 0$. □

With a little more structure on the coefficients of L and M, we can relax the smoothness hypotheses on u.

LEMMA 2.3. *Suppose that there is a positive constant k such that*

$$c \leq k \text{ in } \Omega \tag{2.4a}$$

$$\beta^0 < 0 \text{ on } \mathscr{P}\Omega. \tag{2.4b}$$

If $u \in C^{2,1}(\Omega) \cap C^0(\overline{\Omega})$, *if* $Lu \geq 0$ *in* Ω *and if* $Mu \geq 0$ *on* $\mathscr{P}\Omega$, *then* $u \leq 0$ *in* Ω.

PROOF. Set $v = e^{-(k+1)t}u$, so that $Lv = e^{-(k+1)t}Lu + (k+1)v \geq (k+1)v$ and $Mv = e^{-(k+1)t}Mu - (k+1)\beta^{n+1}v \geq -(k+1)\beta^{n+1}v$. If v has a positive maximum at some $X \in \mathscr{P}\Omega$, then $Mv(X) \leq \beta^0(X)v(X) < 0 \leq -(k+1)\beta^{n+1}(X)v(X)$, contradicting our boundary condition for v. If v has a positive maximum at some $X \in \Omega$, then $v_t(X) = 0$, $Dv(X) = 0$, $D^2v(X)$ is nonpositive definite, and $c(X)v(X) \leq kv(X)$, so $Lv(X) \leq kv(X) < (k+1)v(X)$. Hence v cannot have a positive maximum in Ω.

It follows that if v has a positive maximum, it must occur on $\partial\Omega \setminus \mathscr{P}\Omega$, say at X_0. In this case, set $t_i = t_0 - 1/i$, $\Omega_i = \{X \in \Omega : t < t_i\}$, and $M_i = \sup_{\Omega_i} v$. If i is sufficiently large so that Ω_i is non-empty and $M_i > 0$, choose x_i so that $v(X_i) = M_i$, noting from the first case that v cannot have a positive maximum in $\Omega_i \cup \mathscr{P}\Omega_i$. Then $v(X_i) \to v(X_0)$, so there is a convergent subsequence $(X_{i(j)})$ and $v(X_{i(j)}) \to v(X_0)$ as $j \to \infty$. Using X^* to denote the limit of the sequence $(X_{i(j)})$, we have $v(X^*) = v(X_0)$, so $X^* \in \partial\Omega \setminus \mathscr{P}\Omega$. But then, there is a positive ε such that $Q(X^*, \varepsilon) \subset \Omega$ and hence there is an integer j such that $v(X_{i(j)}) = M_{i(j)}$ with $X_{i(j)} \in Q(X^*, \varepsilon)$. We then have $v_t(X_{i(j)}) \geq 0$, $Dv(X_{i(j)}) = 0$, $D^2v(X_{i(j)})$ is nonpositive definite, and $cv(X_{i(j)}) < kv(X_{i(j)})$. As before, these conditions show that the restriction of v to $\Omega_{i(j)}$ cannot attain a positive maximum at $X_{i(j)}$. Therefore $v \leq 0$, and hence $u \leq 0$ in Ω. □

More generally, we can get an upper bound on u over Ω from a lower bound on Mu just over $\mathscr{P}\Omega$.

THEOREM 2.4. *Suppose $u \in C^{2,1}(\Omega) \cap C^0(\overline{\Omega})$ for some Ω with $t \geq 0$ in Ω, and suppose conditions* (2.2) *and* (2.4) *hold. If $Lu \geq 0$ in Ω and if $Mu \geq \beta^0\Phi$ on $\mathscr{P}\Omega$ for some nonnegative constant Φ, then $u(X) \leq e^{kt}\Phi$ for all $X \in \Omega$.*

PROOF. Set $v = u - e^{kt}\Phi$ and apply Lemma 2.3 to v, noting that $e^{kt} \geq 1$ on $\mathscr{P}\Omega$. □

Crucial tools in the remainder of this book are the following comparison principle and uniqueness result.

COROLLARY 2.5. *Suppose u and v are in $C^{2,1}(\Omega) \cap C^0(\overline{\Omega})$ and conditions* (2.4a,b) *hold. If $Lu \geq Lv$ in Ω and if $Mu \geq Mv$ on $\mathscr{P}\Omega$, then $u \leq v$ in Ω. If $Lu = Lv$ in Ω, and if $Mu = Mv$ on $\mathscr{P}\Omega$, then $u = v$ in Ω.*

PROOF. The inequality follows by applying Lemma 2.3 to $u - v$, and the equality is then clear. □

By taking $\beta \equiv 0$ and $\beta^0 \equiv -1$, we have that $Lu \geq (=)Lv$ in Ω and $u \leq (=)v$ on $\mathscr{P}\Omega$ implies $u \leq (=)v$ in Ω.

As we shall see in the next section, determining the set of points at which u can attain a nonnegative maximum is quite delicate. Related to this issue is the question of continuous attainment of boundary values for solutions of various boundary value problems. We shall address this issue further in the next chapter.

2. The strong maximum principle

For most applications to the first initial-boundary value problem (in which the values of the solution are prescribed on the parabolic boundary of the domain), the weak maximum principle from Theorem 2.4 (with $\beta \equiv 0$ and $\beta^0 \equiv -1$) suffices; however, it is often useful to know that the maximum can't occur inside Ω unless the solution is constant on a suitable set. We shall prove this result via an *a priori* estimate that plays a critical role in our discussion of the weak Harnack inequality from Section VII.10.

LEMMA 2.6. *Let α and R be positive constants and set*

$$Q = \{X \in \mathbb{R}^{n+1} : |x| < R, -\alpha R^2 < t < 0\}. \tag{2.5}$$

Suppose there are positive constants λ, Λ, Λ_1, and Λ_2 such that

$$a^{ij}\xi_i\xi_j \geq \lambda|\xi|^2 \text{ for all } \xi \in \mathbb{R}^n, \tag{2.6a}$$

$$\mathcal{T} \leq \Lambda, \tag{2.6b}$$

$$|b| \leq \Lambda_1, \tag{2.6c}$$

$$|c| \leq \Lambda_2 \tag{2.6d}$$

in Q, and suppose that $Lu \leq 0$ and $u \geq 0$ in Q. Let h and ε be positive constants with $\varepsilon < 1$, and suppose that

$$u \geq h \text{ on } \{|x| < \varepsilon R, t = -\alpha R^2\}. \tag{2.7}$$

Then there is a positive constant κ, determined only by α, λ, Λ, and $\Lambda_1 R + \Lambda_2 R^2$, such that

$$u \geq \varepsilon^\kappa \frac{h}{2} \text{ on } \{|x| < R/2, t = 0\}. \tag{2.8}$$

PROOF. Set

$$\psi_0 = \frac{(1-\varepsilon^2)}{\alpha}(t + \alpha R^2) + \varepsilon^2 R^2, \; \psi_1 = \max\{\psi_0 - |x|^2, 0\},$$

and $\psi = \psi_1^2\psi_0^{-q}$ for some $q \geq 2$ to be chosen. Then $\psi \in C^{2,1}(\tilde{Q})$ for

$$\tilde{Q} = \{X \in \mathbb{R}^{n+1} : |x|^2 < \psi_0, -\alpha R^2 < t < 0\},$$

and

$$L\psi = \psi_0^{-q}[8a^{ij}x^ix^j - 4\psi_1\mathcal{T} - 4\psi_1 b\cdot x + c\psi_1^2 + \frac{1-\varepsilon^2}{\alpha}\left(\frac{q\psi_1^2}{\psi_0} - 2\psi_1\right)]$$

in $\tilde{Q}$. Now we set

$$F_1 = \frac{2}{\alpha} + 8\lambda + 4\Lambda + 4\Lambda_1 R + \Lambda_2 R^2$$

and $\xi = \psi_1/\psi_0$. Noting that $0 \leq \psi_0 \leq R^2$ and $0 \leq \varepsilon \leq 1$, we see that

$$L\psi \geq \psi_0^{1-q}[\frac{(1-\varepsilon^2)q}{\alpha}\xi^2 - F_1\xi + 8\lambda].$$

By choosing

$$q = 2 + \frac{\alpha}{32(1-\varepsilon^2)} F_1^2 \lambda,$$

we see that $L\psi \geq 0$ in Q.

Now we note that

$$\psi(x, -\alpha R^2) = (\varepsilon^2 R^2 - |x|^2)^2 (\varepsilon^2 R^2)^{-q} \leq (\varepsilon R)^{-2q+4} \tag{2.9}$$

if $(x, -\alpha R^2) \in \mathscr{P}\tilde{Q}$. Now set $v = h(\varepsilon R)^{2q-4}\psi$. Since $\psi(X) = 0$ if $X \in \mathscr{P}\tilde{Q}$ and $t > -\alpha R^2$, we infer from (2.7) and (2.9) that $v \leq u$ on $\mathscr{P}\tilde{Q}$. Applying Corollary 2.5 then yields $v \leq u$ in $\tilde{Q}$ and hence

$$u(x,0) \geq h(\varepsilon R)^{2q-4}\psi \geq \frac{h}{2}\varepsilon^{2q-4} \tag{2.10}$$

for $|x| \leq R/2$ because

$$\psi(x,0) = (R^2 - |x|^2)^2 R^{-2q} \geq \frac{9}{16} R^{-2q+4} \tag{2.11}$$

if $|x| \leq R/2$. For $\kappa = 2q-4$, (2.10) easily implies (2.8). □

We are now ready to state and prove the strong maximum principle. For $X_0 \in \overline{\Omega}$, we define $S(X_0)$ to be the set of all $X_1 \in \overline{\Omega} \setminus \mathscr{P}\Omega$ such that there is a continuous function $g\colon [0,1] \to \overline{\Omega} \setminus \mathscr{P}\Omega$ such that $g(0) = X_0$, $g(1) = X_1$ and the t-component of g is nonincreasing.

THEOREM 2.7. *Suppose conditions* (2.6a–d) *hold in* Ω *and* $c \leq 0$ *in* Ω. *If* $Lu \geq 0$ *in* Ω *and* $u(X_0) = \max_{\overline{\Omega}} u \geq 0$ *for some* $X_0 \in \overline{\Omega} \setminus \mathscr{P}\Omega$, *then* $u \equiv u(X_0)$ *in* $S(X_0)$.

PROOF. Set $M = \max_{\overline{\Omega}} u$. As a first step, we show that, if $Y \in \overline{\Omega} \setminus \mathscr{P}\Omega$, if r is chosen so that $Q(Y,3r) \subset \Omega$, and if there is a $Y_1 \in Q(Y,r)$ such that $u(Y_1) < M$, then $u(Y) < M$. to prove this implication, we assume without loss of generality that $y = 0$ and $s = 0$. Then we set $R = 2r$, $\alpha = -s_1/R^2$, and $h = (M - u(Y_1))/2$. Because u is continuous, there is a constant $\varepsilon \in (0,1)$ such that $M - u(X) \geq h$ if $|x| \leq \varepsilon R$ and $t = -\alpha R^2$. Then Lemma 2.6 applied to $M - u$ gives $M - u(Y) \geq \varepsilon^\kappa h/2$ and hence $u(Y) < M$.

We now prove the contrapositive of this theorem. In other words, we show that if $u(X_1) < M$ for some $X_1 \in S(X_0)$, then $u(X_0) < M$. To this end, we let g be the function in the definition of $S(X_0)$ and we define

$$S = \{\sigma \in [0,1] : u(g(\sigma)) < M\}.$$

Our goal is to show that $S = [0,1]$. Clearly, S is a relatively open subset of $[0,1]$ which is non-empty. Hence, we only have to show that S is closed, so let $\sigma_0 \in \overline{S}$. Then there is a positive constant r so that $Q(g(\sigma_0), 3r) \subset \Omega$ and also there is a point $Y \in \overline{Q_0} \setminus \mathscr{P}Q_0$ such that $u(Y) < M$, where $Q_0 = Q(g(\sigma_0), r)$. Since u is continuous, there is a point $Y \in Q_0$ with $u(Y_0) < M$, and, from the first part of the proof of this theorem, we infer that $u(g(\sigma_0)) < M$. Hence S is closed and the contrapositive is proved. □

Related to the strong maximum principle is the so-called boundary point lemma, which loosely states that at a boundary maximum of a non-constant solution, the inner normal derivative must be strictly negative. In fact, the geometry of the domain near a boundary maximum also plays a role.

LEMMA 2.8. *Suppose there are positive constants η, λ, Λ, Λ_1, Λ_2, and R such that conditions* (2.6a–d) *hold in the lower parabolic frustum*

$$PF = PF(R,\eta,Y) = \{X : |x-y|^2 + \eta^2(s-t) < R^2, t < s\} \tag{2.12}$$

for some $Y \in \mathbb{R}^{n+1}$. Suppose also that $u \in C^{2,1}(PF)$ with $Lu \geq 0$ in PF and that there is $X_1 = (x_1, s)$ with $|x_1 - y| = R$ such that

$$u(X_1) \geq u(X) \text{ for all } X \in PF \tag{2.13a}$$

$$u(X_1) > u(X) \text{ for all } X \in PF \text{ with } |x-y| \leq \frac{R}{2}, \tag{2.13b}$$

$$c(X)u(X_1) \leq 0 \text{ for all } X \in PF. \tag{2.13c}$$

If $\nu = (y-x_1)/|y-x_1|$ and $\beta = (\nu,0)$, then $\beta \cdot \partial u(X_1) < 0$ in the sense defined in Section II.1, and $u < u(X_1)$ in PF.

PROOF. Set $r = r(X) = (|x-y|^2 + \eta^2(s-t))^{1/2}$, define

$$P = \{X \in PF : |x-y| > \frac{R}{2}\},$$

and note that $\mathscr{P}P = S_1 \cup S_2$ for $S_1 = \{r = R, t \leq s\}$ and $S_2 = \{|x-y| = \frac{1}{2}R, t \leq s\}$. For α a positive constant to be chosen, we consider the function $v = \exp(-\alpha r^2) - \exp(-\alpha R^2)$. Then a simple calculation shows that

$$\begin{aligned} Lv &= e^{-\alpha r^2}[4\alpha^2 a^{ij}(x_i-y_i)(x_j-y_j) - \alpha(2\mathscr{T} + 2b\cdot(x-y) + \eta^2) \\ &\quad + c(1-e^{-\alpha(R^2-r^2)})] \\ &\geq e^{-\alpha r^2}[\alpha^2\lambda R^2 - \alpha(2\Lambda + 2\Lambda_1 R + \eta^2) - \Lambda_2] \end{aligned}$$

in P. We now set

$$\Lambda_0 = \Lambda + \Lambda_1 R + \Lambda_2 R^2. \tag{2.14}$$

(This constant will be useful later in this chapter.) Choosing $\alpha = g/R^2$ with g sufficiently large (depending only on $(\eta^2 + \Lambda_0)/\lambda$), we can make $Lv \geq 0$ in P. Now $u - u(X_1) \leq 0$ on S_1, and $u - u(X_1) < 0$ on S_2. Hence there is a positive constant ε such that $u - u(X_1) < -\varepsilon$ on S_2. Since $0 \leq v \leq 1$ in $\overline{P}$ and $v = 0$ on S_1, it follows that

$$\begin{aligned} L(u - u(X_1) + \varepsilon v) &\geq -cu(X_1) \geq 0 && \text{in } P, \\ u - u(X_1) + \varepsilon v &\leq 0 && \text{on } \mathscr{P}P. \end{aligned}$$

The weak maximum principle Lemma 2.3 implies that $u - u(X_1) \le -\varepsilon v$ in P, so $u < u(X_1)$ in P, and then evaluating the directional derivative at X_1 yields

$$\beta \cdot \partial u(X_1) \le -\varepsilon v \cdot Dv(X_1) = -2\alpha\varepsilon \exp(-\alpha r(X_1)^2/R^2)\,|x_1 - y|\,.$$

Since this last expression is negative, we are done. □

More generally, we have

$$\limsup_{X \to X_1} \frac{u(X) - u(X_1)}{|X - X_1|} < 0,$$

as long as $t \le s$ and the angle (in $\mathbb{R}^{n+1}$) between $X - X_1$ and v is bounded above away from $\pi/2$. Obviously, the lemma remains true if we assume that $Lu \ge 0$ in Ω and conditions (2.13a,b,c) hold with Ω in place of PF for any domain Ω such that PF is a subset of Ω. In fact, the assumption of the existence of an interior parabolic frustum can be relaxed, but some geometric condition on the domain is needed. Note that (2.13c) follows from any of the three conditions: $c \equiv 0$ in PF, $u(X_1) = 0$, or $c \le 0$ and $u(X_1) \ge 0$.

From Lemma 2.8 we can infer uniqueness and comparison theorems for the oblique derivative problem. To facilitate their statements, we introduce some further terminology. We say that Ω satisfies an *interior paraboloid condition at* X_0 if there is a point $Y \in \mathbb{R}^{n+1}$ and a constant R such that the parabolic frustum PF defined by (2.12) is a subset of Ω and $\{X_0\} = \mathscr{P}(PF) \cap \mathscr{P}\Omega$. If Ω satisfies an interior paraboloid condition at X_0, we say that a vector $\beta \in \mathbb{R}^{n+1}$ *points strongly into* Ω at $X_0 \in \mathscr{P}\Omega$ if there is a positive ε such that $X_0 + h\beta \in PF$ for all $h \in (0, \varepsilon)$. Also, we define $B\Omega$ to be the set of all points $X_0 \in \mathscr{P}\Omega$ such that there is a positive R with $Q((x_0, t_0 + R^2), R) \subset \Omega$, $C\Omega = (\overline{B\Omega} \cap \mathscr{P}\Omega) \setminus B\Omega$, and $S\Omega = \mathscr{P}\Omega \setminus (B\Omega \cup C\Omega)$.

THEOREM 2.9. *Suppose conditions* (2.6a-d) *hold in* Ω *and* $c \le 0$ *in* Ω, *and suppose that* Ω *satisfies an interior paraboloid condition at each point of* $S\Omega$. *Suppose also that* β *is a vector field such that* $\beta(X_1)$ *points strongly into* Ω *at* X_1 *for each* $X_1 \in S\Omega$ *and* $\beta^0 \le 0$ *on* $S\Omega$. *Then if*

$$Lu \ge Lv \text{ in } \Omega \tag{2.15a}$$

$$Mu \ge Mv \text{ on } S\Omega \tag{2.15b}$$

$$u \le v \text{ on } B\Omega \cup C\Omega, \tag{2.15c}$$

then $u \le v$ *in* Ω. *Moreover for any* $f \in C(\Omega)$, *any* $\psi \in C(S\Omega)$ *and any* $\varphi \in C(B\Omega \cup C\Omega)$, *there is at most one solution of the initial-boundary value problem*

$$Lu = f \text{ in } \Omega, \tag{2.16a}$$

$$Mu = \psi \text{ on } S\Omega, \tag{2.16b}$$

$$u = \varphi \text{ on } B\Omega \cup C\Omega. \tag{2.16c}$$

PROOF. To see that $u \le v$ in Ω if (2.15) holds, we note that $u - v$ can only attain a positive maximum on $S\Omega$, say at X_1. From Lemma 2.8, $M(u-v) < \beta^0(u-v) \le 0$ at X_1 while the boundary condition implies that $M(u-v) \ge 0$. Hence $u \le v$. The uniqueness of solutions of (2.16) follows immediately from this comparison principle. □

In fact, we shall see in Chapter V that Theorem 2.9 is valid also if c and β^0 are only assumed to be bounded from above provided we strengthen the hypotheses on $S\Omega$ and M just a little. Of course if $c \le 0$ and $\beta^0 < 0$, then the conclusion of Theorem 2.9 follows immediately from Corollary 2.5.

3. *A priori* estimates

The maximum principle also provides simple pointwise bounds for solutions of the equation $Lu = f$ in a large variety of domains. Because so little information about the coefficients is used, these estimates are very useful in studying nonlinear equations. For simplicity, we write $a^+ = \max\{a, 0\}$ and $a^- = \max\{-a, 0\}$ for any real number (or real-valued function) a.

THEOREM 2.10. *Suppose conditions* (2.2) *and* (2.4a) *are satisfied in some domain* Ω *such that* $I(\Omega) \subset (0,T)$ *for some* $T > 0$, *and suppose that* $u \in C^0(\overline{\Omega}) \cap C^{2,1}(\Omega)$. *If* $Lu \ge f$ *in* Ω, *then*

$$\sup_{\Omega} u \le e^{(k+1)T}(\sup_{\mathscr{P}\Omega} u^+ + \sup_{\Omega} f^-). \tag{2.17}$$

If $Lu = f$ *in* Ω, *then*

$$\sup_{\Omega} |u| \le e^{(k+1)T}(\sup_{\mathscr{P}\Omega} |u| + \sup_{\Omega} |f|). \tag{2.18}$$

PROOF. Suppose first that $Lu \ge f$, and set $v = \exp((k+1)t)$ and

$$F = (\sup_{\mathscr{P}\Omega} u^+ + \sup_{\Omega} f^-).$$

Because

$$Lv = -(k+1)e^{(k+1)t} + ce^{(k+1)t} \le -1$$

in Ω, it follows that

$$L(u - Fv) \ge -\sup f^- + F \ge 0 \text{ in } \Omega,$$
$$u - Fv \le \sup u^+ - F \le 0 \text{ on } \mathscr{P}\Omega,$$

so the maximum principle implies that $u - Fv \le 0$ in Ω. Hence (2.17) is proved. If $Lu = f$, we prove (2.18) by applying (2.17) to u and $-u$. □

As an alternative to Theorem 2.10, we can give an estimate for u which is independent of T if $c \le 0$.

THEOREM 2.11. *Suppose conditions (2.6a,c) hold, suppose $c \le 0$ in Ω, and suppose there is a positive constant R such that $|x| \le R$ for any $X \in \Omega$. If $Lu \ge f$, then*

$$\sup_\Omega u \le \sup_{\mathscr{P}\Omega} u^+ + C(\Lambda_1/\lambda, R)\sup_\Omega(f^-/\lambda). \tag{2.19}$$

If $Lu = f$, then

$$\sup_\Omega |u| \le \sup_{\mathscr{P}\Omega} |u| + C(\Lambda_1/\lambda, R)\sup_\Omega(|f|/\lambda). \tag{2.20}$$

PROOF. Set $v = e^{2\alpha R} - e^{\alpha(x^1+R)}$ for $\alpha = 1 + \Lambda_1/\lambda$. Then

$$Lv = -(\alpha^2 a^{11} + \alpha b^1)e^{\alpha(x^1+R)} + cv \le -\alpha(\alpha\lambda - \Lambda_1) \le -\lambda$$

in Ω. Hence, for $F = \sup(f^-/\lambda)$, we have $L(u - \sup_{\mathscr{P}\Omega} u^+ - Fv) \ge 0$ in Ω and $u - \sup_{\mathscr{P}\Omega} u^+ - Fv \le 0$ on $\mathscr{P}\Omega$. The maximum principle then implies (2.19), and (2.20) is proved in a similar fashion. □

It is also possible to prove various modulus of continuity estimates for solutions of linear equations. Here we give two simple examples which will prove useful in later chapters.

THEOREM 2.12. *Suppose conditions (2.6a–d) are satisfied in a domain Ω, let $X_0 \in \mathscr{P}\Omega$ and suppose there is a parabolic frustum PF of the form (2.12) with $\eta = 1$ for some $Y \in \mathbb{R}^{n+1}$ and $R > 0$ such that $\overline{PF} \cap \overline{\Omega} = \{X_0\}$. If $u \in C^0(\overline{\Omega}) \cap C^{2,1}(\Omega)$ is a solution of $Lu = f$ in Ω, $u = \varphi$ on $\mathscr{P}\Omega$ for some $\varphi \in C^{2,1}(\overline{\Omega})$ and $f \in C^0(\overline{\Omega})$, then there is a constant K determined only by λ, Λ, Λ_1, Λ_2, n, and R such that*

$$|u(X) - u(X_0)| \le K\exp((k+1)t_0)(F + \Phi)\,|X - X_0| \tag{2.21}$$

for any $X \in \Omega$ with $t = t_0$, where $F = \sup|f|$ and $\Phi = \sup\left|D^2\varphi\right| + |\varphi_t| + |D\varphi| + |\varphi|$.

PROOF. Define r by $r^2 = |x-y|^2 + (t_0 - t)$, define v by

$$v(x,t) = \exp(-\alpha R^2) - \exp(-\alpha r^2) + 4(t_0 - t)/R^2,$$

for α a positive constant to be chosen, and set $\Omega^* = \Omega \cap \{t_0 - R^2 < t < t_0\}$. From the same calculation as in the proof of Lemma 2.8, we see that $Lv \le -1$ in Ω^* if $\alpha = g/R^2$ with g taken sufficiently large, determined by Λ_0/λ, n.

Next, we observe that $L\varphi \le K_1\Phi$ in Ω^* for some constant K_1 determined by Λ, Λ_1, Λ_2, and n, and we set $F_1 = \exp((k+1)t_0)[F + (1+K_1)\Phi]$. It follows that

$$L(u - \varphi - F_1 v) \ge 0 \text{ in } \Omega^*.$$

Now on $\Omega \cap \{t = t_0 - R^2\}$, we have $v \ge 4$, and Theorem 2.10 implies that $u \le F_1$, so that

$$u - \varphi - F_1 v \le 0 \text{ on } \mathscr{P}\Omega^*.$$

The maximum principle now implies that $u-\varphi \le F_1 v$ on $\Omega_0 = \Omega \cap \{t=t_0\}$, and a similar argument gives $u-\varphi \ge -F_1 v$ on Ω_0. Hence, for $X=(x,t_0)\in\Omega_0$ with $|x-x_0|\le R$, we have

$$\begin{aligned}|u(X)-u(X_0)| &= |u(X)-\varphi(X_0)| \le |u(X)-\varphi(X)| + |\varphi(X)-\varphi(X_0)| \\ &\le \Phi|x-x_0| + F_1 v(x,t_0) \\ &\le \left(\Phi + \sup_{|x-x_0|\le R} F_1 |Dv(x,t_0)|\right)|x-x_0| \\ &\le (\Phi + 4F_1\alpha R)|x-x_0| .\end{aligned}$$

On the other hand, if $X\in\Omega_0$ and $|x-x_0|\ge R$, then $|u(X)-u(X_0)| \le 2F_1 \le 2(F_1/R)|x-x_0|$. Therefore (2.21) holds with $K = 1+4\alpha K_1 R + (2/R)$. □

In fact the proof of Theorem 2.12 shows that (2.21) holds as long as $X=(x,t)$ with $t\le t_0$, and it is possible to prove (2.21) for $t>t_0$ as well. (See Theorem 3.26.) It will be more useful, though, to use this theorem in its present form, which implies an estimate for $|Du(X_0)|$ (if this derivative exists). We shall see in later chapters that an interior modulus of continuity estimate is also valid, but this result is much deeper. On the other hand, given an interior modulus of continuity estimate with respect to x, the following maximum principle argument of Gilding provides an interior modulus of continuity estimate with respect to t.

THEOREM 2.13. *Suppose conditions* (2.2) *and* (2.6b,c) *hold in*

$$Q = \{|x-x_0|<R\}\times(t_1, t_1+\frac{R^2}{4\Lambda_0}), \tag{2.22}$$

with $c\equiv 0$ in Q and $\Lambda_0 = \Lambda+\Lambda_1 R$, for some point $(x_0,t_1)\in\mathbb{R}^{n+1}$ and some positive constant R, and let $u\in C^{2,1}(Q)$ solve $Lu=f$ in Q. If there is a constant ω such that

$$|u(x,t)-u(x_0,t)|\le\omega \tag{2.23}$$

for all $x\in\mathbb{R}^n$ with $|x-x_0|<R$, then

$$|u(x_0,t)-u(x_0,t_1)| \le 2\omega + \frac{2R^2}{\Lambda_0}\sup_Q|f| . \tag{2.24}$$

PROOF. Set

$$s = \sup_{t_1<t<t_1+R^2/(4\Lambda_0)} |u(x_0,t)-u(x_0,t_1)|$$

and define

$$v^{\pm} = (\sup_Q|f| + \frac{2s\Lambda_0}{R^2})(t-t_1) + \frac{s}{R^2}|x-x_0|^2 + \omega \pm (u-u(x_0,t_1)).$$

A simple calculation shows that

$$Lv^{\pm} = -\sup_Q |f| - \frac{2s\Lambda_0}{R^2} + \frac{2s}{R^2}\mathscr{T} + \frac{2s}{R^2} b\cdot(x-x_0) \pm f \le 0$$

in Q while

$$v^{\pm} \ge \frac{s}{R^2}|x-x_0|^2 + \omega \pm (u - u(x_0,t_1))$$

on $\mathscr{P}Q$. We have $v^{\pm}(x,t_1) \ge \omega \pm (u(x,t_1) - u(x_0,t_1)) \ge 0$ by (2.23), and, if $|x-x_0| = R$, we have

$$\begin{aligned} v^{\pm}(x,t) &\ge s + \omega \pm (u(X) - u(x_0,t_1)) \\ &= s + \omega \pm (u(X) - u(x_0,t)) \pm (u(x_0,t) - u(x_0,t_1)) \ge 0. \end{aligned}$$

It follows that $v^{\pm} \ge 0$ on $\mathscr{P}Q$. We now apply the maximum principle to $v^{\pm}$ on Q to see that $v^{\pm} \ge 0$ on Q. Evaluating this inequality at $x = x_0$ and taking the supremum over all t gives

$$s \le (\sup|f| + \frac{2s\Lambda_0}{R^2})\frac{R^2}{4\Lambda_0} + \omega,$$

and a simple rearrangement of this inequality yields (2.24). □

Theorem 2.13 is easily extended to cylinders of arbitrary heights and to operators with $c \ne 0$.

COROLLARY 2.14. *Suppose, in Theorem 2.13, that* $c \ne 0$ *and that*

$$Q = \{|x-x_0| < R\} \times (t_1, t_1 + hR^2) \tag{2.25}$$

for some positive constant h. Then

$$|u(x_0,t) - u(x_0,t_1)| \le (2 + 4h\Lambda_0)(\omega + \frac{R^2}{\Lambda_0}\sup_Q[|cu| + |f|]). \tag{2.26}$$

PROOF. Define L_0 by $L_0u = Lu - cu$. Then u satisfies the hypotheses of the corollary with $c = 0$ and f replaced by $f - cu$. Hence it suffices to prove the result with $c = 0$. In this case, if $h \le 1/(4\Lambda_0)$, the proof of Theorem 2.13 gives the desired result. Otherwise, we partition the cylinder into k subcylinders of the form $Q_i = \{|x-x_0| < R\} \times (t_i, t_{i+1})$ with $i = 1, \dots, k$ for k the smallest integer greater than or equal to $h\Lambda_0$. Applying Theorem 2.13 in each cylinder and summing the resulting inequalities gives (2.26). □

Typically, Corollary 2.14 is applied to a problem in which a modulus of continuity with respect to x is known. In particular, suppose that a gradient bound is known, that is, $|Du| \le K$ in some cylinder $\{|x-x_0| < \rho\} \times (t_1, t_1 + h\rho^2)$ for some positive h and ρ. If $t_2 \in (t_1, t_1 + h\rho^2)$, then we take $R = [(t_2 - t_1)/h]^{1/2}$ to see that

$$|u(x_0,t_2) - u(x_0,t_1)| \le C(2K + \sup|f|)\,|t_2 - t_1|^{1/2}. \tag{2.27}$$

In other words, u is Hölder continuous with exponent $\frac{1}{2}$ with respect to t. We shall use a more sophisticated version of this estimate in later chapters. Two improvements are that we consider also behavior near the parabolic boundary and that we shall use an integrated form of our equation and estimates.

Notes

One of the most important differences between our discussion of the weak maximum principle and the usual ones (see, for example, [89, Chapter 2] or [193, Section 3.2]) is that we only assume u to be a solution of the equation in Ω. Generally, authors assume u to be a solution in $\overline{\Omega}\setminus\mathscr{P}\Omega$, which allows some simplifications of the arguments. In addition, for linear equations with smooth coefficients, a solution of the equation in Ω is also a solution in the larger set (see Chapters IV and V), but many applications are easier with the stronger version of the maximum principle proved here.

The strong maximum principle for parabolic equations is due to Nirenberg [270], who adapted the proof of Hopf's strong maximum principle for elliptic equations. Our approach to the strong maximum principle is based on [235, Corollary 7.3], which is quite different from Nirenberg's or the one in [115]. Lemma 2.6 is due to Krylov and Safonov [180, Lemma 1.3], but the application to the strong maximum principle first appears in [235]. The ideas in the proof also apply to oblique derivative problems (see [235] and also Exercise 2.8).

Theorem 2.13 is taken from Gilding's paper [102], which is based on earlier work of Kruzhkov [164]. Our statement is slightly different, but the proof is the same.

Further discussion of the maximum principle can be found in the book of Protter and Weinberger [278, Chapter 3].

Exercises

2.1 Show that if (a^{ij}) is a nonnegative definite matrix and (b_{ij}) is a nonpositive definite matrix, then the sum $a^{ij}b_{ij}$ is nonpositive.

2.2 Let Ω be a bounded domain in $\mathbb{R}^{n+1}$ and let σ be a relatively open subset of $S\Omega$. Prove that the problem

$$Lu = f \text{ in } \Omega, Mu = \psi \text{ on } \sigma, u = \varphi \text{ on } \mathscr{P}\Omega\setminus\sigma$$

has at most one solution provided $c \le 0$, $\beta^0 \le 0$, Ω satisfies an interior paraboloid condition at every point of σ, and β points strongly into Ω at every point of σ.

2.3 (Degenerate parabolic equations) Suppose that L is given by

$$Lu = -r(X)u_t + a^{ij}(X)D_{ij}u + b^i(X)D_iu + c(X)u$$

with $r \geq 0$ and (a^{ij}) satisfying

$$a^{ij}(X)\xi_i\xi_j \geq \lambda(X)|\xi|^2.$$

If also $r+\lambda > 0$ in $\overline{\Omega} \setminus \mathscr{P}\Omega$ and there are constants k and K such that $c \leq kr$ and $|b| \leq K\lambda$ in $\overline{\Omega} \setminus \mathscr{P}\Omega$, prove that the maximum principle (Theorem 2.4) is still valid. Also prove analogs of Theorems 2.10 and 2.11 for such an operator.

2.4 State and prove analogs of Theorems 2.10 and 2.11 if the boundary condition is given in terms of Mu rather than u.

2.5 (Integro-differential equations I) Let $\Omega \subset \mathbb{R}^{n+1}$, let $(\Sigma, \mathscr{M}, \pi)$ be a σ-finite measure space, and let $j : \Omega \times \Sigma \to \mathbb{R}^n \setminus \{0\}$ and $j_1 : \Omega \times \Sigma \to \mathbb{R}$ be functions such that $X + j(X,\xi) \in \Omega$ for any $(X,\xi) \in \Omega \times \Sigma$ and $j_1 \geq 0$. Define the integral operator $\mathscr{I}$ by

$$\mathscr{I}u(X) = \int_\Sigma [u(x+j(X,\xi),t) - u(X)]j_1(X,\xi)d\pi. \tag{2.28}$$

Suppose also that L and M satisfy (2.2) and (2.4). Show that if $u \in C^{2,1}(\Omega) \cap C(\overline{\Omega})$ satisfies $Lu + \mathscr{I}u \geq 0$ in Ω, $Mu \geq 0$ on $\mathscr{P}\Omega$, then $u \leq 0$ in Ω. Show also that if $Lu + \mathscr{I}u = f$ in Ω, $Mu = \psi$ on $\mathscr{P}\Omega$, then $|u| \leq C(\sup|f| + \sup|\psi|)$ in Ω. (Hint: Imitate the proof of Lemma 2.3, noting that $\mathscr{I}u \leq 0$ at a maximum point. See [95].)

2.6 (Integro-differential equations II) Let σ be a bounded function on Ω and define $\mathscr{I}$ by

$$\mathscr{I}u(X) = \int_0^t u(x,\tau)\sigma(x,\tau)\, d\tau. \tag{2.29}$$

If L and M satisfy (2.2) and (2.4) and $u \in C^{2,1}(\Omega) \cap C(\overline{\Omega})$ satisfies $Lu + \mathscr{I}u \geq 0$ in Ω, $Mu \geq 0$ on $\mathscr{P}\Omega$, then $u \leq 0$ in Ω. Show also that if $Lu + \mathscr{I}u = f$ in Ω, $Mu = \psi$ on $\mathscr{P}\Omega$, then $|u| \leq C(\sup|f| + \sup|\psi|)$ in Ω. (Hint: Use Exercise 2.5 and Gronwall's inequality.)

2.7 Show that Lemma 2.8 is true if the lower parabolic frustum is replaced by a region of the form $\{|x-y|^{1+\alpha} + \eta^2(s-t) < R^{1+\alpha}, t < s\}$. (See [150, 208].)

2.8 (Oblique derivative problems in nonsmooth domains [235, Section 7]) We say that Ω satisfies an *interior parabolic cone condition* at $X_0 \in \mathscr{P}\Omega$ if there are positive constants R, ω_0, and ω_0 and a unit vector $\gamma \in \mathbb{R}^n$ such that every $X \in Q(X_0,R)$ such that

$$\gamma \cdot (x-x_0) < \omega_0[|x-x_0|^2 - |\gamma \cdot (x-x_0)|^2]^{1/2} + \omega_1|t-t_0|^{1/2}$$

is also in Ω. By rotating the axes, we may assume that $\gamma' = 0$, $\gamma^n = 1$, and $X_0 = 0$, so we have

$$\{X \in Q(0,R) : x^n < \omega_0|x'| + \omega_1|t|^{1/2}\} \subset \Omega.$$

We also assume that there are constants $\mu_1 \ge 0$ and $\varepsilon \in (0,1)$ such that $|\beta'| \le \mu_1 \beta^n$ on $S\Omega \cap Q(0,R)$ and that $\omega_0 \mu_1 \le 1-\varepsilon$.
(a) Show that there are constants A and α_1 such that, if x_1^n satisfies

$$x_1^n \ge (A-\alpha_1)\omega_0 R$$

and if ω_0 and ω_1 are sufficiently small, then

$$G(x) = |x'|^2 + |x^n - x_1^n|^2$$

satisfies

$$\beta \cdot DG \le \varepsilon^4 A \frac{\omega_0}{R} \beta^n$$

on $S\Omega \cap \tilde{Q}(R)$, where

$$\tilde{Q}(R) = \{X \in \mathbb{R}^{n+1} : G(X) < 1, -\alpha R^2 < t < 0\}$$

and α is sufficiently small.
(b) Suppose Ω satisfies an interior parabolic cone condition at a point $X_0 \in S\Omega$ and that β has modulus of obliqueness δ at X_0. Assuming that α, ω_0, and ω_1 are sufficiently small, show that Lemma 2.6 remains true for $Q = \tilde{Q}(R) \cap \Omega$ if $Mu \le 0$ on $S\Omega$. (Hint: use $G(x)$ in place of $|x|^2$.)
(c) Prove a corresponding strong maximum principle and uniqueness theorem for the oblique derivative problem under these hypotheses.

Note that we obtain, in place of the boundary point lemma, a related result (called the Lemma on the Inner Derivative by Nadirashvili [263]; see also [152]), which states that, in every neighborhood of a maximum point, there is a point at which the directional derivative $\beta \cdot Du$ is strictly negative. In this way, we can relax the smoothness assumption on the boundary but then we don't have the precise pointwise behavior of the derivative at the maximum point. A more careful analysis (as in [235]) shows that the smallness conditions on α, ω_0 and ω_1 can be removed.

CHAPTER III

INTRODUCTION TO THE THEORY OF WEAK SOLUTIONS

Introduction

In studying parabolic equations, it is sometimes useful to have a notion of solution that allows for functions which are not as smooth as those considered in Chapter II. In this chapter, we start to develop this notion for equations of the special form

$$-u_t + a^{ij}D_{ij}u = f, \tag{3.1}$$

where (a^{ij}) is a constant, positive definite matrix. (A more complete discussion of the theory, which applies to equations with variable coefficients, is given in Chapter VI.) In fact, our hypotheses are enough to guarantee that the weak solutions considered here are actually $C^{2,1}$ (and therefore solutions in the sense of Chapter II), but it will be very convenient to use weak solutions in developing the existence theory which is completed in the next two chapters. To motivate the notion of weak solution, we observe that if u solves (3.1) in the cylinder $Q = Q(X_0, R)$ and if $\zeta \in C^1(\overline{Q})$ with $\zeta(X) = 0$ if $|x - x_0| = R$, then an integration by parts gives

$$\int_Q (u_t\zeta + a^{ij}D_juD_i\zeta - f\zeta)\,dX = 0. \tag{3.2}$$

We shall use (3.2) as the basis of our definition of a weak solution because it makes sense as long as u has continuous x and t derivatives and u is bounded. In fact, we shall want to make sense of (3.2) when the derivatives are only L^2, not necessarily defined everywhere. To emphasize the nature of this definition, we also write (3.1) as

$$-u_t + D_i(a^{ij}D_ju) = f \tag*{(3.1)$'$}$$

Note that we could also integrate by parts with respect to t; however, the resulting theory is more involved and we defer its discussion to Chapter VI.

The basic theory of these weak derivatives is given in Section III.1. The functional analysis used to derive our existence results appears in Section III.2. Section III.3 contains applications to weak solutions of simple elliptic and parabolic equations in small balls. Section III.4 is concerned with the solution of parabolic equations in more general domains.

1. The theory of weak derivatives

Before studying weak derivatives of general functions, we discuss some standard properties of the L^p spaces. For notational convenience, we consider functions defined in a subset of $\mathbb{R}^N$; the dimension N will represent either n or $n+1$.

The first property we discuss is a special approximation of L^p functions by smooth ones. To this end, let φ be a nonnegative $C^1(\mathbb{R}^N)$ function with $\varphi(z) = 0$ if $|z| > 1$ and

$$\int_{\mathbb{R}^N} \varphi(z)\,dz = 1.$$

Two examples are φ_1 and φ_2 defined by

$$\varphi_1(z) = k_1(N)(1-|z|^2)^2, \varphi_2(z) = k_2(N)\exp(\frac{1}{|z|^2-1})$$

for $|z| < 1$ and $\varphi_1(z) = \varphi_2(z) = 0$ for $|z| \geq 1$. The constants $k_1(N)$ and $k_2(N)$ are chosen to give an integral of 1. (Note that φ_1 is C^1 but not C^2 while φ_2 is C^∞.) We define the *mollification* $u(x;\tau)$ of $u \in L^1_{\mathrm{loc}}(\mathbb{R}^N)$ by

$$u(x;\tau) = \int_{\mathbb{R}^N} u(x-\tau y)\varphi(y)\,dy, \tag{3.3}$$

and observe that $u(x;0) \equiv u(x)$. If $u \in L^1(\Omega)$ for some measurable subset Ω of $\mathbb{R}^N$, we can define $u(x;\tau)$ for all $(x,\tau) \in \mathbb{R}^{N+1}$ via (3.3) provided we extend u to be zero outside of Ω. If u is continuous on $\overline{\Omega}$ (or equivalently, if u is uniformly continuous on Ω), we can extend it to all of $\mathbb{R}^N$ by Exercise 3.1. More generally, we can define $u(x;\tau)$ for $u \in L^1_{\mathrm{loc}}(\Omega)$ if Ω is open, $x \in \Omega$, and $|\tau| < \mathrm{dist}(x, \partial\Omega)$. If u is in an appropriate space, then $u(\cdot;\tau) \to u$ as $\tau \to 0$ in that space. For example, if u is continuous, then the convergence is uniform. More precisely, we have the following result.

LEMMA 3.1. *If Ω is open, and $u \in C^0(\Omega)$, then $u(\cdot;\tau) \to u$ as $\tau \to 0$ uniformly on compact subsets of Ω. If $u \in C^0(\overline{\Omega})$, then the convergence is uniform in Ω.*

PROOF. Let Ω_1 be a bounded open subset of Ω with $h = \mathrm{dist}(\Omega_1, \partial\Omega) > 0$. If $|\tau| < h$ and $x \in \Omega_1$, then

$$\begin{aligned} |u(x;\tau) - u(x)| &= \left| \int [u(x-\tau y) - u(x)]\varphi(y)\,dy \right| \\ &\leq \int |u(x-\tau y) - u(x)|\,\varphi(y)\,dy \\ &\leq \sup_{|z|\leq|\tau|} |u(x-z) - u(x)|. \end{aligned}$$

Since u is uniformly continuous on Ω_1, it follows that this supremum tends to zero (as $\tau \to 0$) uniformly with respect to $x \in \Omega_1$. A similar argument applies if $u \in C^0(\overline{\Omega})$ after extension to all of $\mathbb{R}^N$. □

When $u \in L^p$, the convergence takes place in L^p.

LEMMA 3.2. *If $u \in L^p_{loc}(\Omega)$ for $1 \le p < \infty$, then $u(\cdot;\tau) \to u$ as $\tau \to 0$ in $L^p(\Omega_1)$ for any Ω_1 as in the proof of Lemma 3.1. If $u \in L^p(\Omega)$ for $1 \le p < \infty$, then $u(\cdot;\tau) \to u$ as $\tau \to 0$ in $L^p(\Omega)$ and $|u(\cdot;\tau)|_p \le |u|_p$.*

PROOF. With h as before and $|\tau| < h$, we have

$$\begin{aligned}
\int_{\Omega_1} |u(x;\tau) - u(x)|^p\, dx &= \int_{\Omega_1} \left| \int_{\mathbb{R}^N} [u(x-\tau y) - u(x)]\varphi(y)\, dy \right|^p dx \\
&\le \int_{\Omega_1} \int_{\mathbb{R}^N} |u(x-\tau y) - u(x)|^p \varphi(y)\, dy\, dx \\
&= \int_{\mathbb{R}^N} \int_{\Omega_1} |u(x-\tau y) - u(x)|^p \varphi(y)\, dx\, dy \\
&\le \sup_{|z| \le |\tau|} \int_{\Omega_1} |u(x-z) - u(x)|^p\, dx
\end{aligned}$$

by Hölder's inequality and Fubini's theorem. Since L^p functions are continuous with respect to translation, this supremum tends to zero as $\tau \to 0$, and the first part of the lemma is proved.

For the second part, we extend u to an $L^p(\mathbb{R}^N)$ function and argue as before to deduce the estimate on the L^p norms. □

Now we define weak derivatives. For any multi-index α, and any functions u and v in $L^1_{loc}(\Omega)$, we say that $v = D^\alpha u$ and we call v *the weak α-derivative of u* if

$$\int_\Omega \zeta v\, dx = (-1)^\alpha \int_\Omega u D^\alpha \zeta\, dx$$

for all $\zeta \in C_0^{|\alpha|}(\Omega)$. We leave it to the reader to show (in Exercise 3.2) that there is no more than one function V satisfying this definition. Another useful observation is that the weak derivative of the mollification is the same as the mollification of the weak derivative.

LEMMA 3.3. *Let $u \in L^1_{loc}(\Omega)$, let α be a multi-index, and suppose that the weak derivative $D^\alpha u = v$ exists. If $\varphi \in C^{|\alpha|}$, then*

$$D^\alpha u(x;\tau) = v(x;\tau). \tag{3.4}$$

PROOF. By direct calculation, we have

$$\begin{aligned}
D^\alpha u(x;\tau) &= D^\alpha \left(\int u(x-\tau y)\varphi(y)\,dy \right) \\
&= D^\alpha \left(\tau^{-N} \int u(z)\varphi((x+z)/\tau)\,dz \right) \\
&= (-1)^{|\alpha|}\tau^{-|\alpha|-N} \int u(z) D^\alpha \varphi((x+z)/\tau)\,dz \\
&= \tau^{-N} \int v(z)\varphi((x+z)/\tau)\,dz = v(x;\tau).
\end{aligned}$$

□

Next, we show that a function is weakly differentiable if and only if it can be suitably approximated. To state this result precisely, we define

$$\Omega_h = \{\operatorname{dist}(x,\partial\Omega) > h\}$$

for $h < \operatorname{diam}\Omega$. We note that if Ω is bounded, then for any $h < \frac{1}{2}\operatorname{diam}\Omega$, there is a function $\zeta_h \in C^\infty(\overline{\Omega})$ with support in Ω_h and $\zeta_h \equiv 1$ in Ω_{2h}. (See Exercise 3.3.)

THEOREM 3.4. *Let u and v be in $L^1_{loc}(\Omega)$ and let α be a multi-index. Then $v = D^\alpha u$ if and only if there is a sequence $(u_m) \subset C^\infty(\Omega)$ such that $u_m \to u$ and $D^\alpha u_m \to v$ in $L^1_{loc}(\Omega)$.*

PROOF. If $v = D^\alpha u$, then the sequence is given by $u_m = \zeta_{1/m} u(\cdot;1/m)$, where $\zeta_{1/m}$ is the function corresponding to $\Omega_{1/m}$ described above. Conversely, if the approximating sequence (u_m) is given, then an integration by parts gives

$$(-1)^{|\alpha|} \int_\Omega u_m D^\alpha \zeta\,dx = \int_\Omega D^\alpha u_m \zeta\,dx.$$

Sending $m \to \infty$ gives

$$(-1)^{|\alpha|} \int_\Omega u D^\alpha \zeta\,dx = \int_\Omega v\zeta\,dx,$$

and therefore $v = D^\alpha u$. □

Theorem 3.4 also implies that the product rule $D_i(uv) = uD_iv + vD_iu$ holds provided u, v, uv, and $uD_iv + vD_iu$ are all in L^1_{loc}. Moreover if u is locally Lipschitz in the direction e_j parallel to the x^j-axis, then D_ju exists and is in L^∞_{loc}. (See Exercise 3.4.) Furthermore, the difference quotients of a weakly differentiable function converge to the appropriate derivative.

LEMMA 3.5. *If D_iu exists, then*

$$\lim_{h\to 0} \int \frac{u(x+he_i)-u(x)}{h}\zeta(x)\,dx = \int D_iu\zeta\,dx \tag{3.5}$$

for all $\zeta \in C_0^1$. *If* $D_i u \in L^p(\Omega)$ *for some* $p \geq 1$, *then*

$$\left| \frac{u(x+he_i)-u(x)}{h} \right|_{p;\Omega_h} \leq |D_i u|_p. \tag{3.6}$$

Moreover, if there is a constant K such that

$$\left| \frac{u(x+he_i)-u(x)}{h} \right|_{p;\Omega_h} \leq K \tag{3.7}$$

for all (sufficiently small) h, then $D_i u \in L^p$ *and*

$$|D_i u|_p \leq K. \tag{3.8}$$

PROOF. To prove (3.5), we suppose h is so small that ζ has support in Ω_{2h} and define the operators $\Delta^{\pm}$ by $\Delta^{\pm} v(x) = (v(x \pm he_i) - v(x))/h$. It follows that

$$\int \Delta^+ u\zeta\, dx = -\int u\Delta^- \zeta\, dx.$$

The right side of this equation converges to $-\int uD_i\zeta\, dx$ because $u \in L^1(\operatorname{supp}\zeta)$ and this integral is just $\int D_i u\zeta\, dx$.

To prove (3.6), we note that

$$\begin{aligned}
\int_{\Omega_h} |\Delta^+ u|^p\, dx &= \lim_{\tau\to 0} \int_{\Omega_h} |\Delta^+ u(x;\tau)|^p\, dx \\
&= \lim_{\tau\to 0} \int_{\Omega_h} \left| \int_0^1 D_i u(x+the_i;\tau)\, dt \right|^p dx \\
&\leq \limsup_{\tau\to 0} \int_{\Omega_h} \int_0^1 |D_i u(x+the_i;\tau)|^p\, dt\, dx \\
&= \limsup_{\tau\to 0} \int_0^1 \int_{\Omega_h} |D_i u(x+the_i;\tau)|^p\, dx\, dt \\
&\leq \int_0^1 \int_{\Omega_h} |D_i u(x+the_i)|^p\, dx\, dt \leq \int_\Omega |D_i u|^p\, dx.
\end{aligned}$$

(Here, we used Lemma 3.2 in proving the second to the last inequality.)

Finally, if (3.7) holds, we fix $\zeta \in C_0^1$ and H such that the support of ζ is contained in Ω_H. By the weak compactness of bounded sets in $L^p(\Omega_H)$, there is a function $v \in L^p(\Omega_H)$ and a sequence (h_j) such that $h_j \to 0$ as $j \to \infty$ such that $|v|_{p;\Omega_H} \leq K$ and

$$\lim_{j\to\infty} \int_{\Omega_H} \Delta_j u\zeta\, dx = \int_{\Omega_H} v\zeta\, dx,$$

where $\Delta_j u(x) = (u(x+h_j e_i) - u(x))/h_j$. On the other hand, we have

$$\lim_{j\to\infty} \int_{\Omega_H} \Delta_j u\zeta\, dx = \lim_{j\to\infty} -\int_{\Omega_H} u \frac{\zeta(x-h_j e_i)-\zeta(x)}{h_j}\, dx = -\int_{\Omega_H} uD_i\zeta\, dz.$$

Exercise 3.5 shows that $v = D_i u$. □

We can also prove a chain rule for weakly differentiable functions.

LEMMA 3.6. *If $f \in C^1(\mathbb{R})$, $f' \in L^\infty(\mathbb{R})$, and u and D_iu are in $L^1_{loc}(\Omega)$, then $D_i(f \circ u)$ exists and $D_i(f \circ u) = f'(u)D_iu$.*

PROOF. Let (u_m) be a sequence of smooth functions such that $u_m \to u$ and $D_iu_m \to D_iu$ in $L^1_{\text{loc}}(\Omega)$, and let M be a nonnegative constant such that $|f'| \le M$. Then, for all sufficiently small positive h,

$$\lim_{m\to\infty} \int_{\Omega_h} |f(u_m) - f(u)|\,dx \le \lim_{m\to\infty} \int_{\Omega_h} M\,|u_m - u|\,dx = 0,$$

and

$$\int_{\Omega_h} \left|f'(u_m)D_iu_m - f'(u)D_iu\right|\,dx$$
$$\le \int_{\Omega_h} \left|f'(u_m)\right| |D_iu_m - D_iu|\,dx + \int_{\Omega_h} \left|f'(u_m) - f'(u)\right| |D_iu|\,dx.$$

Next, we have

$$\int_{\Omega_h} \left|f'(u_m)\right| |D_iu_m - D_iu|\,dx \le M \int_{\Omega_h} |D_iu_m - D_iu|\,dx \to 0$$

as $m \to \infty$. Since

$$\left|f'(u_m) - f'(u)\right| |D_iu| \le 2M\,|D_iu|,$$

and $|f'(u_m) - f'(u)|\,|D_iu|$ converges to zero almost everywhere in Ω_h, it follows that

$$\int_{\Omega_h} \left|f'(u_m) - f'(u)\right| |D_iu|\,dx \to 0$$

as well. Hence $f(u_m) \to f(u)$ and $D_i(f \circ u_m) = f'(u_m)D_iu_m \to f'(u)D_iu$ in L^1_{loc}. It follows from Theorem 3.4 that $f(u)$ is differentiable and $D_i(f \circ u) = f'(u)D_iu$. □

In fact, Lemma 3.6 remains valid as long as f is just globally Lipschitz (see [348, Section 2.1]) in which case f' exists almost everywhere because Lipschitz functions are absolutely continuous. For our purposes we only need to consider a special class of Lipschitz, non-C^1 functions.

LEMMA 3.7. *If D_iu exists, then so does D_iu^+, and*

$$D_iu^+ = \begin{cases} D_iu & \text{if } u > 0 \\ 0 & \text{if } u \le 0, \end{cases} \tag{3.9a}$$

$$D_iu^- = \begin{cases} 0 & \text{if } u \ge 0 \\ D_iu & \text{if } u < 0, \end{cases} \tag{3.9b}$$

and

$$D_i|u| = \begin{cases} D_iu & \text{if } u > 0 \\ 0 & \text{if } u = 0 \\ -D_iu & \text{if } u < 0. \end{cases} \tag{3.9c}$$

PROOF. To prove (3.9a), define f_ε by

$$f_\varepsilon(z) = \begin{cases} (z^2+\varepsilon^2)^{1/2} - \varepsilon & \text{if } z > 0 \\ 0 & \text{if } z \leq 0. \end{cases}$$

It follows from Lemma 3.6 that

$$\int_\Omega f_\varepsilon(u) D_i\zeta\, dx = -\int_{\{u>0\}} \zeta \frac{uD_iu}{(u^2+\varepsilon^2)^{1/2}}\, dx$$

for any $\zeta \in C_0^1(\Omega)$, and we can send ε to zero on both sides of this equation to obtain

$$\int_\Omega u^+ D_i\zeta\, dx = -\int_{\{u>0\}} \zeta D_iu\, dx.$$

Hence we have proved (3.9a). The other two results now follow because $u^- = -(-u)^+$ and $|u| = u^+ + u^-$. □

COROLLARY 3.8. *If D_iu exists, then $D_iu = 0$ almost everywhere on $\{u = c\}$ for any constant c.*

PROOF. Apply (3.9c) to $u - c$. □

We also need the following weak compactness result for bounded subsets of $W^{1,p}(\Omega)$, where we define $W^{k,p}(\Omega)$ (k a positive integer and $p \in [1,\infty]$) to be the set of all functions u such that $D^\alpha u \in L^p$ for any multi-index α with $|\alpha| \leq k$. (By definition, $D^\alpha u = u$ when $|\alpha| = 0$.) A norm on $W^{k,p}$ is given by

$$|u|_{k,p} = \left(\sum_{|\alpha|\leq k} \int_\Omega |D^\alpha u|^p\, dx \right)^{1/p}.$$

LEMMA 3.9. *If (u_m) is a bounded sequence in $W^{1,p}(\Omega)$ with $1 < p < \infty$, then there is a function $u \in W^{1,p}$ and a subsequence $(u_{m(k)})$ such that $u_{m(k)} \to u$ weakly in $W^{1,p}$.*

PROOF. The weak compactness of L^p implies that there are functions u, $v_1, \ldots, v_n$ and a subsequence $(u_{m(k)})$ such that $u_{m(k)} \to u$ and $D_iu_{m(k)} \to v_i$ weakly in L^p. The weak convergence of these subsequences guarantees that $D_iu = v_i$. □

It will be useful to have a concept of boundary values for $W^{1,2}$ functions. We define $W_0^{1,2}$ to be the closure in the $W^{1,2}$ norm of C_0^∞ (or, equivalently, C_0^1), and we say that $u = \varphi$ on $\partial\Omega$ for $\varphi \in W^{1,2}$ if $u - \varphi \in W_0^{1,2}$.

We close this section with Poincaré's inequality. Further discussion of this inequality appears in Exercise 3.6 and Section VI.4.

LEMMA 3.10. *Suppose $u \in W_0^{1,2}(\Omega)$ and* $\operatorname{diam}\Omega \le 2R$ *for some positive R. Then*

$$\int_\Omega u^2\,dx \le R^2 \int_\Omega |Du|^2\,dx. \tag{3.10}$$

PROOF. Suppose first that $u \in C_0^1$ and $|x^1| < R$ for $x \in \Omega$. Then

$$u(x) = \begin{cases} \int_{-R}^{x^1} D_1u(t,x^2,\dots,x^N)\,dt & \text{if } x^1 < 0 \\ \int_R^{x^1} D_1u(t,x^2,\dots,x^N)\,dt & \text{if } x^1 \ge 0. \end{cases}$$

Now square this equation, divide by R^2 and integrate with respect to x^1 from $-R$ to 0. Using x' as an abbreviation for $(x^2,\cdots,x^N)$, we see that

$$\begin{aligned}\int_{-R}^0 \left(\frac{u(x)}{R}\right)^2 dx^1 &= \int_{-R}^0 \left(\frac{1}{R}\int_{-R}^{x^1} D_1u(t,x')\,dt\right)^2 dx^1 \\ &\le \int_{-R}^0 \left(\frac{1}{R}\int_{-R}^{0} |Du(t,x')|\,dt\right)^2 dx^1 \\ &= R\left(\frac{1}{R}\int_{-R}^{0} |Du(t,x')|\,dt\right)^2 \\ &\le \int_{-R}^0 |Du(t,x')|^2\,dt = \int_{-R}^0 |Du(x)|^2\,dx^1.\end{aligned}$$

Adding the corresponding inequality obtained by integrating for positive x^1 yields

$$\int_{-R}^R u(x)^2\,dx^1 \le R^2\int_{-R}^R |Du(x)|^2\,dx^1.$$

If we now integrate this inequality with respect to x' and recall that u and D_1u vanish outside Ω, we infer (3.10) for any $u \in C_0^1$. The result for general $u \in W_0^{1,2}$ now follows by approximation. □

The proof of this lemma gives a more general result, which we will use to study strong solutions in Chapter VII.

COROLLARY 3.11. *Suppose $u \in W^{1,2}(\Omega)$ with $u = 0$ on some subset σ of $\partial\Omega$. Suppose also that there are $R > 0$ and a unit vector β such that, for every $x \in \Omega$, there are $r \in (0,R)$ and $x_0 \in \sigma$ with $x = x_0 + r\beta$ and $x + \rho\beta \in \Omega$ for every $\rho \in (0,r)$. Then* (3.10) *holds.*

2. The method of continuity

To prove the existence of solutions to various linear boundary value problems, we use some simple functional analysis. This section discusses the relevant facts.

A map T from a metric space (X,ρ) to itself is called a *contraction map* if there is a constant $\theta \in (0,1)$ such that $\rho(Tx,Ty) \leq \theta\rho(x,y)$ for all x and y in X. We say that T has a fixed point $x \in X$ if $Tx = x$. Our first result asserts the existence of a fixed point for contraction maps in complete metric spaces.

THEOREM 3.12. *A contraction map T on a complete metric space (X,ρ) has a unique fixed point.*

PROOF. Choose $x_0 \in X$ and define a sequence (x_i) inductively by $x_{i+1} = Tx_i$. If $k > m$ are positive integers, then

$$\begin{aligned}\rho(x_k,x_m) &\leq \sum_{i=k+1}^{m} \rho(x_{i-1},x_i) = \sum_{i=k+1}^{m} \rho(T^{i-1}x_0,T^{i-1}x_1) \\ &\leq \sum_{i=k+1}^{m} \theta^{i-1}\rho(x_0,x_1) \leq \frac{\theta^k}{1-\theta}\rho(x_0,x_1).\end{aligned}$$

Since this last expression tends to zero as $k \to \infty$, it follows that (x_i) is a Cauchy sequence and hence has a limit x. Moreover, $\rho(x,Tx) = \lim \rho(x_i,x_{i+1}) = 0$, so x is a fixed point. Finally if y is another fixed point, then

$$\rho(x,y) = \rho(Tx,Ty) \leq \theta\rho(x,y),$$

which implies that $x = y$. □

If $\mathscr{V}_1$ and $\mathscr{V}_2$ are two normed linear spaces, a map $T : \mathscr{V}_1 \to \mathscr{V}_2$ is *bounded* if its norm $|T|$ defined by

$$|T| = \sup_{\substack{x\in\mathscr{V}_1 \\ x\neq 0}} \frac{|Tx|}{|x|}$$

is finite. A linear map is continuous if and only if it is bounded. Our next theorem, known as the method of continuity, asserts that two suitably related bounded linear maps are either both invertible or both non-invertible.

THEOREM 3.13. *Let $\mathscr{B}$ be a Banach space, let $\mathscr{V}$ be a normed linear space, and let L_0 and L_1 be two bounded linear maps from $\mathscr{B}$ into $\mathscr{V}$. For $h \in [0,1]$, define $L_h = hL_1 + (1-h)L_0$, and suppose there is a constant C such that*

$$|x|_{\mathscr{B}} \leq C|L_h x|_{\mathscr{V}} \tag{3.11}$$

for all $h \in [0,1]$. Then L_0 is invertible if and only if L_1 is invertible.

PROOF. Note first that L_h is injective for all h, so we need only prove the surjectivity of the appropriate maps. Denote by S the set of all $h \in [0,1]$ such that L_h is surjective and suppose that S is nonempty. Choose $s \in S$, fix $v \in \mathscr{V}$, let $h \in [0,1]$, and define $T : \mathscr{B} \to \mathscr{B}$ by

$$Tx = L_s^{-1} v + (h-s) L_s^{-1} (L_0 - L_1) x.$$

A straightforward calculation shows that x is a fixed point for T if and only if $L_h x = v$. In addition,

$$|Tx - Ty| = \left|(h-s) L_s^{-1}(L_0 - L_1)(x-y)\right| \le |h-s| \frac{1}{C} |L_0 - L_1|\, |x-y|$$

for all x and y in $\mathscr{B}$. It follows that T is a contraction for $|h-s| \le \delta = \frac{C}{2|L_0 - L_1|}$, and then T has a unique fixed point for $|h-s| \le \delta$ and $0 \le h \le 1$ by Theorem 3.12. It follows that $h \in S$ for any such h and hence that $S = [0,1]$. □

3. Problems in small balls

We now introduce a fixed positive definite matrix (a^{ij}) with constant entries and eigenvalues bounded by the positive constants λ and Λ. In other words,

$$\lambda |\xi|^2 \le a^{ij} \xi_i \xi_j \le \Lambda |\xi|^2$$

for all $\xi \in \mathbb{R}^n$. For nonnegative constants ε and η, we define the operator $L_{\varepsilon,\eta}$ by

$$L_{\varepsilon,\eta} u = D_i(a^{ij} D_j u) + (\varepsilon u_t)_t - \eta u_t,$$

which we consider on the ball $B = B(R) = \{|x|^2 + |t|^2 < R^2\}$. We say that u is a $W^{1,2}$ *solution* of the problem

$$L_{\varepsilon,\eta} u = f \text{ in } B,\ u = \varphi \text{ on } \partial B \tag{3.12}$$

if $u \in W^{1,2}(B)$, $u - \varphi \in W_0^{1,2}(B)$ and

$$\int a^{ij} D_j u D_i \zeta + \varepsilon u_t \zeta_t + \eta u_t \zeta \, dX = \int f \zeta \, dX \tag{3.12$'$}$$

for all $\zeta \in C_0^2(B)$, where we suppress the set of integration B. We call ζ a *test function*. Note that if u is a $W^{1,2}$ solution of (3.12), then (3.12)$'$ holds for all $\zeta \in W_0^{1,2}$. Moreover, for any $\zeta \in C_0^2(B)$, we have

$$\int a^{ij} D_j u D_i \zeta \, dX = -\int u a^{ij} D_{ij} \zeta \, dX,$$

so that there is no loss of generality in assuming that (a^{ij}) is a symmetric matrix. (As we shall see later, such an assertion need not be true if (a^{ij}) is not a constant matrix.) Our first step is to show that (3.12) has a unique solution for all appropriate f and φ.

LEMMA 3.14. *Suppose that $\varepsilon > 0$. If $f \in L^2$ and $\varphi \in W^{1,2}$, then* (3.12) *has a unique $W^{1,2}$ solution.*

PROOF. If u solves (3.12), we note that $v = u - \varphi$ solves

$$L_{\varepsilon,\eta} v = g \text{ in } B,\ v \in W_0^{1,2}(B) \tag{3.13}$$

for $g = f - L_{\varepsilon,\eta}\varphi$. Conversely if v solves (3.13) with this choice of g, then $u = v + \varphi$ solves (3.12). Hence it suffices to show that (3.13) is uniquely solvable for any $g \in L^2$.

We first consider the special case that $\eta = 0$. In this case, we define an inner product on $W_0^{1,2}$ by

$$\langle w, z\rangle = \int a^{ij} D_j w D_i z + \varepsilon w_t z_t \, dX. \tag{3.14}$$

Since $W_0^{1,2}$ is a Hilbert space with respect to this metric, we can define a bounded linear functional F on $W_0^{1,2}$ by

$$F(z) = \int gz dX.$$

The Riesz representation theorem then gives a unique $v \in W_0^{1,2}$ such that $F(z) = \langle v, z\rangle$, which is the unique solution of (3.13) in this case.

We now use the method of continuity to solve (3.13) for $\eta > 0$. To this end, we use $\|\cdot\|$ to denote the norm associated with the inner product $\langle,\rangle$, and we define $\mathscr{L}_h = L_{\varepsilon,h\eta}$ for $h \in [0,1]$. By what was already proved, we know that $\mathscr{L}_0$ is invertible and that $\mathscr{L}_0$ and $\mathscr{L}_1$ are bounded. If $\mathscr{L}_h w = g$, we can use the test function w to see that

$$\|w\|^2 = \int a^{ij} D_i w D_j w + \varepsilon w_t^2 \, dX = h\eta \int w w_t \, dX + \int wg \, dX.$$

But $\int w w_t \, dX = \int \frac{1}{2}(w^2)_t \, dX = 0$, so

$$\|w\|^2 \leq \int wg \, dX \leq \theta \int w^2 \, dX + \frac{1}{\theta} \int g^2 \, dX$$

for any $\theta > 0$. Now Poincaré's inequality gives

$$\int w^2 \, dX \leq R^2 \int |Dw|^2 + w_t^2 \, dX \leq R^2 \left(\frac{1}{\lambda} + \frac{1}{\varepsilon}\right) \|w\|^2,$$

so choosing θ small enough gives $\|w\|^2 \leq C\|g\|^2$, and therefore $\mathscr{L}_0$ and $\mathscr{L}_1$ satisfy the hypotheses of Theorem 3.13. Therefore $\mathscr{L}_h$ is invertible for all $h \in [0,1]$, and the invertibility of $\mathscr{L}_1$ is equivalent to the solvability of (3.13). □

Let us note here that the argument presented in Lemma 3.14 is easily modified to handle any bounded domain Ω in place of B.

Next, we need a simple version of the maximum principle for these operators.

LEMMA 3.15. *Let $\varphi \in C^0(\bar{B}) \cap W^{2,2}(B)$ satisfy the inequality $\varphi \leq M$ on ∂B for some constant M. If $u \in W^{1,2}$ satisfies*

$$0 \geq \int [a^{ij} D_j u D_i v + \varepsilon u_t v_t - \eta v u_t] \,, dX \tag{3.15}$$

for all $v \in C_0^2$ and if $u - \varphi \in W_0^{1,2}$, then $u \leq M$ in B.

PROOF. As already noted, we can take as test function v in (3.15) any function in $W_0^{1,2}$. We take $v = (u - M)^+$ and note that $v \in W_0^{1,2}$ by Exercise 3.7. It follows that

$$0 \geq \int [a^{ij} D_j u D_i v + \varepsilon u_t v_t - \eta v u_t] \, dX.$$

Since $v u_t = \frac{1}{2}(v^2)_t$, we have

$$0 \geq \int [a^{ij} D_j u D_i v + \varepsilon u_t v_t] \, dX.$$

The integrand here is nonnegative and hence it must be zero. Therefore $Dv = 0$ and $v_t = 0$, and then Exercise 3.8 implies that v is a constant, which must be zero. □

For simplicity, we shall write $L_{\varepsilon,\eta} u \geq 0$ in B if (3.15) is satisfied.

We now use Lemma 3.15 to prove an important *a priori* estimate on solutions of (3.12) for sufficiently regular data. Some of the steps in deriving this estimate will reappear in the quasilinear theory.

LEMMA 3.16. *Suppose that $\varphi \in C^2(\bar{B})$, $f \in W^{1,\infty}(B)$, and $\varepsilon \in (0,1]$, and let Φ and F be constants such that*

$$|D\varphi| \leq \Phi, |D^2\varphi| + |\varphi_t| + |\varphi_{tt}| \leq \Phi, \; |f| + |Df| + |f_t| \leq \lambda F. \tag{3.16}$$

If $R \leq \lambda/2$ and $\eta \in [0,1]$, then any $W^{1,2}$ solution of (3.12) *satisfies the estimate*

$$|Du| + |u_t| \leq C(n, \lambda, \Lambda, R)(\Phi + F). \tag{3.17}$$

PROOF. Define $w(X) = R^2 - |x|^2 - t^2$ and write L for $L_{\varepsilon,\eta}$ and B for $B(R)$. Then

$$Lw = -2\mathscr{T} - 2\varepsilon + 2\eta t \leq -2n\lambda + 2\eta R \leq -2n\lambda + 2R \leq -\lambda$$

and

$$|L\varphi| \leq (2n\Lambda + \varepsilon + \eta)\Phi \leq k_1 \lambda \Phi$$

in B, where $k_1 = (2n\Lambda + 2)/\lambda$. Hence, $v = (u - \varphi) - (k_1\Phi + F)w$ satisfies $Lv \geq 0$ in B and $v \leq 0$ on ∂B, so $v \leq 0$ in B by Lemma 3.15. Therefore $u - \varphi \leq (k_1\Phi + F)w$, and a similar argument gives $-(u - \varphi) \leq (k_1\Phi + F)w$, so $|u - \varphi| \leq (k_1\Phi + F)w$.

From this inequality, we can derive a boundary Lipschitz estimate. For $X \in B$ and $Y \in \partial B$ and $\|X-Y\| = (|x-y|^2+(t-s)^2)^{1/2}$, we have

$$\begin{aligned}|u(X)-u(Y)| &= |u(X)-\varphi(Y)| \le |u(X)-\varphi(X)|+|\varphi(X)-\varphi(Y)| \\ &\le (k_1\Phi+F)w(X)+2\Phi\|X-Y\| \\ &\le 2[(k_1R+1)\Phi+FR]\,\|X-Y\|.\end{aligned}$$

Hence, setting $K = 2[(k_1R+1)\Phi+FR]$, we have

$$|u(X+\tau)-u(X)| \le K\|\tau\| \tag{3.18}$$

for any vector $\tau \in \mathbb{R}^{n+1}$ and any $X \in \partial B_\tau$, where $B_\tau = \{Y \in B : Y+\tau \in B\}$, such that B_τ is nonempty.

We use (3.18) to prove the interior estimate. Fix τ so that B_τ is nonempty and define $v^\pm(X) = \pm(u(X+\tau)-u(X)) - K\|\tau\| - F\|\tau\|\,w(X)$. A simple calculation shows that $Lv^\pm \ge 0$ in B_τ and $v^\pm \le 0$ on ∂B_τ, so $v^\pm \le 0$ in B_τ. Hence

$$|u(X+\tau)-u(X)| \le (K+FR^2)\,\|\tau\|$$

for all $X \in B_\tau$. Then Lemma 3.5 implies that $|Du|+|u_t| \le K+FR^2$, which implies (3.17). □

Now we specialize to $\eta = 1$ and send $\varepsilon \to 0$, noting that the estimates derived in Lemma 3.16 are independent of ε. For simplicity, we set $H = L_{0,1}$, which is just the heat operator when $a^{ij} = \delta^{ij}$.

THEOREM 3.17. *If $\varphi \in C^2(\overline{B})$, $f \in W^{1,\infty}(B)$, and $R \le \lambda/2$, then there is a unique $W^{1,2}$ solution u of*

$$Hu = f \text{ in } B, u = \varphi \text{ on } \partial B. \tag{3.19}$$

Moreover, $u \in W^{1,\infty}$ and (3.17) *holds.*

PROOF. Let u_ε be the unique $W^{1,2}$ solution of

$$L_{\varepsilon,1}u_\varepsilon = f \text{ in } B,\ u_\varepsilon = \varphi \text{ on } \partial B$$

given by Lemma 3.14. Then Lemma 3.16 implies that the family $(u_\varepsilon)_{0<\varepsilon\le 1}$ is uniformly bounded and equicontinuous, so there is a uniformly convergent sequence $(u_{\varepsilon(m)})$; we write u for the limit of this sequence. By Lemma 3.16, this limit is also Lipschitz, satisfying (3.17), and hence $u \in W^{1,2}$. In addition, for any $\zeta \in C^2_0$, we have

$$\begin{aligned}\int f\zeta\,dX &= \int [a^{ij}D_ju_{\varepsilon(m)}D_i\zeta + \varepsilon(m)u_{\varepsilon(m),t}\zeta_t + u_{\varepsilon(m),t}\zeta]\,dX \\ &= -\int u_{\varepsilon(m)}[a^{ij}D_{ij}\zeta + \varepsilon(m)\zeta_{tt} - \zeta_t]\,dX.\end{aligned}$$

Now we send $m \to \infty$ and use the uniform boundedness of the sequence $(u_{\varepsilon(m)})$ to see that

$$\int f\zeta\,dX = -\int u[a^{ij}D_{ij}\zeta - \zeta_t]\,dX = \int [a^{ij}D_juD_i\zeta + u_t\zeta]\,dX.$$

Hence u solves (3.19), and the theorem is proved. □

For further existence results, we study the interior regularity of solutions of (3.19). It is important for the present considerations that these estimates not depend on the C^2 estimates for φ, and we only need to consider the case $f \equiv 0$. A key element is the function w from Lemma 3.16.

LEMMA 3.18. *If u is a bounded $W^{1,2}$ solution of $Hu = 0$, then u is a $C^{2,1}$ solution. If $|u| \le M$ in B, then*

$$w^2|Du|^2 + w^4|u_t|^2 \le CM^2. \tag{3.20}$$

PROOF. Fix $\tau \in (-R/2, R/2)$ and set $v(X) = u(X;\tau)$. For $r \in (R/2, R - |\tau|)$, we have $v \in C^\infty(B(r))$ and hence, for $\zeta \in C_0^2(B(r))$, we have

$$\begin{aligned}\int a^{ij}D_jvD_i\zeta\,dX &= \int_{B(r)}\int_{\mathbb{R}^{n+1}} a^{ij}D_ju(X - \tau Y)\varphi(Y)\,dY D_i\zeta(X)\,dX \\ &= \int_B\int_{\mathbb{R}^{n+1}} a^{ij}D_ju(X)\varphi(Y)D_i\zeta(X + \tau Y)\,dY\,,dX \\ &= \int_B a^{ij}D_juD_i\zeta_\tau\,dX,\end{aligned}$$

where we have used the abbreviation $\zeta_\tau = -\zeta(\cdot;\tau)$. Similarly,

$$\int v_t\zeta\,dX = \int u_t\zeta_\tau\,dX,$$

so $Hv = 0$ in $B(R - |\tau|)$ in the weak sense. It follows that $Hv = 0$ in the classical sense as well. (See Exercise 3.9.)

To prove (3.20), we use an idea that goes back to Bernstein [16], namely that $|Dv|^2$ is a subsolution of a certain differential inequality. We implement this idea by defining $W = r^2 - |x|^2 - t^2$ with $R/2 < r < R - |\tau|$ and computing

$$\begin{aligned}H(W^2|Dv|^2) = H(W^2)|Dv|^2 &+ 2[a^{ij}D_{ijk}v - D_kv_t]D_kvW^2 \\ &+ 2a^{ij}D_{ik}vD_{jk}vW^2 + 4a^{ij}D_{ik}vD_kvD_j(W^2).\end{aligned}$$

Now we note that

$$H(W^2) = 2WH(W) + 2a^{ij}D_iWD_jW \ge 2WH(W) = -4\mathcal{T}W + 4tW$$

and that $a^{ij}D_{ijk}v - D_kv_t = 0$. A simple application of Cauchy's inequality now gives us

$$H(W^2|Dv|^2) \ge -C_1(\Lambda, n, R)|Dv|^2.$$

In addition,

$$H(v^2) = 2vHv + 2a^{ij}D_ivD_jv \ge 2\lambda|Dv|^2,$$

so we can apply the maximum principle to $W^2|Dv|^2 + (C_1/\lambda)v^2$ in B_r to find that

$$W^2|Dv|^2 \le \frac{C_1}{\lambda}M^2.$$

Choosing suitable constants C_2 and C_3, we can then apply the maximum principle to $W^4|D^2v|^2 + C_2W^2|Dv|^2 + C_3v^2$ and obtain

$$W^4|D^2v|^2 \le C(C_1/\lambda, C_2, C_3)M^2.$$

Hence from the differential equation $Hv = 0$, we also obtain

$$W^4|v_t|^2 \le C(C_1/\lambda, C_2, C_3)M^2.$$

In a similar fashion, we can derive local L^∞ bounds for any derivative of v. Sending $\tau \to 0$ gives the corresponding bounds for the derivatives of u. Hence

$$W^2|Du|^2 + W^4|u_t|^2 \le CM^2,$$

with C independent of r. Sending $r \to R$ completes the proof. □

In fact, such an estimate also holds for the inhomogeneous equation $Hu = f$. To prove this statement, we first define a slightly more general concept than the $W^{1,2}$ solutions already considered. We say that u is a *continuous weak solution* of (3.12) if $\varphi \in C^0(\bar{B})$, $u \in W^{1,2}_{\text{loc}}(B) \cap C^0(\bar{B})$, u satisfies (3.12)′ for all $\zeta \in C_0^2(B)$, and $u = \varphi$ on ∂B.

THEOREM 3.19. *If* $\varphi \in C^0(\overline{B})$, $f \in W^{1,\infty}(B)$, *and* $R \le \lambda/2$, *then there is a unique continuous weak solution* u *of* (3.19). *In addition,*

$$w^2|Du|^2 + w^4|u_t|^2 \le C(\lambda, \Lambda, n, R)(\sup|\varphi|^2 + \sup(|f|^2 + |Df|^2 + |f_t|^2)). \quad (3.21)$$

PROOF. Let (φ_m) be a sequence of C^2 functions converging uniformly to φ and write u_m for the solution of $Hu_m = f$ in B, $u_m = \varphi_m$ on ∂B. Then $H(u_m - u_k) = 0$ in B and $u_m - u_k = \varphi_m - \varphi_k$ on ∂B for any m and k. Hence the sequence (u_m) is uniformly Cauchy on ∂B, so the maximum principle implies that it is also uniformly Cauchy in B. Applying Lemma 3.16 to $u_m - u_k$ shows that $Du_m - Du_k$ and $u_{m,t} - u_{k,t}$ tend uniformly to zero on compact subsets of B, so the limit function u is a weak solution of (3.19), which is unique by Corollary 2.5 since the difference of any two solutions is a classical solution of $Lu = 0$ in B, $u = 0$ on $\mathcal{P}B$.

To prove (3.21), we write v for the solution of $Hv = 0$ in B, $v = \varphi$ on ∂B given by the first part of the proof. Then Lemma 3.18 and the maximum principle imply that

$$w^2|Dv|^2 + w^4|v_t|^2 \le C(\sup|\varphi|)^2.$$

Since $H(u - v) = f$ in B and $u - v = 0$ on ∂B, it follows from Theorem 3.17 that

$$w^2|D(u-v)|^2 + w^4|(u-v)_t|^2 \le |D(u-v)|^2 + |(u-v)_t|^2 \le CF^2.$$

Combining these two inequalities gives (3.21). □

A corresponding result is true for solutions of $Hu = f$ in lower half-balls $B^- = \{X \in B : t < 0\}$.

COROLLARY 3.20. *If $u \in W^{1,2} \cap C(B^-)$ is a $W^{1,2}$ solution of $Hu = f$ in B^-, then*

$$w^2 |Du|^2 + w^4 |u_t|^2 \leq C(\lambda, \Lambda, n, R)(\sup |u|^2 + \sup(|f|^2 + |Df|^2 + |f_t|^2)). \quad (3.21)'$$

PROOF. Fix $r \in (0, R)$ and $\varepsilon \in (0, R - r)$, let $u_{r,\varepsilon}$ be the restriction of u to $B^-_{r,\varepsilon} = \{X \in B : t < -\varepsilon, |x|^2 + t^2 < r^2\}$, and write φ for a continuous extension of $u_{r,\varepsilon}$ to $B(r)$ and F for a $W^{1,\infty}$ extension of f to $B(r)$ with $\sup |\varphi| \leq \sup |u|$, $\sup |F| \leq \sup |f|$, and $\sup |DF| + \sup |F_t| \leq \sup |Df| + \sup |f_t|$ as given by Exercise 3.1. Then the solution v of $Hv = F$ in $B(r)$, $v = \varphi$ on $\partial B(r)$ satisfies the estimate

$$\begin{aligned} W^2 |Dv|^2 + W^4 |v_t|^2 &\leq c(n, \lambda, \Lambda, R)(\sup |\varphi|^2 + \sup(|F|^2 + |DF|^2 + |F_t|^2)) \\ &\leq C(\sup |u|^2 + \sup(|f|^2 + |Df|^2 + |f_t|^2)) \end{aligned}$$

for $W = r^2 - t^2 - |x|^2$. Restricting this inequality to $B^-_{r,\varepsilon}$ and noting that $u = v$ in $B^-_{r,\varepsilon}$ by the maximum principle, we see that

$$W^2 |Du|^2 + W^4 |u_t|^2 \leq C(\sup |u|^2 + \sup(|f|^2 + |Df|^2 + |f_t|^2))$$

in $B^-_{r,\varepsilon}$. Now take the limit as $\varepsilon \to 0$ and $r \to R$ to infer (3.21)′. □

A final useful fact is the following parabolic version of Harnack's first convergence theorem.

COROLLARY 3.21. *If (u_m) is a bounded sequence of $W^{1,2}$ solutions to $Hu_m = f$ in B and if $f \in C^1$, then there is a subsequence $(u_{m(k)})$ which converges uniformly in $B(R/2)$ to a solution u of $Hu = f$ in $B(R/2)$.*

PROOF. See Exercise 3.10. □

4. Global existence and the Perron process

We are now ready to study the Dirichlet problem for the operator H in fairly general domains, although, strictly speaking, we shall study a slightly different problem, in which Dirichlet data are only prescribed on part of the boundary. For Ω a bounded open subset of $\mathbb{R}^{n+1}$, $\varphi \in C(\overline{\Omega})$, and $f \in C^1(\overline{\Omega})$, we consider the problem

$$Hu = f \text{ in } \Omega, \ u = \varphi \text{ on } \mathscr{P}\Omega. \quad (3.22)$$

We say that $v \in C(\overline{\Omega})$ is a *subsolution* of (3.22) if $v \leq \varphi$ on $\mathscr{P}\Omega$ and if for any ball $B = B(R) \subset \Omega$ with $R \leq \lambda/2$, the solution $\overline{v}$ of

$$H\overline{v} = f \text{ in } B, \ \overline{v} = v \text{ on } \partial B$$

is greater than or equal to v in B. *Supersolutions* are defined by reversing the inequalities for v, that is, $v \geq \varphi$ on $\mathscr{P}\Omega$ and $v \geq \overline{v}$. The Perron process generates a solution as the supremum of all subsolutions. To see that this object really is a solution, we verify certain properties of subsolutions and supersolutions. The first is a comparison theorem based on the strong maximum principle Theorem 2.7.

LEMMA 3.22. *If w is a supersolution and v is a subsolution, then $w \ge v$ in Ω.*

PROOF. If not, then $g = v - w$ has a positive maximum value achieved at some point in $\overline{\Omega} \setminus \mathscr{P}\Omega$. By the same argument as in Lemma 2.3, we see that there is a point $X_0 \in \Omega$ and a positive $R \le \lambda/2$ such that $B = B(X_0, R) \subset \Omega$, $g(X_0) = \varepsilon > 0$ and $g(X) \le \varepsilon$ if $X \in B^* = B \cap \{t < t_0\}$. Writing $G = \overline{v} - \overline{w} - \varepsilon$, we have $HG = 0$ in B^*, $G \le 0$ on $\mathscr{P}B^*$, and $G(X_0) \ge 0$, so the strong maximum principle gives $G \equiv 0$ in B^*. It follows that $v - w \equiv \varepsilon$ on $\mathscr{P}B^*$, and hence, since we can take R arbitrarily small, that $v - w \equiv \varepsilon$ in $\{|x - x_0| < r, t = t_0\}$ for some small r. Hence $C'(X_0)$, the subset of $C(X_0)$ on which $v - w = \varepsilon$, is open in $C(X_0)$. Since it is obviously closed, it must be $C(X_0)$. But then $v - w = \varepsilon$ at some point in $\mathscr{P}\Omega$, which contradicts the assumption that $v \le w$ on $\mathscr{P}\Omega$. □

Next, from a subsolution v and a ball $B \subset \Omega$ with $R \le \lambda/2$, we construct a new subsolution V, called the *lift of* v, as follows: $V = v$ in $\overline{\Omega} \setminus B$ and $V = \overline{v}$ in B. To show that V is a subsolution, let $B(r)$ be another ball in Ω with $r \le \lambda/2$, and let v_1 solve

$$Hv_1 = f \text{ in } B(r),\ v_1 = V \text{ on } \partial B(r),$$

Since $V \ge v$, it follows that $v_1 \ge v$ in $B(r)$. Then $v_1 \ge v = V$ on $B(r) \cap \partial B$ and, by definition, $v_1 = V$ on $\partial B(r) \cap B$. Therefore $v_1 \ge V$ on $\partial(B \cap B(r))$, so the maximum principle gives $v_1 \ge V$ in $B \cap B(r)$. Since $v_1 \ge v$ in $B(r)$, we also have $v_1 \ge V$ in $B(r) \setminus B$ and hence $v_1 \ge V$ in $B(r)$. In addition if v_1 and v_2 are subsolutions, then so is $v_3 = \max\{v_1, v_2\}$ because $\overline{v}_3 \ge \overline{v}_1 \ge v_1$ and $\overline{v}_3 \ge \overline{v}_2 \ge v_2$.

We now construct our solution by the Perron process. If we use S to denote the set of all subsolutions of (3.22), we define u by

$$u(X) = \sup_{v \in S} v(X). \tag{3.23}$$

Our next step is to show that this expression gives a solution of the differential equation.

THEOREM 3.23. *The function u defined by* (3.23) *is a solution of the equation $Hu = f$ in Ω.*

PROOF. Note first that $v_0 = -(\sup|f|)t - \sup|\varphi|$ is a subsolution, so u is defined for all $X \in \Omega$ and is bounded from below. In addition, $-v_0$ is a supersolution, so u is also bounded from above. To see that $Hu = f$, we fix $Y \in \Omega$ and $R \le \lambda/2$ so that $B(Y, R) \subset \Omega$, and choose a sequence (v_m) of subsolutions such that $v_m(Y_1) \to u(Y_1)$ as $k \to \infty$ for $Y_1 = (y, s + R/8)$. Next, define $w_m = \max\{v_m, v_0\}$, and let W_m be the lift of w_m relative to the ball $B(Y, R)$. Because $v_m \le w_m \le W_m \le u$, it follows that $W_m(Y_1) \to u(Y_1)$. Hence Corollary 3.20 gives a subsequence $(W_{m(k)})$

which converges uniformly in $B(Y,R/2)$ to a solution w of $Hw = f$ in $B(Y,R/2)$ with $w(Y_1) = u(Y_1)$. If Y_2 is any point in $B(Y,R/8)$, choose the sequence $(v_m^{(1)})$ so that $v_m^{(1)}(Y_2) \to u(Y_2)$ and define $w_m^{(1)} = \max\{w_m, v_m^{(1)}\}$ so $w_m^{(1)}(Y_1) \to w^{(1)}(Y_1)$ and $w_m^{(1)}(Y_2) \to w^{(1)}(Y_2)$. Taking a convergent subsequence, we have a function $w^{(1)}$ with $Hw^{(1)} = f$ in $B(Y,R/4)$, $w^{(1)} \geq w$ and $w^{(1)}(Y_1) = w(Y_1)$. It follows from the strong maximum principle that $w = w^{(1)}$ in $B(Y,R/8)$. Since Y_2 was arbitary, it follows that $w \equiv u$ in $B(Y,R/8)$. Hence u is a solution of $Hu = f$ in a neighborhood of any point in Ω, and therefore $Hu = f$ in Ω. □

Of course, the gradient bound, Theorem 3.19, gives a modulus of continuity at each point of Ω, and Corollary 3.20 shows that

$$\overline{u}(X_0) = \lim_{\substack{X\to X_0\\ t\leq t_0\\ X\in\Omega}} u(X)$$

exists for any $X_0 \in \partial\Omega \setminus \mathcal{P}\Omega$. If $X_0 \in \partial\Omega \setminus \mathcal{P}\Omega$ and if there is a positive ε such that $B(X_0,\varepsilon) \cap \mathcal{P}\Omega = \emptyset$, then the existence of $\overline{u}(X)$ for any $X \in B(X_0,\varepsilon) \cap \partial\Omega$ implies that $u(X) \to u(X_0)$ as $X \to X_0$ and $X \in \Omega$, so u is continuous at such a point. On the other hand, if $X_0 \in \partial\Omega \setminus \mathcal{P}\Omega$ and there is a sequence (X_k) in $\mathcal{P}\Omega$ with $X_k \to X_0$, it is possible to give boundary data φ on $\mathcal{P}\Omega$ so that the Perron solution from Theorem 3.23 is discontinuous at X_0. For simplicity, we assume that $f \equiv 0$. If R is chosen so that $Q(X_0,R) \subset \Omega$, then the closure of $Q(X_0,R/2)$ and $\mathcal{P}\Omega$ are disjoint and hence there is a continuous function φ which vanishes on $\mathcal{P}\Omega \cap \{t < t_0\}$ and $\varphi \equiv 1$ on $\mathcal{P}\Omega \cap Q(X_0,R/2)$. The maximum principle implies that $u(X_0) = 0$, but $\lim_k \varphi(X_k) = 1$, so u does not take on the prescribed boundary value at X_0. For this reason, we assume that for any $X_0 \in \partial\Omega \setminus \mathcal{P}\Omega$, there is a positive ε such that $B(X_0,\varepsilon) \cap \mathcal{P}\Omega = \emptyset$.

Continuity of the Perron solution at points of $\mathcal{P}\Omega$ is not automatically guaranteed. When the solution is continuous at such a point, we can use the maximum principle to prove this continuity. The conditions guaranteeing continuity are geometric in nature and, as in the case for elliptic equations, are tied to so-called barrier functions. Because of the special role of the time variable for parabolic equations, we shall find it convenient to introduce some intermediate notions first. We say that a continuous function g, defined on $\overline{\Omega}$, is a *global barrier* at $X_0 \in \mathcal{P}\Omega$ (for the operator H) if g is nonnegative and vanishes only at X_0 and if for any $B \subset \Omega$ with $R \leq \lambda/2$, the solution h of

$$Hh = 0 \text{ in } B, h = g \text{ on } \partial B \tag{3.24}$$

satisfies the inequality $h \leq g$ in B.

We say that $X_0 \in \mathcal{P}\Omega$ is a *regular point for H* if the Perron solution of (3.22) is continuous at X_0 for any $f \in C^1$ and $\varphi \in C^0$. The next lemma shows that the existence of a global barrier at a point is equivalent to that point being regular.

LEMMA 3.24. *X_0 is a regular point for H if and only if there is a global barrier at X_0.*

PROOF. If X_0 is regular, use g to denote the Perron solution of $Hg = -1$ in Ω, $g = |X - X_0|$ on $\mathcal{P}\Omega$ and note that $v \equiv 0$ is a subsolution, so that g is nonnegative. Lemma 3.18 shows that $g \in C^{2,1}(\Omega)$ (because $H(g-t) = 0$) so the strong maximum principle shows that $g > 0$ in $\overline{\Omega} \setminus \{X_0\}$.

Conversely, if g is a barrier, set $K = \sup|f|$ and choose $\varepsilon > 0$. Since φ is continuous, there are positive constants δ and k such that

$$|\varphi(X) - \varphi(X_0)| + K|t - t_0| \le \varepsilon \text{ for } |X - X_0| < \delta,$$
$$|\varphi(X) - \varphi(X_0)| + K|t - t_0| \le kg(X) \text{ for } |X - X_0| \ge \delta.$$

We set

$$v^{\pm} = \varphi(X_0) \pm (K[t - t_0] + \varepsilon + kg),$$

and note that

$$v^{\pm}(X) = \varphi(X) + [\varphi(X_0) - \varphi(X)] \pm (K[t - t_0] + \varepsilon + kg).$$

Then, for $X \in \{|X - X_0| < \delta\} \cap \mathcal{P}\Omega$, we have $v^+(X) \ge \varphi(X) + kg(X) \ge \varphi(X)$, and for $X \in \{|X - X_0| \ge \delta\} \cap \mathcal{P}\Omega$, we have $v^+(X) \ge \varphi(X) + \varepsilon \ge \varphi(X)$. In addition, if $B \subset \Omega$ (with $R \le \lambda/2$), let h_0 solve

$$Hh_0 = f \text{ in } B, h_0 = v^+ \text{ on } \partial B$$

let h_1 solve

$$Hh_1 = 0 \text{ in } B, h_1 = g \text{ on } \partial B,$$

and set $h = h_0 - \varphi(X_0) - K[t - t_0] - \varepsilon$. Then

$$Hh = f + K > 0 \text{ in } B, h = kg \text{ on } \partial B,$$

so the maximum principle and the linearity of H imply that $h \le kh_1 \le kg$ in $B(R)$ and hence $h_0 \le v^+$ in $B(R)$. Therefore v^+ is a supersolution. It follows that $u \le v^+$ in Ω, so

$$\limsup_{X \to X_0} u(X) \le \limsup_{X \to X_0} v^+(X) = \varphi(X_0) + \varepsilon.$$

Similarly v^- is a subsolution, so

$$\liminf_{X \to X_0} u(X) \ge \liminf_{X \to X_0} v^-(X) = \varphi(X_0) - \varepsilon.$$

Since ε is arbitrary, it follows that

$$\varphi(X_0) \le \liminf_{X \to X_0} u(X) \le \limsup_{X \to X_0} u(X) \le \varphi(X_0),$$

and therefore $\lim_{X \to X_0} u(X) = \varphi(X_0)$. In other words, u is continuous at X_0. □

The barrier concept is actually a local one. Writing $\Omega[X_0,r] = \Omega \cap B(X_0,r)$, we call W a *two-sided local barrier* (to distinguish it from the one, defined below, for earlier time) if W is a nonnegative, continuous function on $\overline{\Omega[X_0,r]})$ for some positive r vanishing only at X_0 such that for any $B(R) \subset (\Omega \cap B(X_0,r))$ with $R \leq \lambda/2$, the solution h of

$$Hh = 0 \text{ in } B(R),\ h = W \text{ on } \partial B(R)$$

is less than or equal to W in $B(R)$.

From W, we construct a global barrier rather easily. First, set $A = B(X_0,r) \setminus B(X_0,r/2)$ and $m = \inf_A W$, and define

$$g = \begin{cases} \min\{W,m\} & \text{in } \overline{\Omega} \cap B(X_0,r) \\ m & \text{in } \overline{\Omega} \setminus B(X_0,r). \end{cases}$$

Since m is positive, we need only verify that $h \leq g$ if h solves (3.24). By the maximum principle, if h solves (3.24), then $h \leq m$, so if $h-g$ has a positive maximum M at some X_1, then $g(X_1) = W(X_1) < m$ and $\|X_1 - X_0\| < r$. Hence there is a ball $B^* = B(\rho, X_1)$ such that $\overline{B^*} \subset B(X_0,r) \cap B$, and there is $X_2 \in \partial B^* \cap \{t < t_1\}$ with $(h-W)(X_2) < M$. If h_1 solves

$$Hh_1 = 0 \text{ in } B^*, h_1 = W \text{ on } \partial B^*,$$

then $(h-h_1)(X_1) \geq (h-W)(X_1) = (h-g)(X_1) = M$ because W is a two-sided local barrier. Now, on ∂B^*, we have $h - h_1 = h - W \leq h - g \leq M$, so $h - h_1 \leq M$ in B^* by the maximum principle. Since $(h-h_1)(X_1) = M$, the strong maximum principle implies that $h - h_1 \equiv (h-W)(X_1)$ in $B^* \cap \{t \leq t_1\}$ which contradicts $(h-h_1)(X_2) < M$. Hence $h \leq g$ and therefore g is a global barrier.

More significantly, the conditions for a local barrier only need to be satisfied for $t < t_0$. Now we define $\Omega^-[X_0,r] = B^-(X_0,r) \cap \Omega$, and we call w a *local barrier from earlier time* if w is a nonnegative, continuous function in $\overline{\Omega^-[,X_0,r]}$ for some positive r vanishing only at X_0 such that for any $B^- = B^-(R) \subset B^-(X_0,r) \cap \Omega$ with $R \leq \lambda/2$, the solution h of $Hh = 0$ in B^-, $h = w$ on $\mathscr{P}B^-$ is less than or equal to w in B^-.

For brevity (and to conform with standard usage), we shall say *local barrier* when we mean local barrier from earlier time. To show that existence of a local barrier implies that of a global barrier, we first prove an existence result for circular cylinders.

LEMMA 3.25. *Let R and T be positive numbers, let $Y \in \mathbb{R}^{n+1}$, and set $\Omega = \{|x-y| < R\} \times (s-T,s)$. Then for any $f \in C^1(\overline{\Omega})$ and any $\varphi \in C^0(\overline{\Omega})$, there is a unique, continuous solution of* (3.22).

PROOF. From Theorem 3.23, Lemma 3.24 and the remarks following Theorem 3.23 and Lemma 3.24, it follows that we need only construct a two-sided local barrier at each point $X_0 \in \mathscr{P}\Omega$. If $t_0 = s - T$, then $w = |x-x_0|^2 + 8\Lambda(t-t_0)$

is a two-sided local barrier. If $t_0 > s-T$, we set $x_1 = -y+2x_0$. Then the n-dimensional spheres $\{|x-x_1| = R\}$ and $\{|x-y| = R\}$ are tangent at x_0, and hence

$$g = -\exp(-\gamma[(x-x_1)^2+(t-t_0)^2]) + \exp(-\gamma R^2)$$

is a two-sided local barrier at X_0 for a suitable positive γ. □

We can now show the equivalence of regularity of a boundary point and the existence of a local barrier.

THEOREM 3.26. *X_0 is regular if and only if there is a local barrier at X_0.*

PROOF. If X_0 is regular, then there is a global barrier and hence a local barrier.
If w is a local barrier, we construct a two-sided local barrier W. Set $Q = \{|x-x_0| < r, t_0 < t < t_0+r\}$, and let w_0 be a nonnegative continuous function on $\mathbb{R}^{n+1}$ which agrees with w on $\omega(t_0)$ and vanishes only at X_0. (Such a function exists by virtue of Exercise 3.1.) Then write w_1 for the solution of $Hw_1 = 0$ in Q, $w_1(X) = w_0$ on $\mathscr{P}Q$. It is easy to check that W defined by

$$W(X) = \begin{cases} w(X) & \text{if } t \le t_0 \\ w_1(X) & \text{if } t > t_0 \end{cases}$$

gives a two-sided local barrier. □

We close this chapter with an easy example of a local barrier. Let Ω be a domain and let $X_0 \in \mathscr{P}\Omega$. Define the parabolic frustum (as in Lemma 2.8 but with $\eta = 1$)

$$PF(R,Y) = \{|x-y|^2 + (s-t) < R^2, t < s\},$$

and suppose that for some choice of R and Y with $s = t_0$, we have $\overline{PF(R,Y)} \cap \overline{\Omega} = \{X_0\}$. If r is defined by $r(X) = |x-x_0|^2 + t_0 - t$, then

$$w = -\exp(-\alpha r^2) + \exp(-\alpha R^2) + 4(t_0-t)/R + \varepsilon(|x-x_0|^2 + |t-t_0|^2)$$

is a local barrier provided α is sufficiently large and ε is sufficiently small (depending only on λ, Λ, n, and R).

Notes

The theory of weak derivatives given in this chapter is standard in the theory of partial differential equations. Our exposition is based on that in [101, Chapter 7]. More detailed discussion of weak derivatives can be found in [348, Chapter 2] and [68, Chapter 4].

On the other hand, our approach to the existence of weak solutions is quite different from the usual one. The discussion in Sections Section III.2 and III.3 comes from [228]. The underlying philosophy of what we have called the Perron process is quite simple: first one shows local existence and then global existence of solutions. Alternative detailed descriptions of this process can be found in [69]. The key new observation here for local solvability is the gradient bound of Lemma

3.16. Classical approaches [89, Section 4.8] prove local solvability in cylinders. For example, in the Galerkin method, one shows local solvability for boundary values in some finite-dimensional space and then uses an approximation argument to obtain arbitrary boundary values; such an approach is extremely useful for numerical solutions of such equations. Alternatively, one can use the Green's function for the heat equation in a rectangular parallelepiped (see Exercises 1–13 of [89, Chapter 4]) to show local solvability. The Green's function approach has the disadvantage that one must verify that an appropriate infinite series of functions converges.

Our local solvability result for $L_{\varepsilon,\eta}$ is based on the existence of weak solutions of elliptic operators in [101, Chapter 8]. The special form of the operators considered here leads to considerable technical simplification.

Our use of local solvability in balls has another advantage over other methods: The inference of global solvability from local solvability in arbitrary domains is simplified because we can consider only continuous subsolutions, rather than upper semicontinuous ones. The role of semicontinuity is demonstrated in, e.g., [112, Section 6] or [212, Section 8].

We also note that a necessary and sufficient geometric condition for the regularity of a boundary point was proved by Evans and Gariepy [77]. We refer the reader to that work for a detailed history of this problem.

Exercises

3.1 Show that any function u which is continuous on a compact set K in $\mathbb{R}^N$ can be extended to a uniformly continuous function on all of $\mathbb{R}^N$ with the same modulus of continuity and range. Specifically, if there is a function ω which is continuous, subadditive, and increasing on $[0,\infty)$ with $\omega(0) = 0$ such that

$$|u(x) - u(y)| \leq \omega(|x-y|) \text{ for all } x,y \text{ in } K,$$

then there is a function $U \in C(\mathbb{R}^N)$ with $U = u$ on K, $\min U = \min u$, $\max U = \max u$, and

$$|U(x) - U(y)| \leq \omega(|x-y|) \text{ for all } x,y \text{ in } \mathbb{R}^N.$$

3.2 Show that there is at most one weak α-derivative v for a given function u.

3.3 Let Ω be an open set and let $h < \frac{1}{2}\operatorname{diam}\Omega$. Show that there is a function $\zeta_h \in C^\infty(\overline{\Omega})$ with support in Ω_h and $\zeta_h \equiv 1$ in Ω_{2h}.

3.4 Let u be a function defined on some domain Ω and let K be a positive constant. Show that $|u(x+he_i) - u(x)| \leq K|h|$ for all $x \in \Omega$ and $h \in \mathbb{R}$ such that $x + he_i \in \Omega$ if and only if $D_i u \in L^\infty$ and $|D_i u| \leq K$.

3.5 Let $u \in L^1_{\text{loc}}(\Omega)$, and suppose that for all sufficiently small positive H, there is a function $v_H \in L^1(\Omega_H)$ such that

$$\int_{\Omega_H} v_H \zeta\, dx = -\int_{\Omega_H} u D_i \zeta\, dx$$

for all $\zeta \in C^1_0(\Omega_H)$. Show that the weak derivative $D_i u$ exists in Ω and that $D_i u = v_H$ in Ω_H.

3.6 Prove the following generalization of Poincaré's inequality. Let G be a continuous, convex, increasing function on $[0,\infty)$, let $R > 0$, and let Ω be a domain with $\operatorname{diam}\Omega \le 2R$. If $u \in C^1_0(\Omega)$, then

$$\int_\Omega G\left(\frac{|u|}{R}\right) dx \le \int_\Omega G(|Du|)\, dx.$$

3.7 Show that the function v in Lemma 3.15 is in $W^{1,2}_0$.

3.8 Show that a weakly differentiable function which has all first order derivatives equal to zero is constant.

3.9 Prove that any $C^{2,1}$ weak solution of $Hu = 0$ is a classical solution. (Note that it follows from Lemma 3.18 that any bounded weak solution is a classical solution.)

3.10 Prove Corollary 3.21.

3.11 We define a *tusk* to be a set of the form

$$\{X : t_0 - T < t < t_0, \left|[x - x_0] - (t_0 - t)^{1/2} x_1\right|^2 < R^2(t_0 - t)\}$$

for positive constants R and T, and points $X_0 \in \mathbb{R}^{n+1}$, and $x_1 \in \mathbb{R}^n$. We say that Ω satisfies an *exterior tusk condition* at $X_0 \in \mathscr{P}\Omega$ if there is a tusk whose closure intersects $\overline{\Omega}$ only at X_0. Construct a local barrier at any point X_0 at which Ω satisfies an exterior tusk condition. (See [72,221].)

3.12 Show that the theory of this chapter can be reproduced without the restriction $R \le \lambda/2$ if we replace the ball $B(R)$ by the scaled ellipsoid

$$E(R) = \left\{X : |x|^2 + \frac{t^2}{\lambda^2 R^2} < R^2\right\}.$$

CHAPTER IV

HÖLDER ESTIMATES

Introduction

In this chapter, we prove a large number of estimates for solutions of parabolic equations under various hypotheses. Some of these estimates apply to weak solutions of equations in divergence form and others to classical solutions of equations not necessarily in divergence form. Although some of the critical estimates are on integral norms of the solutions and their derivatives, our primary concern is with pointwise behavior of these functions. Specifically, we study their moduli of continuity, and the appropriate class of moduli is the Hölder class introduced in Section IV.1. To connect the Hölder moduli of continuity to the theory of weak solutions developed in Chapter III, we define the Campanato spaces in Section IV.2 and we explore the relation between Campanato and Hölder spaces. Interior estimates for smooth solutions of parabolic equations (in divergence and non-divergence form) are given in Section IV.3. We prove boundary estimates in a neighborhood of a flat boundary portion with zero boundary data in Section IV.4. In order to deal with curved boundaries and nonzero data, and to study an intermediate regularity situation which is important to our considerations for nonlinear equations, we introduce a parabolic regularized distance in Section IV.5. This regularized distance is used in Section IV.6 to derive estimates in certain weighted Hölder spaces (and for equations with possibly unbounded coefficients) with flat boundaries and zero boundary data. Curved boundaries and nonzero data are the subject of Section IV.7.

An important element of our approach is a simple existence theory, which is based on the existence results from Chapter III. Other approaches to these estimates, most of which do not rely on existence theory, are discussed in the Notes to this chapter. In addition, there are two significant connections made throughout this chapter. First, we use estimates for equations in divergence form to infer estimates for equations in nondivergence form. Also, we use estimates for equations with Dirichlet boundary data to infer estimates for equations with oblique derivative data and, conversely, we use estimates for equations with oblique derivative data as the basis for estimates with Dirichlet data.

1. Hölder continuity

For $\alpha \in (0,1]$, we say that a function f defined on $\Omega \subset \mathbb{R}^{n+1}$ is *Hölder continuous at X_0 with exponent α* if the quantity

$$[f]_{\alpha;X_0} = \sup_{X \in \Omega \setminus \{X_0\}} \frac{|f(X) - f(X_0)|}{|X - X_0|^{\alpha}} \tag{4.1}$$

is finite. If the quantity defined by (4.1) is finite for $\alpha = 1$, we also say that f is *Lipschitz* at X_0. Clearly if f is Hölder continuous at a point, it is continuous there. If the semi-norm

$$[f]_{\alpha;\Omega} = \sup_{X_0 \in \Omega} [f]_{\alpha;X_0}$$

is finite, we say that f is *uniformly Hölder continuous in* Ω, and if f is uniformly Hölder continuous on any Ω' with compact closure in Ω, we say that f is *locally Hölder continuous in* Ω. Since functions are Lipschitz at any point of differentiability, we see that Hölder continuity can be thought of as a sort of fractional differentiability. In addition, our definition is in accord with the basic rule from the heat equation that "two x-derivatives are equivalent to one t-derivative" because the exponents with respect to x in the definition of $[f]$ are twice those with respect to t. This equivalence makes the definition of higher order Hölder semi-norms slightly more complicated than in the elliptic case. First, for $\beta \in (0,2]$, we define

$$\langle f \rangle_{\beta;X_0} = \sup \left\{ \frac{|f(x_0,t) - f(X_0)|}{|t - t_0|^{\beta/2}} : (x_0,t) \in \Omega \setminus \{X_0\} \right\},$$

and

$$\langle f \rangle_{\beta;\Omega} = \sup_{X_0 \in \Omega} \langle f \rangle_{\beta;X_0}.$$

Then, for any $a > 0$, we write $a = k + \alpha$, where k is a nonnegative integer and $\alpha \in (0,1]$, and we define

$$\langle f \rangle_{a;\Omega} = \sum_{|\beta|+2j=k-1} \langle D_x^{\beta} D_t^{j} f \rangle_{\alpha+1},$$

$$[f]_{a;\Omega} = \sum_{|\beta|+2j=k} [D_x^{\beta} D_t^{j} f]_{\alpha},$$

$$|f|_{a;\Omega} = \sum_{|\beta|+2j \le k} \sup \left| D_x^{\beta} D_t^{j} f \right| + [f]_a + \langle f \rangle_a.$$

It is easy to verify that $|\cdot|_a$ defines a norm on $H_a(\Omega) = \{f : |f|_a < \infty\}$ which makes $H_a(\Omega)$ a Banach space. Moreover, the inclusion of the term $\langle f \rangle_a$ in the definition of $|f|_a$ (and the corresponding term in other norms) is really superfluous since it is generally true that $\langle f \rangle_a \le C|f|_0 + [f]_a$. See Corollary 4.10 and Exercise 4.1 for details. The reason for including this term in the definition is that $f \in H_a$ if and only if, for each $X_0 \in \Omega$, there is a polynomial $P(X;X_0)$ such that

$|P(X;X_0)-f(X)| \le C|X-X_0|^a$. This characterization, which is studied in more detail in Chapters XII and XIV, is the basis for proving some sharp regularity results for nonlinear equations.

Certain weighted norms will also be useful. To define these norms, we write d for the function defined by

$$d(X_0) = \inf\{|X-X_0| : X \in \mathscr{P}\Omega \text{ and } t < t_0\}$$

and we set $d(X,Y) = \min\{d(X), d(Y)\}$. For simplicity of notation, we define

$$[f]_0 = [f]_0^* = \operatorname*{osc}_\Omega f, |f|_0 = |f|_0^* = \sup_\Omega |f|.$$

If $b \ge 0$, we define

$$|f|_0^{(b)} = \sup_\Omega d^b |f|,$$

and if $b < 0$, we define

$$|f|_0^{(b)} = (\operatorname{diam}\Omega)^b \sup_\Omega |f|.$$

If $a > 0$ and $a + b \ge 0$, we define

$$[f]_a^{(b)} = \sup_{X \ne Y \text{ in } \Omega} \sum_{|\beta|+2j=k} d(X,Y)^{a+b} \frac{\left|D_x^\beta D_t^j f(X) - D_x^\beta D_t^j f(Y)\right|}{|X-Y|^\alpha},$$

$$\langle f\rangle_a^{(b)} = \sup_{\substack{X \ne Y \text{in } \Omega \\ x=y}} \sum_{|\beta|+2j=k-1} d(X,Y)^{a+b} \frac{\left|D_x^\beta D_t^j f(X) - D_x^\beta D_t^j f(Y)\right|}{|X-Y|^{1+\alpha}},$$

$$|f|_a^{(b)} = \sum_{|\beta|+2j\le k} \left|D_x^\beta D_t^j f\right|_0^{(|\beta|+2j+b)} + [f]_a^{(b)} + \langle f\rangle_a^{(b)}$$

and if $a+b<0$, we define $[f]_a^{(b)}$ and $\langle f\rangle_a^{(b)}$ by replacing $d(X,Y)$ with $\operatorname{diam}\Omega$. We also define $|f|_a^* = |f|_a^{(0)}$, and note that (using the obvious definitions) H_a^* and $H_a^{(b)}$ are also Banach spaces. Further weighted spaces will be defined as needed.

A key element in our analysis is an interpolation inequality for Hölder spaces. We first demonstrate some special cases of this inequality.

LEMMA 4.1. *Let* $0 \le \alpha < \beta \le 1$ *and* $\sigma \in (0,1)$. *Then*

$$[u]^*_{\sigma\alpha+(1-\sigma)\beta} \le ([u]^*_\alpha)^\sigma ([u]^*_\beta)^{1-\sigma} \tag{4.2a}$$

$$\langle u\rangle^*_{\sigma\alpha+(1-\sigma)\beta} \le (\langle u\rangle^*_\alpha)^\sigma (\langle u\rangle^*_\beta)^{1-\sigma} \tag{4.2b}$$

$$|Du|_0^{(1)} \le 5([u]_0)^{\alpha/(1+\alpha)}([u]_0 + [u]^*_{1+\alpha})^{1/(1+\alpha)}. \tag{4.2c}$$

PROOF. Set $U_\tau = [u]^*_\tau$ for $\tau \geq 0$, and write d for $d(X,Y)$. Then

$$\begin{aligned}|u(X)-u(Y)| &= |u(X)-u(Y)|^\sigma\, |u(X)-u(Y)|^{1-\sigma}\\ &\leq (d^{-\alpha}U_\alpha\,|X-Y|^\alpha)^\sigma (d^{-\beta}U_\beta\,|X-Y|^\beta)^{1-\sigma}.\end{aligned}$$

If we multiply both sides of this inequality by $(d/|X-Y|)^{\alpha\sigma+\beta(1-\sigma)}$ and then take the supremum over all $X \neq Y$ in Ω, we obtain (4.2a). A similar argument gives (4.2b).

To prove (4.2c), fix $X_1 \in \Omega$ and let $\varepsilon \in (0, \frac{1}{2}]$ be a constant to be further specified. Then take x_2 so that $x_1 - x_2$ is parallel to the vector $Du(X_1)$ and $|x_1 - x_2| = \varepsilon d(X_1)$; it follows that $X_2 = (x_2, t_1) \in \Omega$. By the mean value theorem, there is x_3 on the line segment between x_1 and x_2 (so $X_3 = (x_3, t_1) \in \Omega$ as well) such that

$$Du(X_3)\cdot(x_1 - x_2) = u(X_1) - u(X_2).$$

Also, from our choice of X_2, we have

$$\begin{aligned}|Du(X_1)| &= Du(X_1)\cdot\frac{x_1-x_2}{|x_1-x_2|}\\ &\leq Du(X_3)\cdot\frac{x_1-x_2}{|x_1-x_2|} + |Du(X_3)-Du(X_1)|.\end{aligned}$$

Now our choice of X_3 gives

$$\begin{aligned}|Du(X_1)| &\leq \frac{u(X_1)-u(X_2)}{|x_1-x_2|} + |Du(X_3)-Du(X_1)|\\ &\leq \frac{U_0}{\varepsilon d(X_1)} + U_{1+\alpha}\frac{|x_3-x_1|^\alpha}{d(X_1,X_3)^{1+\alpha}}.\end{aligned}$$

Since $d(X_1) \geq 2|x_1 - x_2|$, it follows that $d(X_1) \geq 2|x_1 - x_3|$, and hence $\frac{1}{2}d(X_1) \leq d(X_1, X_3) \leq d(X_1)$. Therefore, we have

$$|Du(X_1)| \leq \frac{U_0}{\varepsilon d(X_1)} + 2^{1+\alpha}\frac{\varepsilon^\alpha U_{1+\alpha}}{d(X_1)}. \tag{4.3}$$

Now we multiply this inequality by $d(X_1)$ and take the supremum over all $X_1 \in \Omega$. If $U_0/U_{1+\alpha} < 2^{-1-\alpha}$, we choose $\varepsilon = (U_0/U_{1+\alpha})^{1/(1+\alpha)}$ and note that $2^{1+\alpha} \leq 4$ to infer (4.2c). On the other hand, if $U_0/U_{1+\alpha} \geq 2^{-1-\alpha}$, we choose $\varepsilon = \frac{1}{2}$. Then

$$|Du|_0^{(1)} \leq 2U_0^{\alpha/(1+\alpha)}[U_0^{1/(1+\alpha)} + U_{1+\alpha}^{1/(1+\alpha)}].$$

The proof is completed by noting that $a^p + b^p \leq 2(a+b)^p$ if a, b, and p are nonnegative with $p \leq 1$. □

From the elements given in Lemma 4.1, we can infer a general interpolation inequality which is crucial to our derivation of the Hölder estimates.

PROPOSITION 4.2. *Let $0 \le a < b$ and let $\sigma \in (0,1)$. Then there is a constant C determined only by b such that*

$$|u|^*_{\sigma a+(1-\sigma)b} \le C(|u|^*_a)^\sigma(|u|^*_b)^{1-\sigma}. \tag{4.4}$$

PROOF. If $b \le 1$, the result is clear from Lemma 4.1. We consider several cases if $1 < b \le 2$. For simplicity, we write RHS for $C(|u|^*_a)^\sigma(|u|^*_b)^{1-\sigma}$, τ for $\sigma a + (1-\sigma)b$, and U_r for $|u|^*_r$.

Suppose first that $a \ge 1$. In this case, we estimate all quantities involving Du by an easy modification of (4.2a) and then use (4.2b) to estimate the term $\langle u\rangle^*_\tau$.

Next we suppose that $a = 0$. If $\tau = 1$, we use Lemma 4.1 to infer that $|Du|^{(1)}_0 + \langle u\rangle^*_1 \le RHS$. To see that $[u]^*_1 \le RHS$, we fix $X_1 \ne X_2$ in Ω. If also $|X_1 - X_2| \ge \frac{1}{2}d(X_1,X_2)$, then

$$|u(X_1)-u(X_2)| \le 2U_0 \le 4(U_0)^\sigma(U_b)^{1-\sigma}\frac{|X_1-X_2|}{d(X_1,X_2)}.$$

If, instead, $|X_1 - X_2| < \frac{1}{2}d(X_1,X_2)$, then

$$\begin{aligned}|u(X_1)-u(X_2)| &\le |u(X_1)-u(x_2,t_1)| + |u(x_2,t_1)-u(X_2)| \\ &\le \left[|Du|^{(1)}_0 + \langle u\rangle^*_1\right]\frac{|X_1-X_2|}{d(X_1,X_2)}\end{aligned}$$

because $\frac{1}{2}d(X_1,X_2) \le d(x_3,t_1) \le 2d(X_1,X_2)$ if x_3 is on the line segment joining x_1 and x_2. Combining these inequalities gives (4.4) in this case. If $\tau < 1$, we note that

$$U_\tau \le CU_0^\tau U_1^{1-\tau} \le CU_0^\tau(U_b^{1/b}U_0^{(b-1)/b})^{1-\tau} = CU_0^\sigma U_b^{1-\sigma}.$$

If $\tau > 1$, then we proceed as in the case $\tau < 1$ except that we interpolate U_τ between U_1 and U_b and then estimate U_1 in terms of U_0 and U_b.

Next, we suppose that $0 < a < 1$. If $\tau = 1$, we imitate the proof of (4.3) using $|u(X_1) - u(X_2)| \le U_a(|X_1 - X_2|/d)^a$ to find that

$$d(X_1)|Du(X_1)| \le \frac{U_a}{\varepsilon^{1-a}} + 4U_b\varepsilon^{b-1}.$$

The choice $\varepsilon = \min\{(U_a/U_b)^{1/(b-a)}, \frac{1}{2}\}$ gives $|Du|^{(1)}_0 \le CU_a^\sigma(U_b)^{1-\sigma}$ and then (4.4) follows as in the case $a = 0$. If $\tau < 1$, we proceed as in the case $a = 0$, and $\tau < 1$. The case $\tau > 1$ is obtained by combining the cases $a = 1$ and $\tau = 1$.

When $b > 2$, we proceed by induction from our previous cases. □

2. Campanato spaces

The Hölder spaces of the previous section are defined in terms of pointwise properties of their member functions. Surprisingly, an equivalent definition can be given by integral properties of these functions. In this section, we prove a weak version of this equivalence which is useful for applications to parabolic equations.

For $X_0 \in \Omega$, $R > 0$, and $w \in L^1(\Omega)$, we write $\Omega[X_0,R] = \Omega \cap Q(X_0,R)$, and we define the *mean value* $\{w\}_{X_0,R}$ by

$$\{w\}_{X_0,R} = \frac{1}{|\Omega[X_0,R]|} \int_{\Omega[X_0,R]} w\,dX.$$

For $q \geq 0$, we define the *Campanato space* $\mathscr{L}^{1,q}(\Omega)$ to be the set of all functions $w \in L^1(\Omega)$ such that there is a constant W_q for which

$$\int_{\Omega[Y,r]} |w - \{w\}_{Y,r}|\,dX \leq W_q r^q \tag{4.5}$$

for all $Y \in \Omega$. It is easy to check that $\mathscr{L}^{1,q}$ consists only of constant functions if $q > n+4$. Moreover, for any constant L, if we omit the domain of integration $\Omega[Y,r]$ from all integrals, we have

$$\int |w - \{w\}_{Y,r}|\,dX \leq \int |w - L|\,dX + \int |\{w\}_{Y,r} - L|\,dX.$$

Since $\int |\{w\}_{Y,r} - L|\,dX = |\int (w-L)\,dX|$, it follows that

$$\int |w - \{w\}_{Y,r}|\,dX \leq 2\int |w - L|\,dX \tag{4.6}$$

and therefore (4.5) is true if and only if there are constants L and C such that

$$\int |w - L|\,dX \leq Cr^q \tag*{(4.5)$'$}$$

for all $Y \in \Omega$ and all $r > 0$. The equivalence between Hölder spaces and Campanato spaces is proved quite simply in one direction. If $0 < \alpha \leq 1$, then it is easy to check that $H_\alpha \subset \mathscr{L}^{1,n+2+\alpha}$. In the other direction, we first prove a slightly weaker result.

LEMMA 4.3. *Let $f \in L^1(Q(X_0,2R))$ and suppose there are positive constants $\alpha \leq 1$ and H along with a function g defined on $Q(X_0,2R) \times (0,R)$ such that*

$$\int_{Q(Y,r)} |f(X) - g(Y,r)|\,dX \leq Hr^{n+2+\alpha} \tag{4.7}$$

for any $Y \in Q(X_0,R)$ and any $r \in (0,R)$. Then there is a constant $c(n,\alpha)$ such that

$$[f]_{\alpha;Q(X_0,R)} \leq cH. \tag{4.8}$$

PROOF. Write $Q(r)$ for $Q(X_0,r)$. Let $\varphi \in C^1(\mathbb{R}^n)$ and $\eta \in C^1(\mathbb{R})$ be nonnegative functions with

$$\int_{\mathbb{R}^n} \varphi(x)\,dx = \int_{\mathbb{R}} \eta(s)\,ds = 1$$

and $\varphi(x) = 0$ if $|x| > 1$, $\eta(s) = 0$ if $s \notin (0,1)$. We use K and L to denote the maximum of $|D\varphi|$ and $|\eta'|$, respectively. For $(Y,\tau) \in Q(R) \times (0,R)$, we define

$$F(Y,\tau) = \int_{\mathbb{R}^n}\int_{\mathbb{R}} f(y+\tau z, s-\tau^2\sigma)\varphi(z)\eta(\sigma)\,dz\,d\sigma \tag{4.9}$$

and note that

$$F_y(Y,\tau) = -\frac{1}{\tau}\int_{\mathbb{R}^n}\int_{\mathbb{R}} f(y+\tau z, s-\tau^2\sigma)D\varphi(z)\eta(\sigma)\,dz d\sigma.$$

Since $\int D\varphi(z)\,dz = 0$, we also have

$$\int_{\mathbb{R}^n}\int_{\mathbb{R}} g(Y,\tau)D\varphi(z)\eta(\sigma)\,dz d\sigma = 0$$

and hence

$$\begin{aligned}|F_y(Y,\tau)| &\le \frac{1}{\tau}\int_{\mathbb{R}^n}\int_{\mathbb{R}} |f(y+\tau z,s-\tau^2\sigma)-g(Y,\tau)|\,|D\varphi(z)|\,\eta(\sigma)\,d\sigma\,dz\\ &\le \frac{K}{\tau}\int_{|z|\le 1}\int_{\mathbb{R}} |f(y+\tau z,s-\tau^2\sigma)-g(Y,\tau)|\,\eta(\sigma)\,d\sigma\,dz\\ &= \frac{K}{\tau^{n+3}}\int_{Q(Y,\tau)} |f(X)-g(Y,\tau)|\,\eta((s-t)/\tau^2)\,dX\\ &\le HKL\tau^{\alpha-1}\end{aligned}$$

because $0 \le \eta \le L$. Similarly, $|F_s(Y,\tau)| \le HKL\tau^{\alpha-2}$ and $|F_\tau(Y,\tau)| \le (n+K+2+2L)H\tau^{\alpha-1}$.

Now, fix Y and Y_1 in $Q(R)$ and set $\tau = |Y-Y_1|$. If $\varepsilon \in (0,\tau)$, then

$$\begin{aligned}|F(Y,\varepsilon)-F(Y_1,\varepsilon)| &\le |F(Y,\varepsilon)-F(Y,\tau)| + |F(Y,\tau)-F(y_1,s,\tau)|\\ &\quad + |F(y_1,s,\tau)-F(Y_1,\tau)| + |F(Y_1,\tau)-F(Y_1,\varepsilon)|\\ &\le (n+K+2+2L)H\int_\varepsilon^\tau \rho^{\alpha-1}\,d\rho\\ &\quad + KLH\tau^{\alpha-1}|y-y_1| + HKL\tau^{\alpha-2}|s-s_1|\\ &\quad + (n+K+2+2L)H\int_\varepsilon^\tau \rho^{\alpha-1}\,d\rho\\ &\le C(n,K,L,\alpha)H\tau^\alpha.\end{aligned}$$

Therefore F is uniformly Hölder continuous. Since $F(\cdot,\varepsilon) \to f$ in L^1 as $\varepsilon \to 0$, and since we can choose φ and η so that K depends only on n and $L \le 3$, we find that f is also Hölder and (4.8) is satisfied. □

3. Interior estimates

To prove our Hölder estimates, we start with a sharper version of Lemma 3.18.

LEMMA 4.4. *Let (A^{ij}) be a constant matrix and suppose there are positive constants Λ and λ such that*

$$\Lambda|\xi|^2 \ge A^{ij}\xi^i\xi^j \ge \lambda|\xi|^2 \tag{4.10}$$

for all $\xi \in \mathbb{R}^n$. *Let* u *be a* $W^{1,2}$ *solution of*

$$u_t = D_i(A^{ij}D_ju) \qquad \text{in } Q(R) \tag{4.11}$$

with $|u| \le M$ *in* $Q(R)$ *for some constant* M. *Then there is a constant* $C(n,\lambda,\Lambda)$ *such that*

$$\operatorname*{osc}_{Q(r)} u \le C\frac{r}{R}M \tag{4.12}$$

for $r \in (0,R)$.

PROOF. From Lemma 3.18, we know that $u \in C^\infty$ so each component of Du is a solution of (4.11). Now we set

$$\zeta(X) = (R^2 - |x-x_0|^2)^+(R^2 - (t_0 - t))^+.$$

Since $\zeta \le R^4$, $|D\zeta| \le 2R^3$, and $|D^2\zeta| \le CR^2$, it follows that

$$L(\zeta^2|Du|^2) \ge -C_1(n,\lambda,\Lambda)R^6|Du|^2.$$

As in Lemma 3.18, we deduce that

$$\zeta^2|Du|^2 \le C_1M^2R^6/(2\lambda) \tag{4.13}$$

Since (4.12) is clear for $r \ge R/2$, we now suppose that $r < R/2$. Then $\zeta \ge R^4/8$ in $Q(r)$ so $|Du| \le CM/R$ in $Q(r)$. Then (2.27) implies that

$$|u(x_0,t_0) - u(x_0,t)| \le C(M/R)r$$

for any $t \in (t_0 - r^2, t_0)$. Combining this inequality with (4.13) gives (4.12). □

Next, we use Lemma 4.4 to prove an integral estimate for solutions of (4.11).

LEMMA 4.5. *If* (A^{ij}) *satsifies* (4.10) *and if* u *is a* $W^{1,2}$ *solution of* (4.11), *then there is a constant* C *determined only by* n, λ, *and* Λ *such that*

$$\int_{Q(\rho)} u^2\,dX \le C\left(\frac{\rho}{R}\right)^{n+2}\int_{Q(R)} u^2\,dX \tag{4.14a}$$

$$\int_{Q(\rho)} |u - \{u\}_{X_0,\rho}|^2\,dX \le C\left(\frac{\rho}{R}\right)^{n+4}\int_{Q(R)} |u - \{u\}_{X_0,R}|^2\,dX \tag{4.14b}$$

for all $\rho \in (0,R)$.

PROOF. From the proof of Lemma 3.18, we see that $u \in C^\infty(Q(R))$, so $U = [u]^{(1+n/2)}_{0,Q(R/2)}$ is finite. Now take $X_1 \in Q(R/2)$ with $d(X_1)^{1+n/2}|u(X_1)| \ge U/2$, and write d for $d(X_1)$. For $\theta \in (0,1/4)$ to be chosen, Lemma 4.4 implies that

$$\operatorname*{osc}_{Q(X_1,\theta d)} u \le C\theta \sup_{Q(X_1,d/2)} u \le C\theta|u(X_1)| \le C_0\theta d^{-1-n/2}U.$$

Now take $\theta = 1/4(1+C_0)$ and note that

$$|u(X)| \ge |u(X_1)| - \operatorname*{osc}_{Q(X_1,\theta d)} u \ge \frac{3}{4}|u(X_1)|$$

for $X \in Q(X_1, \theta d)$. It follows that

$$U^2 \le 4d^{n+2}\frac{16}{9}(\inf_{Q(X_1,\theta d)}|u|)^2$$
$$\le 8\theta^{-n-2}\int_{Q(X_1,\theta d)} u^2\,dX \le C\int_{Q(R)} u^2\,dX.$$

Hence if $\rho < R/4$, we have

$$\int_{Q(\rho)} u^2 \le C\rho^{n+2}\sup_{Q(\rho)} u^2 \le C\rho^{n+2}\sup_{Q(R/4)} u^2$$
$$\le C\left(\frac{\rho}{R}\right)^{n+2} U^2 \le C\left(\frac{\rho}{R}\right)^{n+2}\int_{Q(R)} u^2\,dX.$$

If $\rho \ge R/4$, then (4.14a) is clear.

To prove (4.14b), we assume that $\rho < R/8$ since otherwise the inequality is clear. In this case, Lemma 4.4 and (4.14a) (with $u - \{u\}_{Y,R}$ in place of u) imply that

$$\sup_{Q(R/4)} |Du|^2 \le CR^{-2}\sup_{Q(R/2)} |u - \{u\}_{Y,R}|^2$$
$$\le CR^{-n-2}\int_{Q(R)} |u - \{u\}_{Y,R}|^2\,dX.$$

Now (2.27) implies that

$$\sup_{Q(\rho)} \left|u - \{u\}_{Y,\rho}\right|^2 \le C\rho^2\sup_{Q(R/4)} |Du|^2$$

because $Q(2\rho) \subset Q(R/4)$. The proof is completed by noting that

$$\int_{Q(\rho)} \left|u - \{u\}_{Y,\rho}\right|^2 dX \le C(n)\rho^{n+2}\sup_{Q(\rho)} \left|u - \{u\}_{Y,\rho}\right|^2$$

and combining these inequalities. □

Note that, as an intermediate result, we have shown that

$$\sup_{Q(R/2)} u^2 \le CR^{-n-2}\int_{Q(R)} u^2\,dX$$

for solutions of (4.11). In Chapter VI, we shall prove a similar estimate for weak solutions to a much larger class of equations.

Another important element in our estimates is the following iteration lemma.

LEMMA 4.6. *Let ω and σ be increasing functions on an interval $(0, R_0]$ and suppose there are positive constants α, δ, and τ with $\tau < 1$ and $\delta < \alpha$ such that*

$$r^{-\delta}\sigma(r) \le s^{-\delta}\sigma(s) \qquad \text{if } 0 < s \le r \le R_0 \tag{4.15a}$$

and

$$\omega(\tau r) \le \tau^\alpha \omega(r) + \sigma(r) \qquad \text{if } 0 < r \le R_0. \tag{4.15b}$$

Then there is a constant $C = C(\alpha, \delta, \tau)$ *such that*

$$\omega(r) \le C[\left(\frac{r}{R_0}\right)^\alpha \omega(R_0) + \sigma(r)]. \tag{4.16}$$

PROOF. Since (4.16) is obvious for $r \ge \tau R_0$, we assume that $0 < r < \tau R_0$ and write k for the smallest integer such that $\tau^k R_0 > r$. By induction, (4.15a,b) imply that

$$\omega(r) \le \omega(\tau^k R_0) \le \tau^{k\alpha}\omega(R_0) + \tau^{-\delta}\sum_{j=0}^{k} \tau^{(\alpha-\delta)j}\sigma(\tau^k R_0).$$

This finite sum is a finite geometric series, which we can estimate by the infinite series. Hence

$$\omega(r) \le \tau^{k\alpha}\omega(R_0) + \frac{\sigma(\tau^k R_0)}{\tau^\delta - \tau^\alpha}.$$

In addition, $\tau^k R_0 \le r/\tau < R_0$, so

$$\sigma(\tau^k R_0) \le \sigma(r/\tau) \le \tau^{-\delta}\sigma(r).$$

These last two inequalities give (4.16) for $C = \max\{\tau^{-\alpha}, \frac{\tau^{-\delta}}{\tau^\delta - \tau^\alpha}\}$. □

In practice, one of two other conditions is used to infer (4.15b) for increasing functions ω and σ. If there are constants K and β with $\beta > \delta$ for which

$$\omega(\tau r) \le K\tau^\beta \omega(r) + \sigma(r) \tag{4.15)$'$}$$

for all sufficiently small τ, then (4.15b) holds with any $\alpha \in (\delta, \beta)$ and τ chosen so that $K\tau^\beta \le \tau^\alpha$. Alternatively, if there are constants τ and ε in $(0,1)$ for which

$$\omega(\tau r) \le \varepsilon\omega(r) + \sigma(r), \tag{4.15)$''$}$$

then (4.15b) holds with $\alpha = \log_\tau \varepsilon$.

The estimates of Lemma 4.5 form the basis of Campanato's approach to Hölder regularity. We illustrate their use with a simple estimate for an inhomogeneous equation with constant coefficients:

$$-u_t + D_i(A^{ij}D_j u) = D_i f^i \tag{4.17}$$

for some vector function f. By this equation, we mean that

$$\int u_t\zeta + A^{ij}D_i\zeta D_j u - f^i D_i\zeta \, dX = 0$$

for all $\zeta \in C_0^2$.

THEOREM 4.7. *Suppose u is a continuous weak solution of* (4.17) *in some cylinder $Q(Y,2R)$ with A^{ij} satisfying* (4.10) *and $[f]_{\alpha;Q(Y,2R)} \le F$ for some positive constants F, λ, and Λ. Then*

$$[Du]_{\alpha,Q(Y,R)} \le C(n,\alpha,\lambda,\Lambda)(R^{-n-2-2\alpha}\int_{Q(Y,2R)} |Du|^2\,dX + F^2)^{1/2}. \tag{4.18}$$

PROOF. Fix $X_1 \in Q(Y,R)$ and $r \in (0,R)$, write $Q(r)$ for $Q(X_1,r)$, and let v be the solution of

$$-v_t + D_i(A^{ij}D_jv) = 0 \text{ in } Q(r), v = u \text{ on } \mathscr{P}Q(r)$$

given by Lemma 3.25. Since Lemma 3.25 only shows that $Dv \in L^2_{\text{loc}}(Q(r))$, our next step is to show that $Dv \in L^2(Q(r))$. For this, we set $w = u - v$, and for $\varepsilon > 0$, we use $(w-\varepsilon)^+$ as test function in the equations for u and v. Subtracting these equations and noting that $\int f^i(X_1)D_i(w-\varepsilon)^+\,dX = 0$, we find that

$$\int [(w-\varepsilon)^+w_t + a^{ij}D_i(w-\varepsilon)^+D_jw]\,dX = \int [f^i - f^i(X_1)]D_i(w-\varepsilon)^+\,dX.$$

Now we observe that $(w-\varepsilon)^+w_t = \frac{1}{2}[(w-\varepsilon)^+]^2{}_t$, so

$$\int (w-\varepsilon)^+w_t\,dX = \frac{1}{2}\int_{B(r)\times\{t_1\}} [(w-\varepsilon)^+]^2\,dx \ge 0,$$

and $D_i(w-\varepsilon)^+D_jw = D_i(w-\varepsilon)^+D_j(w-\varepsilon)^+$, so Schwarz's inequality implies that

$$\int |D(w-\varepsilon)^+|^2\,dX \le C\int |f - f(X_1)|^2\,dX \le CF^2r^{n+2+2\alpha}.$$

Now we send $\varepsilon \to 0$, noting that the uniform L^2 bound on $D(w-\varepsilon)^+$ shows that $Dw^+ \in L^2$ and

$$\int |Dw^+|^2\,dX \le CF^2r^{n+2+2\alpha}.$$

A similar argument applies to Dw^- and hence

$$\int |Dw|^2\,dX \le CF^2r^{n+2+2\alpha}. \tag{4.19a}$$

In particular, $Dv \in L^2(Q(r))$. Next, Lemma 4.4 applied to each component of Dv gives

$$\int_{Q(\rho)} |Dv - \{Dv\}_\rho|^2\,dX \le C\left(\frac{\rho}{r}\right)^{n+4}\int_{Q(r)} |Dv - \{Dv\}_r|^2\,dX. \tag{4.19b}$$

It follows that

$$\int_{Q(\rho)} |Du - \{Dv\}_\rho|^2\,dX \le C\left(\frac{\rho}{r}\right)^{n+4}\int_{Q(r)} |Du - \{Dv\}_r|^2\,dX$$
$$+ CF^2r^{n+2+2\alpha},$$

and therefore ω and σ defined by

$$\omega(r) = \int_{Q(r)} |Du - \{Dv\}_r|^2 \, dX, \ \sigma(r) = CFr^{n+2+2\alpha}$$

satisfy (4.15a) and (4.15)′ with $\beta = n+4$ and $\delta = n+2+2\alpha$. Hence Lemma 4.6 gives

$$\int_{Q(r)} |Du - \{Dv\}_r|^2 \, dX \le C[(R^{-n-2-2\alpha} \int_{Q(R)} |Du|^2 \, dX + F^2]r^{n+2+2\alpha}$$

because $\int_{Q(R)} |\{Dv\}_R|^2 \, dX \le \int_{Q(R)} |Dv|^2 \, dX \le C \int_{Q(R)} |Du|^2 \, dX$, and Lemma 4.3 completes the proof. □

As we shall see in Chapter VI, it is possible to show (4.19a) directly, so the approximating functions $(w \mp \varepsilon)^{\pm}$ are not really necessary to this proof.

It is possible to derive all of our Hölder estimates from Theorem 4.7 via suitable interpolation inequalities and perturbation arguments in small cylinders. We shall take a slightly different view and modify the method of proof in that theorem. Although we cannot avoid the interpolation inequalities entirely, we can eliminate the use of more complicated ones than Proposition 4.2. We also use a weighted Morrey space $M^{(b)}_{p,\delta}$ for $p \ge 1$, $\delta \in [0, n+2]$, and $b \ge n+2-\delta$, which is the set of all $u \in L^1_{\text{loc}}(\Omega)$ such that

$$\|u\|^{(b)}_{p,\delta} = \sup_{\substack{Y \in \Omega \\ 0 \le r \le d(Y)/2}} \left(r^{-\delta} d(Y)^{pb+\delta-n-2} \int_{Q(Y,r)} |u|^p \, dX \right)^{1/p}$$

is finite.

THEOREM 4.8. *Let Ω be a domain in $\mathbb{R}^{n+1}$, and let $a^{ij} \in H^{(0)}_\alpha$, $b^i \in H^{(1)}_\alpha$, $c^i \in M^{(1)}_{1,n+1+\alpha}$, $c^0 \in M^{(2)}_{1,n+1+\alpha}$ satisfy*

$$\lambda |\xi|^2 \le a^{ij}\xi_i\xi_j \le \Lambda |\xi|^2, \ [a^{ij}]^{(0)}_\alpha \le A \tag{4.20a}$$

$$|b^i|^{(1)}_\alpha \le B, \tag{4.20b}$$

$$\|c^i\|^{(1)}_{1,n+1+\alpha} \le c_1, \ \|c^0\|^{(2)}_{1,n+1+\alpha} \le c_2, \tag{4.20c}$$

for some positive constants A, B, c_1, c_2, $\alpha < 1$, λ, and Λ. Let f be a vector-valued function in $H^{(1)}_\alpha$ and let $g \in M^{(2)}_{1,n+1+\alpha}$ such that

$$|f|^{(1)}_\alpha \le F, \ \|g\|^{(2)}_{1,n+1+\alpha} \le G \tag{4.21}$$

for constants F and G. If $u \in H^*_{1+\alpha}$ is a weak solution of

$$-u_t + D_i(a^{ij}D_ju + b^iD_iu) + c^iD_iu + c^0u = D_if^i + g \tag{4.22}$$

in Ω, then there is a constant C determined only by A, B, c_1, c_2, n, α, λ, and Λ such that

$$|u|^*_{1+\alpha} \le C(|u|_0 + F + G). \tag{4.23}$$

PROOF. Suppose first that the coefficients b^i, c^i, and c^0 all vanish. Then fix $Y \in \Omega$ and $r < \frac{1}{2}d(Y)$, and let v solve

$$-v_t + D_i(a^{ij}(Y)D_jv) = 0 \text{ in } Q(Y,r),\ v = u \text{ on } \mathscr{P}Q(Y,r).$$

As in Theorem 4.7, we can show that $Dv \in L^2(Q(Y,r))$ and that (4.19b) holds. Hence we only need to estimate the $L^2(Q(r))$ norm of $D(u-v)$. By using the test function $w = u - v$ in the equations for u and v, we obtain from Schwarz's inequality and the arguments from Theorem 4.7 that

$$\int |Dw|^2\,dX \le C\left[\left(\frac{r}{d}\right)^{2\alpha} r^{n+2} \sup_{Q(r)} |Du|^2 + F^2 r^n \left(\frac{r}{d}\right)^{2+2\alpha} + Gr^n \sup|w| \left(\frac{r}{d}\right)^{1+\alpha}\right].$$

To continue, we define

$$U_{1+\alpha} = [u]^*_{1+\alpha},\ U_1 = |Du|^{(1)}_0,\ U_\alpha = U_{1+\alpha} + \langle u\rangle^*_{1+\alpha},$$

and note that

$$|w(X)| \le |u(X) - Du(Y)\cdot(x-y)| + |v(X) - Du(Y)\cdot(x-y)|.$$

Moreover, V defined by $V(X) = v - Du(Y)\cdot(x-y)$ satisfies the same equation as v, so V obeys the maximum principle. Hence we have $|w| \le 2U_\alpha(r/d)^{1+\alpha}$ and

$$\int |Dw|^2\,dX \le C[U_1^2 + F^2 + GU_\alpha]r^n \left(\frac{r}{d}\right)^{2+2\alpha}.$$

From this inequality and the arguments of Theorem 4.7, we infer that

$$[Du]_{\alpha,Q(d/4)} \le C[U_1^2 + F^2 + GU_\alpha]d^{-\alpha-1}$$

and then (by taking the supremum over all $Y \in \Omega$) that

$$U_{1+\alpha} \le C[U_1 + F + (GU_\alpha)^{1/2}]. \tag{4.24}$$

(As in Proposition 4.2, we only need to consider the case $|X-Y| < \frac{1}{2}d(X,Y)$.)

To estimate $\langle u\rangle^*_{1+\alpha}$ in terms of $U_{1+\alpha}$, we fix $Y \in \Omega$ and $r \le \frac{1}{2}d(Y)$, and then define $\eta(x) = c_0(r^2 - |x-y|^2)^+$ with c_0 chosen so that $\|\eta\|_1 = 1$, and

$$U(t) = \int_{B(r)} [u(X) - Du(Y)\cdot(y-x)]\eta(x)\,dx.$$

If $s-r^2 \le t \le s$, we also have

$$\begin{aligned}U(s)-U(t) &= \int_{B(r)} [u(x,s)-u(x,t)]\eta\,dx \\ &= \int [a^{ij}(X)-a^{ij}(Y)]D_ju(X)D_i\eta(x)\,dX \\ &+ \int a^{ij}(Y)[D_ju(X)-D_ju(Y)]D_i\eta(x)\,dX \\ &+ \int [f^i(X)-f^i(Y)]D_i\eta(x)\,dX + \int g(X)\eta(x)\,dX\end{aligned}$$

and therefore

$$|U(s)-U(t)| \le C(n,\alpha)\left(\frac{r}{d}\right)^{\alpha}\frac{1}{d}[AU_1+\Lambda U_{1+\alpha}+F+G].$$

Now we note that

$$|u(y,t)-u(Y)| \le |u(y,t)-U(t)|+|U(t)-U(s)|+|u(Y)-U(s)|,$$

and use the mean value theorem to infer that

$$|u(Y)-U(s)|+|u(y,t)-U(t)| \le CU_{1+\alpha}\left(\frac{r}{d}\right)^{\alpha}\frac{1}{d}.$$

The combination of these estimates with the one for $|U(t)-U(s)|$ implies that

$$|u(y,t)-u(y,s)| \le C(A,n,\alpha,\Lambda)\left(\frac{r}{d}\right)^{\alpha}\frac{1}{d}[U_1+U_{1+\alpha}+F+G]$$

if $|t-s|^{1/2} \le r \le d/2$. Taking the supremum over all Y in Ω and suitable r gives

$$\langle u\rangle^*_{1+\alpha} \le C(U_{1+\alpha}+F+G).$$

We next use this inequality along with (4.24) and Schwarz's inequality to infer that

$$|u|^*_{1+\alpha} \le C(|Du|^{(1)}_0+F+G),$$

and then (4.23) follows via (4.2c).

For nonzero b^i, c^i, and c^0, we set $\varphi^i = f^i + b^iu$ and $\Gamma = g + c^0u + c^iD_iu$. Then

$$\begin{aligned}|\varphi^i|^{(1)}_\alpha &\le F + |b^i|^{(1)}_\alpha |u|_0 + |b^i|^{(1)}_0 [u]^*_\alpha, \\ \|\Gamma\|^{(2)}_{1,n+1+\alpha} &\le G + c_1|Du|^{(1)}_0 + c_2|u|_0,\end{aligned}$$

and u solves $-u_t + D_i(a^{ij}D_ju) = D_i\varphi^i + \Gamma$, so we can apply what we have just proved to infer that

$$|u|^*_{1+\alpha} \le C(|u|_0+F+G)+C([u]^*_\alpha+[u]^*_1).$$

Using the interpolation inequality now gives (4.23). □

It is possible to improve inequality (4.23) by displaying the explicit dependence on the norms of the coefficients. Specifically, we have

$$|u|^*_{1+\alpha} \le C(n,\alpha,\lambda,\Lambda)(K|u|_0 + F + G)$$

with

$$K = [1 + A + c_1]^{(1+\alpha)/\alpha} + (|b^i|^{(1)}_0)^{1+\alpha} + c_2 + [b]^{(1)}_\alpha.$$

Moreover the assumption that $|u|^*_{1+\alpha}$ is finite can be relaxed to just $|u|^*_1$ being finite; see Exercise 4.12. If b^i and c^i are in $M^{(1)}_{2,n+2\alpha}$, and if c^0 and g are in $M^{(2)}_{2,n+2\alpha}$, then this statement is clear from the proof of Theorem 4.8. The utility of our hypotheses will be made clear in Section IV.6.

Estimates for classical solutions of nondivergence structure equations now follow from the estimate for weak solutions of divergence structure equations.

THEOREM 4.9. *Let Ω be a bounded domain in $\mathbb{R}^{n+1}$, and let $a^{ij} \in H^{(0)}_\alpha$ and $b^i \in H^{(1)}_\alpha$ satisfy* (4.20a,b) *for some positive constants A, B, λ, and Λ. Let $c \in H^{(2)}_\alpha$ satisfy*

$$|c|^{(2)}_\alpha \le c_1 \tag{4.25}$$

for some constant c_1, and let $f \in H^{(2)}_\alpha$. If $u \in H^*_{2+\alpha}$ is a solution of

$$-u_t + a^{ij}D_{ij}u + b^iD_iu + cu = f \tag{4.26}$$

in Ω, then there is a constant C determined only by A, B, c_1, n, λ, and Λ such that

$$|u|^*_{2+\alpha} \le C(|u|_0 + |f|^{(2)}_\alpha). \tag{4.27}$$

PROOF. As in Theorem 4.8, we may assume that $b^i = 0$ and that $c = 0$. Then we fix $Y \in \Omega$ and $r < d/4$. We also fix $\tau \in (0,r)$ and write $U = u(\cdot,\tau)$ and $f_1 = f(\cdot,\tau)$, where $u(\cdot,\tau)$ and $f(\cdot,\tau)$ are the mollifications of u and f, respectively. Defining the operator L by $Lw = -w_t + a^{ij}D_{ij}w$ and defining the function f_2 by

$$f_2(X) = \int_{\mathbb{R}^{n+1}} [a^{ij}(X) - a^{ij}(X - \tau Y_1)]D_{ij}u(X - \tau Y_1)\varphi(Y_1)\,dY_1,$$

we have that $LU = f_1 + f_2$ in $Q(Y,r)$ and (by taking φ smooth enough) that $U \in C^\infty$. Hence $u_k = D_kU$ is a weak solution of the equation

$$-(u_k)_t + D_i(a^{ij}(Y)D_ju_k) = D_if^i$$

in $Q(r)$ for

$$f^i = \delta^{ik}[(a^{mj}(Y) - a^{mj}(X))D_{mj}U + f_1 + f_2].$$

A direct calculation gives constants $K_1(|u|^*_{2+\alpha}, \alpha, A, d, n, r)$ and $K_2(n)$ such that $|f_2| \le K_1\tau^\alpha$ and $|D^2U| \le K_2 |D^2u|^{(2)}_0 (r/d)^{2+2\alpha}$ in $Q(r)$ and therefore

$$\int_{Q(\rho)} |D^2U - \{D^2U\}_\rho|^2\, dX \le C\left(\frac{r}{\rho}\right)^{n+4} \int_{Q(r)} |D^2U - \{D^2U\}_r|^2\, dX$$
$$+ C(F + |D^2u|^{(2)}_0) r^n \left(\frac{r}{d}\right)^{2+2\alpha} + CK_1\tau^\alpha r^{n+2}$$

Sending $\tau \to 0$ gives

$$\int_{Q(\rho)} |D^2u - \{D^2u\}_\rho|^2\, dX \le C\left(\frac{r}{\rho}\right)^{n+4} \int_{Q(r)} |D^2u - \{D^2u\}_r|^2\, dX$$
$$+ C(F + |D^2u|^{(2)}_0) r^n \left(\frac{r}{d}\right)^{2+2\alpha},$$

which implies that

$$|Du|^{(1)}_{1+\alpha} \le C(|D^2u|^{(2)}_0 + F),$$

and the differential equation yields

$$|u_t|^{(2)}_\alpha \le C(|D^2u|^{(2)}_\alpha + F).$$

The proof is completed by combining these inequalities with interpolation and the definitions of the norms. □

From this theorem, we conclude (as mentioned in Section IV.1) that the term $\langle u\rangle^*_{2+\alpha}$ is superfluous in the definition of the norm for $H^*_{2+\alpha}$.

COROLLARY 4.10. *If $u \in H^*_{2+\alpha}$ for some $\alpha \in (0,1)$, then*

$$|u|^*_{2+\alpha} \le C(n,\alpha)(|u|_0 + |D^2u|^{(2)}_\alpha + |u_t|^{(2)}_\alpha). \tag{4.28}$$

PROOF. Define $f = -u_t + \Delta u$. Then a direct calculation gives

$$|f|^{(2)}_\alpha \le C(n)\left(|D^2u|^{(2)}_\alpha + |u_t|^{(2)}_\alpha\right),$$

so (4.28) follows from Theorem 4.9. □

More generally, one can show that if $u \in C^{2,1}$ with D^2u and u_t in $H^{(2)}_\alpha$, then $u \in H^*_{2+\alpha}$ and (4.28) holds. If u is merely continuous with $[u]^*_2$ and $|u|_0$ finite, then $u \in H^*_2$ with $|u|^*_2 \le C(n)([u]^*_2 + |u|_0)$. Similar results are true for $u \in H^{(b)}_a$ if $a \ge 2$ and $a + b \ge 0$. The proof of these results is left to the reader in Exercises 4.2–4.4.

4. Estimates near a flat boundary

The ideas already discussed also give estimates of solutions of linear parabolic equations with various boundary conditions. In this section, we consider these estimates in the special case that the boundary lies in a hyperplane of the form $\{x^n = 0\}$. General (sufficiently smooth) curved boundaries are considered in Section IV.7.

First, we need to modify our notation appropriately. For $Y \in \mathbb{R}^{n+1}$ and $R > 0$, we use $Q^+(Y,R)$ and $Q^0(Y,R)$ to denote the set of points X in $Q(Y,R)$ for which $x^n > 0$ and $x^n = 0$, respectively. We can then prove a boundary version of Lemma 4.3.

LEMMA 4.11. *Let $X_0 \in \mathbb{R}^{n+1}$ with $x_0^n = 0$, let $f \in L^1(Q^+(X_0,2R))$. If there are positive constants $\alpha \le 1$ and H and a function g defined on $Q^0(X_0,R) \times (0,R)$ such that*

$$\int_{Q^+(Y,r)} |f - g(Y,r)|\, dX \le H r^{n+2+\alpha} \tag{4.29}$$

for all $Y \in Q^0(X_0,R)$ and all $r \le R$. Then

$$[f]_{\alpha;Q^0(X_0,R)} \le C(n,\alpha)H. \tag{4.30}$$

If also g can be defined on $Q^+(X_0,R/2) \times (0,R)$ so that (4.7) *holds for all $Y \in Q^+(X_0,R/2)$ and all $r < \frac{1}{2}d$, where $d = y^n$, then*

$$[f]_{\alpha;Q^+(X_0,R/2)} \le C(n,\alpha)H. \tag{4.31}$$

PROOF. If (4.29) holds, then we define F via (4.9) with $\varphi(z) = 0$ for $z^n < 0$. The argument of Lemma 4.3 applies unchanged to give (4.30).

If also (4.7) holds, it follows that $|F_y| + |F_\tau| + \tau|F_s| \le CH\tau^{\alpha-1}$ for $0 < \tau < \frac{1}{2}y^n$. To estimate $[f]_\alpha$, we estimate $|f(Y_1) - f(Y_2)|$ for Y_1 and Y_2 in Q^+. There are three cases to consider.

If $|Y_1 - Y_2| < \frac{1}{2}\min\{y_1^n, y_2^n\}$, then Lemma 4.3 implies that

$$|f(Y_1) - f(Y_2)| \le CH|Y_1 - Y_2|^\alpha. \tag{4.32}$$

If $y_1^n > 0 = y_2^n$, $y_2^i = y_1^i$ for $i < n$ and $s_1 = s_2$, we set $x = \frac{1}{2}y_1^n$ and $y' = (y_1^1, \dots, y_1^{n-1})$. Since $y' = (y_2^1, \dots, y_2^{n-1})$, we have

$$\begin{aligned}
|f(Y_1) - f(Y_2)| &\le |F(Y_1,0) - F(Y_1,x)| + |F(Y_1,x) - F(Y_2,0)| \\
&= \left|\int_0^x F_\tau(Y_1,\sigma)\, d\sigma\right| + \left|\int_0^1 (F_n + F_\tau)(y', 2\sigma x, s_1, \sigma x)\, d\sigma\right| \\
&\le CHx^\alpha.
\end{aligned}$$

Now we note that $x = \frac{1}{2}|Y_1 - Y_2|$ to infer (4.32) in this case.

Finally, if $|Y_1 - Y_2| \ge \frac{1}{2}\min\{y_1^n, y_2^n\}$, we set $Y_k' = (y_k^1, \dots, y_k^{n-1}, 0, s_k)$. Then

$$|f(Y_1) - f(Y_2)| \le |f(Y_1) - f(Y_1')| + |f(Y_1') - f(Y_2')| + |f(Y_2') - f(Y_2)|.$$

We estimate the first and last terms on the right side of this inequality via the second case in this proof and the second term via the first case to obtain

$$|f(Y_1)-f(Y_2)| \le CH((y_1^n)^\alpha + |Y_1'-Y_2'|^\alpha + (y_2^n)^\alpha).$$

Since $|Y_1'-Y_2'| \le |Y_1-Y_2|$ and $|Y_1-Y_2| \ge \frac{1}{3}\max\{y_1^n, y_2^n\}$ by the triangle inequality, we have (4.32) in this case as well. □

Our next step is an analog of Lemma 4.4 near the boundary.

LEMMA 4.12. *Let u be a weak solution of*

$$u_t = D_i(A^{ij}D_ju) \text{ in } Q^+(R),\ u=0 \text{ on } Q^0(R), \tag{4.33}$$

and suppose that $|u| \le M$ *in* $Q^+(R)$ *for some nonnegative constant M. If* A^{ij} *satisfies* (4.10), *then there is a constant C determined only by n, λ, and Λ such that*

$$\operatorname*{osc}_{Q^+(\rho)} u \le CM\frac{\rho}{R} \tag{4.34a}$$

$$\sup_{Q^+(R/2)} |Du| \le CM/R \tag{4.34b}$$

for all $\rho \in (0,R)$.

PROOF. Define v by

$$v(X) = \frac{x^n}{R} - \frac{(x^n)^2}{2R^2} + A\frac{|x|^2-t}{R^2}$$

for A a positive constant to be chosen. Then $Lv \le 0$ in $Q^+(R)$ if $A \le \lambda/(2n\Lambda+1)$. In addition $v \ge 0$ on $Q^0(R)$, and $v(x,t) \ge A$ if $t=-R^2$ or $|x|=R$. The maximum principle implies that $\pm u \le (M/A)v$ in $Q^+(R)$ and a direct calculation shows that $v \le (1+2A)\rho/R$ in $Q^+(\rho)$, which proves (4.34a).

To prove our gradient bound, we fix $X \in Q^+(R/2)$ and set $\rho = x^n$. Then (4.13) applied to $u-u(X)$ gives

$$|Du(X)| \le C\rho^{-1} \operatorname*{osc}_{Q(X,\rho/2)} u.$$

If $\rho > R/4$, we immediately infer that

$$|Du(X)| \le C\frac{M}{R}. \tag{4.35}$$

If $\rho \le R/4$, then we set $X' = (x^1,\dots,x^{n-1},0,t)$. Since

$$Q(X,\rho/2) \subset Q^+(X',3\rho/2) \subset Q^+(X',R/2) \subset Q^+(R),$$

we can use (4.34a) (in $Q^+(X',R/2)$) to infer that

$$|Du(X)| \le C\rho^{-1}\frac{\rho}{R} \operatorname*{osc}_{Q^+(X',R/2)} u,$$

so (4.35) holds also in this case. Taking the supremum over all $X \in Q^+(R/2)$ gives (4.34b). □

Using the same argument as in Lemma 4.5 in Q^+ gives the following estimate.

LEMMA 4.13. *If u is a $W^{1,2}$ solution of* (4.33), *then*

$$\int_{Q^+(r)} u^2\,dX \le C(n,\lambda,\Lambda)\left(\frac{r}{R}\right)^{n+4} \int_{Q^+(R)} u^2\,dX \tag{4.36}$$

for all $r \in (0,R)$.

Lemma 4.13 can be applied to any tangential difference quotient of u, and hence by taking limits, to any derivative $D_i u$ with $i < n$. For the corresponding estimate on $D_n u$, we use another application of Lemmata 4.12 and 4.13.

LEMMA 4.14. *Suppose that u is a weak solution of* (4.33) *with* (4.10) *holding. Then*

$$\sup_{Q^+(R/12)} |DD_n u| \le CR^{-2} \sup_{Q^+(R/2)} |u - \{D_n u\}_R x^n| \tag{4.37a}$$

and

$$\int_{Q^+(r)} |D_n u - \{D_n u\}_r|^2\,dX \le C\left(\frac{r}{R}\right)^{n+4} \int_{Q^+(R)} |D_n u - \{D_n u\}_R|^2\,dX. \tag{4.37b}$$

PROOF. Note that we can mollify with respect to t only and then u_t will be continuous. Next, set $U = u - \{D_n u\}_R x^n$ and note that $DD_n u = DD_n U$, $u_t = U_t$. Since U also satisfies (4.33), we may apply Lemma 4.12 to $D_i u$ and infer that

$$\sup_{Q^+(R/8)} \left|D_{ij} U\right| \le CR^{-1} \sup_{Q^+(R/4)} |D_i U| \le CR^{-2} \sup_{Q^+(R/2)} |U|$$

for $i < n$ and $j \le n$.

Since $u_t \to 0$ as $x^n \to 0$, the differential equation gives

$$\limsup_{X \to X'} |D_{nn} u(X)| \le CR^{-2} \sup_{Q^+(R/2)} |U|$$

uniformly for $X' \in Q^0(R/8)$. Now set $\zeta = (R^2 - 64|x - x_0|^2)^+ (R^2 - 64(t_0 - t))^+$ and apply the maximum principle to $\zeta^4 |D_{nn} u|^2 + k_1 \zeta^2 R^6 |DU|^2 + k_2 R^{12} U^2$ for suitable constants k_1 and k_2 to infer that

$$|D_{nn} U| \le CR^{-2} \sup_{Q^+(R/2)} U$$

in $Q^+(R/12)$. Combining these estimates gives (4.37a).

To deduce (4.37b), we first note that

$$\sup_{Q^+(R/2)} |U| \le \frac{R}{2} \sup_{Q^+(R/2)} |D_n U|$$

and then follow the same reasoning as in Lemma 4.5. □

Our next step involves a slightly different set of weighted norms. First we write $d(X) = \operatorname{dist}(X, \mathcal{P}Q \setminus Q^0)$. Then we define $|\cdot|^{(b)}_{a;Q^+\cup Q^0}$ as in Section IV.1, and we write $H^{(b)}_a(Q^+ \cup Q^0)$ for the set of functions with finite norm $|\cdot|^{(b)}_{a;Q^+\cup Q^0}$. We will generally suppress the expression $Q^+ \cup Q^0$ from the notation since there will be no chance of confusion with the notation in the previous section. It is easy to see that the interpolation inequality Proposition 4.2 is also true for these norms. The only modification is in the proof of (4.2c). In case the choice of x_2 given there is not in Q^+, we take x_2 so that $x_2 - x_1$ is parallel to $Du(X_1)$. Using this interpolation inequality allows us to derive a boundary version of Theorem 4.8.

THEOREM 4.15. *Let $R > 0$ and let $a^{ij} \in H^{(0)}_\alpha$, $b^i \in H^{(1)}_\alpha$, $c^i \in M^{(1)}_{1,n+1+\alpha}$, and $c^0 \in M^{(2)}_{1,n+1+\alpha}$ satisfy* (4.20) *for some positive constants A, B, c_1, c_2, λ, and Λ. Let f be a vector-valued function in $H^{(1)}_\alpha$ and let g be a scalar-valued function in $M^{(2)}_{1,n+1+\alpha}$ satisfying* (4.21) *for constants F and G. If $u \in H^*_{1+\alpha}$ solves* (4.33), *then there is a constant $C(A,B,c_1,c_2,n,\lambda,\Lambda)$ such that* (4.23) *holds.*

PROOF. We proceed as in Theorem 4.7. First, fix $X_1 \in Q^0$ and $R > 0$, and let v solve

$$-v_t + D_i(A^{ij}D_jv) = 0 \text{ in } Q^+,\ v = u \text{ on } \mathcal{P}Q^+$$

for $Q^+ = Q^+(r,X_1)$. If we set $F_0 = F + U_1 + (GU_\alpha)^{1/2}$, then

$$\int_{Q^+(\rho)} |D_iu|^2\, dX \le C\left(\frac{\rho}{r}\right)^{n+4} \int_{Q^+(r)} |D_iu|^2\, dX + CF_0^2 r^n \left(\frac{r}{d}\right)^{2+2\alpha}$$

for $i < n$ by applying (4.36) to D_iv and

$$\int_{Q^+(\rho)} |D_nu - \{D_nu\}_\rho|^2\, dX$$
$$\le C\left(\frac{\rho}{r}\right)^{n+4} \int_{Q^+(r)} |D_nu - \{D_nu\}_r|^2\, dX + CF_0^2 r^n \left(\frac{r}{d}\right)^{2+2\alpha}$$

by applying (4.37b) to D_nv. Using these estimates, we find from Lemma 4.6 and Corollary 4.10 that $U_{1+\alpha} \le CF_0$. To estimate $\langle u\rangle^*_{1+\alpha}$, we modify the definition of η by setting $\eta(x) = c_0(r^2 - 9|x-y_1|^2)^+$ where $y_1 = (y^1,\dots,y^{n-1},r/2)$. □

The proof of the boundary version of Theorem 4.9 turns out to be somewhat more complicated. Our approach to this result involves the oblique derivative problem, in which some directional derivative is prescribed on $S\Omega$, in a suitable domain Ω. To this end, let (A^{ij}) be a constant matrix satisfying (4.10), and let β be a constant vector satisfying

$$\beta^n \ge \chi, |\beta'| \le \mu\chi \tag{4.38}$$

for positive constants χ and μ. We fix the constant $\bar{\mu} \in (\mu/(1+\mu^2)^{1/2}, 1)$, and, for $R > 0$ and $X_0' \in \mathbb{R}^{n+1}$ with $x_0^n = 0$, we set

$$x_0 = (x_0^1, \ldots, x_0^{n-1}, -\bar{\mu}R, t_0)$$

and we write $\Sigma^+(X_0, R)$ and $\Sigma^0(X_0, R)$ for the subsets of $Q(X_0, R)$ on which $x^n > 0$ and $x^n = 0$, respectively. We also define $\sigma(X_0, R) = \mathscr{P}\Sigma^+(X_0, R) \setminus \Sigma^0(X_0, R)$. When the point X_0 is clear from context, we omit it from the notation. We say that $u \in C^1(\Sigma^+(R) \cup \Sigma^0(R)) \cap C^0(\overline{\Sigma^+(R)})$ is a solution of

$$-u_t + D_i(A^{ij}D_j u) = f \text{ in } \Sigma^+, \tag{4.39a}$$

$$\beta \cdot Du = \psi \text{ on } \Sigma^0, \; u = \varphi \text{ on } \sigma \tag{4.39b}$$

if u is a weak solution of (4.39a) which satisfies (4.39b) pointwise. For brevity, we define operators L and M by $Lu = -u_t + D_i(A^{ij}D_j u)$ and $Mu = \beta \cdot Du$.

We first show that (4.39) is uniquely solvable in a suitable space by virtue of appropriate estimates. To this end, we define $d_1(X) = R - |x - x_0|$, $d_2(X) = (t - t_0 + R^2)^{1/2}$, and $d = \min\{d_1, d_2\}$. Thus, $d(X) = \text{dist}(X, \sigma)$.

LEMMA 4.16. *Suppose that u solves (4.39) with $\varphi = 0$ and that there are positive constants α, F and G with $\alpha < 1$ such that*

$$|f| \le F d^{\alpha-2} \text{ in } \Sigma^+, \; |\psi| \le \chi \Psi d^{\alpha-1} \text{ on } \Sigma^0. \tag{4.40}$$

Then there is a constant $C(\alpha, \lambda, \mu, \bar{\mu})$ such that

$$|u| \le C(F + \Psi) d^\alpha \qquad \text{in } \Sigma^+. \tag{4.41}$$

PROOF. Set $r = |x - x_0|$, $w_1 = (R - r^2/R)^\alpha$, and $w_2 = d_2^\alpha$. Then $Lw_1 \le 4(\alpha - 1)\alpha\lambda d_1^{\alpha-2}$ and $Mw_1 \le -c(\mu, \bar{\mu})\alpha\chi d_1^{\alpha-1}$ by direct calculation. In addition, $Lw_2 = -(\alpha/2)d_2^{\alpha-1}$ and $Mw_2 = 0$.

If we define w by

$$w = (\frac{F}{4(1-\alpha)\alpha\lambda} + \frac{\Psi}{2\alpha c(\mu, \bar{\mu})})w_1 + \frac{2}{\alpha}(F + G)w_2,$$

it follows that $Lw \le f$ in Σ^+, $Mw \le \chi\Psi$ on Σ^0, and $w \ge 0$ on σ. The comparison principle implies that $w \ge u$ in Σ^+, and hence

$$u \le C(\alpha, \mu)(F + \Psi)(d_1^\alpha + d_2^\alpha).$$

Next we apply the comparison principle to u and $C(F + \Psi)w_1$ in $\{d_1 < d_2\}$ and to u and $C(F + \Psi)w_2$ in $\{d_2 < d_1\}$ to see that $u \le C(F + \Psi)d^\alpha$. Arguing similarly with $-u$ gives (4.41). □

Note that $c(\mu, \bar{\mu}) = \mu(1 - \bar{\mu}^2)^{1/2} - \bar{\mu}$, so the constant C in (4.41) depends only on a lower bound for $\bar{\mu}$.

Next we prove a Hölder estimate in appropriate weighted spaces. Here, we define $|\cdot|_a^{(b)}$ as in Section IV.1 with d defined as above.

LEMMA 4.17. *Under the hypotheses of Lemma 4.16, if $f \in H_\alpha^{(2-\alpha)}$ and $\psi \in H_{1+\alpha}^{(1-\alpha)}$, then*

$$|u|_{2+\alpha}^{(-\alpha)} \leq C(\alpha,\mu,\lambda,\Lambda)(|f|_\alpha^{(2-\alpha)} + |\psi|_{1+\alpha}^{(1-\alpha)}). \tag{4.42}$$

PROOF. We abbreviate $H = |f|_\alpha^{(2-\alpha)} + |\psi|_{1+\alpha}^{(1-\alpha)}$. From Lemma 4.16 and the definitions of the norms, we have $|u| \leq CHd^\alpha$. To estimate the derivatives of u near Σ^0, we set $v = Mu - \psi$, and note that $Lv = D_i(\beta^i f - A^{ij}D_j\psi)$ in Σ^+, $v = 0$ on Σ^0. Now fix $X_1 \in \Sigma^0(R)$ and $r < \frac{1}{2}d(X_1)$, and set $Q^+(r) = \Sigma^+(Xx_1', r)$. Then we use the proof of Theorem 4.15 to infer that

$$\int_{Q^+(r)} |D_k v|^2\,dX \leq C(H^2 + d^{-n+2-2\alpha}\int_{Q^+(d/2)} |D_k v|^2\,dX)r^{n+2+2\alpha}d^{-4}$$

for $k < n$ and

$$\int_{Q^+(r)} |D_n v - \{D_n v\}_r|^2\,dX$$
$$\leq C(H^2 + d^{-n+2-2\alpha}\int_{Q^+(d/2)} \left|D_n v - \{D_n v\}_{d/2}\right|^2 dX)r^{n-2+2\alpha}d^{-4}.$$

To estimate the integrals on the right sides of these inequalities, we first multiply the differential equation for v by $\eta^4 v$ for the usual cut-off function η and then integrate by parts. Writing $B^+ = \{x : |x - (x_1', -\bar{\mu})| < 1, x^n > 0\}$ and $Q^+ = Q^+(d)$, we find that

$$\int_{B^+\times\{t_1\}} v^2\eta^4\,dx + \int_{Q^+} \eta^4 |Dv|^2\,dX \leq C(d^{-2}\int_{Q^+} \eta^2 v^2\,dX + H^2 d^{n-2+2\alpha})$$
$$\leq C(d^{-2}\int_{Q^+} \eta^2 |Du|^2\,dX + H^2 d^{n-2+2\alpha}).$$

We now estimate the integral of $\eta^2|Du|^2$ by multiplying the differential equation for u by $\eta^2 u$. An integration by parts yields

$$\frac{1}{2}\int_{B^+\times\{t_1\}} \eta^2 u^2\,dx = \int_{Q^+} \eta^2(A^{ij}D_{ij}u + f) + 2\eta\eta_t u^2\,dX.$$

Next, we define the matrix (B^{ij}) by

$$B^{ij} = \begin{cases} A^{ij} & \text{for } i < n,\ j < n \\ A^{nn}\beta^j/\beta^n & \text{for } i = n,\ j \leq n \\ A^{in} + A^{ni} - A^{nn}\beta^i/\beta^n & \text{for } i < n,\ j = n. \end{cases}$$

Since the difference between (A^{ij}) and (B^{ij}) is an anti-symmetric matrix, it follows that $A^{ij}D_{ij}u = B^{ij}D_{ij}u$, and hence an integration by parts yields

$$\int_{Q^+} \eta^2 u A^{ij} D_{ij} u \, dX = -\int_{Q^+} 2\eta u B^{ij} D_j u D_j \eta + \eta^2 B^{ij} D_i u D_j u \, dX$$
$$- \int_{Q^0} \eta^2 u (A^{nn}/\beta^n) g \, dx' \, dt.$$

It follows by simple rearrangement that

$$\int_{Q^+} \eta^2 |Du|^2 \, dX \le CH^2 d^{n+2\alpha}.$$

Hence

$$\int_{Q^+} \eta^4 |Dv|^2 \, dX \le CH^2 d^{n-2+2\alpha}$$

and therefore

$$|v|^{(1-\alpha)}_{1+\alpha} \le CH. \tag{4.43}$$

Now we use the partial differential equation $Lu = f$ on Σ^0. Defining

$$C^{ij} = A^{ij} - A^{nj}\frac{\beta^i}{\beta^n} - A^{in}\frac{\beta^j}{\beta^n} + A^{nn}\frac{\beta^i\beta^j}{(\beta^n)^2},$$
$$B^i = (A^{ni} + A^{in})/\beta^n - A^{nn}\beta^i/(\beta^n)^2$$

for $i, j < n$, we see that $A^{ij}D_{ij}u = C^{ij}D_{ij}u + B^i D_i v - B^i D_i \psi$. In addition for any vector $\xi \in \mathbb{R}^{n-1}$, we have $C^{ij}\xi^i\xi^j = A^{ij}\Xi^i\Xi^j$ for the vector $\Xi = (\xi, -\beta\cdot\xi/\beta^n)$. It follows that u satisfies the uniformly parabolic equation $-u_t + C^{ij}D_{ij}u = f_1$ on Σ^0 for $f_1 = f - B^i D_i v + B^i D_i \psi$, and (4.43) implies that $|f_1|^{(2-\alpha)}_{\alpha,\Sigma^0} \le CH$. It follows by applying Theorem 4.9 to u on Σ^0 that $|D_k u|^{(1-\alpha)}_{1+\alpha,\Sigma^0} \le CH$ for $k = 1, \dots n-1$. Now, fix $k \in \{1, \dots, n-1\}$, set $h_k(X) = D_k u(x^1, \dots x^{n-1}, 0, t)$ and define $w = D_k u - h_k$. It follows that $Lw = D_i(\delta^{ik}(f - f^*))$ in Σ^+, $w = 0$ on Σ^0, where $f^*(X) = f(x^1, \dots, x^{n-1}, 0, t)$. Hence (4.43) holds for $v = w$ as well. Using the original differential equation to estimate u_t yields (4.42). □

From this estimate, we infer our first uniqueness and existence theorem for problem (4.39).

PROPOSITION 4.18. *Under the hypotheses of Lemma 4.17, there is a unique $H^{(-\alpha)}_{2+\alpha}$ solution u of* (4.39) *with $\varphi = 0$. This solution satisfies the estimate* (4.42).

PROOF. Suppose first that $A^{ij} = \delta^{ij}$ and $\beta^i = \delta^{in}$. If also $\psi = 0$ and $\varphi \ne 0$, we obtain a solution by extending f and φ as even functions of x^n and solving the Dirichlet problem $Lu = f$ in $\Sigma(R)$ and $u = \varphi$ on $\partial\Sigma(R)$, where $\Sigma(R)$ is the set of all X such that X or $(x', -x^n, t)$ is in $\Sigma^+(R) \cup \Sigma^0(R)$. If $\psi \neq 0$ is a $H_{2+\alpha}$ function and $\varphi = 0$, we define U by

$$U(X) = \int_0^{x^n} \psi(x', y, t)\, dy,$$

and then use the previous case to solve $Lv = f - LU$ in Σ^+, $Mv = 0$ on Σ^0, $v = -U$ on σ. It follows that $u = v + U$ solves (4.39) with $\varphi = 0$. If $\psi \neq 0$ is only $H_{1+\alpha}^{(1-\alpha)}$, we can approximate ψ by $H_{2+\alpha}$ functions with uniformly bounded $H_{1+\alpha}^{(1-\alpha)}$ norms, and then (4.42) implies the solvability with the correct estimate.

For general A^{ij} and β, we use the method of continuity to infer the conclusion of this proposition. □

Proposition 4.18 leads to another useful existence result.

COROLLARY 4.19. *If $f = 0$, $\psi = 0$ and φ is merely continuous, then there is a unique solution of* (4.39).

PROOF. Extend φ to all of $\mathbb{R}^{n+1}$ as a continuous function and then approximate uniformly on $\overline{\Sigma^+}$ by smooth functions φ_m. If v_m solves

$$Lv_m = -L\varphi_m \text{ in } \Sigma^+,\ Mv_m = M\varphi_m \text{ on } \Sigma^0,\ v_m = 0 \text{ on } \sigma,$$

then $u_m = v_m + \varphi_m$ solves

$$Lu_m = 0 \text{ in } \Sigma^+,\ Mu_m = 0 \text{ on } \Sigma^0, u_m = \varphi_m \text{ on } \sigma.$$

The maximum principle implies that the sequence (u_m) converges uniformly to some limit function u and Lemma 4.17 implies that this sequence also converges in $H_{2+\alpha}^{(-\alpha)}$, so u satisfies (4.42) and is the unique solution of (4.39). □

It is now straightforward to infer the usual estimates needed to prove Hölder estimates.

LEMMA 4.20. *Let u be a weak solution of $Lu = 0$ in $\Sigma^+(R)$, $Mu = 0$ on $\Sigma^0(R)$. Then*

$$\int_{\Sigma^+(r)} u^2\, dX \leq C(n, \lambda, \Lambda) \left(\frac{r}{R}\right)^{n+4} \int_{\Sigma^+(R)} u^2\, dX \tag{4.44a}$$

$$\int_{\Sigma^+(r)} |u - \{u\}_r|^2\, dX \leq C(n, \lambda, \Lambda) \left(\frac{r}{R}\right)^{n+4} \int_{\Sigma^+(R)} |u - \{u\}_R|^2\, dX \tag{4.44b}$$

for all $r \in (0, R)$.

PROOF. With ζ as in Lemma 4.4, it's easy to check that

$$v = \zeta^2 [\sum_{k<n} |D_k u|^2 + (\beta \cdot Du/\chi)^2] + C\frac{u^2}{R^2}$$

satisfies $Lv \geq 0$ in Σ^+ for a suitable positive constant C. Moreover, because $Mv = 2M\zeta/\zeta$ on Σ^0, we also have $Mv > 0$ on Σ^0. It follows from the maximum principle that $\sup_{\Sigma^+} v \leq \sup_\sigma v \leq CM_0^2/R^2$ for $M_0 = \sup|u|$. Now we write

$$|D_n u| = \left|\beta \cdot Du - \sum_{k<n} \beta^k D_k u\right| /\beta^n \leq |\beta \cdot Du| /\chi + \bar{\mu} |D'u|$$

to see that (4.13) holds in this case. Next, we imitate the proof of Theorem 2.13, noting that $Mv^\pm < 0$ on Σ^0, to obtain $|u(X) - u(X_0)| \leq C(M_0/R)r$ if $X \in Q(X_0,R)$ as in Lemma 4.4. The proof is completed as in Lemma 4.5. □

We are now in a position to prove boundary estimates for the oblique derivative problem.

THEOREM 4.21. *Let $u \in H^*_{1+\alpha}$ be a weak solution of* (4.22) *in Q^+ and*

$$\beta \cdot Du + \beta^0 u = \psi \tag{4.45}$$

on Q^0 with coefficients satisfying (4.20), (4.38), *and*

$$|\beta|^*_\alpha \leq B_1, |\beta^0|^{(1)}_\alpha \leq B_2. \tag{4.46}$$

If also (4.21) *holds and if $|\psi|^{(1)}_\alpha \leq \Psi\chi$, then*

$$|u|^*_{1+\alpha} \leq C(|u|_0 + F + G + \Psi). \tag{4.47}$$

PROOF. Fix $Y_0 \in Q^0$, choose $r < \bar{\mu}R$, set $X_0 = (y'_0, -\bar{\mu}r, s_0)$, and set $\beta_0 = \beta(X_0)$ and $\psi_0 = \psi(X_0)$. If v is the solution of (4.39) with $A^{ij} = a^{ij}(X_0)$, β replaced by β_0, $f = 0$, $\psi = 0$, and φ replaced by $u - \psi_0(\beta_0 \cdot (x - x_0))/|\beta_0|^2$, then Lemma 4.20, applied to $D_k v$ for $k < n$, gives

$$\int_{\Sigma^+(\rho)} |D'v - \{D'v\}_\rho|^2\, dX \leq C\left(\frac{\rho}{r}\right)^{n+4} \int_{\Sigma^+(r)} |D'v - \{D'v\}_r|^2\, dX,$$

and Lemma 4.13 applied to $\beta_0 \cdot Dv$ gives

$$\int_{\Sigma^+(\rho)} |\beta_0 \cdot Dv|^2\, dX \leq C\left(\frac{\rho}{r}\right)^{n+4} \int_{\Sigma^+(r)} |\beta_0 \cdot Dv|^2\, dX.$$

For $v_1(X) = v(X) + \psi_0(\beta_0 \cdot (x - x_0))/|\beta_0|^2$, we can use $u - v_1$ as test function in the weak forms of the equations for v_1 and u to obtain

$$\int_{\Sigma^+(r)} |Du - Dv_1|^2\, dX \leq C(F^2 + GU_\alpha + U_1^2 + \Psi U_1)r^n \left(\frac{r}{d}\right)^{2+2\alpha},$$

and hence

$$\int_{\Sigma^+(\rho)} |D'u - \{D'v_1\}_\rho|^2\, dX \leq C\left(\frac{\rho}{r}\right)^{n+4} \int_{\Sigma^+(r)} |D'u - \{D'v_1\}_r|^2\, dX$$
$$+ C(F^2 + GU_\alpha + U_1^2 + \Psi U_1)r^n \left(\frac{r}{d}\right)^{2+2\alpha},$$

and

$$\int_{\Sigma^+(\rho)} |\beta_0 \cdot Du - \psi_0|^2 \, dX \le C\left(\frac{\rho}{r}\right)^{n+4} \int_{\Sigma^+(r)} |\beta_0 \cdot Du - \psi_0|^2 \, dX$$
$$+ C\beta_0^n (F^2 + GU_\alpha + U_1^2 + \Psi U_1) r^n \left(\frac{r}{d}\right)^{2+2\alpha}$$

Now we set $h(r) = (\varphi - \beta_0' \cdot \{D'u\}_r)/\beta_0^n$ and note that

$$\beta_0 \cdot Du = \beta_0^n D_n u + \beta_0' \cdot \{D'u\}_r + \beta' \cdot (D'u - \{D'u\}_r).$$

It follows that

$$\int_{\Sigma^+(\rho)} |D_n u - h(\rho)|^2 \, dX$$
$$\le C\left(\frac{\rho}{r}\right)^{n+4} \int_{\Sigma^+(r)} |D_n u - h(r)|^2 + \left|D'u - \{D'u\}_r\right|^2 dX$$
$$+ C(F^2 + GU_\alpha + U_1^2 + \Phi U_1) r^n \left(\frac{r}{d}\right)^{2+2\alpha}$$

From these estimates, we see that

$$|Du|^{(1)}_{\alpha, Q^0} \le C(F + G + \Psi + |Du|_0^{(1)}).$$

Combining this estimate with the interior estimates already proved via Lemma 4.11 gives the desired result. □

It is now a simple matter to prove $H_{2+\alpha}$ boundary regularity with Dirichlet boundary conditions.

THEOREM 4.22. *Let $u \in H^*_{2+\alpha}$ solve* (4.26) *in Q^+ and suppose that $u = 0$ on Q^0. If a^{ij}, b^i, c and f satisfy conditions* (4.20a,b) *and* (4.25), *then there is a constant C determined only by A, B, c_1, n, λ, and Λ such that* (4.27) *holds.*

PROOF. As before, we estimate $|D_k u|^{(1)}_{1+\alpha}$ for $k < n$ by using the boundary condition $D_k u = 0$ on Q^0. To estimate the same norm for $v = D_n u$, we note that $\beta \cdot Dv = h$ on Q^0 for $\beta^k = (a^{kn} + a^{nk})$, $\beta^n = a^{nn}$, and $h = f - b^n D_n u$. Theorem 4.21 gives the desired estimate on v. □

For the oblique derivative problem, we obtain the boundary analog of Theorem 4.9 by similar reasoning.

THEOREM 4.23. *Let $u \in H^*_{2+\alpha}$ satisfy* (4.26) *in Q^+ and* (4.45) *on Q^0. Suppose that a^{ij}, b^i, and c satisfy* (4.20a,b) *and* (4.25), *and that β and β^0 satisfy* (4.38) *and*

$$|\beta|^{(1)}_{1+\alpha} \le B_1 \chi, \; |\beta^0|^{(2)}_{1+\alpha} \le B_0 \chi. \tag{4.48}$$

Then there is a constant C determined only by A, B, B_0, B_1, c_1, n, λ, and Λ such that

$$|u|^*_{2+\alpha} \le C(|u|_0 + |f|^{(2)}_\alpha + |\psi|^{(1)}_{1+\alpha}/\chi + |\varphi|^*_{2+\alpha}). \tag{4.49}$$

PROOF. Now we use Lemma 4.20 to estimate $D_k u$, Theorem 4.15 to estimate $\beta \cdot Du$ and then the algebra from Theorem 4.21 to infer the appropriate estimate on Du. □

5. Regularized distance

To deal with curved boundaries and non-zero Dirichlet boundary data, and to discuss the intermediate Schauder theory, we now introduce a regularized distance, that is, a function which is comparable in size to the parabolic distance to $\mathscr{P}\Omega$ and smooth inside Ω. In this section we prove the existence of such a function and examine some of its properties.

To define regularized distance, let $\Omega \subset \mathbb{R}^{n+1}$ and define d in all of $\Omega^* = \mathbb{R}^n \times I(\Omega)$ as follows. First we define

$$d_0(X) = \min_{\substack{Y \in \mathscr{P}\Omega \\ s \le t}} |X - Y|$$

and then

$$d(X) = \begin{cases} d_0(X) & \text{if } X \in \Omega \\ -d_0(X) & \text{if } X \notin \Omega. \end{cases}$$

We then say that $\rho \in C^{2,1}(\Omega^* \setminus \mathscr{P}\Omega) \cap H_1(\Omega^*)$ is a *regularized distance* for Ω if there are positive constants R_1 and R_2 such that $R_1 \le \rho/d \le R_2$ on $\Omega^* \setminus \mathscr{P}\Omega$. We shall not consider regularized distances for arbitrary domains Ω, but only a special class of domains and we shall work locally. To distinguish between cylinders in spaces of different dimension, we write Q' for cylinders in $\mathbb{R}^n$ and Q for cylinders in $\mathbb{R}^{n+1}$. Now let $X_0 \in S\Omega$ and suppose that there are positive constants A and δ, a coordinate system Y with origin at X_0 and a function $f \in H_1(Q'(4\delta))$ with $|f|_1 \le A$ such that

$$\Omega \cap Q(4\delta) = \{Y \in Q(4\delta) : y^n > f(y', s)\}.$$

For φ and η as in the proof of Lemma 4.3, we define $L = 4(A^2+1)^{1/2}$ and $K = \int |\eta'(\sigma)|\, d\sigma$. We then set

$$F(Y,\tau) = y^n - \int_{\mathbb{R}} \int_{\mathbb{R}^n} f(y' - \frac{\tau}{L} z, s - \frac{\tau^2}{2(1+K)^2 L^2} \sigma) \varphi(z) \eta(\sigma)\, dz\, d\sigma.$$

It is easy to check that $F \in C^2(\Omega^* \times (\mathbb{R} \setminus \{0\})$ and that $|F_\tau| \le \frac{2A}{L} < \frac{1}{2}$, and hence

$$|F(Y,\tau_1) - F(Y,\tau_2)| \le \frac{1}{2} |\tau_1 - \tau_2|. \tag{4.50a}$$

Another simple calculation gives

$$|F(Y_1,\tau) - F(Y_2,\tau)| \le \frac{L}{4} |Y_1 - Y_2|. \tag{4.50b}$$

It follows that $F(Y,0)>0$, $F(Y,4\delta)\le 2\delta$, and $F(Y,\tau)-\tau$ is strictly increasing with respect to $\tau\in(0,4\delta)$ for any $Y\in\Omega\cap Q(4\delta)$. Hence for any $Y\in\Omega\cap Q_n(4\delta)$, there is a unique number $\tau\in(0,4\delta)$ such that $\tau=F(Y,\tau)$. We denote this τ by $\rho(Y)$. From (4.50), we have

$$\begin{aligned}|\rho(Y_1)-\rho(Y_2)|&\le|F(Y_1,\rho(Y_1))-F(Y_2,\rho(Y_1))|\\&\quad+|F(Y_2,\rho(Y_1))-F(Y_2,\rho(Y_2))|\\&\le\frac{L}{4}|Y_1-Y_2|+\frac{1}{2}|\rho(Y_1)-\rho(Y_2)|.\end{aligned}$$

Therefore $|\rho(Y_1)-\rho(Y_2)|\le L|Y_1-Y_2|/2$, so $\rho\in H_1$. More calculation shows that $|F_y|\le L/2$ and that $|F_s|+|F_{yy}|+|F_{\tau\tau}|\le CL/\tau$, and so

$$|D\rho|\le\frac{L}{2}\text{ and }|D^2\rho|+|\rho_t|\le C\frac{L}{\rho}.$$

In addition, if we define $g(Y)=y^n-f(y',s)$, we have $g(Y)-\rho(Y)=F(Y,0)-F(Y,\rho(Y))$ so $|g-\rho|\le\frac{1}{2}\rho$ and therefore

$$\frac{1}{2}\rho\le g\le\frac{3}{2}\rho.$$

All that is left to show that ρ is a regularized distance is that g is comparable to d. For this, we fix $Y\in\Omega\cap Q(4\delta)$, set $Y_1=(y',f(y',s),s)$ (so that $Y_1\in\mathscr{P}\Omega$), and use Z to denote the point on $\mathscr{P}\Omega$ such that $d(Y)=|Z-Y|$. Then $d(Y)\le|Y-Y_1|=g(Y)$ and $g(Y)=g(Y)-g(Z)\le(1+A)|Y-Z|=(1+A)d(Y)$. It follows that

$$\frac{1}{2+2A}\rho\le d\le\frac{3}{2}\rho,$$

so ρ is a regularized distance in $\Omega\cap Q_n(4\delta)$. Moreover, we have $D_nF=1$, so that

$$D_n\rho=\frac{1}{1-F_\tau}\ge\frac{2}{3}. \tag{4.51}$$

If, in addition, $f\in H_{1+\alpha}$ for some $\alpha\in(0,1]$ with $|f|_{1+\alpha}\le A_\alpha$, then it follows that $|F_{yy}|+|F_s|+|F_{\tau\tau}|+|F_{y\tau}|\le CA_\alpha\tau^{\alpha-1}$ and therefore

$$|\rho_s|+|D^2\rho|\le CA_\alpha\tau^{\alpha-1},|\rho|_{1+\alpha}\le CA_\alpha.$$

Also we note that the change of variables $\xi'=y'$, $\sigma=s$, $\xi^n=\rho(Y)$ is a one-to-one map from $\Omega\cap Q(2\delta)$ onto a set of the form

$$\{0<\xi^n<h(\xi',\sigma),|\xi'|<2\delta,-4\delta^2<\sigma<0\}$$

with $h>c(A)(4\delta^2-|\xi'|^2)$.

We have two tasks remaining in this section: to describe appropriate weighted norms, and to discuss the behavior of these norms under the change of variables just described.

To define these norms, we set $Q=Q^+(0,R)$ for a fixed constant R. Then, we take numbers $a>0$ and $b\ge -a$ and write $a=k+\alpha$ with k a nonnegative integer and $\alpha \in (0,1]$. We then define $|\cdot|_a^{(b)}$ as in Section IV.1 with d replaced by x^n and $\operatorname{diam}\Omega$ replaced by R. A simple modification of the proof of Proposition 4.2 shows that whenever $c \geq a > 0$, $d \geq b \geq -a$ and $\sigma \in (0,1)$, there is a constant K determined only by c such that

$$|f|_{\sigma a+(1-\sigma)c}^{(\sigma b+(1-\sigma)d)} \leq K(|f|_a^{(b)})^{\sigma}(|f|_c^{(d)})^{1-\sigma}. \tag{4.52}$$

Our next step is an extension lemma to convert arbitrary boundary and initial data to zero data. To simplify the statement of our extension lemma, we introduce some notation. Fix $R > 0$ and write σ and β for the subsets of $\mathscr{P}Q^+(R)$ on which $x^n = 0$ or $t = -R^2$, respectively and set $\omega = \sigma \cup \beta$. We also write $V(\tau)$ ($\tau \in (0,1]$) for the set of functions defined on Q^+ which vanish for $|x| > \tau R$.

LEMMA 4.24. *Let $a \geq b > 0$ and $0 < \tau < \tau' \leq 1$, and suppose that $u \in H_b(\omega) \cap V(\tau)$. If $b \leq 1$, then there is a function $U \in H_a^{(-b)} \cap V(\tau')$ such that*

$$U = u \qquad \text{on } \omega, \tag{4.53a}$$

and there is a constant $C(a,\tau,\tau')$ such that

$$|U|_a^{(-b)} \leq C|u|_b. \tag{4.53b}$$

If $b > 1$ and if there are functions $u_j \in H_{b-j}(\sigma)$ and $v_k \in H_{b-2k}(\beta)$ for all non-negative integers $j < b$ and $k < b/2$ such that

$$\lim_{t\to 0} D_t^k u_j(x',0,t) = \lim_{\xi\to 0} D_n^j v_k(x',\xi,0) \tag{4.54}$$

for all x' with $|x'| < R$, then there is a function $U \in H_a^{(-b)}$ satisfying

$$D_n^j U = u_j \text{ on } \sigma,\ D_t^k U = v_k \text{ on } \beta, \tag{4.55a}$$

and

$$|U|_a^{(-b)} \leq C(a,\tau,\tau')[\sum_j |u_j|_{b-j} + \sum_k |v_k|_{b-2k}]. \tag{4.55b}$$

PROOF. Suppose first that $b \leq 1$. Then we may assume that $u(\cdot,0) \in H_b(\mathbb{R}^n)$ by considering the even extension of u and defining u to be zero outside $\{|x| < R\}$.) Now we set

$$U_1(X) = \int_{\mathbb{R}^n} u(x - t^{1/2}\frac{y}{\tau'-\tau},0)\varphi(y)\,dy$$

for φ a mollifier. A simple calculation shows that

$$\begin{aligned}|U_{1,t}(X)| &= \frac{1}{2t}\left|\int [u(x - t^{1/2}\frac{y}{\tau'-\tau},0) - u(x)]D_i(y^i\varphi(y))\,dy\right| \\ &\leq C|u|_b t^{(b-2)/2}.\end{aligned}$$

Similar reasoning shows that $\left|D_x^\beta D_t^j U_1\right| \le C|u|_b t^{(b-|\beta|-2j)/2}$ for $|\beta|>0$ or $j>0$, so $|U_1|_a^{(-b)} \le C|u|_b$. Next we define $u_2 = u - U_1$ on β and extend u_2 to $\{x^n=0\}$ to be zero where $|x|>R$ and as an even function of t. We then set

$$U_2(X) = \int_{\mathbb{R}}\int_{\mathbb{R}^{n-1}} u_2(x' - x^n\frac{y'}{\tau'-\tau}, 0, t-(x^n)^2\frac{s}{\tau'-\tau})\varphi(y')\eta(s)\,dy'\,ds.$$

Again we find that $|U_2|_a^{(-b)} \le C|u|_b$. If $\eta \in C_0^\infty(Q(\tau' R))$ with $\eta \equiv 1$ in $Q(\tau R))$, it follows that U defined by $U(X) = \eta(X)[U_1(X)+U_2(X)-U_2(x,0)]$ is the desired function.

If $1<b\le 2$, we only need to find a function $U_4 \in H_a^{(-b)}$ such that $U_4 = 0$ on ω and $D_nU_4 = v$ on σ for $v \in H_{b-1}(\sigma)$. Such a function is constructed by letting $U_5 \in H_{a-1}^{(1-b)}$ satisfy $U_5 = v$ on ω and taking $U_4 = x^n U_5$.

The case $b>2$ is proved similarly. □

6. Intermediate Schauder estimates

In this section, we consider estimates for solutions of the equation (4.26) in an intermediate situation. We allow the functions b^i, c, and f to blow up as x^n or t approaches 0, but we still prove good estimates on u and its first spatial derivatives near the boundary. Our first estimate is for solutions of (4.26) satisfying a Dirichlet boundary condition.

PROPOSITION 4.25. *Let $u \in H_{2+\alpha}^{(-b)}(Q^+(R)) \cap V(\tau)$ for some constants $\delta \in (1,2)$, $\alpha \in (0,1)$, $\tau \in (0,1)$, and $R>0$, and suppose u is a solution of* (4.26) *with $u=0$ on Q^0. If* (4.20a) *holds, if*

$$|b^i|_\alpha^{(2-\delta)} \le B,\ |c|_\alpha^{(2-\delta)} \le c_1 \tag{4.56}$$

for some constants A, B, and c_1 and if there is a continuous increasing function ζ with $\zeta(0)=0$ such that

$$|a^{ij}(X)-a^{ij}(Y)| \le \zeta(|X-Y|/R) \tag{4.57}$$

for all X and Y in Q^+, then there is a constant $C(A,B,c_1,n,\delta,\lambda,\Lambda,\zeta)$ such that

$$|u|_{2+\alpha}^{(-\delta)} \le C(|u|_0 + |f|_\alpha^{(2-\delta)}). \tag{4.58}$$

PROOF. Define

$$g(X) = [a^{ij}(X_1)-a^{ij}(X)]D_{ij}u(X) - b^i(X)D_iu(X) - c(X)u(X) + f(X),$$

extend g and u to be zero for $t<0$ or $|x|>R$ and let $X_1 \in Q^0$. It follows that u is a weak solution of $-u_t + D_i(a^{ij}(X_1)D_ju) = g$ in $Q^+(r,X_1)$ for any $r<R$. Moreover,

$$\int_{Q^+(r,X_1)} |g(X)|\,dX \le Cr^{n+\delta}(\zeta(\frac{r}{R})U_1 + U_1 + F)$$

for $F = |f|_\alpha^{(2-\delta)}$, $U_1 = |Du|_0^{(1-\delta)} + |u|_0^{(1-\delta)}$, and $U_2 = \left|D^2u\right|_0^{(2-\delta)}$. Then the proof of Theorem 4.15 shows that $U_\delta \le C\zeta(\frac{r}{R})U_2 + C(U_1+F)$ for $U_\delta = |u|_\delta^{(-\delta)}$, and the proof of Theorem 4.9 gives us $U_2 \le CU_\delta$. Hence

$$U_\delta \le C\zeta(\frac{r}{R})U_\delta + C(U_1+F),$$

and therefore, by taking r/R small enough, $U_\delta \le C(U_1+F)$. Then (4.52) gives $U_\delta \le C(|u|_0+F)$ and the proof of Theorem 4.9 is easily modified to give $|u|_{2+\alpha}^{(-\delta)} \le C(U_\delta+F)$. Combining these final two estimates gives (4.58). □

Similar reasoning (with Theorem 4.21 in place of Theorem 4.15) gives the corresponding estimate for the oblique derivative problem.

PROPOSITION 4.26. *Let $u \in H_{2+\alpha}^{(-\delta)}(Q^+(R)) \cap V(\tau)$ for some constants $\delta \in (1,2)$, $\alpha \in (0,1)$, $\tau \in (0,1)$, and $R > 0$, and suppose u is a solution of* (4.26) *with $\beta^i D_i u + \beta^0 u = \psi$ on ω. Suppose also that* (4.20a) *and* (4.56) *hold for some constants A, B, and c_1 and that there is a continuous increasing function ζ with $\zeta(0) = 0$ such that* (4.57) *is satisfied for all X and Y in Q^+. If also there is a constant B_1 such that*

$$|\beta^i|_{\delta-1} + |\beta^0|_{\delta-1} \le \chi B_1, \tag{4.59}$$

then there is a constant $C(A,B,B_1,c_1,n,\delta,\lambda,\Lambda,\zeta)$ such that

$$|u|_{2+\alpha}^{(-\delta)} \le C(|u|_0 + |f|_\alpha^{(2-\delta)} + |\psi|_{\delta-1}). \tag{4.60}$$

7. Curved boundaries and nonzero boundary data

From the estimates of the previous sections in this chapter, we can prove analogous results for solutions of the various initial-boundary value problems. This section is devoted to this final part of the program. Since the proofs rely on the invariance of certain relations between norms under a change of variables, our first step is to define a suitable class of domains in which such a change of variables can be used to transform to the simple situations previously considered.

For $\delta \ge 1$, we say that $\mathscr{P}\Omega \in H_\delta$ if $B\Omega \subset \{t = t^*\}$ for some t^* and if there is $\varepsilon > 0$ such that for any $X_0 \in S\Omega$, there are a function $f \in H_\delta(Q'(\varepsilon,0))$ and a coordinate system Y centered at X such that

$$\Omega \cap Q(\varepsilon,0) = \{Y \in Q(\varepsilon,0) : y^n > f(y',s)\}$$

in this coordinate system. Using the regularized distance ρ constructed in Section IV.5, we define the change of variables $\Phi(X) = (y',\rho(X),s)$ which maps $\Omega \cap N$ onto $Q^+(\varepsilon) \cap \{t > t_0 - t^*\}$ for some parabolic neighborhood N of X_0. A simple calculation shows that

$$|\Phi|_a^{(-\delta)} + |\Phi^{-1}|_a^{(-\delta)} \le C(a,\delta,|f|_\delta,\varepsilon),$$

and therefore

$$|u|_a^{(b)} \le C_1(a,\delta,|f|_\delta,\varepsilon)\left|u\circ\Phi^{-1}\right|_a^{(b)} \le C_2(a,\delta,|f|_\delta,\varepsilon)\,|u|_a^{(b)}$$

for any u with compact support in $\overline{\Omega}\cap N$ and constants a and b satisfying $a\ge 0$, $b\ge\max\{-a,-\delta\}$. Hence if $\mathscr{P}\Omega\in H_\delta$, we can define $H_a^{(b)}$ for this range of a and b locally: We introduce a partition of unity (χ_i) subordinate to the finite cover of $\overline{\Omega}$ by sets of the form $\Omega\cap N$ and a single set S_0 on which $d>\varepsilon/2$, and write χ_0 for the function supported on S_0. If we define $\|u\|_a^{(b)}=\sum\left|(\chi_i u)\circ\Phi_i^{-1}\right|_a^{(b)}+|\chi_0 u|_a$, it follows that $\|\cdot\|$ and $|\cdot|$ are equivalent norms.

To see the effect of this change of variables on the differential equation and boundary conditions, we first suppose that $\mathscr{P}\Omega\in H_{2+\alpha}$. We write $Y=\Phi(X)$ and defining $v(Y)=u(X)$ and $\partial_i=\partial/\partial y^i$. Then (since $t=s$), we have

$$u_t=v_s+y^k_t\partial_k v,\ D_iu=D_iy^k\partial_k v,$$
$$D_{ij}u=D_iy^kD_jy^m\partial_{km}v+D_{ij}y^k\partial_k v,$$

and hence

$$-u_t+a^{ij}D_{ij}u+b^iD_iu+cu=-v_s+A^{ij}\partial_{ij}v+B^i\partial_iv+Cv,$$

where

$$A^{ij}(Y)=a^{km}(X)D_ky^iD_my^j,$$
$$B^i(Y)=a^{km}(X)D_{km}y^i+b^k(X)D_ky^i-y^i_t,\ C(Y)=c(X).$$

It follows that the hypotheses of Theorem 4.22 are invariant under an $H_{2+\alpha}$ change of variables.

For equations in divergence form, we have

$$-u_t+D_i(a^{ij}D_ju+b^iu)+c^iD_iu+c^0u=-v_s+\partial_i(A^{ij}\partial_jv+B^iv)+C^i\partial_iv+C^0v$$

for

$$A^{ij}=a^{km}D_ky^iD_my^j,\ B^i=b^kD_ky^i,$$
$$C^i=c^kD_ky^i-a^{mj}\partial_k(D_my^k)D_jy^i-y^i_t,\ C^0=c^0-b^m\partial_k(D_my^k).$$

Therefore the hypotheses of Theorem 4.15 are invariant under an $H_2^{(-1-\alpha)}$ change of variables.

From this change of variables, we infer the following global estimates. We also define the Morrey space $M^{p,\delta}$ to be the set of all functions $u\in H_1$ with finite norm

$$\|u\|_{p,\delta}=\sup_{\substack{Y\in\Omega\\ r<\operatorname{diam}\Omega}}\left(r^{-\delta}\int_{\Omega[Y,r]}|u|^p\,dX\right)^{1/p}.$$

THEOREM 4.27. *Let $u \in H_{1+\alpha}$ be a weak solution of* (4.22) *in some domain Ω with $\mathscr{P}\Omega \in H_{1+\alpha}$ and suppose that $u = \varphi$ on $\mathscr{P}\Omega$ for some $\varphi \in H_{1+\alpha}(\Omega)$. Suppose also that a^{ij} and b^i are in H_α, and that c^i and c^0 are in $M^{1,n+1+\alpha}$ with*

$$|a^{ij}|_\alpha \le A, |b^i|_\alpha \le B, \tag{4.61a}$$

$$\|c^j\| \le c_1 \tag{4.61b}$$

for $j = 0, \dots, n$. Suppose also that the first inequality in (4.20a) *holds. If $f \in H_\alpha$ and $g \in M^{1,n+1+\alpha}$, then*

$$|u|_{1+\alpha} \le C(A, B, c_1, n, \lambda, \Lambda, \Omega)(|u|_0 + |f|_\alpha + \|g\|_{1,n+1+\alpha} + |\varphi|_{1+\alpha}). \tag{4.62}$$

PROOF. By considering $u - \varphi$ in place of u, we may assume $\varphi = 0$ on $\mathscr{P}\Omega$. If we set $F = |f|$ and $G = \|g\|$, then for each $X_0 \in \mathscr{P}\Omega$, there is a parabolic neighborhood N such that

$$|u|^*_{1+\alpha;(N\cap\Omega)\cup T} \le C(|u|_0 + F + G),$$

where $T = N \cap \mathscr{P}\Omega$, by Theorem 4.15. Hence, there is a $\varepsilon = \varepsilon(X_0) \le 1$ such that $\Omega[2\varepsilon, X_0] \cap \Omega$ is a subset of N, so that

$$\varepsilon^{1+\alpha} |u|_{1+\alpha;\Omega[\varepsilon]} \le C(|u|_0 + F + G).$$

We now cover $\mathscr{P}\Omega$ by a finite number of cylinders $Q(X_i, \varepsilon(X_i))$ and set $\sigma = \frac{1}{4}\min_i \varepsilon(X_i)$. It follows that $\overline{\Omega}$ can be covered by a finite number of cylinders $Q(X_i, \sigma)$ with

$$|u|^*_{1+\alpha;\Omega[X_i,\sigma]} \le C(|u|_0 + F + G),$$

and hence

$$|Du(X) - Du(Y)| \le C(|u|_0 + F + G)\,|X - Y|^\alpha$$

as long as $|X - Y| \le \sigma$. A similar argument yields

$$|u(x,t) - u(x,s)| \le C(|u|_0 + F + G)\,|t - s|^{(1+\alpha)/2}$$

for all (x,t) and (x,s) in Ω with $t - \sigma^2 < s < t$. In addition, we have

$$|Du|_0 \le C(|u|_0 + F + G),$$

and (4.62) follows from these three inequalities. □

For equations in nondivergence form, a similar argument gives the following result.

THEOREM 4.28. *Let $u \in H_{2+\alpha}$ be a solution of* (4.26) *in Ω and $u = \varphi$ on $\mathscr{P}\Omega$ with $\mathscr{P}\Omega \in H_{2+\alpha}$ and $\varphi \in H_{1+\alpha}(\Omega)$. Suppose also that a^{ij}, b^i, and c are in H_α with a^{ij} and b^i satisfying* (4.61a) *and c satisfying $|c|_\alpha \le c_1$. If $f \in H_\alpha$, then*

$$|u|_{2+\alpha} \le C(A, B, c_1, n, \lambda, \Lambda, \Omega)(|u|_0 + |f|_\alpha + |\varphi|_{2+\alpha}). \tag{4.63}$$

Note that under the hypotheses of this theorem, φ and f satisfy a compatibility condition because u_t is prescribed in two different ways along $C\Omega$. For the case of a cylindrical domain $\Omega = D \times (0,T)$, if the data are $u = \varphi$ on $\partial D \times (0,T)$ and $u(\cdot,0) = \psi$ on $\overline{D}$, the condition is just that $\varphi = \psi$ and $\varphi_t = a^{ij}D_{ij}\psi + b^iD_i\psi + c\psi - f$ on $\partial D \times \{0\}$. For a general domain with $H_{2+\alpha}$ boundary, if $u = \varphi$ on $S\Omega$ and $u = \psi$ on $B\Omega$, then the condition is $\varphi = \psi$ and

$$\varphi_t - \frac{\rho_t}{|D\rho|^2}D\rho \cdot D\varphi = a^{ij}D_{ij}\psi + (b^i - \frac{\rho_t}{|D\rho|^2}D_i\rho)D_i\psi + c\psi - f$$

on $C\Omega$, where ρ is a regularized distance to the boundary. Note that $\varphi_t - \frac{\rho_t}{|D\rho|^2}D\rho \cdot D\varphi$ can be written as $v \cdot (D\varphi, \varphi_t)$ for some $(n+1)$-dimensional vector v which is tangent to $S\Omega$, hence the compatibility condition is independent of the choice of extension of φ. It is also easily checked that $\rho_t/|D\rho|$ and $D\rho/|D\rho|$ are independent of the choice of regularized distance, since they are just components of the $n+1$-dimensional normal vector to $S\Omega$. Equivalently, if φ is given on all of $\mathscr{P}\Omega$, the compatibility condition can be written as

$$-\varphi_t + a^{ij}D_{ij}\varphi + b^iD_i\varphi + c\varphi = f$$

on $C\Omega$, where φ is extended into Ω as an $H_{2+\alpha}$ function , and this condition is independent of the particular extension.

We also obtain estimates for the intermediate case.

THEOREM 4.29. *Let* $1 < \delta < 2$ *and suppose* $\mathscr{P}\Omega \in H_\delta$. *Let* $u \in H^{(-\delta)}_{2+\alpha}$ *be a solution of* (4.26) *and* $u = \varphi$ *on* $\mathscr{P}\Omega$ *with* a^{ij}, b^i *and* c *satisfying* (4.20a), (4.56) *and*

$$|a^{ij}(X) - a^{ij}(Y)| \le \zeta(|X-Y|) \tag{4.64}$$

for some continuous increasing function ζ *with* $\zeta(0) = 0$. *If* $|f|^{(2-\delta)}_\alpha \le F$, *then there is a constant* C *determined only by* A, B,c_1, n, δ, ζ, λ, Λ, *and* Ω *such that*

$$|u|^{(-\delta)}_{2+\alpha} \le C(|u|_0 + F + |\varphi|_\delta). \tag{4.65}$$

PROOF. Again, we may assume that $\varphi \equiv 0$. Then we use the partition of unity (χ_i) from the definition of $\|\cdot\|$. Then $u_i = \chi_i u$ satisfies the equation $Lu_i = f_i$ with

$$f_i = \chi_i f + (a^{kj}D_{kj}\chi_i + b^kD_k\chi_i)u + 2a^{kj}D_k\chi_iD_ju.$$

It follows that

$$|u|^{(-\delta)}_{2+\alpha} \le C(|u|_0 + F + |Du|^{(2-\delta)}_\alpha)$$

and then (4.52) gives (4.65). □

Analogous reasoning also gives us global estimates for the oblique derivative problem, in which the condition $u = \varphi$ on $\mathscr{P}\Omega$ is replaced by

$$u = \varphi \text{ on } B\Omega,\ \beta^iD_iu + \beta^0u = \psi \text{ on } S\Omega. \tag{4.66}$$

We recall that the boundary condition in (4.66) is oblique if there is a positive constant χ such that

$$\beta\cdot\gamma\geq\chi \text{ on } S\Omega. \tag{4.67}$$

THEOREM 4.30. *Suppose that, in the hypotheses of Theorem 4.27, the condition $u=\varphi$ on $\mathcal{P}\Omega$ is replaced by (4.66) and that (4.67) holds for some positive χ. Suppose also that condition (4.59) holds with $\delta=1+\alpha$. If $|\psi|_\alpha\leq\chi\Psi$ and $|\varphi|_{1+\alpha}\leq\Phi$, then*

$$|u|_{1+\alpha}\leq C(A,B,B_1,c_1,c_2,n,\alpha,\lambda,\Lambda,\Omega)(|u|_0+F+G+\Psi+\Phi). \tag{4.68}$$

THEOREM 4.31. *Suppose that, in the hypotheses of Theorem 4.28, the condition $u=\varphi$ on $\mathcal{P}\Omega$ is replaced by (4.66) and that (4.67) holds for some positive χ. If $|\beta^i|_{1+\alpha}+|\beta^0|_{1+\alpha}\leq B_1\chi$, $|\psi|_{1+\alpha}\leq\chi\Psi$, and $|\varphi|_{2+\alpha}\leq\Phi$, then*

$$|u|_{2+\alpha}\leq C(A,B,B_1,c_1,c_2,n,\alpha,\lambda,\Lambda,\Omega)(|u|_0+F+\Psi+\Phi). \tag{4.69}$$

THEOREM 4.32. *Suppose that, in the hypotheses of Theorem 4.30, the condition $u=\varphi$ is replaced by (4.66) and that (4.67) holds for some positive χ. If condition (4.59) holds and if $|\psi|_{\delta-1}\leq\chi\Psi$ and $|\varphi|_\delta\leq\Phi$, then*

$$|u|^{(-b)}_{2+\alpha}\leq C(A,B,B_1,b,c_1,n,\alpha,\lambda,\Lambda,\Omega)(|u|_0+F+\Phi+\Psi). \tag{4.70}$$

8. Two special mixed problems

We now examine two problems in which a Dirichlet condition is prescribed on part of $S\Omega$ and an oblique derivative condition is prescribed on the rest of $S\Omega$. Rather than develop a full theory of such problems, we only look at two particular such problems.

First, for β a vector-valued function satisfying (4.38), $R>0$ and $\bar{\mu}$ as in Section IV.4, we look at

$$-u_t+a^{ij}D_{ij}u=f \text{ in } \Sigma^+, \tag{4.71a}$$

$$\beta\cdot Du=\psi \text{ on } \Sigma^0,\ u=\varphi \text{ on } \sigma, \tag{4.71b}$$

which is just (4.39) when a^{ij} and β are constants. If a^{ij}, β, f, ψ, and φ are smooth enough, the methods of Section IV.4 show that this problem has a unique solution. Here we are concerned with the smoothness of u near σ^0, the subset of σ on which $x^n=0$ and $t>0$.

LEMMA 4.33. *Let u be a solution of (4.71) with β and ψ in $H_\alpha(\Sigma^0)$ and $\varphi\in H_{1+\alpha}(\sigma)$ for some $\alpha\in(0,1)$. Suppose also that there are positive constants B, F, λ, Λ, Ψ, and Φ such that*

$$\lambda|\xi|^2\leq a^{ij}\xi_i\xi_j\leq\Lambda|\xi|^2, \tag{4.72a}$$

$$|f|\leq\lambda Fd^{\alpha-1}, \tag{4.72b}$$

$$|\beta|_\alpha\leq B\chi,\ |\psi|_\alpha\leq\Psi\chi,\ |\varphi|_{1+\alpha}\leq\Phi, \tag{4.72c}$$

where $d(X) = \operatorname{dist}(X, S\Sigma^+)$. Then there are constants $\alpha_0(n,\Lambda/\lambda,\mu,\bar{\mu}) > 0$ and $C(B,n,R,\lambda,\Lambda,\mu,\bar{\mu})$, and a vector-valued function v such that $\alpha \le \alpha_0$ implies that

$$|u(X) - \varphi(X_1) - v(X_1)\cdot(x - x_1)| \le \frac{C}{\alpha}|x - x_1|^{1+\alpha}[F + \Psi + \Phi] \tag{4.73}$$

for any $X_1 \in \sigma^0$.

PROOF. First, we note that (4.73) implies that u is differentiable at any $X_1 \in \sigma^0$ and that $v(X_1) = Du(X_1)$. Moreover, there is a unique vector $v(X_1)$ such that $\beta(X_1)\cdot v(X_1) = \psi(X_1)$ and $v(X_1)\cdot\tau(X_1) = D\varphi(X_1)\cdot\tau(X_1)$ for any vector $\tau(X_1)$ tangential to Σ^0 at X_1. In addition $v \in H_\alpha(\sigma^0)$ with $|v|_{1+\alpha} \le C\Psi$. Specifically, writing γ^* for the unit inner normal to Σ^+, we have $v = D\varphi + V_0\gamma^*$ for some function $V_0 \in H_\alpha(\sigma^0)$ with $|V_0|_\alpha \le C\Psi$.

Next, we note that V_0 can be extended as an H_α function to all of σ^+, the closure of the subset of $S\Sigma^+$ on which $x^n > 0$ and $t > 0$, so that $V_0 = D\varphi\cdot\gamma^*$ wherever $t = 0$. Then Lemma 4.24 gives a function U such that $U = \varphi$ and $DU\cdot\gamma^* = V_0$ on σ^+. By considering $u - U$ in place of u, we may assume that $\varphi \equiv 0$ on σ^+ and that $\psi \equiv 0$ on σ^0.

Now, we fix $X_1 \in \sigma^0$ and rotate and translate the (x', t) coordinates so that X_1 is the origin and the tangent plane to σ^+ at X_1 has the equation $[\cos\theta_0]x^1 + [\sin\theta_0]x^n = 0$ for $\theta_0 = \arccos\bar{\mu}$. Setting

$$r = [(x^1)^2 + (x^n)^2]^{1/2}, \theta = \arctan(x^n/x^1)$$

(these are just polar coordinates in the x^1x^n-plane), we introduce the function $w_1 = r^{1+\alpha}f(\theta)$ with $f \in C^2[0,\theta_0]$ to be further specified. A direct calculation shows that there is a symmetric 2×2 matrix-valued function (b^{ij}) such that

$$Lw_1 = r^{\alpha-1}[b^{11}(1+\alpha f) + 2b^{12}\alpha f' + b^{22}(f'' + (1+\alpha)f)]$$

in Σ^+ and

$$Mw = r^\alpha[(1+\alpha)\beta^1 f + \beta^n f']$$

on Σ^0. Now we choose $\theta_1 \in (\theta_0, \arctan\mu)$ and set $\delta = \theta_1/\theta_0$ and

$$f(\theta) = \cos((1+\delta)\theta - \theta_1).$$

Then f is increasing and $f(0) = \cos\theta_1 > 0$, so f is bounded from below by a positive constant. In addition,

$$Lw_1 \le r^{\alpha-1}\lambda\left[2\alpha\frac{\Lambda}{\lambda} + (\alpha-\delta)\cos\theta_0\right]$$

and

$$Mw_1 \le r^\alpha\chi\cos\theta_1[(1+\alpha)\mu - \tan\theta_1],$$

so there are constants α_0 and C_0 determined only by n, Λ/λ, μ, and $\bar{\mu}$ such that $Lw_1 \le -C_0\lambda r^{\alpha-1}$ and $w_1 \ge C_0 r^{\alpha+1}$ in Σ^+, $Mw \le -C_0\chi r^\alpha$ on Σ^0.

We also define

$$w_2 = (x^n)^{1+\alpha}, w_3 = (R^2 - |x - x_0|^2)^{1+\alpha},$$

and

$$w = A_1(\Psi + F)w_1 - \frac{F}{\alpha}w_2 - A_2Fw_3,$$

with A_1 and A_2 positive constants to be chosen. Since

$$Lw_3 \geq C(\bar{\mu}, R)\alpha\lambda(R^2 - |x - x_0|^2)^{\alpha-1} - C(n, R, \Lambda/\lambda),$$

we can choose first A_2 and then A_1 so that $Lw \leq Lu$ and $w \geq 0$ in Σ^+. Since also $Mw_2 \leq 0$, we can increase A_1 so that $Mw \leq Mu$ on Σ^0, and then the maximum principle gives $u \leq w$ in Σ^+. Similarly, $-u \leq w$, and then (4.73) follows easily. □

Next, we assume that β is a constant vector satisfying (4.38) and we take $\bar{\mu}$ as in Section IV.4. Now, for $X_0' \in \mathbb{R}^{n+1}$ with $x_0^n = 0$, and positive numbers $r < R$, we write X_1' and X_2' for elements of $\mathbb{R}^{n+1}$ with $x_1^n = x_2^n = 0$ and $t_1 = t_2 = t_0$ such that $|x_0' - x_1| = R$ (recall that $x_1^n = -\bar{\mu}R$) and

$$x_0' - x_1' = \left(1 - \frac{r}{R}\right)(x_2' - x_0').$$

We then set

$$\begin{aligned}
\Sigma^+(X_0; R, r) &= \Sigma^+(X_1, R) \cap \Sigma^+(X_2, R), \\
\Sigma^0(X_0; R, r) &= \Sigma^0(X_1, R) \cap \Sigma^0(X_2, R), \\
\sigma(X_0; R, r) &= \mathscr{P}\Sigma^+(X_1; R, r) \setminus \Sigma^0(X_0; R, r), \\
\sigma^*(X_0; R, r) &= \sigma(X_1, R) \cap \sigma(X_2, R).
\end{aligned}$$

The arguments leading to Proposition 4.18 then imply that there is a unique solution of (4.39) with $\phi \equiv 0$. In trying to prove higher regularity of the solution, we note that the condition $\phi \equiv 0$ implies that $Du = 0$ on σ^*, so we would generally need

$$\lim_{\substack{X \to X^* \\ X \in \Sigma^0}} \psi(X) = 0.$$

For our applications, this condition will be trivially satisfied by taking $\psi \equiv 0$ as well. We then have the following analog of Lemma 4.33.

LEMMA 4.34. *Let u be a solution of* (4.39) *with β constant, $\psi \equiv 0$, and $\varphi \equiv 0$. Suppose also that there are positive constants F, λ, and Λ such that* (4.72a,b) *hold with $d(X) = \operatorname{dist}(X, S\Sigma^+)$. Then there are constants $\alpha_0(n, \Lambda/\lambda, \mu, \bar{\mu}) > 0$ and $C(n, R, r, \lambda, \Lambda, \mu, \bar{\mu})$ such that $\alpha \leq \alpha_0$ implies that*

$$|u(X)| \leq \frac{C}{\alpha}|x - x_1|^{1+\alpha}F \tag{4.74}$$

for any $X_1 \in \sigma^0 \cup \sigma^*$.

PROOF. First, we argue as in Lemma 4.33 that (4.74) holds for $X_1 \in \sigma^0$ and then a similar argument applies for $X_1 \in \sigma^*$ once we note that $w_1 \geq 0$ rather than $Mw_1 \leq -C_0\lambda r^\alpha$. □

Notes

The Campanato spaces used in this chapter are parabolic versions of spaces introduced by Campanato [39] for the study of elliptic equations [40]; the parabolic versions are a special case of a more general concept due to DaPrato [59]. Campanato also used the parabolic versions of these spaces to study parabolic equations [41]. Our proofs of the estimates in this chapter are simplifications of his. The mollification proof that Hölder and Campanato spaces are equivalent given here is based on the proof by Trudinger [311] (see Section 3.1 there). It is quite different from the original proof of DaPrato [59], who used an iteration scheme to estimate the mean values of the function and then a variant of the Lebesgue differentiation theorem to infer that these mean value estimates imply the corresponding ones for the function itself.

The estimates in Lemma 4.5 are a crucial part of the Campanato space approach. His proof of these estimates is based on the Sobolev imbedding theorem, which we shall visit in Chapter VI. Our proof is completely different and comes from the author's work [229, 230], which seems to be the first time such an approach was used although it is similar to DiBenedetto's proof of intrinsic Harnack inequalities from Hölder estimates for solutions of degenerate parabolic equations [62].

The proof of Theorem 4.7 (and the proofs of all estimates for inhomogeneous equations with variable coefficients from the corresponding ones for homogeneous equations with constant coefficients) is essentially Campanato's except for some technical details. Our boundary estimates and the use of hypothesis (4.20b) come from [229]. A somewhat stronger hypothesis for the term g appears in [343]. In fact, Hölder estimates of this type are usually proved directly for equations in general form. Theorem 4.9 was first proved by Ciliberto [54] for the case $n = 1$ and extended to arbitrary dimensions by Barrar [13, 14].

Boundary estimates for the Cauchy-Dirichlet problem were proved by Ciliberto [54] ($n = 1$), who considered only cylindrical domains, and by Friedman [87] ($n \geq 1$), who considered arbitrary domains with $H_{2+\alpha}$ boundary. Estimates for the oblique derivative problem were first given by Kamynin and Maslennikova [155].

Regularized distance was introduced (in an elliptic setting) by Calderón and Zygmund [38, Lemma 3.2] via a very different construction; see also [268]. We have followed [208], in which both a regularized distance and a parabolic regularized distance were given. The extension lemma, Lemma 4.24, is adapted from [100].

Our notation and results on intermediate Schauder theory, taken from [213], are modeled on the corresponding elliptic results of Gilbarg and Hörmander [100].

Our study of the oblique derivative problem, starting after Theorem 4.15, comes from [207]. Earlier related results for parabolic equations are due to Kamynin [143–145, 147, 148] and Krüger [162] although these authors did not consider the parameters a and b to be independent. It should be noted that the regularity of $\mathscr{P}\Omega$ can be further relaxed without affecting the regularity of the solution. If $\mathscr{P}\Omega \in H_1$, then the solution is in $H_{2+\alpha}^{-1-\theta}$ for some $\theta \in (0,1)$ (see [221, 236]) and Theorem 4.30 is also true for $\mathscr{P}\Omega \in H_{1+\alpha}$ (see [221, 236] and, for the elliptic case only, [286]).

The estimate for the mixed problem in Section IV.8 is essentially due to Azzam and Kreyszig [11, Lemma 1], who assumed a more general geometric configuration but a more restrictive set of conditions for a^{ij} and β, specifically that $\beta^i = a^{ij}\gamma_j$. Our proof is a slight modification of theirs; more general mixed problems are studied in [222].

There are many different methods for deriving the Hölder estimates. Representation of solutions via Green's functions was the original technique, used in [13, 14, 54, 87, 88, 213, 296]. The Campanato approach was used in [41, 229]. More recently, Safonov [282] has proved the Schauder estimates using the Harnack inequality of Krylov and Safonov [180]; we shall discuss this approach in Chapter XIV. Trudinger [311] has used mollification to prove Schauder estimates for elliptic equations, and that technique is easily modified to deal with parabolic equations. Brandt [24] and Knerr [159] used the maximum principle and some clever comparison functions to obtain Schauder type estimates. Using the theory of viscosity solutions, Wang [333, 334] has given a still different proof.

Many of these Hölder estimates (with continuity in space and time for the derivatives) remain true even when the coefficients of the equation have no assumed continuity in time. Complete details of this situation can be found in [24, 73, 159, 229] and the references therein. In the exercises, we consider some of the extensions, which use essentially the proofs given in the text but require some additional existence theorems.

On the other hand, the results of this chapter generally fail if the Hölder modulus of continuity is replaced by an arbitrary one. The sharp condition on the modulus of continuity is that it be Dini. If the coefficients a^{ij}, b^i, c, and f are continuous with modulus of continuity ω satisfying

$$\int_0^\delta \frac{\omega(s)}{s}\,ds < \infty$$

for some positive constant δ and if there is a constant $\alpha \in (0,1)$ such that $s^{-\alpha}\omega(s)$ is a decreasing function of s, then u_t and D^2u are continuous with modulus of continuity ω_1 defined by

$$\omega_1(\sigma) = \int_0^\sigma \frac{\omega(s)}{s}\,ds.$$

The proof of this fact (see [251,252,297]) shows that

$$\int_{Q(r)} \left|D^2u - \{D^2u\}_r\right| \, dX \le Cr^{n+2}\omega(r),$$

so there is an isomorphism with respect to the Campanato-Dini norms. Some related results (for equations in divergence form) appear in [32].

Furthermore, the Dirichlet and oblique derivative boundary conditions can be replaced by other ones. For example, the Ventsell′ boundary condition, in which a combination of second tangential derivatives and first tangential and normal derivatives is prescribed on $S\Omega$, was studied by Apushkinskaya and Nazarov [6]. There is also a large literature on nonlocal boundary conditions. We refer to [142, 149] for a discussion of some nonlocal boundary conditions and to [233] for a discussion of periodic solutions.

Exercises

4.1 Prove Corollary 4.10 without invoking any estimates for solutions of parabolic differential equations.

4.2 Show that if $u \in C^{2,1}$ and if the right side of (4.28) is finite, then so is the left side, and hence $u \in H^*_{2+\alpha}$.

4.3 Show that if u is continuous with $[u]^*_2 + |u|_0$ finite then $u \in H^*_2$ with $|u|^*_2 \le C([u]^*_2 + |u|_0)$.

4.4 Prove an analog of Exercise 4.2 with $H^{(b)}_a$ in place of $H^*_{2+\alpha}$ provided $a+b \ge 0$ and a is not an integer. What regularity of $\mathscr{P}\Omega$ is needed for these results?

4.5 Prove the analogs of Theorems 4.9, 4.28, and 4.31 for $H_{k+\alpha}$. Specifically, show the following estimates.

(a) Suppose that $a^{ij} \in H^*_{k+\alpha}$, $b^i \in H^{(1)}_{k+\alpha}$, $c \in H^{(2)}_{k+\alpha}$. If $Lu = f$ with $f \in H^{(2)}_{k+\alpha}$, then

$$|u|^*_{k+2+\alpha} \le C(|u|_0 + |f|^{(2)}_{k+\alpha}).$$

(b) Suppose that a^{ij}, b^i, and c are in $H_{k+\alpha}$, $\mathscr{P}\Omega \in H_{k+2+\alpha}$, and $\varphi \in H_{k+2+\alpha}$. If $Lu = f$ in Ω and $u = \varphi$ on $\mathscr{P}\Omega$, then

$$|u|_{k+2+\alpha} \le C(|u|_0 + |\varphi|_{k+2+\alpha} + |f|_{k+\alpha}).$$

(c) Suppose L, f, and Ω are as in part (b), β^j ($j \ge 0$) and ψ are in $H_{k+1+\alpha}$, and $\varphi \in H_{k+2+\alpha}(B\Omega)$. If $Lu = f$ in Ω, $Mu = \psi$ on $S\Omega$, and $u = \varphi$ on $B\Omega$ with $f \in H_{k+\alpha}$, then

$$|u|_{k+2+\alpha} \le C(|u|_0 + |\varphi|_{k+2+\alpha} + |\psi|_{k+1+\alpha} + |f|_{k+\alpha}).$$

4.6 Suppose that A^{ij} in Theorem 4.7 depends on t (but not on x) and that f satisfies the weaker condition

$$|f(x,t)-f(x_1,t)| \le F\,|x-x_1|^\alpha$$

for all (x,t) and $(x_1,t) \in Q(Y,2R)$. Assuming the existence of the solution v, show that (4.18) still holds.

4.7 Suppose that μ is a Radon measure defined on Ω such that

$$\mu(Q(Y,r)) \le Gr^n(r/d(Y))^{1+\alpha}$$

for all $Y \in \Omega$ and $r < \frac{1}{2}d(Y)$. If $u \in H^*_{1+\alpha}$ is a weak solution of

$$-u_t + D_i(a^{ij}D_{ij}u + b^i u) + c^i D_i u + c^0 u = D_i f^i + g + \mu,$$

that is,

$$\int_\Omega (u_t - c^0 u - g)\zeta + (a^{ij}D_j u + b^i u - f^i)D_i\zeta\,dX = \int_\Omega \zeta\,d\mu$$

for all $\zeta \in C^1_0$ and if the coefficients satisfy (4.20) and (4.25), then (4.23) holds. (Compare with [231, 280].)

4.8 Let $(\Sigma, \mathscr{M}, \pi)$ be a σ-finite measure space, and let j, j_1, and $\mathscr{I}$ be as in Exercise 2.5. Suppose also that there are constants c_3 and J_0 and a nonnegative function $j_0 \in L^1(\Sigma)$ such that

$$|j_1(\cdot,\xi)|_\alpha \le c_3, |j(\cdot,\xi)|_\alpha \le j_0(\xi), \tag{4.75a}$$

$$\int_\Sigma j_0(\xi)\,d\pi \le J_0. \tag{4.75b}$$

If u satisfies $Lu + \mathscr{I}u = f$ in Ω, $u = \varphi$ on $\mathscr{P}\Omega$, show (4.63) holds with C depending also on c_3, J, and J_0.

4.9 If $\mathscr{I}$ is defined by (2.29) for $\sigma \in H_\alpha$, show that (4.63) holds with C depending also on $|j_1|_\alpha$.

4.10 Let $\mathscr{L} : H_\alpha \to H_\alpha$ (or, more generally, $\mathscr{L} : H_a \to H_\alpha$ with $0 < a < 2+\alpha$) be a bounded operator. Show that if $u \in H_{2+\alpha}$ is a solution of $Lu + \mathscr{L}u = f$ in Ω, $u = \varphi$ on $\mathscr{P}\Omega$, then (4.63) holds with C depending also on the norm of $\mathscr{L}$. As special cases, infer the results of Exercises 4.8 and 4.9.

4.11 State and prove suitable forms of Exercise 4.10 for divergence structure equations, oblique derivative boundary conditions, and intermediate spaces.

4.12 Show that Theorem 4.8 remains true if $u \in H^*_1$. (Hint: first show that $|u|^*_{1+\delta}$ is finite for $\delta = \alpha/2$, then for $\delta = \alpha[1-2^{-k}]$ for any positive integer k. Finally show that the estimate on $|u|^*_{1+\delta}$ is independent of δ.)

CHAPTER V

EXISTENCE, UNIQUENESS AND REGULARITY OF SOLUTIONS

Introduction

In this chapter, we show the unique solvability of various initial-boundary value problems, and we investigate the smoothness of their solutions. We only consider problems with nondivergence structure differential equations; the theory for divergence structure equations is the topic of Chapter VI. Because of the important connection between existence and uniqueness of solutions, we begin with the uniqueness theorems in Section V.1. Existence for the Cauchy-Dirichlet problem is covered by Sections V.2 (for equations with bounded coefficients) and V.3 (for unbounded coefficients). The oblique derivative problem is the topic for Section V.4, including bounded and unbounded coefficients.

1. Uniqueness of solutions

Our first results are for the Cauchy-Dirichlet problem, in which the solution is prescribed on all of $\mathscr{P}\Omega$, and the operator L is given by

$$Lu \equiv -u_t + a^{ij}D_{ij}u + b^i u + cu. \tag{5.1}$$

When the coefficient c is bounded from above, Corollary 2.5 implies that there is at most one solution of (5.1) with boundary data $u = \varphi$ on $\mathscr{P}\Omega$ for any continuous φ. When the coefficients are unbounded as in Theorem 4.29, we need to assume also that $\mathscr{P}\Omega \in H_\delta$ for some $\delta \in (1,2)$.

LEMMA 5.1. *Let $\mathscr{P}\Omega \in H_\delta$ for some $\delta \in (1,2)$ and suppose there are positive constants B, c_1, λ, and Λ such that*

$$a^{ij}\xi_i\xi_j \geq \lambda|\xi|^2,\ \mathscr{T} \leq \Lambda, \tag{5.2a}$$

$$\left|b^i\right| \leq Bd^{\delta-2},\ c \leq c_1 d^{\delta-2} \tag{5.2b}$$

in Ω, where d is given by $d(X) = \operatorname{dist}(X, \mathscr{P}\Omega)$. If $u \in C(\overline{\Omega}) \cap C^{2,1}(\Omega)$ is a solution of $Lu = f$ in Ω, $u = \varphi$ on $\mathscr{P}\Omega$, then there is a constant $C(B, c_1, \delta, \lambda, \Lambda, n, \Omega)$ such that

$$|u|_0 \leq C(|f|_0^{(2-\delta)} + |\varphi|_0). \tag{5.3}$$

PROOF. Let ρ be a regularized distance for Ω, let $R = \sup\rho$, and choose $\alpha \in (0, \delta - 1)$. Then $w_1 = R^{1+\alpha} - \rho^{1+\alpha}$ satisfies

$$\begin{aligned} Lw_1 &= (1+\alpha)\rho^\alpha\rho_t - (1+\alpha)\alpha\rho^{\alpha-1}a^{ij}D_i\rho D_j\rho \\ &\quad - (1+\alpha)[\rho^\alpha(a^{ij}D_{ij}\rho + b^iD_i\rho)] + c(R^{1+\alpha} - \rho^{1+\alpha}) \\ &\le -(1+\alpha)\alpha\lambda\rho^{\alpha-1}|D\rho|^2 + Cd^{\delta-2}. \end{aligned}$$

It follows from (4.51) that there is a positive constant ε_0 such that $|D\rho| > 1/2$ on the set $\{d < t/2, d < \varepsilon_0\}$, and therefore there are positive constants C_1 and η such that, for all sufficiently small ε, we have $Lw_1 \le -\eta d^{\alpha-1}$ wherever $d < \min\{t^{1/2}, \varepsilon\}$ and $Lw_1 \le C_1 d^{\delta-2}$ elsewhere. Similarly, for $w_2 = t^{(1+\alpha)/2}$ and ε sufficiently small, we have $Lw_2 \le -\eta d^{\alpha-1}$ wherever $d = t^{1/2} < \varepsilon$ and $Lw_2 \le C_1 d^{\delta-2}$ elsewhere. (There is no loss of generality in taking the same C_1 and η here.)

With k a positive constant to be determined, we set $w_3 = e^{kt}[w_1 + w_2]$. It follows that

$$Lw_3 \le e^{kt}[-\eta + C_1\varepsilon^{\delta-\alpha-1}]d^{\alpha-1}$$

wherever $d < \varepsilon$ and

$$Lw_3 \le e^{kt}[-k + C_1\varepsilon^{\delta-2}]$$

wherever $d \ge \varepsilon$. We now take ε sufficiently small so that $C_1\varepsilon^{\delta-\alpha-1} \le \eta/2$ and then $k = C_1\varepsilon^{\delta-2} + \eta\varepsilon^{\alpha-1}$ to infer that

$$Lw_3 \le -\frac{\eta}{2}d^{\alpha-1} \le -\eta_1 d^{\delta-2}$$

in Ω for some positive η_1, which we may take to be less than one.

Finally, set $F = |\varphi|_0 + |f|_0^{(2-\delta)}$. Then, for any $\theta > 0$, we see that $w = (\theta + 2F/\eta_1)w_3$ satisfies $L(\pm u) > Lw$ in Ω and $\pm u < w$ on $\mathscr{P}\Omega$. From Lemma 2.1, we infer that $\pm u < w$ in Ω. Sending $\theta \to 0$ in this inequality yields

$$|u|_0 \le 2\frac{F}{\eta_1}\sup w_3$$

for any $\eta > 0$, and hence (5.3) holds with $C = 2(F/\eta_1)\sup w_3$. □

COROLLARY 5.2. *Suppose that L and Ω are as in Lemma 5.1. If $Lu = Lv$ in Ω and $u = v$ on $\mathscr{P}\Omega$, then $u = v$ in Ω.*

For the oblique derivative problem, we use a similar argument to prove an L^∞ bound and uniqueness. Here the operator M is defined by $Mu = \beta \cdot Du + \beta^0 u$. We shall guarantee that β points into $\overline{\Omega}$ by assuming that $\mathscr{P}\Omega \in H_\delta$ for some $\delta \in (1,2)$ and that

$$\beta \cdot \gamma \ge \chi \tag{5.4}$$

for some positive constant χ, where $\gamma(X_0)$ is the unit inner normal to the boundary of $\Omega(t_0) = \{x \in \mathbb{R}^n : (x, t_0) \in \Omega\}$ at X_0.

LEMMA 5.3. *Let $\mathscr{P}\Omega \in H_\delta$ for some $\delta \in (1,2)$, and suppose that $u \in C(\overline{\Omega}) \cap C^{2,1}(\Omega)$ is a solution of $Lu = f$ in Ω, $Mu = \psi$ on $S\Omega$, and $u = \varphi$ on $B\Omega$. If conditions* (5.2) *and* (5.4) *hold and if $\beta^0 \leq c_2\chi$ on $S\Omega$, then*

$$|u|_0 \leq C(|\varphi|_0 + |\psi|_0/\chi + |f|_0^{(2-\delta)}) \tag{5.5}$$

with C depending on B, c_1, c_2, n, δ, λ Λ, and Ω.

PROOF. Suppose first that $\beta^0 \leq -\chi$ on $S\Omega$. With w_3 as in the proof of Lemma 5.1, we have $Mw_3 \leq -C_4\chi$ on $S\Omega$ for some positive constant C_4 because $D\rho = \gamma$ on $S\Omega$. Therefore $L(\pm u) > Lw$ in Ω, $M(\pm u) > Mw$ on $S\Omega$ and $\pm u \leq w$ on $B\Omega$ for

$$w = |\varphi|_0 + [c_1|\varphi|_0 + |f|_0^{(2-\delta)} + |\psi|_0/(C_4\chi)]w_3,$$

and hence (5.5) holds in this case.

For the general case of $\beta^0 \leq c_2\chi$, we define set $K = c_2 + 1$, $\Gamma = \beta^0 - K\beta\cdot\gamma$, and $v = e^{K\rho}u$. It follows that $\beta^i D_i v + \Gamma v = \psi$ on $S\Omega$ and that $\Gamma \leq -\chi$. In addition, v satisfies a differential equation with coefficients satisfying (5.2), so we can apply the previous case of this theorem to v and hence infer (5.5) again. □

COROLLARY 5.4. *Suppose L, M, and Ω are as in Lemma 5.3. If $Lu = Lv$ in Ω, $Mu = Mv$ on $S\Omega$, and $u = v$ on $B\Omega$, then $u = v$ in Ω.*

2. The Cauchy-Dirichlet problem with bounded coefficients

Now we prove that the Cauchy-Dirichlet problem (that is, the problem in which the solution is prescribed on $\mathscr{P}\Omega$) has a unique solution when the coefficients of the differential operator are in H_α by adapting the arguments from Chapter III. The key element is the solvability of the Dirichlet problem in a space-time ball. We shall prove this solvability for variable coefficients after proving the corresponding result for cylinders. We start with a basic existence result.

LEMMA 5.5. *Let R, T, α, and δ be positive constants with $\alpha, \delta < 1$. Let $X_0 \in \mathbb{R}^{n+1}$, and set $\Omega = \{|x - x_0| < R, t_0 - T < t < t_0\}$. Suppose that $a^{ij} = \delta^{ij}$, and b^i and c are zero. Then for any $f \in H_\alpha^{(2-\delta)}$, there is a unique continuous weak solution of*

$$Lu = f \text{ in } \Omega,\ u = 0 \text{ on } \mathscr{P}\Omega. \tag{5.6}$$

Moreover, $u \in H_{2+\alpha}^{(-\delta)}$ and there is a constant C determined only by n, R, T, α, and δ such that

$$|u|_{2+\alpha}^{(-\delta)} \leq C|f|_\alpha^{(2-\delta)}. \tag{5.7}$$

PROOF. Suppose first that $f \in C^1(\overline{\Omega})$. Then there is a unique solution u of (5.6) by Lemma 3.25. Using mollification and Theorem 4.9, we find that $u \in C^{2,1}(\Omega)$. Now we set $F = |f|_\alpha^{(2-\delta)}$, and define functions w_1 and w_2 by

$$w_1(X) = (R^2 - |x - x_0|^2)^\delta,\ w_2(X) = (t - t_0 + T)^{\delta/2}.$$

A simple calculation shows that there are positive constants c_1 and c_2 such that $w = c_1w_1 + c_2w_2$ satisfies $Lw \le -d^{\delta-2}$ in Ω and hence $|u| \le CFw$ in Ω. Next, we define

$$\Omega_1 = \{X \in \Omega : |x - x_0| < \frac{1}{2}R, (t - t_0 + T)^{1/2} > d\},$$
$$\Omega_2 = \{X \in \Omega : (t - t_0 + T)^{1/2} < R - |x - x_0|\},$$
$$\Omega_3 = \{X \in \Omega : (t - t_0 + T)^{1/2} > R - |x - x_0|, |x - x_0| > \frac{1}{2}R\}.$$

It follows that $|u| \le CFd^\delta$ in Ω_3 and on $\mathscr{P}\Omega_i$ for $i = 1, 2$. On Ω_1, we have $Rd \le w_1 \le 2Rd$, so applying the maximum principle to $\pm u + Cw_i$ gives $|u| \le CFd^\delta$ in Ω_i for $i = 1, 2$. Hence $|u| \le CFd^\delta$ in Ω and then Theorem 4.9 gives (5.7).

Next, we suppose only that $f \in H_\alpha$. A simple mollification argument shows that there is a sequence (f_m) from $C^1(\overline{\Omega})$ such that $f_m \to f$ and $|f_m|_\alpha^{(2-\delta)} \le C|f|_\alpha^{(2-\delta)}$. If u_m denotes the solution of (5.6) corresponding to f_m, we see that (5.7) holds for each u_m (with the norm of f on the right side). The maximum princple along with the uniform convergence of the sequence (f_m) shows that the sequence (u_m) converges uniformly in Ω to a limit function u which must be the unique solution of (5.6) and which satisfies (5.7).

For arbitrary $f \in H_\alpha^{(2-\delta)}$, we set $f_m = (\operatorname{sgn} f)\min\{|f|, m\}$. Since each f_m is in H_α with the $H_\alpha^{(2-\delta)}$ norms uniformly bounded, we see that each u_m satisfies (5.7). Hence, for any $\varepsilon > 0$, there is $d_0 > 0$ such that $|u_m| < \varepsilon$ on $\{d \le d_0\}$ for all m. Furthermore, there is a positive integer J such that $f_m = f$ on $\{d > d_0\}$ for all $m \ge J$. The maximum principle implies that $|u_m - u_k| \le \varepsilon$ on $\{d > d_0\}$ if $m, k \ge J$ and hence (u_m) converges uniformly to the solution u of (5.6) and that u satisfies (5.7). □

The method of continuity allows us to extend Lemma 5.5 to arbitrary operators with smooth coefficients in cylinders.

THEOREM 5.6. *Let R and T be positive numbers, let $X_0 \in \mathbb{R}^{n+1}$, and set $\Omega = \{|x - x_0| < R, t_0 - T < t < t_0\}$. Let (a^{ij}) be a matrix-valued function, let (b^i) be a vector-valued function and let c be a scalar-valued function on Ω and suppose that (a^{ij}), b^i and c are in H_α for some $\alpha \in (0,1)$ satisfying* (5.2a) *and*

$$|a^{ij}|_\alpha \le A, \ |b^i|_\alpha \le B, \ |c|_\alpha \le c_1, \tag{5.8}$$

for positive constants A, B, c_1, λ and Λ. If $\delta \in (0,1)$ and $f \in H_\alpha^{(2-\delta)}$, then there is a unique solution $u \in C^{2,1}(\Omega) \cap C^0(\overline{\Omega})$ of (5.6)*, where L is defined by* (5.1)*. Moreover, $u \in H_{2+\alpha}^{(-\delta)}$, and* (5.7) *holds with C determined only by A, B, c_1, n, R, T, λ, and Λ.*

PROOF. With w_1 and w_2 as before, we find that there are positive constants k_1, k_2, k_3, and R_0 such that

$$Lw_1 \leq \begin{cases} -k_1 d^{\delta-2} & \text{if } R_0 < r < R, d = R - r \\ k_2 d^{\delta-2} & \text{otherwise,} \end{cases}$$

and

$$Lw_2 \leq \begin{cases} -k_2 d^{\delta-2} & \text{if } r \leq R_0 \text{ or } d = (t - t_0 + T)^{1/2} \\ 0 & \text{otherwise.} \end{cases}$$

It follows that there are positive constants k_3 and k_4 such that $w = k_3 w_1 + k_4 w_2$ satisfies $Lw \leq -d^{\delta-2}$ in Ω. The proof of Lemma 5.5 now shows that (5.7) holds for any solution of (5.6), and then the method of continuity implies the unique solvability of (5.6) for all such f by virtue of (5.6) and Lemma 5.5. □

Next, we prove unique solvability for the Dirichlet problem in balls.

THEOREM 5.7. *Let $B = \{|x - x_0|^2 + |t - t_0|^2 < R^2\}$, and let L be defined by* (5.1). *Suppose also that R is so small that*

$$2R + 2\sup|b|R + R^2 \sup c^+ \leq \lambda. \tag{5.9}$$

Then for any $f \in H_\alpha(B)$, there is a unique solution of $Lu = f$ in B, $u = 0$ on ∂B.

PROOF. For $\varepsilon \in (0, 1)$, set

$$B_\varepsilon = B \cap \{|t - t_0| < (1 - \varepsilon)R\},$$
$$Q = \{|y| < 1, 0 < s < 2(1 - \varepsilon)R\},$$

and define a map from B_ε to Q by

$$y = (R^2 - |t - t_0|^2)^{1/2}(x - x_0), \; s = t - t_0 + (1 - \varepsilon)R.$$

Since this map is smooth, we can define an operator L_ε in Q by $L_\varepsilon v(Y) = Lu(X)$, where $v(Y) = u(X)$. Defining also $g_\varepsilon(Y) = f(X)$, we see that there is a unique solution v_ε of $L_\varepsilon v_\varepsilon = g_\varepsilon$ in Q, $v_\varepsilon = 0$ on $\mathscr{P}Q$. It follows that the corresponding u_ε (defined by $u_\varepsilon(X) = v_\varepsilon(Y)$) satisfies $Lu_\varepsilon = f$ in B_ε, $u_\varepsilon = 0$ on $\mathscr{P}B_\varepsilon$.

Now set $W = R^2 - |x - x_0|^2 - |t - t_0|^2$. Then $0 \leq W \leq R^2$ in B, so

$$\begin{aligned} LW &= 2(t - t_0) - 2\mathscr{T} - 2b^i[x^i - x_0^i] + cW \\ &\leq 2R - 2\lambda + 2\sup|b|R + R^2 \sup c^+ \leq -\lambda \end{aligned}$$

in B. From the maximum principle, we have $|u_\varepsilon| \leq (\sup|f|/\lambda)W$ in B_ε, so there is a uniform pointwise bound on u_ε. In addition, for any fixed $\eta \in (0, 1)$ and any $\varepsilon \in (0, \eta)$, we have $|u_\varepsilon| \leq 2\eta R^2(\sup|f|)/\lambda$ on $\{t - t_0 = -(1 - \eta)R\}$. Applying the maximum principle to $u_\varepsilon - u_\zeta$ on a fixed B_η with ε and ζ in $(0, \eta)$, we conclude that the family (u_ε) is uniformly Cauchy on B_η as $\varepsilon \to 0$. It follows that this family has a limit u which is continuous in $\overline{B}$. Moreover, according to Theorem 5.6, each v_ε is in $H_{2+\alpha}^{(-\delta)}$ for any $\delta \in (0, 1)$. Hence $u_\varepsilon \in H_{2+\alpha}^*(B_\varepsilon)$ and we can apply Theorem 4.9 to infer that the limit function is in this space and hence is the unique solution of $Lu = f$ in B, $u = 0$ on ∂B. □

To show the existence of solutions to $Lu = f$ with arbitrary continuous boundary values in a ball, we follow the proof of Theorem 3.19.

COROLLARY 5.8. *If L, B and f are as in Theorem 5.7, then for any continuous φ, there is a unique solution of $Lu = f$ in B, $u = \varphi$ on ∂B.*

PROOF. If φ is smooth, we set $g = f - L\varphi$ and write v for the solution of $Lv = g$ in B, $v = 0$ on ∂B, and then $u = v + \varphi$ is the desired solution. In general, we approximate φ by a sequence of smooth functions (φ_m) and use the maximum principle along with Theorem 4.9 to conclude that the corresponding sequence (u_m) converges uniformly to the solution. □

To study the Cauchy-Dirichlet problem in an arbitrary domain, we follow Section III.4. Let $f \in H_\alpha$, let $\varphi \in C^0(\overline{\Omega})$, and define the operator L by (5.1) with coefficients in H_α. Suppose that $c \leq 0$. We then say that $v \in C(\overline{\Omega})$ is a *subsolution* of

$$Lu = f \text{ in } \Omega,\ u = \varphi \text{ on } \mathscr{P}\Omega \tag{5.10}$$

if $v \leq \varphi$ on $\mathscr{P}\Omega$ and if, whenever R is so small that $2R(1+\sup|b|) \leq \lambda$ and $B \subset \Omega$, the solution $\overline{v}$ of $L\overline{v} = f$ in B, $\overline{v} = v$ on ∂B is greater than or equal to v in B. We then write S for the set of all subsolutions of (5.10) and define u by

$$u(X) = \sup_{v \in S} v(X). \tag{5.11}$$

The argument used to prove Theorem 3.23 then implies the following theorem provided we use the Schauder-type estimates of Theorem 4.9 in place of Theorem 3.19. (The hypothesis $c \leq 0$ is used to apply the strong maximum principle.)

THEOREM 5.9. *If u is given by* (5.11), *then $u \in H^*_{2+\alpha}$ and $Lu = f$ in $\overline{\Omega} \setminus \mathscr{P}\Omega$.*

The study of boundary behavior is analogous to that in Chapter III. For $X_1 \in \mathbb{R}^{n+1}$ and $R > 0$, we write $B(R)$ for the space-time ball $B(X_1, R)$ and $B^-(R)$ for the set of all $X \in B(R)$ with $t < t_1$, and we set $\Omega^-[X_1, R] = \Omega \cap B^-(R)$. We then say that w is a *local barrier at $X_0 \in \mathscr{P}\Omega$ (from earlier time)* if there is a positive constant r such that w is nonnegative and continuous in $\overline{\Omega^-[X_0, R]}$, w vanishes only at X_0, and, for all $B(R) \subset Omega^-[X_0, r]$ with $2R(1+\sup|b|) \leq \lambda$, the solution h of

$$Lh = -1 \text{ in } B^-(R), h = w \text{ on } \mathscr{P}B^-(R)$$

is greater than or equal to w in $B^-(R)$. As before, the existence of a local barrier implies that u is continuous at X_0.

THEOREM 5.10. *If u is defined by* (5.11), *if φ is continuous at X_0, and if there is a local barrier at X_0, then u is continuous at X_0 and $u(X_0) = \varphi(X_0)$.*

Note that if there is a parabolic frustum $PF = \{|x-y|^2 + (t_0 - t) < R^2, t < t_0\}$ such that $\overline{PF} \cap \overline{\Omega} = \{X_0\}$ (see the last page of Chapter III), then a barrier exists. In particular, if $\mathscr{P}\Omega \in H_2$, then the u given by (5.11) is a $C^{2,1}(\Omega) \cap C(\overline{\Omega})$ solution of (5.10).

It is also a simple matter to relax the condition $c \leq 0$ to $c \leq k$ for some positive constant k. If we define L_k, f_k, and φ_k by

$$L_k v = Lv - kv, f_k(X) = e^{-kt} f(X), \varphi_k(X) = e^{-kt} \varphi(X),$$

then there is a unique solution u_k to $L_k u_k = f_k$ in Ω and $u_k = \varphi_k$ on $\mathscr{P}\Omega$. A simple calculation shows that $u(X) = e^{kt} u_k(X)$ solves (5.10).

For $H_{2+\alpha}$ boundary data, smoothness up to the boundary is an easy consequence of regularity results for a simpler problem. Suppose that $\Omega = Q^+(X_0, R)$ for some X_0 with $x_0^n = 0$, suppose that $\varphi \in C(\mathscr{P}\Omega)$, and that $\varphi \in H_{2+\alpha}(Q^0)$. Our first step is to show that u is smooth up to Q^0 for the special operator L_0 defined by $L_0 v = -v_t + \delta^{ij} D_{ij} v$.

LEMMA 5.11. *Suppose $L_0 u = f$ in Q^+, $u = \varphi$ on $\mathscr{P}Q^+$ with $\varphi \in C(\overline{\Omega})$, $f \in H_\alpha(Q^+)$, and $\varphi = 0$ on Q^0. Then $u \in H^*_{2+\alpha}(Q^+ \cup Q^0)$.*

PROOF. Extend u, f and φ as odd functions of x^n to all of $Q(X_0, R)$, and let v be the solution of $L_0 v = f$ in Q, $v = \varphi$ on $\mathscr{P}Q$. The maximum principle applied to $w(X) = v(X) - v(x', -x^n, t)$ shows that v is odd, so $v = 0$ on Q^0. Hence u and v solve the same boundary value problem so $u = v$. Since $v \in H^*_{2+\alpha}(Q)$, we have the desired regularity for u. □

Next, we prove $H^{(-\delta)}_{2+\alpha}(Q^+ \cup Q^0)$ estimates with zero boundary data on $\sigma = \mathscr{P}Q^+ \setminus Q^0$.

LEMMA 5.12. *If $L_0 u = f$ in Q^+ and $u = \varphi$ on $\mathscr{P}Q^+$ with $\varphi \in H^{(-\delta)}_{2+\alpha}$, $f \in H_\alpha$, and $\varphi = 0$ on σ, then $u \in H^{(-\delta)}_{2+\alpha}$ and*

$$|u|^{(-\delta)}_{2+\alpha} \leq C(n, R, \alpha, \delta)(|f|_\alpha + |\varphi|^{(-\delta)}_{2+\alpha}). \tag{5.12}$$

PROOF. The barrier of Lemma 5.5 shows that $|u| \leq C(|f|_\alpha + |\varphi|^{(-\delta)}_{2+\alpha}) d^\delta$ in Q^+, and then the proof of Lemma 5.5 gives (5.12). □

By reasoning as in Theorem 5.6, we infer the existence of solutions to (5.10) in case $\Omega = Q^+$ which are smooth up to Q^0 if φ is smooth on Q^0. By the uniqueness theorem, these solutions are the same as the ones given by the Perron process, but the proof here does not assume the existence of a solution.

THEOREM 5.13. *Suppose $\varphi \in C(\overline{Q^+}) \cap H^*_{2+\alpha}(Q^0)$ and $f \in H_\alpha$. Then there is a unique $C(\overline{\Omega}) \cap C^{2,1}$ solution of (5.10) for $\Omega = Q^+$ and L given by (5.1). Moreover, $u \in H^*_{2+\alpha}$, and*

$$|u|^*_{2+\alpha} \leq C(A, B, c_1, n, R, \alpha, \delta, \lambda, \Lambda)(|f|_\alpha + |\varphi|^*_{2+\alpha}). \tag{5.13}$$

PROOF. Suppose first that $\varphi = 0$ and fix $\delta \in (0,1)$. Then we can proceed as before to infer that any solution $u \in H_{2+\alpha}^{(-\delta)}$ obeys the estimate (5.7). From this estimate and the existence of solutions for $L = L_0$, the method of continuity shows that there is a unique solution $u \in H_{2+\alpha}^{(-\delta)}$ with $u = 0$ on $\mathscr{P}Q^+$ for any $f \in H_\alpha^{(2-\delta)}$. Now suppose that $\varphi \in H_{2+\alpha}^{(-\delta)}$. If v is the solution of $Lv = f - L\varphi$ in Q^+, $v = 0$ on $\mathscr{P}Q^+$ given by the method of continuity, then $u = v + \varphi$ solves (5.10). For general φ, we proceed by approximation. □

An alternative approach is to show that there is a solution of $Lu = f$ in B^*, $u = \varphi$ on ∂B^* for $B^* = \{X \in B(R) : x^n > 0\}$ and then imitate the proof of Theorem 5.10 to show that there is a solution of (5.10) in $H_{2+\alpha}$. We shall use this approach in Section V.4 to study the oblique derivative problem.

We close this section by proving that the solution of (5.10) is globally smooth if the data are smooth and if the compatibility condition from Section IV.7 is imposed.

THEOREM 5.14. *Suppose $\mathscr{P}\Omega \in H_{2+\alpha}$, $\varphi \in H_{2+\alpha}$, and $f \in H_\alpha$ for some $\alpha \in (0,1)$. If the coefficients of L satisfy* (5.2a) *and* (5.8), *then there is a unique solution $u \in C(\overline{\Omega}) \cap C^{2,1}(\Omega)$ of* (5.10). *If also $L\varphi = f$ on $C\Omega$, then $u \in H_{2+\alpha}$ and*

$$|u|_{2+\alpha} \le C(A,B,c_1,n,\alpha,\Omega)(|f|_\alpha + |\varphi|_{2+\alpha}). \tag{5.14}$$

PROOF. The existence and uniqueness of the solution follow from the remarks after Theorem 5.10. To prove (5.14), we use Theorems 4.28 and 2.10 to infer that it suffices to show that $u \in H_{2+\alpha}$. We prove this regularity locally in a neighborhood of an arbitrary point X_0 of $\overline{\Omega}$. For $X_0 \in \overline{\Omega} \setminus \mathscr{P}\Omega$, the smoothness follows from Theorem 5.9, and for $X_0 \in S\Omega$, it follows from Theorem 5.13 and a simple change of variables.

To study the cases $X_0 \in B\Omega$ and $X_0 \in C\Omega$, we first use Lemma 4.24 along with a partition of unity and a change of variables to see that φ can be taken so that $L\varphi = f$ on $B\Omega \cup C\Omega$. Hence, after subtracting φ from u, we may assume that $\varphi \equiv 0$ and that $f = 0$ on $B\Omega$. With this assumption, we first suppose that $X_0 \in B\Omega$ and take r so small that $Q(2r,(x_0,t_0+r^2)) \cap \{t > t_0\} \subset \Omega$. We can then extend u, f, and the coefficients of L to all of $Q(2r) = Q((x_0,t_0+r^2),2r)$ to be independent of t for $t \le t_0$; in particular $u(X) = f(X) = 0$ for $t \le t_0$. Theorem 5.10 gives a unique solution v of $Lv = f$ in $Q(2r)$, $v = u$ on $\mathscr{P}Q(2r)$ and uniqueness implies that $u = v$. Theorem 5.10 then implies that $u \in H_{2+\alpha}(Q(r))$, which is the desired regularity. For $X_0 \in C\Omega$, we imitate the above argument with Theorem 5.13 in place of Theorem 5.10. □

3. The Cauchy-Dirichlet problem with unbounded coefficients

If the coefficients of the equation are unbounded as in Section IV.6, we prove existence of suitably smooth solutions via an approximation argument. Suppose that the coefficients of L satisfy the conditions (5.2a) and

$$|a^{ij}|_{\alpha}^{(0)} \le A,\ |b^i|_{\alpha}^{(2-\delta)} \le B,\ |c|_{\alpha}^{(2-\delta)} \le c_1 \tag{5.15}$$

for some constants A, B, c_1, α,and δ with $0 < \alpha < 1 < \delta < 2$, and suppose that there is a continuous increasing function ζ with $\zeta(0) = 0$ such that

$$|a^{ij}(X) - a^{ij}(Y)| \le \zeta(|X-Y|) \tag{5.16}$$

for all X and Y in Ω. Suppose also that $\mathscr{P}\Omega \in H_\delta$ and $\varphi \in H_\delta(\mathscr{P}\Omega)$, and write ρ for a regularized distance in Ω. For $\varepsilon > 0$, we define $\Omega(\varepsilon) = \{X \in \Omega : \rho > \varepsilon, t > \varepsilon^2\}$. Extend φ to an $H_{2+\alpha}^{(-\delta)}$ function and write u_ε for the solution of $Lu_\varepsilon = f$ in $\Omega(\varepsilon)$, $u_\varepsilon = \varphi$ on $\mathscr{P}\Omega(\varepsilon)$. From the results of the previous section, $u_\varepsilon \in H_{2+\alpha}(K)$ for any compact subset K of $\overline{\Omega}(\varepsilon) \setminus C\Omega(\varepsilon)$. Since $v_\varepsilon = u_\varepsilon - \varphi$ solves $Lv_\varepsilon = g$ in $\Omega(\varepsilon)$, $v_\varepsilon = 0$ on $\mathscr{P}\Omega(\varepsilon)$ for the bounded function $g = f - L\varphi$, it follows that $|v_\varepsilon| \le C(\varepsilon)t$ in $\Omega(\varepsilon)$ and hence $v_\varepsilon \in H_{2+\alpha}^{(-\delta)}(\Omega(\varepsilon))$. From Theorem 4.29, we obtain a uniform estimate on $|u_\varepsilon|_{2+\alpha}^{(-\delta)}$, which depends on $|u_\varepsilon|_0$, but this norm is estimated via Lemma 5.1. An application of the maximum principle shows that $u_\varepsilon \to u$ uniformly as $\varepsilon \to 0$, and hence we have proved the following theorem.

THEOREM 5.15. *Suppose conditions* (5.2a), (5.15), *and* (5.16) *are satisfied for some $\alpha \in (0,1)$ and $\delta \in (1,2)$. If $\mathscr{P}\Omega \in H_\delta$, $f \in H_\alpha^{(2-\delta)}$, and $\varphi \in H_\delta$, then there is a unique solution of* (5.10), *which is in $H_{2+\alpha}^{(-\delta)}$. In addition, there is a constant $C(A, B, c_1, n, \lambda, \Lambda, \zeta)$ such that*

$$|u|_{2+\alpha}^{(-\delta)} \le C(|f|_{\alpha}^{(2-\delta)} + |\varphi|_\delta). \tag{5.17}$$

4. The oblique derivative problem

Existence of solutions to the oblique derivative problem is proved by appropriate modification of the Perron arguments, especially the proof of Theorem 3.19. The major difference is in the proof of local existence. For $R > 0$ and $\tau \in (0,1)$ to be chosen, we set $x_0 = (0, \dots, 0, -\tau R)$ and

$$E^+ = \{X \in \mathbb{R}^{n+1} : |x - x_0|^2 + t^2 < R^2, x^n > 0, t > 0\}.$$

We then write Σ, σ, and E^0 for the subsets of ∂E^+ on which $t = 0$, $|x - x_0|^2 + t^2 = R^2\}$, and $x^n = 0$, respectively, and we consider the boundary value problem

$$Lu = f \text{ in } E^+,\ Mu = \psi \text{ on } E^0,\ u = \varphi \text{ on } \sigma \cup \Sigma, \tag{5.18}$$

where M is defined by (2.3) for some β which points into $\overline{\Omega}$. We suppose that $\beta^0 \le 0$ and that there is a nonnegative constant μ such that $|\beta'| \le \mu\beta^n$. Noting that $d(X,\sigma) = (R^2 - t^2)^{1/2} - |x - x_0|$, we see that $W = R^2 - |x - x_0|^2 - t^2$ satisfies the inequalities $\bar{\mu}Rd \le W \le 2Rd$ and

$$MW \le 2\beta^n R(-\bar{\mu} + \mu(1 - \bar{\mu}^2)^{1/2}) \le -C(\mu)\beta^n R$$

provided $\bar{\mu} > \mu/(1+\mu^2)^{1/2}$. Hence, if we take $w_1 = W^\delta$ and $w_2 = 0$, we can imitate the proof of Lemma 5.5 to obtain the following result.

LEMMA 5.16. *Let α and δ be constants in* $(0,1)$. *If $f \in H_\alpha^{(2-\delta)}(E^+ \cup E^0)$ and $\psi \in H_{1+\alpha}^{(1-\delta)}(E^0)$, then there is a unique solution u of* (5.18) *with $L = L_0$ and $\varphi \equiv 0$. Moreover, $u \in H_{2+\alpha}^{(-\delta)}$, and*

$$|u|_{2+\alpha}^{(-\delta)} \le C(R,\alpha,\delta,n,\mu)(|f|_\alpha^{(2-\delta)} + |\psi|_{1+\alpha}^{(1-\delta)}). \tag{5.19}$$

The reasoning which leads to Corollary 5.8 then gives the following existence theorem.

COROLLARY 5.17. *Suppose L and R satisfy the hypotheses of Corollary 5.8 and M is defined by* (2.3) *with $\beta^0 \le 0$ and $|\beta'| \le \mu\beta^n$. Suppose also that*

$$|\beta^j|_{1+\alpha} \le B_1\chi \tag{5.20}$$

for $j = 0,\dots,n$. Then for all $f \in H_\alpha$, $\psi \in H_{1+\alpha}$, and $\varphi \in C^0(\overline{E^+})$, there is a unique solution u of (5.18). *Moreover, $u \in H_{2+\alpha}^*$ and there is a constant C determined only by A, B, B_1, c_1, n, R, and α such that*

$$|u|_{2+\alpha}^* \le C(|f|_\alpha + |\psi|_{1+\alpha}/\chi + |\varphi|_0). \tag{5.21}$$

To remove the restriction that c and β^0 be nonpositive, we proceed as in the remarks after Theorem 5.10 except that now we define

$$L_{m,k}v = Lv - m(a^{in} + a^{ni})D_iv + (2m^2a^{nn}mb^n - k)v,$$
$$M_{m,k}v = Mv - m\beta^n v,$$
$$f_{m,k} = \exp(-kt - mx^n)f,\ \psi_{m,k} = \exp(-kt - mx^n)\psi,$$
$$\varphi_{m,k} = \exp(-kt - mx^n)\varphi.$$

By first choosing $m \ge \sup(\beta^0/\beta^n)^+$ and then k sufficiently large, and finally R small enough, we see that (5.18) is solvable in this case as well. A suitable $H_{2+\alpha}$ change of variables then gives local solvability in some neighborhood of any point on $S\Omega \cup C\Omega$. Now the proof of Theorem 5.14 establishes the following existence result for the oblique derivative problem.

THEOREM 5.18. *Let Ω be a domain with $\mathscr{P}\Omega \in H_{2+\alpha}$ for some $\alpha \in (0,1)$, let the coefficients of L be in H_α and let the coefficients of M be in $H_{1+\alpha}$ and suppose conditions* (5.2a), (5.4), (5.8), (5.20) *and*

$$|\beta| \le \mu\beta\cdot\gamma, \tag{5.22}$$

are satisfied. Then for any $f \in H_\alpha$, $\psi \in H_{1+\alpha}$ and $\varphi \in C(B\Omega \cup C\Omega)$, there is a unique solution of $Lu = f$ in Ω, $Mu = \psi$ on $S\Omega$, $u = \varphi$ on $B\Omega$. If also $\varphi \in H_{2+\alpha}(B\Omega)$ and $M\varphi = \psi$ on $C\Omega$, then $u \in H_{2+\alpha}$ and there is a constant C determined only by A, B, B_1, c_1, n, R, α, and Ω such that

$$|u|_{2+\alpha} \le C(|f|_\alpha + |\psi|_{1+\alpha}/\chi + |\varphi|_{2+\alpha}). \tag{5.23}$$

Finally, a straightforward approximation argument gives the corresponding result for the oblique derivative problem with intermediate regularity assumptions.

THEOREM 5.19. *Let* $\mathscr{P}\Omega \in H_\delta$ *for some* $\delta \in (1,2)$, *suppose that the coefficients of* L *satisfy the hypotheses of Theorem 5.14, and that the coefficients of* M *satisfy* (5.4), (5.22) *and* $|\beta^j|_{\delta-1} \le B_1\chi$ *for* $j = 0,\dots,n$. *If* $f \in H_\alpha^{(2-\delta)}$, $\psi \in H_{\delta-1}$, *and* $\varphi \in C(B\Omega \cup C\Omega)$, *then there is a unique solution of* $Lu = f$ *in* Ω, $Mu = \psi$ *on* $S\Omega$, $u = \varphi$ *on* $B\Omega$. *If also* $\varphi \in H_\delta(B\Omega)$ *and* $M\varphi = \psi$ *on* $C\Omega$, *then* $u \in H_{2+\alpha}^{(-\delta)}$ *and there is a constant* C *determined only by* A, B, B_1, c_1, n, R, α, δ, *and* Ω *such that*

$$|u|_{2+\alpha}^{(-\delta)} \le C(|f|_{2+\alpha}^{(2-\delta)} + |\psi|_{\delta-1}/\chi + |\varphi|_\delta). \tag{5.24}$$

Notes

The uniqueness theorems in Section V.1 are taken from [212] although the proofs are not. Existence, uniqueness, and regularity for the Cauchy-Dirichlet problem with bounded coefficients are standard (see [89, Theorem 3.7; 183, Theorem IV.5.2]), especially Theorems 5.10 and 5.14. Our proof is based on ideas of Michael [255] and van der Hoek [112]. The proof of regularity near $B\Omega$ and $C\Omega$ in Theorem 5.14 are taken from [230]. Theorem 5.15, in a slightly different form, first appears in [212, Theorem 11.3]; see also [162, Satz V.4].

Theorem 5.18 was first proved by Kamynin and Maslennikova [155] as an offhand remark (see note 2 of that work). The discussion of this result in [89, Chapter 5] is rather cursory; in particular, that work does not give an existence result for $H_{2+\alpha}$ solutions. The corresponding result in [183, Theorem 5.2] is more detailed. The intermediate Schauder theory of Theorem 5.19 comes from [212]. In fact, the hypotheses on the coefficients can be further relaxed (see [89, Theorem 5.2]; the elliptic analogs of these results are in [236, Section 6] and they are easily extended to the parabolic setting) yielding a continuous solution of the initial-boundary value problem which satisfies the boundary condition only in the sense defined prior to Lemma 2.2. In fact, if $\mathscr{P}\Omega \in H_1$, then a complete theory can still be developed which yields $H_{1+\theta}$ solutions for some $\theta > 0$ (see [221, 236]).

Construction of suitable barriers [11, 222] gives corresponding results in suitable weighted spaces for the mixed boundary problem, in which oblique derivative data are prescribed on part of $S\Omega$ and Dirichlet data are prescribed elsewhere. In particular, Lemma 5.16 and Corollary 5.17 are valid for $\bar{\mu} = 0$ provided δ is taken sufficiently small depending on μ. Some of these results will be important in our study of nonlinear problems, but we shall prove them only in Σ^+ as in Lemma 4.16.

There are a large number of alternative proofs of the existence results in this chapter. Their applicability is limited only by the specific nature of the solution sought. For example, the original method of Fourier (separation of variables) can be used if the coefficients are time-independent, if the domain is cylindrical, and if one is familiar with the theory of eigenvalue problems for the associated elliptic operator. Variations on Galerkin's method are also applicable under suitable hypotheses, but the coefficients may depend on time and the domain need not be cylindrical. As a special case of Galerkin's method, we mention the finite element method, which consists of subdividing Ω into small subdomains and approximating the problem by a finite dimensional one. Details of these alternative methods can be found in Sections III.14, III.15, III.16, III.17, III.18, and IV.16 of [183] or in Sections 3, 4, 6, 7, 8, and 9 of [115].

Exercises

5.1 Suppose that $(\Sigma, \mathscr{M}, \pi)$, j, j_1 and $\mathscr{I}$ are as in Exercise 4.6. Suppose also that $\mathscr{P}\Omega$, φ, L and f satisfy the hypotheses of Theorem 5.14. Show that the problem $Lu + \mathscr{I}u = f$ in Ω, $u = \varphi$ on $\mathscr{P}\Omega$ has a unique solution $u \in H_{2+\alpha}$.

5.2 If $\mathscr{I}$ is replaced by $\mathscr{L}$ as in Exercise 4.7, show that the result of Exercise 5.1 is still true.

5.3 Extend Exercise 5.2 to problems with oblique derivative boundary conditions and also to equations as in Theorem 5.15.

5.4 (See [160]) Let P be the operator defined on $C^{2,1}$ by

$$Pu = -u_t + a^{ij}(X,u,Du)D_{ij}u + a(X,u,Du)$$

and suppose the matrix (a^{ij}) is positive definite on $\Omega \times \mathbb{R} \times \mathbb{R}^n$. Suppose also that there is a constant $\alpha \in (0,1)$ such that $\mathscr{P}\Omega \in H_{1+\alpha}$ and a^{ij} and a are in $H_{1+\alpha}(K)$ for any bounded subset K of $\Omega \times \mathbb{R} \times \mathbb{R}^n$, and define the operator J which maps $H_{2+\alpha}^{(-1-\alpha)}$ to itself by $u = Jv$ if u solves the linear initial-boundary value problem

$$-u_t + a^{ij}(X,0,0)D_{ij}u = f(X) \text{ in } \Omega, \; u = 0 \text{ on } \mathscr{P}\Omega,$$

where

$$f(X) = [a^{ij}(X,0,0) - a^{ij}(X,v,Dv)]D_{ij}v - a(X,v,Dv).$$

(a) Show that there are constant C_1 and C_2 determined only by Ω, a^{ij}, and a such that $|v|_{2+\alpha}^{(-1-\alpha)} \le C_2$ implies

$$|u|_{2+\alpha}^{(-1-\alpha)} \le C_1 \, |v|_2^{(-1-\alpha)} + \frac{1}{2}C_2.$$

(b) Infer that there is a positive constant ε_0 (determined by the same quantities as were C_1 and C_2) such that J maps $\mathscr{M}(\varepsilon)$ into itself for $\varepsilon < \varepsilon_0$, where $\mathscr{M}(\varepsilon)$ is the set of all $v \in H_{2+\alpha}^{(-1-\alpha)}$ with $|v|_{2+\alpha}^{(-1-\alpha)} \le C_2$.
(c) Show also that there are positive constants ε_1 and θ with $\theta < 1$ such that $|Jv_1 - Jv_2| \le \theta \, |v_1 - v_2|$ for v_1 and v_2 in $\mathscr{M}(\varepsilon)$ if $\varepsilon \le \varepsilon_1$.

(d) Use the contraction mapping theorem to conclude that the problem $Pu = 0$ in $\Omega(\varepsilon)$, $u = 0$ on $\mathscr{P}\Omega(\varepsilon)$ is solvable for ε sufficiently small.

(e) Generalize this result to nonzero initial-boundary conditions.

5.5 Repeat Exercise 5.4 with the zero boundary condition replaced by the nonlinear boundary condition $G(X,u,Du) = 0$ on $S\Omega$ subject to the conditions $G_p \cdot \gamma > 0$ on $S\Omega \times \mathbb{R} \times \mathbb{R}^n$, $G \in H_{2+\alpha}(K)$ for any bounded subset K of $S\Omega \times \mathbb{R} \times \mathbb{R}^n$, and $G(X,0,0) = 0$ on $C\Omega$. How should this last condition be modified for nonzero initial data?

CHAPTER VI

FURTHER THEORY OF WEAK SOLUTIONS

Introduction

In Chapter III, we introduced the notion of weak solutions by multiplying the differential equation by a test function, integrating the resultant equation and then integrating by parts. For the purposes of that chapter, it suffices to consider weak solutions which have weak derivatives with respect to t. In fact, the solutions of the equations appearing in that chapter are classical by virtue of the results in Chapters IV and V. In this chapter, we study equations in divergence form which have only bounded measurable coefficients. There is no reason to expect such equations to have solutions with bounded derivatives with respect to space or time, and therefore we need a notion of solution which is somewhat weaker than those previously given.

We begin in Section VI.1 by defining weak solutions and showing that the Cauchy-Dirichlet problem always has a unique weak solution. Section VI.2 studies the higher differentiability of weak solutions, in particular the existence of second spatial derivatives for weak solution of equations with suitable smooth coefficients. We then prove some integral inequalities, valid for arbitrary functions, in Sections VI.3 and VI.4. These inequalities are used in Sections VI.5–VI.8 to study some important qualitative properties of weak solutions which are crucial to our study of nonlinear parabolic equations. The oblique derivative problem is the subject of Sections VI.9 and VI.10, and Section VI.11 discusses a mixed boundary value problem which is needed in Chapter VII. Existence of weak solutions to various initial-boundary value problems is described in Section VI.12 and an alternative approach to the qualitative properties, based on DeGiorgi's method, is given in Section VI.13.

1. Notation and basic results

The basic space in which we seek solutions is V, the set of all $u \in L^2(\Omega)$ such that $Du \in L^2(\Omega)$, $u(\cdot,t) \in L^2(\omega(t))$ for all $t \in I(\Omega)$, and the norm on V defined by

$$\|u\|_V^2 = \int_\Omega |Du|^2\, dX + \sup_{t\in I(\Omega)} \int_{\omega(t)} u^2\, dx$$

is finite. We write $C^1_{\mathscr{P}}$ for the set of all functions in $C^1(\overline{\Omega})$ which vanish on $S\Omega$ and V_0 for the closure of $C^1_{\mathscr{P}}$ in the norm of V. For f^i and g in $L^2(\Omega)$ and $\varphi \in L^2(B\Omega)$, we then say that u is a weak solution of

$$Lu = D_i f^i + g \text{ in } \Omega,\ u = 0 \text{ on } S\Omega,\ u = \varphi \text{ on } B\Omega \tag{6.1}$$

if $u \in V_0$ and

$$\begin{aligned}\int_{\omega(\tau)} uv\,dx - \int_{\Omega(\tau)} uv_t\,dX \\ + \int_{\Omega(\tau)} (a^{ij}D_ju + b^iu)D_iv - (c^iD_iu + c^0u)v\,dX \\ = \int_{\Omega(\tau)} -vg + f^iD_iv\,dX + \int_{B\Omega} \varphi v\,dx\end{aligned}$$

for all $v \in C^1_{\mathscr{P}}$ and almost all $\tau \in I(\Omega)$. An easy approximation argument shows that we may take $v \in W^{1,2}_{\mathscr{P}}(\Omega)$, that is, v is the limit in $W^{1,2}$ of $C^1_{\mathscr{P}}$ functions.

The key to studying weak solutions in a cylinder is the *Steklov average* v_h of a function v for a nonzero constant h, defined by

$$v_h(X) = \frac{1}{h}\int_t^{t+h} v(x,s)\,ds.$$

For $h > 0$ this expression gives an average of v over later times, and for $h < 0$ it gives an average over earlier times. In a cylinder $\Omega = \omega \times (0,T)$, the proofs of Lemmata 3.2 and 3.3 show that $v_h \to v$ in $L^p(\omega \times (0, T - \delta))$ if $v \in L^p$ and that $Dv_h \to Dv$ in $L^p(\omega \times (0, T - \delta))$ if $Du \in L^p$ provided $0 < \delta < T$ and $h \to 0^+$. In addition, if $v \in V_0$, then $v_h \to v$ in $V_0(\omega \times (0, T - \delta))$ with the same restrictions on δ and h. (If $h \to 0^-$, we replace $\omega \times (0, T - \delta)$ by $\omega \times (\delta, T)$.)

We are now ready to prove our basic estimate in cylinders.

THEOREM 6.1. *Let $\Omega = \omega \times (0,T)$ for some $\omega \subset \mathbb{R}^n$ and some $T > 0$ and let u be a solution of* (6.1) *with f^i and $g \in L^2(\Omega)$ and $\varphi \in L^2(B\Omega)$. If a^{ij}, b^i, c^i and c^0 are in L^∞ and if there are positive constants λ, Λ, and Λ_1 such that*

$$a^{ij}\xi_i\xi_j \geq \lambda|\xi|^2 \tag{6.2a}$$

for all $\xi \in \mathbb{R}^n$,

$$|a^{ij}| \leq \Lambda\lambda, \tag{6.2b}$$

$$\frac{1}{\lambda^2}\left(\sum |b^i - c^i|^2\right) + \frac{|c^0|}{\lambda} \leq \Lambda_1^2, \tag{6.2c}$$

then there is a constant C determined only by λ, Λ_1, and n such that

$$\|u\|_V \leq Ce^{CT}(\|f^i\|_2 + \|g\|_2 + \|\varphi\|_2). \tag{6.3}$$

PROOF. Let $\tau \in (0,T)$ and $h \in (0,T-\tau)$. Suppose that $\eta \in W^{1,2}_{\mathscr{P}}(\omega \times (0,T-h))$ with $\eta(X) = 0$ if $t = 0$ or $t = T-h$, and set $v = \eta_{-h}$, $w = \eta_t$. A simple calculation shows that $v_t = w_{-h}$ and hence

$$\begin{aligned}\int_\Omega uv_t\,dX &= \frac{1}{h}\int_\omega\int_0^T u(X)\int_{t-h}^t w(x,s)\,ds\,dt\,dx\\ &= \frac{1}{h}\int_\omega\int_{-h}^{T-h}\int_s^{s+h} u(X)w(x,s)\,dt\,ds\,dx\\ &= \int_\Omega u_h(X)w(X)\,dX\end{aligned}$$

because $w(x,s) = 0$ for $s < 0$ and $s > T-h$. Therefore integration by parts yields

$$\int_\Omega uv_t\,dX = -\int_\Omega u_{ht}\eta\,dX.$$

(Here u_{ht} means the derivative with respect to t of u_h.) It follows that

$$\begin{aligned}\int_\Omega (a^{ij}D_ju + b^iu - f^i)_h D_i\eta\,dX = &\int_\Omega (c^iD_iu + c^0u - g)_h\eta\,dX\\ &-\int_\Omega u_{ht}\eta\,dX\end{aligned}$$

for any $\eta \in W^{1,2}_{\mathscr{P}}$ which vanishes for $t > T-h$.

Our next step is to show that this identity remains true if η is replaced by $u_h\chi(t)$, where $\chi(t) = 0$ if $t > \tau$ and $\chi(t) = 1$ if $t < \tau$. To this end, we fix k a sufficiently large integer and write z_k for the continuous function which is linear on the intervals $(0,1/k)$ and $(\tau - 1/k, \tau)$, is zero on the intervals $(-\infty,0)$ and (τ,∞) and is one on $(1/k, \tau - 1/k)$. Since $\eta_k = u_hz_k$ is an admissible test function, we can take the limit as $k \to \infty$ to infer that

$$\begin{aligned}\int_{\Omega(\tau)} (a^{ij}D_ju + b^iu + f^i)_h D_iu_h\,dX = &\int_{\Omega(\tau)} (c^iD_iu + c^0u - g)_h u_h\,dX\\ &-\int_{\Omega(\tau)} u_{ht}u_h\,dX,\end{aligned}$$

where $\Omega(\tau) = \omega \times (0,T)$. Now we can integrate with respect to t and infer that

$$\int_{\Omega(\tau)} u_{ht}u_h\,dX = \frac{1}{2}\int_{\omega(\tau)} u_h(X)^2\,dx - \frac{1}{2}\int_{\omega(0)} u_h(X)^2\,dx.$$

With this evaluation, we can send $h \to 0$ to conclude that

$$\begin{aligned}\frac{1}{2}\int_{\omega(\tau)} u^2\,dx + &\int_{\Omega(\tau)} (a^{ij}D_ju + b^iu + f^i)D_iu\,dX\\ &= \int_{\Omega(\tau)} (c^iD_iu + c^0u - g)u\,dX + \frac{1}{2}\int_{B\Omega}\varphi^2\,dx.\end{aligned} \tag{6.4}$$

Now (6.2) implies that

$$\int_{\omega(\tau)} u^2\,dx + \int_{\Omega(\tau)} |Du|^2\,dX \le C\Big(\int_\Omega [|f|^2 + g^2]\,dX + \int_{\Omega(\tau)} u^2\,dx + \int_{B\Omega} \varphi^2\,dx\Big).$$

If we set $H(\tau) = \int_{\omega(\tau)} u^2\,dx$ and $K = C[\|f\|_2^2 + \|g\|_2^2 + \|\varphi\|_2^2]$, we can rewrite this inequality as

$$H(\tau) \le C\int_0^\tau H(t)\,dt + K,$$

so Gronwall's inequality implies that

$$\int_0^\tau H(t)\,dt \le \frac{K}{C}(e^{C\tau} - 1),$$

and hence $(\|u\|_V)^2 \le (C + e^{CT})K$, which implies (6.3). □

Note that Theorem 6.1 remains valid if we replace $|c^0|$ by $(c^0)^+$ in (6.2c). Moreover, the dependence of the estimate in (6.3) on T can be eliminated if an additional condition is assumed, namely,

$$\int_\Omega (b^i - c^i)D_i v + 2c^0 v\,dX \le 0$$

for almost all $\tau \in (0,T)$ and all nonnegative $v \in C^1_{\mathscr{P}}$ provided we replace $\|g\|_2$ by the norm

$$\|g\|_{2,1} = \int_0^T \Big(\int_\omega g^2\,dx\Big)^{1/2} dt.$$

It's now simple to obtain an existence and uniqueness result for cylindrical domains.

THEOREM 6.2. *If the coefficients of L, f^i, g, and φ satisfy the hypotheses of Theorem 6.1, then there is a unique solution u of* (6.1) *for any domain $\omega \subset \mathbb{R}^n$.*

PROOF. Let (f^i_m) and (g_m) be sequences of C^2 functions which converge to f^i and g, respectively, in $L^2(\Omega)$ and let (a^{ij}_m), (b^i_m), (c^i_m), and (c^0_m) be sequences of C^2 functions which are uniformly bounded and which converge almost everywhere to a^{ij}, b^i, c^i, and c^0, respectively. By using a mollification, we see that the approximations can be chosen to satisfiy the hypotheses of Theorem 6.1 (including (6.2a)) for all m. Now we define the operator L_m on $C^{2,1}$ by

$$L_m w = -w_t + a^{ij}_m D_{ij} w + (D_j(a^{ji}_m) + b^i_m + c^i_m)D_i w + (D_i b^i_m + c^0_m)w,$$

and use ρ to denote a regularized distance for ω. By Sard's theorem, there is a sequence (ε_m) tending to zero such that $D\rho$ never vanishes on $\{\rho = \varepsilon_m\}$. Hence if we define $\omega_m = \{\rho > \varepsilon_m\}$, then $\partial\omega_m \in H_2$ and there is a unique classical solution u_m of

$$L_m u_m = D_i f^i_m + g_m \text{ in } \Omega_m, u_m = 0 \text{ on } \mathscr{P}\Omega_m, u = \varphi_m \text{ on } B\Omega_m, \tag{6.5}$$

where $\Omega_m = \omega_m \times (0,T)$ by virtue of Theorem 5.6 and (φ_m) is a sequence of $H_2(\omega)$ functions which vanish for $\rho < 2\varepsilon_m$. Since $u_m \in H_{1+\alpha}$ for any $\alpha \in (0,1)$, it follows that u_m is a weak solution of (6.5), that is,

$$-u_{m,t} + D_i(a^{ij}_m D_j u_m + b^i_m u) + c^i_m D_i u + c^0 u = D_i f^i_m + g_m,$$

$u_m = \varphi_m$ on $B\Omega_m$, and $u_m \in V_0(\Omega_m)$. Inequality (6.3) implies that $\|u_m\|_{V(\Omega_m)}$ is uniformly bounded and hence there is a subsequence $(u_{m(k)})$ which converges weakly in V to a limit function u. It's easy to check that u is a weak solution of (6.1). Theorem 6.1 implies that u satisfies (6.3) and that u is the unique solution. □

For noncylindrical domains, we use a somewhat different approach. In such domains, u_h is not an appropriate test function because it may not vanish on $S\Omega$. Several techniques have been developed to study weak solutions in noncylindrical domains (see the Notes to this chapter) but only one gives the results of Theorems 6.1 and 6.2 (with the same regularity assumptions on the coefficients of the problem) under a weak regularity hypotheses on $\mathscr{P}\Omega$. The basis of the proof is the following version of Hardy's inequality.

LEMMA 6.3. *Suppose $\mathscr{P}\Omega \in H_1$ and $u \in V_0$. Then there is a constant C determined only by Ω such that*

$$\int_\Omega \left(\frac{u}{d}\right)^2 dX \le C \int_\Omega |Du|^2\, dX. \tag{6.6}$$

PROOF. First suppose that $u \in C^1_{\mathscr{P}}$. Let $X_0 \in S\Omega$ and choose a parabolic neighborhood N of X_0 and a coordinate system centered at X_0 such that $S\Omega \cap N$ is the graph of a function $f \in H_1(Q'(\delta))$ for some positive δ. We then introduce coordinates Y by the equations

$$Y' = X', y^n = \rho(X),$$

where ρ is the regularized distance on Ω. By decreasing δ or N as necessary, we may assume that $\Omega \cap N$ is the same as $Q^+ = Q^+(\delta)$ in Y-coordinates. It follows that

$$\int_{\Omega\cap N} \left(\frac{u}{d}\right)^2 dX \le C \int_{Q^+} \left(\frac{u}{y^n}\right)^2 dY.$$

Now the one-dimensional Hardy's inequality (Exercise 6.1) shows that

$$\int_0^R \left(\frac{u(Y',y^n)}{y^n}\right)^2 dy^n \le 2 \int_0^R \left(\frac{\partial u(Y',y^n)}{\partial y^n}\right)^2 dy^n$$

for any $R > 0$, in particular for $R = (\delta - |Y'|^2)^{1/2}$. With this choice for R, we integrate this inequality over $Q'(\delta)$ to infer that

$$\int_{Q^+} \left(\frac{u}{y^n}\right)^2 dY \le 2 \int_{Q^+} \left|\frac{\partial u}{\partial y}\right|^2 dY \le C \int_{\Omega\cap N} |Du|^2\, dX.$$

An easy covering argument using the compactness of $S\Omega \cup C\Omega$ show that

$$\int_{\{d<\delta\}} \left(\frac{u}{d}\right)^2 dX \le C \int_{\{d<\delta\}} |Du|^2 \, dX.$$

Next, we infer from Poincaré's inequality that

$$\int_{\{d\ge\delta\}} \left(\frac{u}{d}\right)^2 dX \le \delta^{-2} \int_{\Omega} u^2 \, dX \le C \int_{\Omega} |Du|^2 \, dX,$$

and (6.6) follows by combining these last inequalities.

For arbitrary $u \in V_0$, the result follows by approximation. □

We are now ready to prove the analog of Theorem 6.1 in H_1 domains. The idea of the proof is to work locally in the Y-coordinates of Lemma 6.3 and then follow the proof of Theorem 6.1 in those coordinates. Lemma 6.3 will be used to verify that the appropriate functions are integrable.

THEOREM 6.4. *Let $\mathscr{P}\Omega \in H_1$ and let u be a solution of* (6.1) *with f^i and $g \in L^2(\Omega)$ and $\varphi \in L^2(B\Omega)$. If a^{ij}, b^i, c^i and c^0 are in L^∞ and if there are positive constants λ, Λ, and Λ_1 such that conditions* (6.2a,b,c) *are satisfied, then there is a constant C determined only by λ, Λ, Λ_1, and n such that* (6.3) *holds.*

PROOF. Fix X_0, δ, N and Y as in the proof of Lemma 6.3. Then the Jacobian determinant $\det(\partial X/\partial Y) = 1/D_n\rho$ and $\psi_t = \psi_s + (\partial\psi/\partial y^n)\rho_t$ for $\psi \in C^1$. To simplify notation, we write Q for $Q^+(\delta)$, $\omega(1) = B^+(0,\delta) \times \{-\delta^2\}$ and $\omega(2) = B^+(0,\delta) \times \{0\}$. Then, with $A^{ij} = a^{km} D_k y^i D_m y^j$, $B^i = b^k D_k y^i$, $C^i = c^k D_k y^i$, $F^i = f^k D_k y^i$, and $\partial_i = \partial/\partial y^i$, we have

$$\begin{aligned}
\int_Q u\psi_s \frac{1}{D_n\rho} \, dY &+ \int_Q u\partial_n\psi \frac{\rho_t}{D_n\rho} \, dY - \int_Q (A^{ij}\partial_j u + B^i u - F^i)\partial_i\psi \frac{1}{D_n\rho} \, dY \\
&+ \int_Q (C^i\partial_i u + c^0 u - g)\psi \frac{1}{D_n\rho} \, dY \\
&= \int_{\omega(1)} u\psi \frac{1}{D_n\rho} \, dy - \int_{\omega(2)} u\psi \frac{1}{D_n\rho} \, dy
\end{aligned}$$

for any $\psi \in C^1_{\mathscr{P}}(N \cap \Omega)$. By virtue of Hardy's inequality, we can relax the regularity on ψ. We define the norm

$$\|\psi\|_2^* = \left(\int_\Omega |\psi|^2 + |D\psi|^2 + d^2 |\psi_t|^2 \, dX\right)^{1/2},$$

and we suppose that ψ is in the closure of $C^1_{\mathscr{P}}(N \cap \Omega)$ with respect to this norm. As in Theorem 6.1, we replace ψ by ψ_{-h} with $h < \delta^2/2$ and reduce the time interval of integration to $(-\delta^2 + h, -h)$. Writing $Q(h)$ for $B^+(\delta) \times (-\delta^2 + h, -h)$, we see that

$$I_h = \int_{Q(h)} u(\psi_{-h})_s \frac{1}{D_n\rho} \, dY = -\int_{Q(h)} \psi_s \left(\frac{u}{D_n\rho}\right)_h dY.$$

Now we set

$$v = \frac{u}{D_n\rho}, \psi = v_h\zeta$$

with $\zeta \in C^1_{\mathscr{P}}(N\cap\Omega)$ to be further specified. Then, we can integrate by parts to see that

$$I_h = -\tfrac{1}{2}\int_{Q(h)} v_h^2\zeta_s\,dY + \tfrac{1}{2}\int_{B^+\times\{-\delta^2+h\}} v_h^2\zeta\,dy - \tfrac{1}{2}\int_{B^+\times\{-h\}} v_h^2\zeta\,dy.$$

Now we set $\zeta = \eta D_n\rho$ with $\eta \in C^1_{\mathscr{P}}(N\cap\Omega)$ to infer that

$$I_h \to -\tfrac{1}{2}\int_Q u^2[\eta_s\frac{1}{D_n\rho} + \eta\left(\frac{1}{D_n\rho}\right)_s]\,dY + \tfrac{1}{2}\int_{\omega(1)} u^2\frac{1}{D_n\rho}\,dy - \tfrac{1}{2}\int_{\omega(2)} u^2\frac{1}{D_n\rho}\,dy.$$

Using this information in the weak form of the differential equation implies that

$$\tfrac{1}{2}\int_{\omega(2)} u^2\eta\frac{1}{D_n\rho}\,dy - \tfrac{1}{2}\int_{\omega(1)} u^2\eta\frac{1}{D_n\rho}\,dy - \int_Q (A^{ij}\partial_i u + B^i u - F^i)\partial_i(u\eta)\frac{1}{D_n\rho}\,dY + \int_Q [(C^i\partial_i u + c^0 u - g)u\eta]\frac{1}{D_n\rho}\,dY = I,$$

where

$$I = -\tfrac{1}{2}\int_Q \eta\partial_n(u^2)\frac{\rho_t}{D_n\rho}\,dY - \int_Q u^2\partial_n\eta\frac{\rho_t}{D_n\rho}\,dY - \tfrac{1}{2}\int_Q u^2\eta_s\frac{1}{D_n\rho}\,dY + \tfrac{1}{2}\int_Q u^2\eta\left(\frac{1}{D_n\rho}\right)_s\,dY.$$

Integrating the first integral in the definition of I by parts, we see that

$$I = \tfrac{1}{2}\int_Q u^2 S(Y)\,dY$$

with

$$S(Y) = \partial_n\left(\eta\frac{\rho_t}{D_n\rho}\right) - 2\partial_n\eta\frac{\rho_t}{D_n\rho} - \frac{\eta_s}{D_n\rho} + \eta\left(\frac{1}{D_n\rho}\right)_s.$$

Since $\partial_n = \frac{1}{D_n\rho}D_n$, it follows that $S(Y) = -\eta_t/D_n\rho$. Therefore, back in the X-coordinates, we have

$$\frac{1}{2}\int_{\Omega(\tau)} \eta u^2\,dx + \int_{\Omega(\tau)} (a^{ij}D_j u + b^i u + f^i)D_i(u\eta)\,dX = \int_{\Omega(\tau)} (c^i D_i u + c^0 u - g)(u\eta)\,dX + \int_{B\Omega} \varphi^2\eta\,dx.$$

Now we take a suitable partition of unity on Ω to infer that (6.4) holds, and then the proof is exactly like that of Theorem 6.1. □

Of course the remarks after Theorem 6.1 apply equally well when $\mathscr{P}\Omega \in H_1$.

The proof of Theorem 6.2 is easily adapted to the noncylindrical setting to yield the following existence and uniqueness result.

THEOREM 6.5. *If the coefficients of L, f^i, g, and φ satisfy the hypotheses of Theorem 6.1, then there is a unique solution u of* (6.1) *for any Ω with $\mathscr{P}\Omega \in H_1$.*

2. Differentiability of weak solutions

The remainder of this chapter is a consideration of improved regularity of weak solutions. Later sections will discuss pointwise properties of weak solutions (such as boundedness, Hölder continuity, and a strong maximum principle), but for now we give conditions under which u has weak derivatives D^2u and u_t.

THEOREM 6.6. *Let u be a weak solution of $Lu = f$ with $f \in L^2$, and suppose the coefficients of L satisfy* (6.2) *and*

$$\|Da^{ij}\|_\infty + \|Db^i\|_\infty \le K \tag{6.2)$'$}$$

for some nonnegative constant K. Then for any $\Omega' \subset\subset \Omega$, we have D^2u and u_t are in $L^2(\Omega')$. In addition, there is a constant C depending only on $\operatorname{dist}(\Omega', \partial\Omega)$, *$K$, λ, Λ, Λ_1, n, and T such that*

$$\|D^2u\|_{2,\Omega'} + \|u\|_{2,\Omega'} \le C(\|Du\|_2 + \|f\|_2). \tag{6.7}$$

PROOF. Set $g = (b^i + c^i)D_iu + (D_ib^i + c^0)u - f$, so that we can write the equation as $-u_t + D_i(a^{ij}D_ju) = g$. The idea now is to differentiate this equation with respect to x. Since this differentiation is not really permitted, we use a slightly different approach. Fix $X_0 \in \Omega'$ and $R < \frac{1}{2}d(X_0, \partial\Omega)$. Then for $m \in \{1, \dots, n\}$ and $h \in (0, R/2)$, set $\bar{u}(X) = u(X) - \underline{u(x + he_m, t)}$. Using the obvious version of this notation, we have that $-u_t + D_j(\overline{a^{ij}D_ju}) = \bar{g}$ and $\overline{a^{ij}D_ju} = \bar{a}^{ij}D_ju + a^{ij}(x + he_m, t)D_j\bar{u}$. Writing $f^i = \bar{a}^{ij}D_ju$ allows us to write

$$-\bar{u}_t + D_i(a^{ij}D_j\bar{u} + f^i) = \bar{g}.$$

Now we use the test function $\eta^2\hat{u}$ in the weak form of this equation, where $\eta \in C^2(\overline{\Omega})$ is supported in $Q(2R)$ with $0 \le \eta \le 1$ and $\hat{u}$ is a Steklov average of $\bar{u}$. As in the proof of Theorem 6.1, we infer that

$$\int_{Q(2R)} \eta^2 |D\bar{u}|^2 \, dX \le C \int_{Q(2R)} [|f^i|^2 + |\bar{g}|^2 + |\bar{u}|^2] \, dX,$$

and then we divide this inequality by h and send $h \to 0$. It follows that

$$\int_{Q(2R)} \eta^2 |DD_mu|^2 \, dX \le C \int_{Q(2R)} [|Du|^2 + |f|^2] \, dX.$$

By taking η to be an element of a suitable partition of unity in Ω', we find that

$$\int_{\Omega'} |D^2u|^2\, dX \leq C \int_{\Omega} [|Du|^2 + |f|^2]\, dX.$$

To show that u_t is in L^2, we mollify u and note that, for $\tau < \frac{1}{2}\operatorname{dist}(\Omega', \partial\Omega)$, the mollified function $v = u(\cdot, \tau)$ satisfies the equation

$$-v_t = [D_i(a^{ij}D_ju + b^iu) + c^iD_iu + c^0u - f](X, \tau)$$

in Ω' and we can estimate the $L^2(\Omega')$ norm of right side of this equation by the right side of (6.7). □

Note that it is a simple matter to replace the term $\|Du\|_2$ by $\|u\|_2$ in the right side of (6.7). In addition, it follows by induction that if a^{ij}, b^i, and f^i have bounded spatial derivatives of order less than or equal to some positive integer k, and if c^i, c^0 and g have bounded spatial derivatives of order less than or equal to $k-1$, then the solution u has spatial derivatives of all orders less than or equal to $k+2$ in L^2_{loc} and u_t has k spatial derivatives in L^2_{loc}. A similar argument gives higher regularity with respect to t corresponding to regularity in t of the coefficients.

3. Sobolev inequalities

For further estimates on weak solutions, we return to the theory of weak derivatives to investigate additional properties. The basic result is the Sobolev imbedding theorem.

THEOREM 6.7. *If $u \in W^{1,p}(\mathbb{R}^n)$ with $1 \leq p < n$, then $u \in L^q(\mathbb{R}^n)$ for $q = np/(n-p)$ and*

$$\|u\|_q \leq C(n,p)\, \|Du\|_p. \tag{6.8}$$

PROOF. We first suppose that $u \in C_0^1$ and $p = 1$, so $q = n/(n-1)$. By induction on n, we shall show that

$$\|u\|_{n/(n-1)} \leq \prod_{i=1}^{n} \|D_iu\|_1^{1/n}. \tag{6.9}$$

For $n = 2$, we have

$$
\begin{aligned}
\int u^2(x)\,dx &= \int\int u^2(x^1,x^2)\,dx^1\,dx^2 \\
&\le \int\int \max_{t^1\in\mathbb{R}} |u(t^1,x^2)| \max_{t^2\in\mathbb{R}} |u(x^1,t^2)|\; dx^1\,dx^2 \\
&= \max_{t^1\in\mathbb{R}} \int |u(t^1,x^2)|\; dx^2 \max_{t^2\in\mathbb{R}} \int |u(x^1,t^2)|\; dx^1 \\
&\le \left(\int D_1\,|u(t^1,x^2)|\; dt^1\,dx^2\right)\left(\int D_2\,|u(x^1,t^2)|\; dx^1\,dt^2\right) \\
&= \prod_{i=1}^{2} \|D_i u\|_1\,,
\end{aligned}
$$

which is just (6.9).

For $n > 2$, we set $t = (x',t^n)$ and $q = (n-1)/(n-2)$. Then we have

$$
\begin{aligned}
\int |u|^{n/(n-1)}(x)\,dx &\le \int\int [\max_{t^n\in\mathbb{R}} |u(t)|^{1/(n-1)}]\,|u(x)|\,dx'\,dx^n \\
&\le \int\left(\int \max_{t^n} |u(t)|\,dx'\right)^{1/(n-1)}\left(\int |u(x)|^q\,dx'\right)^{1/q} dx^n \\
&= \left(\int \max_{t^n} |u(t)|\,dx'\right)^{1/(n-1)} \int\left(\int |u(x)|^q\,dx'\right)^{1/q} dx^n \\
&\le \left(\int D_n\,|u(t)|\,dx'\,dt^n\right)^{1/(n-1)} \int \prod_{i=1}^{n-1}\left(\int D_i\,|u(x)|\,dx'\right)^{1/(n-1)} dx^n \\
&\le \prod_{i=1}^{n} \|D_i u\|_1^{1/(n-1)}\,,
\end{aligned}
$$

which is, again, (6.9). We now use the geometric-arithmetic mean inequality to infer (6.8) for $u \in C_0^1$ with $C = 1/n$ when $p = 1$. A simple approximation argument shows that the inequality holds for arbitrary $u \in W^{1,1}$.

When $p > 1$, we set $v = u^r$ for $r = (np-p)/(n-p)$ and $q = p/(p-1)$. Applying (6.8) to v then yields

$$
\begin{aligned}
\left(\int v^{n/(n-1)}\,dx\right)^{(n-1)/n} &\le \frac{1}{n}\int |Dv|\,dx = \frac{r}{n}\int u^{r-1}\,|Du|\,dx \\
&\le \frac{r}{n}\left(\int |u|^{q(r-1)}\,dx\right)^{1/q}\left(\int |Du|^p\,dx\right)^{1/p} \\
&= \frac{r}{n}\left(\int v^{n/(n-1)}\,dx\right)^{(p-1)/p}\left(\int |Du|^p\,dx\right)^{1/p}.
\end{aligned}
$$

By rearrangement, we have

$$\left(\int v^{n/(n-1)}\,dx\right)^{(n-p)/np} \leq \frac{r}{n}\left(\int |Du|^p\,dx\right)^{1/p},$$

which is just (6.8) once we note that $rn/(n-1) = q$. □

When $p > n$, elements of $W^{1,p}$ have Hölder continuous representatives.

THEOREM 6.8. *If $u \in W^{1,p}(\mathbb{R}^n)$ for $p > n$, then $u \in H_\alpha$ for $\alpha = 1 - n/p$, and*

$$[u]_\alpha \leq C(n,p)\,\|Du\|_p. \tag{6.10}$$

PROOF. We first observe that, for $u \in W^{1,p}$, we have

$$\int_{B(r)} |Du|\,dx \leq C(n)\,\|Du\|_p\, r^{n-n/p}$$

for any $r > 0$. Hence, if $u(\cdot;\tau)$ is a mollification of u, and we set $U = \|Du\|_p$, then

$$|Du(x;\tau)| \leq \int_{\{|y|\leq 1\}} |Du(x-\tau y)|\,|\varphi(y)|\,dy \leq C(n)U\tau^{-n/p},$$

and

$$|u_\tau(x;\tau)| \leq \int_{\{|y|\leq 1\}} |Du(x-\tau y)\cdot -y|\,|\varphi(y)|\,dy \leq C(n)U\tau^{-n/p}.$$

The required estimates follows from these two via integration as in Lemma 4.3: Set $\tau = |x-y|$ to see that

$$\begin{aligned} |u(x)-u(y)| &\leq |u(x;0)-u(x;\tau)| + |u(x;\tau)-u(y;\tau)| + |u(y;\tau)-u(y;0)| \\ &\leq CU\int_0^\tau s^{-n/p}\,ds + CU\,|x-y|\,\tau^{-n/p} + CU\int_0^\tau s^{-n/p}\,ds \\ &= C(n,p)U\,|x-y|^\alpha, \end{aligned}$$

which is (6.10). □

The case $p = n$ is more delicate (except for $n = 1$, in which case $W^{1,1}(\mathbb{R})$ is imbedded in C^0), and we leave the details to Exercise 6.3.

It is also possible to derive estimates like Sobolev's in parabolic domains. These estimates will form the basis for our pointwise estimates of solutions of parabolic equations.

THEOREM 6.9. *If $u \in V_0$ for some cylinder $Q = Q(R)$, then*

$$\int_Q |u|^{2(n+2)/n}\,dX \leq C(n)\max_{t\in I(Q)}\left(\int_{B(R)} |u(x,t)|^2\,dx\right)^{2/n}\int_Q |Du|^2\,dX. \tag{6.11}$$

PROOF. Suppose first that $n=1$. Then for any $t \in I(Q)$, we have

$$\begin{aligned}\int u^6\,dx &\le \max|u(\cdot,t)|^4 \int u^2\,dx \le \int |D(u^4)|\,dx \int u^2\,dx \\ &= 4\int |u|^3\,|Du|\,dx \int u^2\,dx \\ &\le 4\left(\int u^6\,dx\right)^{1/2}\left(\int u^2\,dx\right)\left(\int |Du|^2\,dx\right)^{1/2}\end{aligned}$$

by integration and Hölder's inequality.

On the other hand, for $n>1$, there is a constant C (determined only by n) such that

$$\begin{aligned}\int u^{2(n+2)/n}\,dx &\le \left(\int u^2\,dx\right)^{1/n}\left(\int u^{2(n+1)/(n-1)}\,dx\right)^{(n-1)/n} \\ &\le C\left(\int u^2\,dx\right)^{1/n}\left(\int |u|^{(n+2)/n}\,|Du|\,dx\right) \\ &\le C\left(\int u^2\,dx\right)^{1/n}\left(\int u^{2(n+2)/n}\,dx\right)^{1/2}\left(\int |Du|^2\,dx\right)^{1/2}\end{aligned}$$

by the Sobolev imbedding theorem, and Hölder's inequality. Rearranging these two inequalities gives, for any $n \ge 1$,

$$\begin{aligned}\int |u(x,t)|^{2(n+2)/n}\,dx &\le C\left(\int u^2(x,t)\,dx\right)^{2/n}\int |Du(x,t)|^2\,dx \\ &\le C\left(\max_s \int u^2(x,s)\,dx\right)^{2/n}\int |Du(x,t)|^2\,dx.\end{aligned}$$

The proof is completed by integrating this inequality with respect to t. □

More generally, an analogous weighted estimate is true.

COROLLARY 6.10. *Let $u \in V_0(Q)$ and $\lambda \in L^\infty(Q)$. Let $N=2$ if $n=1$, $N>2$, if $n=2$, and $N=n$ if $n>2$. Then*

$$\begin{aligned}\int_Q u^{2(N+2)/N}\lambda\,dX \le C(N)&\left(\max_s \int_B u^2\lambda^{N/2}(x,s)\,dx\right)^{2/N} \\ &\times\left(\int_Q |Du|^2\,dX\right)^{n/N}\left(\int u^2\,dX\right)^{1-n/N}.\end{aligned}\tag{6.12}$$

PROOF. For $n=1$, we have

$$\begin{aligned}\int u^4\lambda\,dx &\le (\max u^2)\int u^2\lambda\,dx \\ &\le 2\left(\int u^2\lambda\,dx\right)\left(\int |Du|^2\,dx\right)^{1/2}\left(\int u^2\,dx\right)^{1/2}\end{aligned}\tag{6.13}$$

by integration and Hölder's inequality. For $n = 2$, Hölder's inequality gives

$$\int u^{2(N+2)/N}\lambda\,dx \le \left(\int u^2\lambda^{N/2}\,dx\right)^{2/N}\left(\int u^{2N/(N-2)}\,dx\right)^{1-2/N},$$

and another application of Hölder's inequalities yields

$$\int u^{2N/(N-2)}\,dx \le \left(\int u^{(4N-4)/(N-2)}\,dx\right)^{2/N}\left(\int u^2 dx\right)^{1-2/N}.$$

The first integral on the right side of this inequality is estimated via the Sobolev imbedding theorem and Hölder's inequality as in Theorem 6.9. In this way, we find that

$$\left(\int u^{2N/(N-2)}\,dx\right)^{1-2/N} \le C\left(\int u^2\,dx\right)^{1-2/N}\left(\int |Du|^2\,dx\right)^{2/N},$$

and hence

$$\begin{aligned}\int u^{2(N+2)/N}\lambda\,dx \\ \le C\left(\int u^2\lambda^{N/2}\,dx\right)^{2/N}\left(\int |Du|^2\,dx\right)^{2/N}\left(\int u^2 dx\right)^{1-2/N}.\end{aligned} \tag{6.14}$$

Finally, for $n > 2$, we have

$$\begin{aligned}\int u^{2(n+2)/n}\,dx &\le \left(\int u^2\lambda^{n/2}\,dx\right)^{2/n}\left(\int u^{2n/(n-2)}\,dx\right)^{1-2/n} \\ &\le C(n)\left(\int u^2\lambda^{n/2}\,dx\right)^{2/n}\left(\int |Du|^2\,dx\right)\end{aligned} \tag{6.15}$$

by Hölder's inequality and the Sobolev imbedding theorem (with $p = 2$). The proof is completed from the appropriate choice of (6.13), (6.14), or (6.15) as in Theorem 6.9. □

In later chapters, we shall need L^p versions of Theorem 6.9 and Corollary 6.10. For the statement, we define $V_0^p(\Omega)$ to be the closure of $C^1_{\mathscr{P}}(\Omega)$ in the norm

$$\|u\|_{V^p} = \|Du\|_p + \max_{t\in I(\Omega)} \|u\|_{p;\omega(t)}\,.$$

THEOREM 6.11. *Let $u \in V_0^p(Q)$ for some $p \ge 1$. Then*

$$\int_Q |u|^{p(n+p)/n}\,dX \le C(n,p)\left(\max_s \int_B |u(x,s)|^p\,dx\right)^{p/n}\int_Q |Du|^p\,dX. \tag{6.16a}$$

Moreover, for any nonnegative $\lambda \in L^\infty(Q)$, *if* $N = p$ *for* $n = 1$, $N > p$ *for* $1 < n \le p$, *and* $N = n$ *for* $n > p$, *then*

$$\int_Q |u|^{p(N+p)/N} \lambda\, dX \le C(n,p) \left(\max_s \int_B |u(x,s)|^p \lambda(x,s)^{N/p}\, dx \right)^{p/N} \times \left(\int_Q |Du|^p\, dX \right)^{n/N} \left(\int_Q |u|^p\, dX \right)^{1-n/N}. \tag{6.16b}$$

PROOF. Inequality (6.16a) is proved in the same way as Theorem 6.9. For (6.16b), we follow the proof of Corollary 6.10, with the case $n = 1$ as in that corollary, $n > p$ as in case $n > 2$ and $1 < n \le p$ as in case $n = 2$. □

4. Poincaré's inequality

In Chapter III, we saw how to estimate the L^p norm of a $W_0^{1,p}$ function in terms of the L^p norm of its gradient. In this section, we prove a similar estimate for $W^{1,p}$ functions without assuming that they vanish on $\partial\Omega$. Instead, we normalize them by subtracting a suitable constant. Such estimates are related to those in Lemma 4.3, and they play a crucial role in our study of the weak Harnack inequality in Section VI.6.

Our first result is a weighted version of Poincaré's inequality, which will be used directly in Section VI.6 to prove the weak Harnack inequality. It will also be the basis of our further investigation of Poincaré's inequality with various mean values.

LEMMA 6.12. *Let* η *be a nonnegative, continuous function in* $\mathbb{R}^n$ *with compact support* Σ *and* $\int_\Sigma \eta = 1$, *suppose that the sets* $\{\eta \ge k\}$ *are convex for* $k < \sup\eta$, *and set* $R = \operatorname{diam}\Sigma$. *If* $u \in W^{1,p}$ *for some* $p \in [1,\infty)$ *and if* $L = \int u\eta\, dx$, *then*

$$\int |u - L|^p \eta\, dx \le C(n)(\sup\eta) R^{n+p} \int |Du|^p \eta\, dx. \tag{6.17}$$

PROOF. By direct calculation and Hölder's inequality, we have

$$\begin{aligned} \int |u-L|^p \eta\, dx &= \int \left| \int [u(x) - u(y)]\eta(y)\, dy \right|^p \eta\, dx \\ &\le \int\int |u(x) - u(y)|^p \eta(x)\eta(y)\, dy dx. \end{aligned} \tag{6.18}$$

Hence we need only estimate the right side of this inequality in terms of the right side of (6.17). To this end, we choose two points $x \ne y$ in the support of η. Since $|x - y| \le R$, we have

$$\begin{aligned} |u(x) - u(y)|^p &= \left| \int_0^1 Du(x + s(y-x)) \cdot (y - x)\, ds \right|^p \\ &\le R^p \int_0^1 |Du(x + s(y-x)|^p\, ds \end{aligned}$$

by Jensen's inequality. Because $\min_{s\in[0,1]}\eta(x+s[x-y]) = \min\{\eta(x),\eta(y)\}$ and this minimum is positive, it follows that

$$|u(x)-u(y)|^p \le R^p \int_0^1 |Du(x(s))|^p\,\eta(x(s))\,ds \frac{1}{\min\{\eta(x),\eta(y)\}}$$

for $x(s) = x+s[x-y]$, and hence

$$|u(x)-u(y)|^p\,\eta(x)\eta(y) \le (\sup\eta)R^p \int_0^1 |Du(x(s))|^p\,\eta(x(s))\,ds.$$

Now we fix $z\in\mathbb{R}^n$ and set $y = x+z$. Integrating this inequality with respect to x then implies that

$$\begin{aligned}\int |u(x)&-u(x+z)|^p\,\eta(x)\eta(x+z)\,dx\\ &\le |\eta|_0 R^p \int\int_0^1 |Du(x+sz)|^p\,\eta(x+sz)\,ds\,dx.\end{aligned}$$

We now simplify the integral on the right side of this inequality:

$$\begin{aligned}\int\int_0^1 &|Du(x+sz)|^p\,\eta(x+sz)\,ds\,dx\\ &= \int_0^1\int_{\mathbb{R}^n} |Du(x+sz)|^p\,\eta(x+sz)\,dx\,ds\\ &= \int_0^1\int_{\mathbb{R}^n} |Du(w)|^p\,\eta(w)\,dw\,ds\\ &= \int |Du(x)|^p\,\eta(x)\,dx,\end{aligned}$$

and therefore

$$\int |u(x)-u(x+z)|^p\,\eta(x)\eta(x+z)\,dx \le (\sup\eta)R^p \int |Du(x)|^p\,\eta(x)\,dx.$$

Finally, we observe that $\eta(x)\eta(x+z) = 0$ if $|z| > R$, so we can integrate this inequality over the set on which $|z|\le R$ to obtain

$$\int\int |u(x)-u(y)|^p\,\eta(x)\eta(y)\,dx\,dy \le C(n)R^n(\sup\eta)R^p \int |Du(x)|^p\,\eta(x)\,dx.$$

The proof is completed by combining this inequality with (6.18). □

To continue our investigation, we now show that the mean value L in (6.17) can be replaced by the weighted average of u with respect to any suitable L^∞ weight.

LEMMA 6.13. *Let Ω be a domain in $\mathbb{R}^n$, let η be a nonnegative continuous function in Ω, and suppose $f\in L^\infty(\Omega)$ with $\int_\Omega f\eta\,dx = 1$. If $u\in L^p(\Omega)$ and if $u_f = \int_\Omega uf\eta\,dx$, then*

$$\int_\Omega |u-u_f|^p\,\eta\,dx \le 2^p \sup|f| \inf_{L\in\mathbb{R}} \int_\Omega |u-L|^p\,\eta\,dx. \tag{6.19}$$

PROOF. From the triangle inequality, we have that

$$\left(\int_\Omega |u-u_f|^p \eta\,dx\right)^{1/p} \le \left(\int_\Omega |u-L|^p \eta\,dx\right)^{1/p} + \left(\int_\Omega |L-u_f|^p \eta\,dx\right)^{1/p}$$

for any $L \in \mathbb{R}$. But

$$\int_\Omega |L-u_f|^p \eta\,dx = |L-u_f|^p = \left|\int_\Omega (L-u) f\eta\,dx\right|^p$$
$$\le \int_\Omega |u-L|^p f\eta\,dx \le \sup|f| \int_\Omega |u-L|^p \eta\,dx$$

by Hölder's inequality and the conditions on f. Combining these two inequalities with the inequality $1 \le \sup|f|$ and taking the infimum on L gives (6.19). □

From the preceding two lemmata, we infer a Poincaré inequality for balls using general mean values.

PROPOSITION 6.14. *Let $u \in W^{1,p}(B)$ and suppose $f \in L^\infty(B)$, where B is a ball with radius R. Then there is a constant C determined only by n and p such that*

$$\int_B |u-u_f|^p\,dx \le CR^p \sup|f| \int_B |Du|^p\,dx, \tag{6.20}$$

where $u_f = |\Omega|^{-1} \int_\Omega uf\,dx$.

PROOF. Let (g_k) be a sequence of continuous functions supported in $\overline{B}$ which increase to 1 in B. (For example $g_k = [(1-|x|^2/R^2)^+]^{1/k}$.) By using

$$\eta_k = \frac{g_k}{\int g_k\,dx}$$

in place of η in Lemma 6.12 and sending $k \to \infty$, we infer that

$$\inf_{L\in\mathbb{R}} \int_B |u-L|^p\,dx \le C(n,p)R^p \int_B |Du|^p\,dx.$$

Combining this inequality with Lemma 6.13 gives (6.20). □

We shall use Proposition 6.14 with $p=1$ in giving an alternative proof of the weak Harnack inequality in Section VI.13.

5. Global boundedness

We now show that weak solutions of $Lu = D_i f^i + g$ which are bounded on $\mathscr{P}\Omega$ are also bounded in Ω if f and g are bounded. An important element of our proof is that the linearity of the operator L is not the relevant structure; instead we use a growth condition in a nonlinear setting. To illustrate these conditions, we write $Lu = D_i f^i + g$ as

$$-u_t + \operatorname{div} A(X,u,Du) + B(X,u,Du) = 0, \tag{6.21}$$

where

$$A^i(X,z,p) = a^{ij}(X)p_j + b^i(X)z - f^i(X),$$
$$B(X,z,p) = c^i(X)p_i + c^0(X)z - g(X).$$

We then say that u is a weak *subsolution* (*supersolution*, *solution*) of (6.21) in Ω if $u \in V$ and if

$$\int_{\omega(\tau)} uv\,dx - \int_{B\Omega} uv\,dx + \int_{\Omega(\tau)} [D_i v A^i(X,u,Du) - vB(X,u,Du) - v_t u]\,dx\,dt \le (\ge, =) 0$$

for all nonnegative $v \in C^1_{\mathscr{P}}$ and almost all $\tau \in I(\Omega)$. (In order to simplify our statements, we assume that $I(\Omega) = (0,T)$ for some positive T.) To see what structure conditions are satisfied by A and B, we first use (6.2a) and Schwarz's inequality to see that

$$p \cdot A(X,z,p) \ge \frac{\lambda}{2}|p|^2 - \frac{1}{2\lambda}(|bz|^2 + |f|^2).$$

(Here we use $|b|$ to denote the Euclidean length of the vector $(b^1,\dots,b^n)$, etc.) Assuming now that f and g are bounded, we infer that there is a constant k such that

$$|f| + |g| \le k\lambda.$$

From (6.2c), it then follows that

$$p \cdot A(X,z,p) \ge \frac{\lambda}{2}|p|^2 - (\Lambda_1^2 + 1)\lambda|z|^2 \tag{6.22a}$$

$$zB(X,z,p) \le \frac{\lambda}{4}|p|^2 + 4(\Lambda_1^2 + 1)\lambda|z|^2 \tag{6.22b}$$

as long as $|z| \ge k$. From these structure conditions, we can infer a bound for u in terms of boundary information. We start by considering the case of a subsolution which is nonpositive on $\mathscr{P}\Omega$.

We also need a suitable definition of boundary inequalities. We say that $u \in V$ satisfies the inequality $u \le K$ on $\mathscr{P}\Omega$ if $(u-K)^+ \in V_0$ and $u \le K$ almost everywhere on $B\Omega$. We also define $u \ge K$ on $\mathscr{P}\Omega$ if $-u \le -K$ on $\mathscr{P}\Omega$.

THEOREM 6.15. *Let A and B satisfy* (6.22a,b) *for $|z| \ge k$, and suppose u is a weak subsolution of* (6.21) *with $u \le 0$ on $\mathscr{P}\Omega$. If Ω is a cylinder or if $\mathscr{P}\Omega \in H_1$, then*

$$\sup_\Omega u \le C(n,\Lambda_1,\lambda)\,\|u\|_2 + 2k. \tag{6.23}$$

PROOF. Suppose first that $u \in W^{1,2}$ and take $v = j(u)$ for j a $C^1(\mathbb{R})$ function with bounded derivative such that $j(s) = 0$ for $s < k$. If we define the function J by

$$J(s) = \int_0^s j(\sigma)\,d\sigma,$$

then we have

$$\begin{aligned} 0 &\geq \int_0^\tau \int_{\Omega(t)} -u_t v + Dv \cdot A(X,u,Du) - vB(X,u,Du)\,dx\,dt \\ &= \int_{\Omega(\tau)} J(u)\,dx + \int_0^\tau \int_{\Omega(t)} j'(u)Du \cdot A - j(u)B\,dx\,dt. \end{aligned}$$

Using (6.22), we find that

$$\begin{aligned} 0 \geq \int_{\Omega(\tau)} J(u)\,dx + \frac{\lambda}{2}\int_0^\tau \int_{\Omega(t)} [j'(u) - j(u)/(2u)]\,|Du|^2\,dX \\ - \lambda(\Lambda_1^2+1)\int_0^\tau \int_{\Omega(t)} [j'(u)u + 4j(u)]\,|u|\,dX. \end{aligned} \tag{6.24}$$

By imitating the argument involving Steklov averages from Theorem 6.1, we see that (6.24) holds for any weak subsolution u with $u \leq 0$ on $\mathscr{P}\Omega$.

To continue, we take j to have the form $j(s) = \chi(s)(s-k)^+$ for some C^1 function χ satisfying

$$0 \leq \frac{\chi'(s)(s-k)}{\chi(s)} \leq k_1 \tag{6.25}$$

for some constant k_1 (and also such that j' is uniformly bounded) and all $s > k$. Then, for $s \geq k$,

$$\begin{aligned} J(s) &= \int_k^s \chi(\sigma)(\sigma - k)\,d\sigma \\ &= \frac{1}{2}\chi(s)(s-k)^2 - \frac{1}{2}\int_k^s \chi'(\sigma)(\sigma-k)^2\,d\sigma \\ &\geq \frac{1}{2}\chi(s)(s-k)^2 - \frac{k_1}{2}\int_k^s \chi(\sigma)(\sigma-k)\,d\sigma, \end{aligned}$$

and hence, after some rearrangement,

$$J(s) \geq \frac{1}{2+k_1}\chi(s)(s-k)^2.$$

Since χ is increasing, we also have

$$j'(s) - j(s)/(2s) \geq \frac{1}{2}\chi(s).$$

Using these estimates in (6.24), we find that

$$\int_{\omega(\tau)} \chi(u)(u-k)^2\,dx + \int_{\Omega(\tau)} \chi(u)\,|Du|^2\,dX \\ \leq C(n,\lambda,\Lambda_1)(1+k_1)^2 \int_{\Omega(\tau)} \chi(u)u^2\,dX. \tag{6.26}$$

We first take

$$\chi(s) = \left[\min\{s,Z\}(1-\frac{k}{s})^+\right]^q$$

for positive constants Z and q at our disposal. It follows from (6.26) that

$$\begin{aligned}\int_\sigma \min\{u,Z\}^q u^2(1-\frac{k}{u})^{q+2}\,dx \\ &\leq C(n,q,\lambda,\Lambda_1)\int_\Sigma \min\{u,Z\}^q u^2(1-\frac{k}{u})^q\,dX \\ &\leq C[\int_\Sigma \min\{u,Z\}^q u^2(1-\frac{k}{u})^{q+2}\,dX \\ &+\int_\Sigma \min\{u,Z\}^q k^2(1-\frac{k}{u})^q\,dX],\end{aligned}$$

where we have abbreviated $\sigma = \{X \in \omega(\tau) : u(X) > k\}$ and $\Sigma = \{X \in \Omega(\tau) : u(X) > k\}$. We now use Gronwall's inequality to infer that

$$\begin{aligned}\int_\Sigma \min\{u,Z\}^{q+2}(1-\frac{k}{u})^{q+2}\,dX &\leq \int_\Sigma \min\{u,Z\}^q u^2(1-\frac{k}{u})^{q+2}\,dX \\ &\leq Ck^2\int_\Sigma \min\{u,Z\}^q(1-\frac{k}{u})^q\,dX,\end{aligned}$$

and hence Young's inequality implies that

$$\int_\Sigma \min\{u,Z\}^{q+2}(1-\frac{k}{u})^{q+2}\,dX \leq C\,|\Omega|\,k^{q+2}.$$

Sending $Z \to \infty$ gives

$$\int_\Omega (u^+)^{q+2}\,dX \leq C(q)k^{q+2} \tag{6.27}$$

for any positive q. A similar argument shows that (6.26) remains valid for arbitrary χ satisfying (6.25) since all the integrals are finite.

Now we fix $q > 1$ and define $\chi(s) = s^{2q-2}[(1-\frac{k}{s})^+]^{(n+2)q-n-2}$. Since χ satisfies (6.25), we infer from (6.26) that

$$\int_\sigma u^{2q}(1-\frac{k}{u})^{(n+2)q-n}\,dx + \int_\Sigma u^{2q-2}(1-\frac{k}{u})^{(n+2)q-n-2}\,|Du|^2\,dX \\ \leq Cq^2\int_\Sigma u^{2q}(1-\frac{l}{u})^{(n+2)q-n-2}\,dX.$$

Now we set $h = (u^q[(1-\frac{k}{u})^+]^{(n+2)q-n})^{1/2}$ and note that

$$|Dh|^2 \le C(n)q^2u^{2q-2}[(1-\frac{k}{u})^+]^{(n+2)q-n-2}|Du|^2.$$

Therefore we have

$$\max_{\tau\in I(\Omega)}\int_{\omega(\tau)} h^2\,dx + \int_\Omega |Dh|^2\,dX \le Cq^4\int_\Omega u^{2q}[(1-\frac{k}{u})^+]^{(n+2)q-n-2}\,dX,$$

and the Sobolev imbedding theorem implies that

$$\left(\int_\Omega u^{2\kappa q}[(1-\frac{k}{u})^+]^{\kappa(n+2)q-\kappa n}\,dX\right)^{1/\kappa} \le Cq^4\int_\Omega u^{2q}[(1-\frac{k}{u})^+]^{(n+2)q-n-2}\,dX$$

for $\kappa = (n+2)/n$. We now rewrite this inequality by setting $w = u^2(1-k/u)^{n+2}$ and $d\mu = (1-k/u)^{-n-2}\,dX$ to see that

$$\left(\int_\Sigma w^{\kappa q}\,d\mu\right)^{1/(\kappa q)} \le C^{1/q}q^{4/q}\left(\int_\Sigma w^q d\mu\right)^{1/q}. \tag{6.28}$$

We now iterate this inequality for $q = 1, \kappa, \kappa^2, \ldots$, and write $I(q)$ for the $L^q(d\mu)$ norm of w. It follows from (6.28) and induction that

$$I(\kappa^j) \le \prod_{m=1}^{j}(C\kappa^{4m})^{a(m)}I(1) = C^{b(j)}\kappa^{\sigma(j)}I(1)$$

for $a(m) = \kappa^{-2m}$,

$$b(j) = \sum_{m=1}^{j} 4a(m), \text{ and } \sigma(j) = \sum_{m=1}^{j} 2ma(m).$$

Since $b(j)$ and $\sigma(j)$ are partial sums of convergent series, it follows that $b(j) \le C(n)$ and $\sigma(j) \le C(n)$, and therefore $I(\kappa^j) \le CI(1)$ for any positive integer j. From Exercise 6.4, we have that $\lim_{j\to\infty} I(\kappa^j) = \sup_\Sigma w$, and hence

$$\sup_\Sigma w \le C\int_\Omega u^2(1-\frac{k}{u})^{n+2}[(1-\frac{k}{u})^+]^{-n-2}\,dX,$$

or

$$\sup_\Omega u^2[(1-\frac{k}{u})^+]^{n+2} \le C\int_\Omega u^2\,dX.$$

The desired inequality follows from this one by considering separately the cases $\sup u \le 2k$ and $\sup u > 2k$. □

Note that (6.27) gives an estimate of $\|u\|_{2;\{u>k\}}$, which is enough to estimate $\sup u$ in terms of the data of the problem. Moreover, this pointwise estimate holds if we only assume a nonnegative upper bound for u on $\mathscr{P}\Omega$.

COROLLARY 6.16. *If, in the hypotheses of Theorem 6.15, we assume that $u \le M$ on $\mathcal{P}\Omega$ for some nonnegative constant M, then*

$$\sup_{\Omega} u \le C(n,\lambda,\Lambda_1,|\Omega|,T)(M+k). \tag{6.29}$$

PROOF. Follow the proof of Theorem 6.15 with $(1-(M+k)/u)^+$ in place of $(1-k/u)^+$. □

In fact, the global L^∞ bound in Corollary 6.16 remains valid if the coefficients of the linear equation (and their counterparts in the quasilinear one) are in the spaces $L^{p,q} = L^p([0,T];L^q(\Omega(t)))$ for a suitable range of p and q; see Exercise 6.5.

6. Local estimates

By modifying the iteration scheme used in the previous section, we can study the local behavior of weak solutions of linear equations also. For this study, a slightly different version of the structure conditions is useful. We note also that $|A| \le (\sum |a^{ij}p_j|^2)^{1/2} + |b|\,|z| + |f|$. Then if $|z| \ge \lambda kR$ for some constants k and R, and if

$$|f| \le \lambda k, |g| \le \lambda k/R,$$

we have

$$p\cdot A \ge \frac{\lambda}{2}|p|^2 - \Lambda_2\lambda\left(\frac{|z|}{R}\right)^2, \tag{6.30a}$$

$$|A| \le n\Lambda|p| + \Lambda_2^{1/2}\lambda\frac{|z|}{R}, \tag{6.30b}$$

$$zB \le \frac{\lambda}{4}|p|^2 + 4\Lambda_2\lambda\left(\frac{|z|}{R}\right)^2, \tag{6.30c}$$

for $\Lambda_2 = (\Lambda_1 R+1)^2$, which will imply our local supremum estimate.

THEOREM 6.17. *Let A and B satisfy* (6.30) *for $|z| \ge kR$ and $X \in Q(2R)$. If u is a subsolution of* (6.21) *in $Q(2R)$, then for any $m > 0$, there is a constant $C(m,n,\lambda,\Lambda,\Lambda_2)$ such that*

$$\sup_{Q(R)} u \le C\left[\left(R^{-n-2}\int_{Q(2R)}(u^+)^m\,dX\right)^{1/m} + kR\right]. \tag{6.31}$$

PROOF. We now let $\eta \in C^1_{\mathcal{P}}(Q(2R))$ and take $\chi \in C^1(0,\infty)$ with $\chi(s) = 0$ if $s \le kR$ and χ' uniformly bounded. By using the test function $v = \eta^2\chi(u)(u-kR)^+$

and conditions (6.30a,b,c) as in Theorem 6.15, we infer that

$$\sup_\tau \int_{B(2R)\times\{\tau\}} \chi(u)\eta^2[(u-kR)^+]^2\,dx + \int_{Q(2R)} \chi(u)\,|Du|^2\,\eta^2\,dX$$
$$\leq C(k_1+1)^2 \int_{Q(2R)} \chi(u)u^2\left(\frac{\eta^2}{R^2}+\eta\eta_t+|D\eta|^2\right)dX$$

provided χ satisfies (6.25). (The argument in Theorem 6.15 is easily modified to remove the assumption that χ' is bounded.)

Next, we define

$$\zeta = \left(1-\frac{|x|^2}{4R^2}\right)^+ \left(1-\frac{t}{4R^2}\right)^+,$$

and take $\eta^2 = \zeta^{\alpha q-n}$ and $\chi(s) = s^{q-2}v^{\alpha q-n}$ for $\alpha = (n+2)/m$ and $v = \zeta(1-\frac{kR}{u})^+$ to find that

$$\sup_\tau \int_{B(2R)\times\{\tau\}} u^q v^{\alpha q-n}\,dx + \int_{Q(2R)} u^{q-2}\,|Du|^2\,v^{\alpha q-n-2}\zeta^2\,dX$$
$$\leq Cq^2R^{-2}\int_{Q(2R)} u^q v^{\alpha q-n-2}\,dX.$$

Now we apply the Sobolev imbedding theorem with $h^2 = u^q v^{\alpha q-n}$ and set $w = uv^\alpha$, $d\mu = [R\zeta(1-kR/u)]^{-n-2}\,dX$ to see again that (6.28) holds with $\Sigma = \{X \in Q(2R) : v(X) > 0\}$.

We iterate as before but with $q = m\kappa^j$ for j a sufficiently large integer to infer that

$$\sup_\Sigma w \leq C(m)\left(\int_\Sigma w^m\,d\mu\right)^{1/m}.$$

The proof is completed by noting that

$$\int_\Sigma w^m\,d\mu \leq \int_{Q(2R)} (u^+)^m\,dX$$

and that $\sup_\Sigma w \geq C\sup_{Q(R)} u$ provided $\sup_{Q(R)} u \geq 2kR$. □

In our further development of local properties of weak solutions, we use the following variant of our nonlinear structure conditions: For $z \geq 0$, we set $\bar{z} = z + kR$ and we suppose that

$$p\cdot A(X,z,p) \geq \frac{\lambda}{2}|p|^2 - 2\Lambda_2\lambda\frac{\bar{z}^2}{R^2}, \tag{6.32a}$$

$$|A(X,z,p)| \leq \Lambda|p| + \Lambda_2^{1/2}\lambda\frac{\bar{z}}{R}, \tag{6.32b}$$

$$\bar{z}B \geq -\varepsilon\lambda|p|^2 - \frac{\lambda\Lambda_2}{\varepsilon}\frac{|\bar{z}|^2}{R^2} \tag{6.32c}$$

for any $\varepsilon \in (0,1)$ and $z \geq 0$. We also use a special geometric situation. Fix $Y \in \Omega$ and $R > 0$ such that $Q(R,Y) \subset \Omega$, and define $\Theta(R) = Q((y, s - 4R^2), R)$. Our weak Harnack inequality can now be stated as follows.

THEOREM 6.18. *Let A and B satisfy* (6.32), *and suppose that $u \in V$ is a supersolution of* (6.21) *which is nonnegative in $Q(4R)$. Then for any $\sigma \in [1, 1 + 2/n)$, there is a constant $C(n, \lambda, \Lambda, \Lambda_2, \sigma)$ such that*

$$\left(R^{-n-2} \int_{\Theta(R)} u^\sigma \, dX\right)^{1/\sigma} \leq C(\inf_{Q(R)} u + kR). \tag{6.33}$$

A key idea of the proof of this theorem is a restatement of (6.33) in terms of the functional Φ defined for real numbers ρ and subsets Σ of Ω by

$$\Phi(\rho, \Sigma) = \left(\frac{1}{|\Sigma|} \int_\Sigma u^\rho \, dX\right)^{1/\rho}$$

This expression is defined for any Σ with positive measure and any $\rho \in \mathbb{R} \setminus \{0\}$. For $\rho \geq 1$, it is the normalized L^ρ norm of u over the set Σ, and, according to Exercise 6.4, it can be extended as a continuous function of $\rho \in [-\infty, \infty]$ by setting

$$\Phi(-\infty, \Sigma) = \inf_\Sigma u, \Phi(\infty, \Sigma) = \sup_\Sigma u,$$
$$\Phi(0, \Sigma) = \exp\left(\frac{1}{|\Sigma|} \int_\Sigma \ln u \, dX\right).$$

Suppressing the dependence of Φ on Σ temporarily, we can write (6.33) as $\Phi(\sigma) \leq C\Phi(-\infty)$ whereas (6.31) can be written as $\Phi(\infty) \leq \Phi(\sigma)$. The crucial difference in the proof here comes from "jumping over zero", that is, comparing Φ for a positive argument with Φ for a negative argument. In fact the proof proceeds essentially in three steps: First $\Phi(\sigma) \leq C\Phi(\rho)$ for any $\rho \in (0, \sigma)$, then there is a positive ρ such that $\Phi(\rho) \leq C\Phi(0)$, and finally $\Phi(0) \leq C\Phi(-\infty)$. (In fact, as is suggested by the form of $\Phi(0)$, we estimate various quantities involving $\ln u$ for several of these steps.)

The first step (estimating $\Phi(\sigma)$ in terms of $\Phi(\rho)$) is quite straightforward.

LEMMA 6.19. *For any $\sigma \in [1, 1 + 2/n)$ and any $\rho \in (0, 1]$ there is a constant C, determined only by n, λ, Λ, Λ_2, ρ, and σ such that*

$$\left(R^{-n-2} \int_{Q(R)} \bar{u}^\sigma \, dX\right)^{1/\sigma} \leq C \left(R^{-n-2} \int_{Q(3R/2)} \bar{u}^\rho \, dX\right)^{1/\rho} \tag{6.34}$$

PROOF. Use the test function $\bar{u}^{-q} \eta^{\alpha q - n}$ with $q \in (0,1)$ and $\alpha \geq 0$ to be further specified and η a nonnegative cut-off function supported in $Q = Q(2R)$. By taking $\varepsilon = \varepsilon(q)$ sufficiently small, we find that

$$\sup_{\tau} \int_{B(2R)\times\{\tau\}} \eta^{\alpha q-n}\bar{u}^{1-q}\,dx + \int_Q \eta^{\alpha q-n}\bar{u}^{-q-1}|D\bar{u}|^2\,dX \leq C(q)R^{-2}\int_Q \eta^{\alpha q-n-2}\bar{u}^{1-q}\,dX,$$

and hence, from the Sobolev imbedding theorem,

$$\left(R^{-n-2}\int_Q \bar{u}^{\kappa(1-q)}\eta^{\alpha\kappa q-n-2}\,dX\right)^{1/\kappa} \leq CR^{-n-2}\int_Q \bar{u}^{1-q}\eta^{\alpha q-n-2}\,dX$$

for $\kappa = 1+2/n$. If $\sigma = \kappa^j\rho$ for some positive integer j, we take $q = 1-\kappa^i\rho$ for $i = 0,\ldots,j-1$ and obtain (6.34) after a finite iteration of this inequality. If $\kappa^j\rho < \sigma < \kappa^{j+1}\rho$ for some positive integer j, we infer (6.34) by finite iteration along with Hölder's inequality provided α is taken sufficiently large. □

The next step (estimating $\Phi(\rho)$ in terms of $\Phi(0)$) is more complicated, requiring an additional iteration scheme.

LEMMA 6.20. *There are positive constants $\rho \in (0,1)$ and C (both determined only by n, λ, Λ, and Λ_2) such that*

$$\left(R^{-n-2}\int_{Q(3R/2)} \left(\frac{\bar{u}}{K}\right)^{\rho}\,dX\right)^{1/\rho} \leq C\exp\left(R^{-n-2}\int_{Q(2R)} \left[\ln^+\left(\frac{\bar{u}}{K}\right)\right]^{1/2}\,dX\right) \tag{6.35}$$

for any positive constant K.

PROOF. Set $v = \ln^+(\bar{u}/K)$ and, for constants $q > 2$, α to be chosen, set $\Gamma = (4q)^{q-1}$. We then use the test function $\eta^{\alpha q-n}(\Gamma + v^{q-1})/\bar{u}$ and note that Young's inequality gives $qv^{q-2} \leq \frac{1}{4}(v^{q-1}+\Gamma)$ and $\Gamma v \leq v^q + (4q)^q$. It follows from the structure conditions (6.32) that

$$\sup_{\tau}\int_{B(2R)\times\{\tau\}} v^q\eta^{\alpha q-n}\,dx + q^2\int_Q v^{q-2}|Dv|^2\eta^{\alpha q-n}\,dX \leq Cq^2R^{-2}\int_Q v^q\eta^{\alpha q-n-2}\,dX + C(4q)^{q+2}R^{-2}\int_Q \eta^{\alpha q-n-2}\,dX.$$

For $d\mu = (\eta R)^{-n-2}\,dX$, the Sobolev imbedding theorem then yields a constant c_1 such that

$$\left(\int_Q (v\eta^{\alpha})^{\kappa q}\,d\mu\right)^{1/\kappa} \leq c_1q^2\left[\int_Q (v\eta^{\alpha})^q\,d\mu + (4q)^q\right].$$

We rewrite this inequality as

$$I(\kappa q) \leq c_1^{1/q}q^{2/q}[I(q) + 4q]$$

with $I(m) = (\int (v\eta^\alpha)^m \, d\mu)^{1/m}$, noting this inequality holds also for $q \in [\frac{1}{2}, 2]$ by Hölder's inequality. Induction then gives

$$I(\kappa^j q) \leq (c_1 q^2)^{a(j)} \kappa^{b(j)} [I(q) + 4q \sum_{i=0}^{j-1} \kappa^i],$$

where

$$a(j) = \frac{1}{q} \sum_{i=0}^{j} \kappa^{-i} \leq \frac{1}{q} \frac{\kappa}{\kappa - 1},$$

and

$$b(j) = \frac{2}{q} \sum_{i=1}^{j-1} \kappa^{-i} \leq \frac{2}{q} \frac{1}{\kappa - 1}.$$

Noting also that

$$\sum_{i=0}^{j-1} \kappa^i \leq \frac{\kappa^j}{\kappa - 1},$$

we see that

$$I(\kappa^j q) \leq c_2(q)[I(q) + \kappa^j q]$$

for any $q \geq \frac{1}{2}$ and positive integer j.

Now let m be a positive integer and choose $q \in [\frac{1}{2}, \frac{\kappa}{2})$ and j a positive integer such that $m = \kappa^j q$. Noting that $c_2(q)$ is uniformly bounded on this interval, we consider two cases. If $q = \frac{1}{2}$, we have $I(m) \leq c_2[I(\frac{1}{2}) + m]$, while if $q > \frac{1}{2}$, we have $I(m) \leq c_2[I(q) + m]$ and

$$I(q) \leq I(\frac{\kappa}{2})^\beta I(\frac{1}{2})^{1-\beta} \leq [c_1^2(I(\frac{1}{2}) + 2)]^\beta I(\frac{1}{2})^{1-\beta} \leq c_1^2(I(\frac{1}{2}) + 2)$$

for some $\beta \in (0, 1)$. It follows from these two cases that

$$I(m) \leq c_3[I(\frac{1}{2}) + m],$$

and therefore (after choosing α appropriately),

$$R^{-n-2} \int_{Q(3R/2)} v^m \, dX \leq c_4^m [(R^{-n-2} \int_{Q(2R)} v^{1/2} \, dX)^2 + m]^m.$$

We now fix $\beta \in (0, 1/e)$ and set $\rho = \beta / c_4$. Multiplying both sides of this inequality by $\rho^m / m!$, summing on m and then using Exercise 6.6 leads to the estimate

$$R^{-n-2} \int_{Q(3R/2)} \exp(\rho v) \, dX \leq C \exp(R^{-n-2} \int_{Q(2R)} v^{1/2} \, dX).$$

Finally, we note that $\exp(\rho v) = (\frac{\bar{u}}{K})^\rho$ if $\bar{u} > K$,

$$\int_{Q(3R/2) \cap \{\bar{u} \leq K\}} \left(\frac{\bar{u}}{K} \right)^\rho dX \leq C(n) R^{n+2},$$

and that $\exp(w) \geq 1$ if $w \geq 0$. □

The next step is to choose K in such a way that the right side of (6.35) can be estimated.

LEMMA 6.21. *Let* $\eta(x) = c_0(1 - |x-y|^2/(9R^2))^+$, *where* c_0 *is chosen so that* $\int_{\mathbb{R}^n} \eta^2\,dx = 1$. *For*

$$K = \exp\left(\int_{B(3R)\times\{s\}} \eta^2 \ln \bar{u}\,dx\right), \tag{6.36}$$

we have

$$\int_{Q(2R)} \left(\ln^+ \frac{\bar{u}}{K}\right)^{1/2} dX \le C(n,\lambda,\Lambda,\Lambda_2)R^{n+2}. \tag{6.37}$$

PROOF. Set $v = \ln(\bar{u}/K)$ and

$$V(t) = \int_{B(3R)\times\{t\}} \eta^2 v\,dx,$$

noting that $V(s) = 0$. If $\bar{u}$ is strictly positive and $u \in W^{1,2}$, then

$$\begin{aligned} V(t_1) - V(t_2) &= -\int_{t_1}^{t_2}\int_{B(3R)} \frac{d}{dt}\left(\ln\frac{\bar{u}}{K}\right)\eta^2\,dX = -\int_{t_1}^{t_2}\int_{B(3R)} \frac{u_t}{\bar{u}}\eta^2\,dX \\ &\le \int_{t_1}^{t_2}\int_{B(3R)} \frac{\eta}{\bar{u}}\left(2D\eta\cdot A - \frac{\eta}{\bar{u}}[Du\cdot A + \bar{u}B]\right)dX \end{aligned}$$

for $t_2 > t_1$, and a simple Steklov averaging argument shows that

$$V(t_1) - V(t_2) \le \int_{t_1}^{t_2}\int_{B(3R)} \frac{\eta}{\bar{u}}\left(2D\eta\cdot A - \frac{\eta}{\bar{u}}[Du\cdot A + \bar{u}B]\right)dX$$

for an arbitrary subsolution u. Using the structure conditions, we find that

$$V(t_1) - V(t_2) + \frac{\lambda}{4}\int_{t_1}^{t_2}\int_{B(3R)} \eta^2 |Dv|^2\,dX \le CR^{-2}\int_{t_1}^{t_2}\int_{B(3R)} 1\,dX,$$

and then Poincaré's inequality (Lemma 6.12) implies that

$$V(t_1) - V(t_2) + c_1R^{-2}\int_{t_1}^{t_2}\int_{B(3R)} \eta^2 |v - V(t)|^2\,dX \le c_2R^{n-2}(t_2 - t_1)$$

for constants c_1 and c_2 determined by n, λ, Λ, and Λ_2.

If V is differentiable, the proof can be completed fairly directly. We set $w = v + c_2R^{-2}(t-s)$ and $W(t) = V(t) + c_2R^{-2}(t-s)$ to infer that there is a constant c_3 such that

$$W(t_1) - W(t_2) + \frac{c_3}{R^2}\int_{t_1}^{t_2}\int_{B(2R)} (w - W(t))^2\,dX \le 0$$

(because $\eta^2 \ge C(n)$ in $B(2R)$). For $\mu > 0$ and $t \in (s - 4R^2, s)$, we write $Q_\mu(t)$ for the set of all points $x \in B(2R)$ such that $w(x,t) > \mu$. Then $W(t) \le 0$ for $t \in (s - 4R^2, s)$, so $w - W(t) > \mu - W(t) > 0$ in $Q_\mu(t)$, and hence

$$W(t_1) - W(t_2) + \frac{c_4}{|Q|}\int_{t_1}^{t_2} |Q_\mu(t)|\,(\mu - W(t))^2\,dt \le 0.$$

Now divide by $t_2 - t_1$ and take the limit as $t_1 \to t_2$, thus obtaining

$$-W'(t)(\mu - W(t))^{-2} + \frac{c_4}{|Q|}|Q_\mu(t)| \le 0.$$

This inequality can now be integrated from $s - 4R^2$ to s. The result is that

$$\begin{aligned} \frac{1}{\mu} &\ge \frac{1}{\mu - W(s)} - \frac{1}{\mu - W(s-4R^2)} \\ &\ge \frac{c_4}{|Q|}\int_{s-4R^2}^{s} |Q_\mu(t)|\, dt = c_4 |\{X \in Q(2R) : w(X) > \mu\}| / |Q| \end{aligned} \tag{6.38}$$

because $W(s) = 0$. Simple integration shows that

$$\begin{aligned} \int_{\{X\in Q:w(X)>1\}} w^{1/2}\, dX &= \frac{1}{2}\int_1^\infty \mu^{-1/2}|\{X \in Q(2R) : w(X) > \mu\}|\, d\mu \\ &\le \frac{|Q|}{2c_4}\int_1^\infty \mu^{-3/2}\, d\mu = C|Q|. \end{aligned}$$

We easily infer (6.37) from this estimate because $(v^+)^{1/2} \le w^{1/2} + 2c_2^{1/2}$ if $w \ge 1$ and $(v^+)^{1/2} \le 1 + 2c_2^{1/2}$ if $w < 1$.

If V is not differentiable, we derive (6.38) in a slightly different manner. By hypothesis, V is continuous, so, for a given positive constant ε, there is a positive δ such that $|t - t_2| < \delta$ implies $|V(t) - V(t_2)| < \varepsilon$. In particular, if $t_2 < t_1 + \delta$, then

$$\begin{aligned} V(t_1) - V(t_2) + c_1 R^{-2}\int_{t_1}^{t_2}\int_{B(3R)} \eta^2 |v - V(t_2)|^2\, dX \\ \le (c_2 + \varepsilon^2 c_1) R^{n-2}(t_1 - t_2). \end{aligned}$$

If we now define

$$w = v + [c_2 + \varepsilon^2 c_1] R^{-2}(t - s)$$

and

$$W(t) = V(t) + [c_2 + \varepsilon^2 c_1] R^{-2}(t - s),$$

then we obtain

$$\frac{W(t_1) - W(t_2)}{(\mu - W(t_2))^2} + \frac{c_4}{|Q|}\int_{t_1}^{t_2} |Q_\mu(t)|\, dt \le 0,$$

and hence, because W is increasing,

$$\frac{c_4}{Q}\int_{t_1}^{t_2} |Q_\mu(t)|\, dt \le \frac{W(t_2) - W(t_1)}{(\mu - W(t_2))(\mu - W(t_1))} = \frac{1}{\mu - W(t_2)} - \frac{1}{\mu - W(t_1)}.$$

We now take $\delta_1 \le \delta$ so that $4R^2$ is an integer multiple of δ_1, say $4R^2 = k\delta_1$, and set $t_2 = s + (j+1)\delta_1$ and $t_1 = s + j\delta_1$ for j between 0 and $k - 1$. Adding the resultant inequalities gives (6.38) and then we repeat the previous argument. (Note that by taking the limit as $\varepsilon \to 0$, we can infer the results with the same constants as before.) □

The estimates for $\Phi(\rho)$ with $\rho < 0$ are proved similarly.

LEMMA 6.22. *Let η be as in Lemma 6.21. For*

$$K = \exp\left(\int_{B(3R)\times\{s-4R^2\}} \eta^2 \ln \bar{u}\, dx\right), \tag{6.36$'$}$$

we have

$$\int_{Q(2R)} (\ln^- \frac{\bar{u}}{K})^{1/2}\, dX \le C(n,\lambda,\Lambda_2)R^{n+2}. \tag{6.37$'$}$$

PROOF. With v as in Lemma 6.21, we define

$$w = v + c_2R^{-2}(t-s+4R^2),\ W(t) = V(t) + c_2R^{-2}(t-s+4R^2),$$

and we write $Q_\mu(t)$ for the set of all $x \in B(2R)$ such that $w(x,t) < \mu$. Then $W \ge 0$ and $W(s-4R^2) = 0$, so, in place of (6.38), we infer that

$$\begin{aligned} -\frac{1}{\mu} &\ge \frac{1}{\mu - W(s)} - \frac{1}{\mu - W(s-4R^2)} \\ &\ge \frac{c_4}{|Q|}\int_{s-4R^2}^{s} \left|Q_\mu(t)\right| dt = c_4 \left|\{X \in Q(2R) : w(X) > \mu\}\right| / |Q|. \end{aligned}$$

From this inequality, we proceed as in Lemma 6.21. □

LEMMA 6.23. *For any positive K, we have*

$$\sup_{Q(R)} \ln(K/\bar{u}) \le C\left(R^{-n-2}\int_{Q(2R)} [\ln^+(\frac{K}{\bar{u}})]^{1/2}\, dX\right)^2. \tag{6.39}$$

PROOF. With $v = \ln^+(K/\bar{u})$, we use the test function $\eta^{\alpha q - n}v^{q-1}/\bar{u}$ and imitate the proof of Lemma 6.19. □

PROOF OF THEOREM 6.18. We apply Lemmata 6.19, 6.20, and 6.21 with $Q((y,s-4R^2),r)$ in place of $Q(Y,r)$ to infer that

$$\left(R^{-n-2}\int_{\Theta(R)} u^\sigma\, dX\right)^{1/\sigma} \le CK$$

for K given by (6.36)$'$. Then Lemmata 6.22 and 6.23 show that

$$CK \le \inf_{Q(R)} \bar{u}$$

for this same choice of K. Combining these two inequalities gives (6.34). □

A stronger form of Theorem 6.18 is sometimes useful in applications. It is proved by a straightforward modification of the proof of that theorem.

COROLLARY 6.24. *Let* (6.32) *hold and suppose that $u \in V$ is a nonnegative supersolution of* (6.21) *in Q. Then for any positive constants θ_i, $i = 1,2,3,4$ and any $\sigma \in [1, 1+2/n)$, there is a constant $C(n,\lambda,\Lambda,\Lambda_2,\theta_i,\sigma)$ such that, if Y_1 and*

Y_2 are points in Q with $s_1-(\theta_1+\theta_3)R^2>s_2$ and $Q(Y_1,(\theta_1+\theta_3)R)\cup Q(Y_2,(\theta_2+\theta_4)R)\subset Q$, then

$$\left(R^{-n-2}\int_{Q(Y_2,\theta_2R)}u^\sigma\,dX\right)^{1/\sigma}\le C(\inf_{Q(Y_1,\theta_1R)}u+kR). \tag{6.33$'$}$$

As in Section VI.5, all of the results in this section remain valid provided the functions b^i, c^i, c^0, f^i, and g lie in appropriate $L^{p,q}$ spaces. See Exercise 6.5.

7. Consequences of the local estimates

Theorems 6.17 and 6.18 imply many useful pointwise properties of weak solutions of parabolic equations. In this section, we discuss a few of them.

The first property is the strong maximum principle for weak subsolutions.

THEOREM 6.25. *Let L satisfy* (6.2) *and*

$$\int_\Omega -b^iD_iv+c^0v\,dX\le 0 \tag{6.40}$$

for all nonnegative $v\in C^1_{\mathscr{P}}$, and let $u\in V$ satisfy $Lu\ge 0$ in Ω. If there is a cylinder $Q=Q(X_0,R)\subset\subset\Omega$ such that $\sup_Q u=\sup_\Omega u\ge 0$, then u is constant in $S(X_0)$, the set of all points $X_1\in\overline{\Omega}\setminus\mathscr{P}\Omega$ such that there is a continuous function $\Gamma:[0,1]\to\overline{\Omega}\setminus\mathscr{P}\Omega$ such that $\Gamma(0)=X_0$, $\Gamma(1)=X_1$, and the t-component of Γ is nonincreasing.

PROOF. We apply Corollary 6.24 to $M-u$, where $M=\sup_\Omega u$. It follows that

$$\int_{Q(Y,r)}(M-u)\,dX\le C\inf_Q(M-u)=0$$

for any cylinder $Q(Y,r)\subset Q$ with $s<t_0$, and hence that $u\equiv M$ in Q. The chaining argument in Theorem 2.7 completes the proof. □

From the strong maximum principle, we also infer the weak maximum principle.

COROLLARY 6.26. *Let L satisfy* (6.2) *and* (6.40), *and let $u\in V$ satisfy $Lu\ge 0$ in Ω. Then $\sup_\Omega u\le\sup_{\mathscr{P}\Omega}u$.*

From Theorems 6.17 and Theorem 6.18, we also conclude a parabolic analog of Harnack's inequality for harmonic functions.

THEOREM 6.27. *Let L satisfy* (6.2). *If $u\in V$ satisfies $Lu=0$ and is nonnegative in $Q(4R)$, then $\sup_{\Theta(R/2)}u\le C\inf_{Q(R)}u$.*

In general, it is not possible to estimate the supremum of u over a cylinder Q in terms of its infimum over the same cylinder. See Exercise 6.7.

A further consequence of the weak Harnack inequality for positive supersolutions is the Hölder continuity of weak solutions of parabolic equations. This result, first proved by Nash, opened up the theory of quasilinear parabolic equations in more than one space dimension.

THEOREM 6.28. *Let L satisfy conditions* (6.2), *and let $u \in V$ be a bounded solution of $Lu = D_i f^i + g$ in $Q(R)$. If f and g are bounded, then u is Hölder continuous. Specifically, if $|u| \le M_0$ in $Q(R)$, and $M_0 |b| + |f| \le k_1$, $M_0 |c^0| + |g| \le k_2$ in $Q(R_0)$, then there are positive constants C and α determined only by n, λ, Λ, and Λ_2 such that*

$$\operatorname*{osc}_{Q(r)} u \le C\left[\left(\frac{r}{R}\right)^{\alpha} \operatorname*{osc}_{Q(R)} u + k_1 r + k_2 r^2\right] \tag{6.41}$$

for $0 < r < R$.

PROOF. Without loss of generality, we may assume that $r < R/4$. Then we define

$$M_1 = \sup_{Q(r)} u, M_4 = \sup_{Q(4r)} u, m_1 = \inf_{Q(r)} u, m_4 = \inf_{Q(r)} u.$$

Then

$$L(M_4 - u) = D_i(M_4 b^i - f^i) + (M_4 c^0 - g),$$
$$L(u - m_4) = D_i(f^i - m_4 b^i) + (g - m_4 c^0).$$

Since $M_4 - u$ and $u - m_4$ are nonnegative in $Q(4r)$, we may apply the weak Harnack inequality to these functions to see that

$$r^{-n-2} \int_{\Theta(r)} (M_4 - u)\, dX \le C_1[(M_4 - M_1) + kr],$$
$$r^{-n-2} \int_{\Theta(r)} (u - m_4)\, dX \le C_1[(m_1 - m_4) + kr]$$

for $k = k_1 + k_2 r$. Adding these inequalities yields

$$M_4 - m_4 \le C_1(M_4 - m_4 + m_1 - M_1 + 2kr),$$

and then, for $\varepsilon = 1 - 1/C_1$ and $\omega(\rho) = \operatorname{osc}_{Q(\rho)} u$, we see that $\omega(r) \le \varepsilon\omega(4r) + 2kr$. This inequality is just (4.15)$''$, so Lemma 4.6 gives the desired result. □

From Exercise 6.5, we see that the coefficients of L can lie in the Lebesgue spaces $L^{p,q}$ for suitable p and q. In fact, we can prove Hölder continuity of weak solutions even if the coefficients lie only in a suitable Morrey space, $M^{p,q}$, defined to be the set of all functions $u \in L^1$ with finite norm

$$\|u\|_{p,q} = \sup_{Q(r)\subset\Omega} \left(r^{-q} \int_{Q(r)} |u|^p \, dX\right)^{1/p}.$$

THEOREM 6.29. *Let a^{ij} satisfy* (6.2a,b), *let $\delta \in (0,1)$, suppose that b^i and f^i are in $M^{2,n+2\delta}$, $c^i \in M^{2,n}$, and c^0 and g are in $M^{1,n+\delta}$, and let u be a weak solution of $Lu = D_i f^i + g$ in $Q(R)$. Then there are constants α and ε_0 determined only by n, λ, and Λ such that, for any $\theta < \alpha$ with $\theta \le \delta$, if $u \in H^*_\theta$, if*

$$B = R^\delta \|b\|_{2,n+2\delta}, C_0 = R^\delta \left\|c^0\right\|_{1,n+\delta} \tag{6.42a}$$

and if there is $r_0 > 0$ such that

$$\int_{Q(X,r)} \left|c^i\right|^2 dX \le \varepsilon_0 r^n \tag{6.42b}$$

for $r \le \min\{r_0, d(X)\}$ and all $X \in Q$, then

$$[u]^*_\theta \le C(B, C_0, n, r_0, \lambda, \Lambda, \theta)(|u|_0 + R^\delta(\|f\|_{2,n+2\delta} + \|g\|_{1,n+\delta}). \tag{6.43}$$

PROOF. Let $M_\theta = [u]^*_\theta$ and set $M_0 = \sup|u|$. Since

$$R^\delta \left\|b^i u\right\|_{2,n+2\delta} \le M_0 B \text{ and } R^\delta \left\|c^0 u\right\|_{1,n+\delta} \le M_0 C_0,$$

we assume without loss of generality that $b^i = 0$ and $c^0 = 0$. We also assume that (6.42b) holds with $\varepsilon_0 \le 1$ and r_0 given. To estimate M_θ, fix $X_0 \in Q(R)$ and take $r \le \min\{\frac{1}{4}d(X_0), r_0\}$. If v solves

$$-v_t + D_i(a^{ij} D_j v) = 0 \text{ in } Q(X_0, r), v = u \text{ on } \mathscr{P}Q(X_0, r).$$

then Theorem 6.28 (along with the argument from Lemma 4.5) implies that there are constants α and C_1 determined only by n, λ, and Λ such that

$$\int_{Q(\rho)} |v - \{v\}_\rho|^2 \, dX \le C_1 \left(\frac{\rho}{r}\right)^{n+2+2\alpha} \int_{Q(r)} |v - \{v\}_r|^2 \, dX \tag{6.44}$$

for $\rho \le r$, where here $Q(\rho) = Q(X_0, \rho)$ and $Q(r) = Q(X_0, r)$. Then we use the test function $w = u - v$ in the weak forms of the equations for u and v to see that

$$\int_{Q(r)} |Dw|^2 \, dX = \int_{Q(r)} |Du - Dv|^2 \, dX$$
$$\le C \int_{Q(r)} [f^i D_i w + (c^i D_i u + g) w] \, dX,$$

and therefore, setting $F = \|f\|_{2,n+2\delta}$ and $G = \|g\|_{1,n+\delta}$,

$$\int_{Q(r)} |Dw|^2 \, dX \le C \int |f|^2 \, dX + C \sup w \int_{Q(r)} \left|c^i D_i u\right| + |g| \, dX$$
$$\le C F^2 r^{n+2\delta}$$
$$+ C \sup w [\varepsilon_0^{1/2} r^{n/2} (\int_{Q(r)} |Du|^2 \, dX)^{1/2} + G r^{n+\delta}].$$

To estimate the integral of $|Du|^2$, we take ζ to be our usual cut-off function in $Q(2r)$ which is bounded away from zero on $Q(r)$, and we set $U = u - \{u\}_r$. By taking the test function $\zeta^2 U$ in the equation for u, we find that

$$\begin{aligned}\int_{Q(r)} |Du|^2\, dX &\le C\int_{Q(2r)} a^{ij} D_i u D_j u \zeta^2\, dX \\ &\le C\int_{Q(2r)} |f^i D_i \zeta U \zeta| + |gU|\zeta^2\, dX \\ &\quad + C\int_{Q(2r)} |c^i D_i u U|\zeta^2\, dX \\ &\quad + C\int_{Q(2r)} a^{ij} D_j u U D_i \zeta\zeta + U^2 \zeta\zeta_t\, dX.\end{aligned}$$

Since $|U| \le 2M_\theta (r/d)^\theta$, it follows that

$$\int_{Q(r)} |Du|^2\, dX \le CM_\theta d^{-\theta}[(G+F)r^{n+\delta+\theta} + M_\theta d^{-\theta} r^{n+2\theta}].$$

Moreover, it is easy to check that $|w| \le CM_\theta (r/d)^\theta$, and hence

$$\int_{Q(r)} |Dw|^2\, dX \le C[(F+G)^2 r^{n+2\delta} + \varepsilon_0^{1/2} (M_\theta d^{-\theta})^2 r^{n+2\theta}].$$

Now we use Poincaré's inequality to see that

$$\int_{Q(r)} w^2\, dX \le C[(F+G)^2 r^{n+2+2\delta} + \varepsilon_0^{1/2} (M_\theta d^{-\theta})^2 r^{n+2+2\theta}]$$

Adding this inequality to (6.44) gives

$$\begin{aligned}\int_{Q(\rho)} |u - \{v\}_\rho|^2\, dX &\le C\left(\frac{\rho}{r}\right)^{n+2+2\alpha} \int_{Q(r)} |u - \{v\}_r|^2\, dX \\ &\quad + C[(F+G)^2 d^{2\delta - 2\theta} + \varepsilon_0^{1/2} (M_\theta d^{-\theta})^2] r^{n+2+2\theta}\end{aligned}$$

and therefore

$$M_\theta \le C[\varepsilon_0^{1/2} M_\theta + (F+G)R^\delta]$$

because $d \le R$. From this estimate, we infer (6.43) by simple algebra. □

8. Boundary estimates

It is possible to define inequalities for functions in V near $\mathscr{P}\Omega$ which allow us to extend our previous results, which were valid in the interior of Ω, to a neighborhood of an arbitrary point in $\mathscr{P}\Omega$. If Δ is a subset of $\mathscr{P}\Omega$, we define $C^1_{\mathscr{P}}(\overline{\Omega}\setminus\Delta)$ to be the set of all functions in $C^1(\overline{\Omega})$ which vanish on $\mathscr{P}\Omega\setminus\Delta$. We then say that $u \le 0$ on Δ if u^+ is the limit in V of a sequence of $C^1_{\mathscr{P}}(\overline{\Omega}\setminus\Delta)$ functions. If u is continuous, this definition is equivalent to the usual pointwise one. We then define $u \le M$ on Δ, for a constant M, to mean that $u - M \le 0$ on Δ and $u \ge M$ to mean that $M - u \le 0$ on Δ. We also define $\inf_\Delta u$ to be the least upper bound of all numbers M such that $u \ge M$ on Δ and $\sup_\Delta u$ is the greatest lower bound of all M such that $u \le M$ on Δ. To simplify our notation, we also define $\Omega[Y,R] = \Omega \cap Q(Y,R)$ and $\mathscr{P}\Omega[Y,R] = \mathscr{P}\Omega \cap Q(Y,R)$ for $R > 0$ and $Y \in \mathbb{R}^{n+1}$, and we supress Y from the notation when it is clear from the context. With these definitions, we have the following analogs of Theorems 6.17 and 6.18.

THEOREM 6.30. *Suppose L, f^i and g satisfy the hypotheses of Theorem 6.17 and that u is a weak solution of $Lu \ge D_i f^i + g$ in Ω. Let $Y \in \mathbb{R}^{n+1}$, let $R > 0$ and set $M = \sup_{\mathscr{P}\Omega[2R]} u^+$. If Ω is a cylinder or if $\mathscr{P}\Omega \in H_1$, then for any $m > 0$, there is a constant $C(n,\lambda,\Lambda,\Lambda_2,m)$ such that*

$$\sup_{\Omega[R]} u \le C[\left(R^{-n-2}\int_{\Omega[2R]} (u^+)^m\, dX\right)^{1/m} + M + kR]. \tag{6.45}$$

PROOF. Replace M by $M + kR$ in the proof of Theorem 6.17. □

THEOREM 6.31. *Suppose L, f^i and g are as in Theorem 6.18 and suppose that u is a weak solution of $Lu \le D_i f^i + g$ in Ω. Let $Y \in \mathbb{R}^{n+1}$ and $R > 0$, and suppose that $u \ge 0$ in $\Omega[4R]$. For $m = \inf_{\mathscr{P}\Omega[2R]} u$, set*

$$u_m(X) = \begin{cases} \inf\{u(X), m\} & \text{if } X \in \Omega \\ m & \text{otherwise.} \end{cases} \tag{6.46}$$

If Ω is as in Theorem 6.30, then for any $\sigma \in [1, 1 + 2/n)$, there is a constant $C(n,\lambda,\Lambda,\Lambda_2,\sigma)$ such that

$$\left(R^{-n-2}\int_{\Theta(R)} (u_m)^\sigma\, dX\right)^{1/\sigma} \le C(\inf_{Q(R)} u_m + kR). \tag{6.47}$$

PROOF. First, we set $\bar{u} = u_m + kR$. In the proof of Lemma 6.19, we use the test function

$$(\bar{u}^{-q} - (m + kR)^{-q})\eta^{\alpha q - n}$$

with $0 < q < 1$ to infer (6.34) again for u_m.

Next, in the proof of Lemma 6.20, we set $v = [\ln(\bar{u}/K) - (\bar{u}/(m + kR))]^+$ and use the test function

$$(v^{q-1} + \Gamma)\left(\frac{1}{\bar{u}} - \frac{1}{m + kR}\right)\eta^{\alpha q - n}.$$

We then obtain

$$\left(R^{-n-2}\int_{Q(3R/2)} \left(\frac{\bar{u}}{K}\right)^r \exp\left(\frac{-\bar{u}}{m + kR}\right) dX\right)^{1/r}$$
$$\le C\exp(R^{-n-2}\int_{Q(2R)} (v^+)^{1/2}\, dX).$$

Since $0 \le \bar{u} \le m + kR$, it follows that $1/e \le \exp(-\bar{u}/(m+kR)) \le 1$, and hence (6.35) is valid in this case as well.

Analogous reasoning to that in Lemmata 6.21–6.23 now implies (6.47). □

Continuity of weak solutions up to $\mathscr{P}\Omega$ does not follow directly from Theorem 6.31 without some geometric restriction on Ω. We shall use a simple one, known merely as condition (A). We say that Ω satisfies *condition (A)* at $X_0 \in \mathscr{P}\Omega$ if there are positive constants $A < 1$ and R such that

$$A\,|Q(X_0,r)| \le |\Omega[X_0,r]| \tag{6.48}$$

for $0 < r < R$. It is easy to check that Ω satisfies condition (A) at each point of $\mathscr{P}\Omega$ if $\mathscr{P}\Omega \in H_1$.

THEOREM 6.32. *Let L, f^i and g be as in Theorem 6.28, let Ω be as in Theorem 6.30, and suppose that Ω satisfies condition (A) at $X_0 \in \mathscr{P}\Omega$. Then there are constants C and α determined only by A, n, λ, Λ, and Λ_2 such that if σ is an increasing function with $\tau^{-\delta}\sigma(\tau)$ decreasing in τ for some $\delta < \alpha$ and $\operatorname{osc}_{\mathscr{P}\Omega[r]} u \le \sigma(r)$, then*

$$\operatorname*{osc}_{\Omega[r]} u \le C\Big(\Big(\frac{r}{R}\Big)^{\alpha} \operatorname*{osc}_{\Omega[R]} u + k_1 r + k_2 r^2 + \sigma(r)\Big) \tag{6.49}$$

for $0 < r < R$.

PROOF. Without loss of generality, we have $r \le R/4$. With

$$M_0 = \sup_{\Omega[R]} |u|,$$

$$M_i = \sup_{\Omega[ir]} u, \quad m_i = \inf_{\Omega[ir]} u$$

for $i = 1, 4$, and

$$m = \inf_{\mathscr{P}\Omega[R]} u,$$

we see from Theorem 6.31 that

$$\frac{1}{\omega_n r^{n+2}} \int_{\Theta(r)} (M_4 - u)_{M_4 - m}\, dX \le C(M_4 - M_1 + kr).$$

On the other hand, we have

$$\frac{1}{\omega_n r^{n+2}} \int_{\Theta(r)} (M_4 - u)_{M_4 - m}\, dX \ge (M_4 - M)\frac{|Q(r)\setminus\Omega|}{|Q(r)|} \ge (1-A)(M_4 - M),$$

so that

$$(1-A)(M_4 - M) \le C(M_4 - M_1 + kr).$$

A similar argument gives

$$(1-A)(m - m_4) \le C(m_1 - m_4 + kr).$$

Adding these estimates and rearranging the resultant inequality leads to

$$M_1 - m_1 \le \gamma(M_4 - m_4) + kr + M - m.$$

The desired result follows from this inequality by virtue of Lemma 4.6. □

In particular, if the restriction of u to $\mathcal{P}\Omega$ is continuous at X_0, then u is continuous at X_0. Moreover, if u is Hölder continuous and if the constants in condition (A) can be taken independent of the point X_0, then u is Hölder continuous in Ω. To simplify the notation, we say that Ω satisfies condition (A) on a subset Δ of $\mathcal{P}\Omega$ if there are positive constants A and R such that (6.48) holds for all $0 < r < R$ and all $X_0 \in \Delta$.

THEOREM 6.33. *Let Δ be a subset of $\mathcal{P}\Omega$ and suppose that Ω satisfies condition (A) on Δ. Under the hypotheses of Theorem 6.32, if there are positive constants K and β such that*

$$\operatorname*{osc}_{\mathcal{P}\Omega[Y,r]} u \le K r^{\beta} \tag{6.50}$$

for any $Y \in \Delta$, then for any compact subset Σ of $\Omega \cup \Delta$, there are positive constants $\alpha(n,\lambda,\Lambda)$ and $C(n,\beta,\theta,\lambda,\Lambda,\Lambda_2,\operatorname{dist}(\Sigma,\mathcal{P}\Omega\setminus\Delta))$ such that

$$|u|_{\theta;\Sigma} \le C(|u|_0 + K + k) \tag{6.51}$$

for $\theta \le \min\{\alpha,\beta\}$.

PROOF. The boundary Hölder estimate of Theorem 6.32 and the interior Hölder estimate of Theorem 6.28 are combined as in Lemma 4.11. □

9. More Sobolev-type inequalities

For the conormal problem in the next section, we shall use extensions of the results in Section VI.4 to functions which need not vanish on $\partial\Omega$ with Ω a subset of $\mathbb{R}^n$. The first such result is a simple extension theorem.

LEMMA 6.34. *Let $1 \le p \le \infty$, and suppose that $u \in W^{1,p}(B^+(R))$ with $u(x) = 0$ if $|x| = R$, then the function $\hat{u}$, defined by*

$$\hat{u}(x) = \begin{cases} u(x',x^n) & \text{if } x^n \ge 0 \\ u(x',-x^n) & \text{if } x^n < 0 \end{cases} \tag{6.52}$$

is in $W_0^{1,p}(B(R))$ and

$$\|D\hat{u}\|_p + \|\hat{u}\|_p \le 2^{1/p}(\|Du\|_p + \|u\|_p). \tag{6.53}$$

PROOF. If $u \in C^1$, then $\hat{u} \in W_0^{1,\infty}$ and $|D\hat{u}(x)| = |Du(x',|x^n|)|$ for $x \in B(R)$. Hence (6.53) holds (as an equality) in this case. The general result follows by an easy approximation argument. □

The next step is a general Sobolev inequality.

THEOREM 6.35. *Let $\partial\Omega \in C^{0,1}$. If $u \in W^{1,p}(\Omega)$ for some $p \in [1,n)$, then there is a constant $C(n,p)$ such that*

$$\|u\|_{np/(n-p)} \leq C(\|Du\|_p + \|u\|_p). \tag{6.54}$$

PROOF. Let (Ω_i) be a finite open cover of $\overline{\Omega}$ such that if Ω_i meets $\partial\Omega$, then $\Omega_i \cap \Omega$ can be mapped by a $C^{0,1}$ homeomorphism onto $B^+(R)$ for some R with $\partial\Omega \cap \Omega_i$ mapped onto $B^0(R)$. If (η_i) is a partition of unity subordinate to this cover, then

$$\|\eta_i u\|_{np/(n-p)} \leq \|\eta_i \hat{u}\|_{np/(n-p)} \leq C\|D(\eta_i \hat{u})\|_p \leq C(\|D(\eta_i u)\|_p + \|\eta_i u\|_p)$$

by Lemma 6.34 and the Sobolev imbedding theorem. The desired result now follows by adding these inequalities. □

Our next estimate relates boundary integrals to interior ones.

LEMMA 6.36. *If $\partial\Omega \in C^{0,1}$ and if $u \in W^{1,1}(\Omega)$, then there is a constant $C(\Omega)$ such that*

$$\int_{\partial\Omega} |u|\, ds \leq C \int_\Omega |u| + |Du|\, dx. \tag{6.55}$$

PROOF. Let (η_i) be a partition of unity as in Theorem 6.35. If η_i is supported in a small neighborhood $\mathcal{N}$ of a point $x_0 \in \partial\Omega$ such that there is a Lipschitz homeomorphism $f : B^+(R) \to \mathcal{N} \cap \Omega$ with $f(B^0) = \mathcal{N} \cap \partial\Omega$, then, for $u_i = \eta_i u$, we have

$$\begin{aligned}
\int_{\partial\Omega} |u_i|\, ds &= \int_{B^0} |u_i(f(y))|\,(1 + |Df(y)|^2)^{1/2}\, dy' \leq C \int_{B^0} |u_i(f(y))|\, dy' \\
&= C \int_{B^+} \frac{\partial\, |u_i(f(y))|}{\partial y^n}\, dy \leq C \int_\Omega |Du_i|\, dx \\
&\leq C \int_\Omega |u| + |Du|\, dx.
\end{aligned}$$

Summing on i gives (6.55). □

For our local estimates, it is useful to have a sharper version of the results in Theorem 6.35 and Lemma 6.36.

THEOREM 6.37. *Let $x_0 \in \partial\Omega$ and suppose that there are positive constants R and K and a function $f \in H_1(B'(R,x_0))$ such that $|Df| \leq K$ and*

$$\Omega[x_0,R] = \{x^n > f(x'), |x - x_0| < R\}, \tag{6.56}$$

Then, for any $p \in [1,n)$, there are constants $C_1(K,n,p)$ and $C_2(K,n)$ such that

$$\|u\|_{np/(n-p)} \leq C_1\, \|Du\|_p \tag{6.54'}$$

for any $u \in W_0^{1,p}(B(R) \cap \overline{\Omega})$ and

$$\int_{\partial\Omega} |u|\, ds \leq C_2 \int_\Omega |Du|\, dx \tag{6.55'}$$

for any $u \in W_0^{1,1}(B(R) \cap \overline{\Omega})$.

PROOF. For both inequalities, we define $v(x) = u(x', x^n - f(x'))$ and note that $v = 0$ on $\{x^n = f(x') + x_0^n + (R^2 - |x' - x_0'|^2)^{1/2}\}$ and that $v \in W^{1,p}(\{0 \leq x^n \leq f(x') + x_0^n + (R^2 - |x' - x_0'|^2)^{1/2}\})$ for (6.54)′ while $v \in W^{1,1}(\{0 \leq x^n \leq f(x') + x_0^n + (R^2 - |x' - x_0'|^2)^{1/2}\})$ for (6.55)′. From Lemma 6.34 applied to v and then the Sobolev imbedding theorem applied to $\hat{v}$, we infer (6.54)′, and integration by parts yields (6.55)′. □

10. Conormal problems

We now examine weak solutions of $Lu = D_i f^i + g$ subject to the boundary condition $\mathscr{M}u = \psi$ on $S\Omega$, where $\mathscr{M}$ is defined by

$$\mathscr{M}u = (a^{ij}D_j u + b^i u - f^i)\gamma_i + b^0 u$$

for a scalar function b^0, where γ is the inner normal to $S\Omega$, and $\Omega = \omega \times (0,T)$ for some domain $\omega \in \mathbb{R}^n$ with $\partial\omega \in C^{0,1}$ and some constant $T > 0$. The weak formulation of $Lu = D_i f^i + g$ in Ω, $\mathscr{M}u = \psi$ on $S\Omega$, $u = \varphi$ on $B\Omega$ is that

$$\int_\Omega [-uv_t + (a^{ij}D_j u + b^i u - f^i)D_i v - (c^i D_i u c^0 u - g)v]\, dX$$

$$= \int_{S\Omega} (b^0 u + \psi) v\, ds\, dt + \int_{\omega(0)} \varphi v\, dx$$

for all $v \in C^1(\overline{\Omega})$ which vanish on $\omega(T)$. It follows from Lemma 6.36 that the integral over $S\Omega$ is defined for all u and v in V.

Some comments are in order concerning this oblique derivative problem. First, the use of cylindrical domains is not essential to the theory, but noncylindrical domains cannot be handled in a simple fashion for the following reason. In deriving our integral identity, we need to integrate by parts to obtain

$$\int_\Omega u_t \eta\, dX = -\int_\Omega u\eta_t + \int_{\partial\Omega} u\eta\Gamma\, d\sigma,$$

where $d\sigma$ denotes surface measure on $\partial\Omega$ and Γ is the t-component of the space-time normal to the surface $\partial\Omega$. Now let us assume that η has support in a small neighborhood of a point in $\partial\Omega$, which we take to be the origin and that, in the support of η, we can write Ω as the graph of the inequality $x^n > g(x', t)$ for some (smooth) function g. Then

$$d\sigma = (1 + |D'g|^2 + g_t^2)^{1/2}\, dx'\, dt, \Gamma = g_t/(1 + |D'g|^2 + g_t^2)^{1/2}.$$

Therefore

$$\int_{\partial\Omega} u\eta\Gamma\, d\sigma = \int u\eta g_t\, dx'\, dt.$$

As this last integral is only defined if $g_t \in L^\infty$ (or if $u\eta$ and g_t are suitably related), the appropriate boundary regularity for a noncylindrical conormal theory would not be one of our H_a spaces. For this reason, we shall investigate the conormal problem only in cylindrical domains. In addition, we note that none of the terms $a^{ij}D_j u\gamma_i$, $b^i u\gamma_i$, $-f^i\gamma_i$ has an intrinsic meaning on $S\Omega$, nor does their sum, because they are all restrictions to $S\Omega$ of $L^2(\Omega)$ functions. Hence the boundary condition can only be defined via the integral identity. Of course if the functions in the problem are all sufficiently smooth, the integral formulation agrees with the pointwise one.

The results proved (in previous sections) for the Cauchy-Dirichlet problem all have analogs for the conormal problems which are proved by similar methods.

THEOREM 6.38. *If $u \in V$ is a weak solution of $Lu = D_i f^i + g$ in Ω, $\mathscr{M}u = \psi$ on $S\Omega$ with $u = \varphi$ on $B\Omega$, then*

$$\|u\|_V \le Ce^{CT}(\|f\|_2 + \|g\|_2 + \|\psi\|_2 + \|\varphi\|_2). \tag{6.57}$$

PROOF. By approximating as in Theorem 6.1, we see that

$$\int_{\omega(\tau)} u^2\,dx + \int_{\Omega(\tau)} |Du|^2\,dX \le C(F^2 + \int_{\Omega(\tau)} u^2\,dX + \int_{S\Omega(\tau)} u^2\,ds\,dt)$$

for $F = \|f\|_2 + \|g\|_2 + \|\psi\|_2 + \|\varphi\|_2$. Lemma 6.36 and Cauchy's inequality then imply that

$$\int_{S\Omega(\tau)} u^2\,ds\,dt \le C(1+\frac{1}{\varepsilon})\int_{\Omega(\tau)} u^2\,dX + C\varepsilon\int_{\Omega(\tau)} |Du|^2\,dX$$

for any $\varepsilon > 0$. By taking ε sufficiently small, we have

$$\int_{\omega\times\{\tau\}} u^2\,dx + \int_\Sigma |Du|^2\,dX \le CF^2 + C\int_\Sigma u^2\,dX,$$

and (6.57) follows by Gronwall's inequality. □

THEOREM 6.39. *If $\varphi \in L^2(\omega)$ and if the coefficients of L and $\mathscr{M}$ are bounded, then there is a unique solution of $Lu = D_i f^i + g$ in Ω, $\mathscr{M}u = \psi$ on $S\Omega$, $u = \varphi$ on $B\Omega$.*

PROOF. We proceed as in Theorem 6.2 with Theorem 5.18 in place of Theorem 5.6, noting that we can approximate so that the approximations to φ and ψ are zero near $C\Omega$. □

For the pointwise estimates, we use the structure from Sections VI.5 and VI.6, noting that the boundary condition $\mathscr{M}u = \psi$ can be written as

$$A(X,u,Du)\cdot\gamma + \Psi(x,u) = 0 \tag{6.58}$$

for $\Psi(X,z) = b^0 u - \psi$. The structure for Ψ is then that

$$z\Psi(X,z) \le (\Lambda_1 + 1)\,|z|^2 \tag{6.59}$$

for $|z| \ge k$, which corresponds to $|\psi| \le k\lambda$ and $b^0 \le \Lambda_1^2\lambda$. Then our global boundedness result takes the following form.

THEOREM 6.40. *Let A, B, and* Ψ *satisfy* (6.22) *and* (6.59) *for* $|z| \geq k$. *If* $u \in V$ *is a weak subsolution of* (6.21), (6.58) *with* $u \leq M$ *on* $B\Omega$, *then*

$$\sup_{\Omega} u \leq C(n,\lambda,\Lambda_1)(M+k). \tag{6.60}$$

PROOF. Following the proof of Theorem 6.15, we first obtain (6.26) with an additional term of

$$C(1+k_1)\int_0^\tau \int_{\partial\omega} \chi(u)(u-k)\,ds\,dt$$

on the right side. Now we use Lemma 6.36, Cauchy's inequality, and (6.25) to infer that

$$\int_{\partial\omega}(u-k)\chi\,ds \leq C(1+\frac{1}{\varepsilon})(1+k_1)^2\int_{\Omega}\chi u^2\,dx + C\varepsilon\int_{\Omega}\chi\,|Du|^2\,dx.$$

It follows that

$$\int_{\omega(\tau)}\chi(u)(u-k)^2\,dx + \int_{\Omega(\tau)}\chi(u)\,|Du|^2\,dX$$
$$\leq C(n,\lambda,\Lambda_1)(1+k_1)^3\int_{\Omega(\tau)}\chi(u)u^2\,dX$$

and the proof of Theorem 6.15 is easily modified to use this inequality in place of (6.26). □

Local estimates are proved by using the same test function arguments as in Section VI.6 and estimating the boundary integrals as in Theorem 6.40.

THEOREM 6.41. *Let A, B and* Ψ *satisfy* (6.30) *and* (6.59) *for* $|z| \geq kR$ *and* $X \in Q[2R]$. *If u is a subsolution of* (6.21), (6.58) *in* $\Omega[2R]$, *then for any* $m > 0$, *there is a constant* $C(m,n,\lambda,\Lambda,\Lambda_2)$ *such that*

$$\sup_{\Omega[R]} u \leq C\left[(R^{-n-2}\int_{\Omega[2R]}(u^+)^m\,dX)^{1/m} + kR\right]. \tag{6.61}$$

THEOREM 6.42. *Let A and B satisfy* (6.32), *let* Ψ *satisfy*

$$\bar{z}\Psi(X,z) \geq -\Lambda_2\lambda\frac{|\bar{z}|^2}{R^2}, \tag{6.59$'$}$$

and suppose that $u \in V$ *is a supersolution of* (6.21), (6.58) *which is nonnegative in* $\Omega[4R]$. *Then for any* $\sigma \in [1, 1+2/n)$, *there is a constant* $C(n,\lambda,\Lambda,\Lambda_2,\sigma,\Omega)$ *such that*

$$\left(R^{-n-2}\int_{\Theta(R)\cap\Omega}u^\sigma\,dX\right)^{1/\sigma} \leq C(\inf_{\Omega[R]} u + kR). \tag{6.62}$$

We also have a strong maximum principle and a weak maximum principle for the conormal derivative problem.

THEOREM 6.43. *Let L satisfy* (6.2) *and suppose* (6.40) *is satisfied for all nonnegative $v \in C^1(\overline{\Omega})$. Suppose also that $b^0 \le 0$, and let $u \in V$ satisfy $Lu \ge 0$ in Ω and $\mathscr{M}u \ge 0$ on $S\Omega$. If there is a point $X_0 \in \overline{\Omega} \setminus B\Omega$ such that*

$$\limsup_{X \to X_0} u = \sup_{\Omega} u \ge 0, \tag{6.63}$$

then u is constant in $S(X_0)$, the set of all points $X_1 \in \overline{\Omega} \setminus \mathscr{P}\Omega$ such that there is a continuous functions $\Gamma : [0,1] \to \overline{\Omega} \setminus \mathscr{P}\Omega$ such that $\Gamma(0) = X_0$, $\Gamma(1) = X_1$, and the t-component of Γ is nonincreasing. In any case, $\sup_{\Omega} u \le \sup_{B\Omega} u^+$.

In addition, the obvious analog of the Harnack inequality, Theorem 6.27, holds for solutions of the conormal problem. We leave its statement to the reader. Of primary importance to the quasilinear theory is the following Hölder estimate.

THEOREM 6.44. *Let L satisfy conditions* (6.2), *and let u be a bounded weak solution of $Lu = D_i f^i + g$ in $\Omega[R]$, $\mathscr{M}u = \psi$ on $S\Omega[R]$. If f, g, and ψ are bounded, then u is Hölder continuous. Specifically, if $|u| \le M_0$, $M_0(|b| + |c^0|) + |f| \le k_1$, and $|g| \le k_2$ in $\Omega[R]$ and if $M_0 |b^0| \le k_2$ on $S\Omega[R]$, then there are positive constants C and α determined only by n, λ, Λ, and Λ_2 such that*

$$\operatorname*{osc}_{\Omega[r]} u \le C\left[\left(\frac{r}{R}\right)^{\alpha} \operatorname*{osc}_{\Omega[R]} u + k_1 r + k_2 r^2\right] \tag{6.64}$$

for $0 < r < R$.

As in the previous sections, the boundedness of the coefficients of the differential equation can be relaxed considerably.

11. A special mixed problem

All of our results have analogs for mixed boundary value problems. For our applications, we only need to examine a simple mixed boundary value problem for weak solutions. In particular, following the notation of Section IV.8, we use Σ^+ to denote either $\Sigma^+(X_0,R)$ or $\Sigma^+(X_0;R,r)$. For a constant matrix (a^{ij}), we define the operators L_0 and $\mathscr{M}_0$ by

$$L_0 u = -u_t + D_i(a^{ij} D_j u)$$

in Σ^+ and

$$\mathscr{M}_0 u = a^{nj} D_j u$$

on Σ^0. Writing C^1_σ for the set of all $C^1(\overline{\Omega})$ functions which vanish on σ, we define V_m to be the closure in V of C^1_σ. We then say that $u \in V_m$ is a weak solution of

$$L_0 u = D_i f^i \text{ in } \Omega,\ \mathscr{M}_0 u = f^n \text{ on } \Sigma^+,\ u = 0 \text{ on } \sigma, \tag{6.65}$$

for $\Omega = \Sigma^+$, if

$$\int_{\omega(\tau)} uv\,dx + \int_{\Omega(\tau)} [-uv_t + (a^{ij} D_j u - f^i) D_i v]\,dX = 0$$

for all $v \in C^1_\sigma$. Here we make explicit use of the observation that $f \in L^2$ does not have a trace on Σ^0. It's then easy to verify the energy estimate.

THEOREM 6.45. *If* $u \in V_m$ *is a weak solution of* (6.65) *and if* (a^{ij}) *satisfies* (6.2a), *then*

$$\|u\|_V \leq \frac{2}{\lambda^{1/2}}\|f\|_2. \tag{6.66}$$

From Lemma 4.34, it follows that (6.65) has a weak solution (which is in $H_{1+\alpha}$ for some $\alpha > 0$) if $f^i \in C^2$ and then the usual approximation argument gives our existence result.

THEOREM 6.46. *For any* $f \in L^2$, *there is a unique weak solution of* (6.65).

Finally, we have the following local maximum principle.

THEOREM 6.47. *Let* a^{ij} *satisfy* (6.2a,b). *If* $u \in W^{1,2}(\Omega)$ *satisfies* $L_0 u \geq 0$ *in* $\Omega[2R]$, $\mathscr{M}_0 u \geq 0$ *on* $\Sigma^0[2R]$, *and* $u \leq 0$ *on* $\sigma[2R]$, *then there is a constant* $C(n,\lambda,\Lambda)$ *such that*

$$\sup_{\Omega[R]} u \leq C\left(R^{-n-2}\int_{\Omega[2R]}(u^+)^2\,dX\right)^{1/2}. \tag{6.67}$$

12. Solvability in Hölder spaces

In Chapter V, we proved the existence of solutions for initial-boundary value problems when the differential equation is in nondivergence form under exactly the hypotheses that led to estimates in one of the spaces $H_{2+\alpha}$ or $H^{(-\delta)}_{2+\alpha}$ (with $\alpha \in (0,1)$ and $\delta \in (1,2)$) in Chapter IV. Chapter IV also contained estimates for solutions of initial-boundary value problems when the differential equation is in divergence form, but there were no corresponding existence results in Chapter V. The main reason for this lack is that the solutions of these problems need not be in $W^{1,2}$, which was the only space available there for weak solutions. In this section, we prove the existence of weak solutions to those problems not already covered in Chapter V.

We begin by noting that the proofs of the theorems in Chapter IV, which were given for $W^{1,2}$ weak solutions, are also valid for weak solutions. In all cases, the proofs are either the same or simpler.

Next, we recall the hypotheses under which estimates were proved in Chapter IV. For $\alpha \in (0,1)$ and a given domain Ω, we assume that a^{ij} and b are in H_α and that $c^j \in M^{1,n+1+\alpha}$ for $j = 0,\dots,n$. We also suppose that

$$[a^{ij}]_\alpha + [b]_\alpha + \|c^j\|_{1,n+1+\alpha} \leq \Lambda_1 \tag{6.68}$$

for some nonnegative constant Λ_1. We then have the following existence, uniqueness, and regularity theorem.

THEOREM 6.48. *Suppose $\mathscr{P}\Omega \in H_{1+\alpha}$ and that the coefficients of L satisfy conditions* (6.2a,b) *and* (6.68). *Then for any $\varphi \in H_{1+\alpha}$, $f \in H_\alpha$ and $g \in M^{1,n+1+\alpha}$, there is a unique H_1 weak solution of $Lu = D_i f^i + g$ in Ω, $u = \varphi$ on $\mathscr{P}\Omega$. Moreover, $u \in H_{1+\alpha}$ and*

$$|u|_{1+\alpha} \le C(n,\alpha,\lambda,\Lambda,\Lambda_1,\Omega)(|\varphi|_{1+\alpha} + |f|_\alpha + \|g\|_{1,n+1+\alpha}). \tag{6.69}$$

PROOF. We first note from Exercise 4.12 that $u \in H_1$ implies $u \in H_{1+\alpha}$.

To prove (6.69), we assume without loss of generality that $I(\Omega) = (0,T)$ for some $T > 0$ and we set $F = |\varphi|_{1+\alpha} + |f|_\alpha + \|g\|_{1,n+1+\alpha}$ and $v = u - \varphi$. Then Theorem 4.8 in $\Omega(\varepsilon)$ gives a constant C_1 such that

$$|v|_{1+\alpha;\Omega(\varepsilon)} \le C_1[|v|_{0;\Omega(\varepsilon)} + F]$$

for any $\varepsilon \in (0,T)$, and hence

$$[v]_{1;\Omega(\varepsilon)} \le C_1[|v|_{0;\Omega(\varepsilon)} + F].$$

Since $v = 0$ on $\mathscr{P}\Omega$, it follows that $|v|_{0;\Omega(\varepsilon)} \le \varepsilon[v]_{1;\Omega(\varepsilon)}$, and therefore, if $\varepsilon C_1 \le 1/2$,

$$|v|_{0;\Omega(\varepsilon)} \le \frac{1}{2}[F + |v|_{0;\Omega(\varepsilon)}],$$

so $|v|_{0;\Omega(\varepsilon)} \le F$ for $\varepsilon C_1 \le 1/2$. With this choice of ε, we have

$$|u|_{1+\alpha;\Omega(\varepsilon)} \le C_2 F.$$

Now we set $\Sigma(k) = \Omega((k+1)\varepsilon) \setminus \Omega(k\varepsilon)$ and apply this same argument in $\Sigma(k)$ to see that

$$|u|_{1+\alpha;\Sigma(k)} \le C_2[F + |u|_{1+\alpha;\Sigma(k-1)}]$$

for $k \ge 1$. Since $\Sigma(1) = \Omega(\varepsilon)$, a finite induction gives (6.69).

Existence now follows from the same argument as in Theorem 6.2 and uniqueness is an immediate consequence of (6.69). □

For oblique derivative problems, the same arguments apply and we infer the following result.

THEOREM 6.49. *Suppose L and Ω are as in Theorem 6.48, and let β^0 and β^i be $H_\alpha(S\Omega)$ functions with $\beta \cdot \gamma \ge \chi$ on $S\Omega$ and $|\beta|_\alpha + |\beta^0|_\alpha \le B_1\chi$ for some positive constant B_1. Then for any $\varphi \in H_{1+\alpha}$, $\psi \in H_\alpha$, $f \in H_\alpha$ and $g \in M^{1,n+1+\alpha}$, there is a unique H_1 weak solution of $Lu = D_i f^i + g$ in Ω, $\beta \cdot Du + \beta^0 u = \psi$ on $S\Omega$, and $u = \varphi$ on $B\Omega$. Moreover, $u \in H_{1+\alpha}$ and*

$$|u|_{1+\alpha} \le C(B_1,n,\alpha,\lambda,\Lambda,\Lambda_1,\Omega)(|\varphi|_{1+\alpha} + |\psi|_\alpha + |f|_\alpha + \|g\|_{1,n+1+\alpha}). \tag{6.69$'$}$$

13. The parabolic DeGiorgi classes

The local estimates of Section VI.6 can be proved without using the Moser iteration scheme. In its place, we use an adaptation of DeGiorgi's original proof of local boundedness and Hölder continuity for solutions of elliptic equations.

Before defining the parabolic DeGiorgi classes $PDG^{\pm}$, we introduce some notation for this section only. First, we define the cylinder

$$Q(Y,R,\tau) = \{X : |x-y| < R, s < t < s+\tau R^2\}.$$

(Note that cylinders with a parameter τ are different from cylinders elsewhere in this book, so $Q(Y,R,1)$ is not the same as $Q(Y,R)$.) Next, if $u \in L^1(\Omega)$ and $Q(Y,R,\tau) \subset \Omega$, we define

$$A^{\pm}_{k,R,\tau}(Y) = \{X \in Q(Y,R,\tau) : \pm(u-k)(X) > 0\}.$$

Similarly, we write

$$A^{\pm}_{k,R}(Y) = \{X \in Q(Y,R) : \pm(u-k)(X) > 0\}.$$

When Y is clear from the context, we shall not include it in our notation. Then the parabolic DeGiorgi class PDG^{+} is the subset of all $u \in V(\Omega)$ for which there are positive constants $\varepsilon \leq \frac{2}{n+2}$ and α, and an increasing function χ such that

$$\begin{aligned} \sup_{s-R^2<t<s} \int_{B(\sigma R)\times\{t\}} [(u-k)^+]^2\,dx + \int_{Q(\sigma R)} \left|D(u-k)^+\right|^2 dX \\ \leq \alpha \frac{1}{(1-\sigma)^2R^2} \int_{Q(R)} \left|(u-k)^+\right|^2 dX \\ + \alpha[\chi(R)^2 + R^{-(n+2)\varepsilon}k^2] \left|A^{+}_{k,R}\right|^{1+\varepsilon-2/(n+2)} \end{aligned} \tag{6.70a}$$

for all $\sigma \in (0,1)$, all $k > 0$ and all $Q(R) = Q(Y,R) \subset \Omega$. We define PDG^{-} somewhat differently. It is the subset of all $u \in V(\Omega)$ for which there are positive constants $\varepsilon \leq \frac{2}{n+2}$ and α, and an increasing function χ such that

$$\begin{aligned} \sup_{s-\sigma R^2<t<s} \int_{B(\sigma R)\times\{t\}} [(u-k)^-]^2\,dx + \int_{Q(\sigma R)} \left|D(u-k)^-\right|^2 dX \\ \leq \alpha \frac{1}{(1-\sigma)^2R^2} \int_{Q(R)} \left|(u-k)^-\right|^2 dX \\ + \alpha[\chi(R)^2 + R^{-(n+2)\varepsilon}k^2] \left|A^{-}_{k,R}\right|^{1+\varepsilon-2/(n+2)} \end{aligned} \tag{6.70b}$$

for all $\sigma \in (0,1)$, all $k > 0$ and all $Q(R) = Q(Y,R) \subset \Omega$ and also

$$\begin{aligned}\sup_{s<t<s+\tau R^2} \int_{B(\sigma R)\times\{t\}} [(u-k)^-]^2\,dx \le &\int_{B(\sigma R)\times\{s\}} [(u-k)^-]^2\,dx \\ &+ \alpha\frac{1}{(1-\sigma)^2R^2}\int_{Q(R,\tau)} \left|(u-k)^-\right|^2 dX \\ &+ \alpha[\chi(R)^2 + R^{-(n+2)\varepsilon}k^2]\left|A^-_{k,R,\tau}\right|^{1+\varepsilon-2/(n+2)}\end{aligned} \tag{6.70c}$$

for all $\sigma \in (0,1)$, all $k > 0$ and all $Q(R,\tau) = Q(Y,R,\tau) \subset \Omega$.

Now we show that weak subsolutions of $Lu = D_i f^i + g$ are in PDG^+ if $|b^i|^2$, $|c^i|^2$, $|c^0|$, $|f^i|^2$, and $|g|$ are in L^q with $q > (n+2)/2$. Let $\zeta \in C^1_{\mathscr{P}}(\overline{Q(R)})$ with $\zeta \equiv 1$ in $Q(\sigma R)$, and $|D\zeta|^2 + |\zeta_t| \le C[(1-\sigma)R]^{-2}$ and $0 \le \zeta \le 1$ in $Q(R)$, and use the test function $\zeta^2[u-k]^+$. Some simple rearrangement along with the Sobolev inequality shows that (6.70a) holds with

$$\varepsilon = \frac{2}{n+2} - \frac{1}{q},\ \chi(R) = \|f\|_{2q} + R^{(n+2)\varepsilon/2}\|g\|_q,$$

and α determined only by n, q, λ, Λ, and

$$R^{(n+2)\varepsilon/2}[\|b^i\|_{2q} + \|c^i\|_{2q}] + R^{(n+2)\varepsilon}\|c^0\|_q.$$

Similarly, supersolutions satisfy (6.70b). The proof of (6.70c) is similar except that ζ is taken as a function of x only.

We start by proving a local maximum principle.

THEOREM 6.50. *If $u \in PDG^\pm(Q(r))$, then*

$$\sup_{Q(r/2)} u^\pm \le C(n,\alpha)\left([r^{-n-2}\int_{Q(r)} (u^\pm)^2\,dX]^{1/2} + \chi(r)r^{(n+2)\varepsilon/2}\right). \tag{6.71}$$

PROOF. Without loss of generality, we assume that $u \in PDG^+$ and we abbreviate $Q^-(R)$ to $Q(R)$. For K a positive constant to be further specified and any positive integer N, we set

$$\begin{aligned}k_N = (1-2^{-N})K,\ R_N = (1+2^{-N})r/2,\\ \overline{R}_N = (R_N + R_{N+1})/2,\ \rho_N = 2^{-N-3}r.\end{aligned}$$

Since $R_N - \overline{R}_N = \rho_N$, we can find functions $\zeta_N \in C^1_{\mathscr{P}}(Q(R_N))$ such that $\zeta_N \equiv 1$ on $Q(\overline{R}_N)$ with $0 \le \zeta_N \le 1$ and $|D\zeta_N| \le 2^{N+3}r$ in $Q(R_N)$. Writing $A(N) = A^+_{k_{N+1},R_N}$,

$A_N = |A(N)|$, $Q_N = Q(R_N)$, $\tilde{Q}_N = Q(\overline{R}_N)$, and $v = \zeta_N(u-k_{N+1})^+$, we have

$$\begin{aligned}
\int_{Q_{N+1}} [(u-k_{N+1})^+]^2\,dX &\le \int_{\tilde{Q}_N} v^2\,dX \\
&\le \left(\int_{\tilde{Q}_N} |v|^{2(n+2)/n}\,dX\right)^{n/(n+2)} A_N^{2/(n+2)} \\
&\le C\left(\sup_t \int_{B(\overline{R}_N)\times\{t\}} v^2\,dx + \int_{\tilde{Q}_N} |Dv|^2\,dX\right) A_N^{2/(n+2)} \\
&\le C\left(\frac{1}{\rho_N^2}\int_{A(N)} |u-k_{N+1}|^2\,dX + r^{-(n+2)\varepsilon}K^2 A_N^{1+\varepsilon-2/(n+2)}\right) A_N^{2/(n+2)} \\
&\le C\left(\frac{1}{\rho_N^2}\int_{Q_N} |(u-k_N)^+|^2\,dX + r^{-(n+2)\varepsilon}K^2 A_N^{1+\varepsilon-2/(n+2)}\right) A_N^{2/(n+2)}
\end{aligned}$$

provided $K \ge \chi R^{(n+2)\varepsilon/2}$. Let us also suppose that

$$K^2 \ge r^{-n-2}\int_{Q(r)} (u^+)^2\,dX$$

and set

$$Y_N = K^{-2}r^{-n-2}\int_{Q_N} [(u-k_N)^+]^2\,dX.$$

It is simple to check that $Y_N \le Y_0$ for all N and that $A_N \le C4^N r^{n+2} Y_N$. Because $Y_0 \le 1$, we infer that

$$Y_{N+1} \le C16^N Y_N^{1+\varepsilon}.$$

Setting $a(N) = [(1+\varepsilon)^N - 1]/\varepsilon$ and $b(N) = (1+\varepsilon)^N$, we see by a simple induction argument that

$$Y_N \le C^{a(N)}16^{(a(N)-N)/\varepsilon} Y_0^{b(N)}$$

and hence $Y_N \to 0$ as $N \to \infty$ if

$$Y_0 \le C_0 = C^{-1/\varepsilon}16^{-1/\varepsilon^2}.$$

Now we take

$$K = \max\{\chi R^{(n+2)\varepsilon/2}, \left(1+\frac{1}{C_0}\right)\left(r^{-n-2}\int_{Q(r)} (u^+)^2\,dX\right)^{1/2}\}$$

to conclude that $Y_N \to 0$ from which (6.71) follows easily. □

Note that only inequality (6.70b) is needed for the supersolution case of Theorem 6.50.

The complete proof of the weak Harnack inequality for nonnegative functions in PG^- requires a large amount of theory which we defer to Chapter VII, so we shall only sketch the proof; however, with the tools already presented, we can prove a Hölder estimate for weak solutions of (6.21). The key to both results is the following estimate.

LEMMA 6.51. *Suppose* $u \in PDG^-(\Omega)$ *is nonnegative in* $Q(X_0, r, \tau) \subset \Omega$ *and set* $\bar{u} = u + \chi r^{(n+2)\varepsilon/2}$. *If there are positive constants* $\mu < 1$ *and* K *such that*

$$|\{x \in B(r) : \bar{u}(x, t_0) < K\}| \leq \mu |B(r)|, \tag{6.72}$$

then for every $\xi \in (\mu^{1/2}, 1)$ *and* $b \in (\mu/\xi^2, 1)$*, there is a constant* $\theta \in (0,1)$ *determined only by* b, n, α, ε, ξ, *and* μ *such that*

$$|\{x \in B(r) : \bar{u}(x, t_0) < (1-\xi)K\}| \leq b|B(r)|, \tag{6.73a}$$

for

$$t_0 < t + \min\{\tau, \theta\} r^2. \tag{6.73b}$$

PROOF. If $K \leq \chi r^{(n+2)\varepsilon/2}$, then (6.72) and (6.73a) hold with $\mu = b = 0$ so we may assume that $K > \chi r^{(n+2)\varepsilon/2}$. Let $\eta = \min\{\tau, \theta\}$ and $k = K - \chi r^{(n+2)\varepsilon/2}$. We abbreviate $A = A^-_{k,r,\eta}$ and $Q = Q(r, \eta)$. Then, for t satisfying (6.73b), we infer from (6.70c) that

$$\begin{aligned} \int_{B(\sigma r) \times \{t\}} |(K - \bar{u})^+|^2 \, dx \leq & \int_{B(r) \times \{t_0\}} |(K - \bar{u})^+|^2 \, dx \\ & + \frac{\alpha}{(1-\sigma)^2 r^2} \int_Q |(K - \bar{u})^+|^2 \, dX \\ & + 2\alpha r^{-(n+2)\varepsilon} K^2 |A|^{1+\varepsilon - 2/(n+2)} \end{aligned}$$

for any $\sigma \in (0,1)$.

Using (6.72) and direct calculation to estimate the terms in this inequality, we see that

$$\begin{aligned} & |\{x \in B(r) : \bar{u}(X) > (1-\xi)K\}| \xi^2 \\ & \qquad \leq |B(r)| [\mu + \frac{\alpha\theta}{(1-\sigma)^2} + 2\alpha\theta^{1-\varepsilon+2/(n+2)} + (1 - \sigma^n)]. \end{aligned}$$

We now take $\sigma < 1$ close enough to 1 that $\mu + (1 - \sigma^n) < b\xi^2$, and then θ small enough that

$$\mu + \frac{\alpha\theta}{(1-\sigma)^2} + 2\alpha\theta^{1-\varepsilon+2/(n+2)} + (1 - \sigma^n) \leq b\xi^2$$

to complete the proof. □

Lemma 6.51 says that if $\bar{u}$ is bounded from below by a given constant on a fixed fraction of $B(r) \times \{t_0\}$, then $\bar{u}$ is bounded from below by some other constant on a fixed fraction of $B(r) \times \{t\}$ provided t is close enough to t_0. Our next step is to show that we can bound $\bar{u}$ from below by some constant on a prescribed fraction of $Q^+(r, \tau)$ for τ sufficiently small.

LEMMA 6.52. *Let u, $\bar{u}$, μ, θ, and ξ be as in Lemma 6.51 and set $b = (1+\mu/\xi^2)/2$. If $Q((x_0,t_0-3\eta r^2),2r,\eta) \subset \Omega$ for some $\eta \in (0,\min\{\theta,\tau\}]$, then for any $\nu \in (0,1)$, there is a positive integer σ determined only by n, α, ε, μ, ξ, θ, and ν such that*

$$|\{X \in Q(r,\eta) : \bar{u}(X) < (1-\xi)^\sigma K\}| \le \nu |Q(r,\eta)|. \tag{6.74}$$

PROOF. With σ to be determined, we have

$$\begin{aligned}\left|\{x \in B(r) : \bar{u}(x,t_0) < (1-\xi)^i K\}\right| &\le |\{x \in B(r) : \bar{u}(x,t_0) < K\}| \\ &\le \mu |B(r)|\end{aligned}$$

for $i = 0,\dots,\sigma-1$, so Lemma 6.51 implies that

$$\left|\{x \in B(r) : \bar{u}(x,t_0) < (1-\xi)^{i+1} K\}\right| \le b|B(r)|.$$

To continue, we fix i and set

$$v = \begin{cases} (1-\xi)^i K - (1-\xi)^{i+1} K & \text{if } \bar{u} > (1-\xi)^i K \\ (1-\xi)^i K - \bar{u} & \text{if } (1-\xi)^i K \ge \bar{u} > (1-\xi)^{i+1} K \\ 0 & \text{if } \bar{u} \le (1-\xi)^{i+1} \end{cases}$$

and

$$f(X) = \begin{cases} \frac{1}{|\{x \in B(r) : v(X) = 0\}|} & \text{if } v(X) = 0 \\ 0 & \text{otherwise} \end{cases}$$

for t satisfying $t_0 < t < t_0 + \eta r^2$. Then Proposition 6.14 applied in $B(r) \times \{t\}$ with $p = 1$ implies that

$$\int_{B(r)\times\{t\}} v\,dx \le C(n,b) r \int_{A_i(t)\setminus A_{i+1}(t)} |Du|\,dx,$$

where $A_i(t)$ denotes the subset of $B(R) \times \{t\}$ on which $\bar{u} < (1-\xi)^i K$. We then integrate this inequality with respect to t and write $A(i)$ for the subset of $Q(r,\eta)$ on which $\bar{u} < (1-\xi)^i K$. Estimating the integral of v from below as in Lemma 6.51, we find that

$$\begin{aligned}\xi(1-\xi)^i K |A(i+1)| &\le C(n,b) r \int_{A(i)\setminus A(i+1)} |Du|\,dX \\ &\le Cr \left(\int_{A(i)} |Du|^2\,dX\right)^{1/2} (|A(i)| - |A(i+1)|)^{1/2}.\end{aligned}$$

To estimate the integral of $|Du|^2$, we use (6.70b) with $\sigma = 1/2$. Since $\left|A^-_{k,2r,\eta}\right| \le C(n,\eta) r^{n+2}$, we infer that

$$\xi(1-\xi)^i K |A(i+1)| \le C(1-\xi)^i K r^{(n+2)/2} [|A(i)| - |A(i+1)|]^{1/2}.$$

Now we set $A_i = r^{-n-2} |A(i)|$ and rewrite this inequality as

$$A_{i+1}^2 \le C(n,\alpha,\varepsilon,\xi,\eta)[A_i - A_{i+1}].$$

If we now add these inequalities for i from 1 to σ and note that $A_{i+1} \geq A_{\sigma+1}$, $A_0 \leq 1$ and $A_{\sigma+1} \geq 0$, we conclude that $\sigma A_{\sigma+1}^2 \leq C$ and therefore

$$|\{X \in Q(r,\eta) : \bar{u} < (1-\xi)^\sigma K\}| \leq C\sigma^{-1/2}|Q(r,\eta)| \leq \nu|Q(r,\eta)|$$

provided σ is sufficiently large. □

Our next step is to show that if $\bar{u}$ is bounded from below on a large enough fraction of $Q(r,\tau)$, then $\bar{u}$ is bounded from below on all of $Q(r/2,\tau)$.

LEMMA 6.53. *Suppose $u \in PDG^-(Q(r,\tau))$ is nonnegative with $\tau < \frac{1}{2}$. Then there is a constant $\bar{\mu} \in (0,1)$ determined only by n, α, ε, and τ such that if*

$$|\{X \in Q(r,\tau) : \bar{u} < K\}| \leq \bar{\mu}|Q(r,\tau)| \tag{6.75}$$

for some $K > 0$, then $\bar{u} \geq K/2$ on $Q((y,s+\tau r^2/4),r/2,\tau)$. Moreover, $\bar{\mu}$ is an increasing function of τ.

PROOF. Fix $Z \in Q^* = Q((y,s+\tau r^2/4),r/2,\tau)$ and apply Theorem 6.50 in $Q(Z,\tau r)$ to infer that

$$\begin{aligned}\sup_{Q(Z,\tau r/2)}(K-\bar{u}) &\leq C[\frac{1}{|Q(\tau r)|}\int_{Q^+(Z,\tau r)}|(K-\bar{u})^+|^2\,dX] \\ &\leq C[\frac{1}{|Q(\tau r)|}\int_{Q^*}|(K-\bar{u})^+|^2\,dX]\end{aligned}$$

since $Q(Z,\tau r) \subset Q^*$. Now we note that $(K-\bar{u})^+ \leq K$ and that $|Q(\tau r)| = \tau^n|Q^*|$. Therefore

$$K - \bar{u}(Z) \leq \sup_{Q(Z,\tau r/2)}(K-\bar{u}) \leq C\tau^{-n/2}\bar{\mu}^{1/2}K.$$

The proof is completed by choosing $\bar{\mu}$ so small that $C\tau^{-n/2}\bar{\mu}^{1/2} \leq 1/4$. □

From these results, we infer the following estimate, which leads easily to a Hölder estimate.

PROPOSITION 6.54. *Suppose u is a weak solution of (6.21) with $0 \leq u \leq M$ in $Q(Y,R,\tau)$. then there are positive constants c_0, θ_0, and ξ_0 with $\xi_0 < 1$ such that $\tau \leq \theta_0$ and $M > c_0\chi r^{(n+2)\varepsilon/2}$ imply that either*

$$\xi_0 M \leq u \leq M \tag{6.76a}$$

or

$$0 \leq u \leq (1-\xi_0)M \tag{6.76b}$$

in $B(R/4) \times (s+\frac{15}{16}\tau R^2, s+\tau R^2)$.

PROOF. Without loss of generality, we may assume that $c_0 \geq 4$ and that

$$\left|\{X \in Q(R,\tau) : u \leq \frac{M}{2}\}\right| \leq \frac{1}{2}|Q(R,\tau)|.$$

Then there is a t_1 in the interval $(s, s+\frac{2}{3}\tau R^2)$ such that

$$\left|\{x \in B(R) : \bar{u}(x,t_1) < \frac{M}{4}\}\right| \leq \frac{3}{4}|B(R)|$$

because, otherwise,

$$\begin{aligned}\left|\{X \in Q(R,\tau) : u \leq \frac{M}{2}\}\right| &\geq \left|\{X \in Q(R,\tau) : u < \frac{M}{4}\}\right| \\ &= \int_s^{s+\tau R^2} \left|\{x \in B(R) : \bar{u}(x,t) < \frac{M}{4}\}\right| dt \\ &\geq \int_s^{s+2\tau R^2/3} \left|\{x \in B(R) : \bar{u}(x,t) < \frac{M}{4}\}\right| dt \\ &> \frac{2}{3}\tau R^2 \frac{3}{4}|B(R)| = \frac{1}{2}|Q(R,\tau)|.\end{aligned}$$

Now for $K = M/4$, $\mu = 3/4$, $\xi = 5/8$, and $b = (1+\mu/\xi^2)/2$, take θ_0 to be the constant θ from Lemma 6.51. Then use $\nu = \bar{\mu}$, the corresponding constant from Lemma 6.53, in Lemma 6.52 to infer that $\bar{u} > (1-\xi)^\sigma M/2$ in $B(R/4) \times (s + \frac{15}{16}\tau R^2, s+\tau R^2)$. This inequality then gives (6.76a) with $c_0 = \max\{16, 4/(1-\xi)^\sigma\}$ and $\xi_0 = 1-(1-\xi)^\sigma/4$. □

It is a simple matter to infer Theorem 6.28 from this result. First, Proposition 6.54 applied to $u - \min_{Q(R,\tau)} u$ gives

$$\operatorname*{osc}_{Q(\rho,\tau)} u \leq C[\left(\frac{\rho}{R}\right)^\alpha \operatorname*{osc}_{Q(R,\tau)} u + \chi\rho^{(n+2)\varepsilon/2}]$$

for some $\alpha > 0$. Then we note that $Q(r) \subset Q(r/\tau, \tau)$ and $Q(R,\tau) \subset Q(R)$ for $r \leq \tau R$.

To prove the weak Harnack inequality for elements of PDG^-, we need the following proposition which compares minima of u on different sets.

PROPOSITION 6.55. *Suppose that $u \in PDG^-(\Omega)$ and that u is nonnegative in some $Q(4r,\tau) \subset \Omega$. Suppose also that $\bar{u} \geq K$ on $B(r) \times \{s\}$ for some positive constant K. With $\mu = 1-4^{-n}$, $\xi = (1+\mu^{1/2})/2$, and $b = (1+\mu/\xi^2)/2$, let θ be the constant from Lemma 6.51 and suppose that $\tau \leq \theta$. Then there is a constant $\delta_0 \in (0,1)$ determined only by n, α, ε and τ such that $\bar{u} \geq \delta_0 K$ in $B(2r) \times (s, s+2\tau r^2)$.*

PROOF. Let $\bar{\mu}$ be the constant from Lemma 6.53 and take σ from Lemma 6.52 corresponding to $v = \bar{\mu}$. Then Proposition 6.54 implies that

$$|\{X \in Q(4r,\tau) : \bar{u} < (1-\xi)^\sigma K\}| \leq \bar{\mu}\,|Q(4r,\tau)|,$$

and then Lemma 6.53 gives the desired inequality with $\delta_0 = (1-\xi)^\sigma/2$. □

From this lemma we infer the basic estimate for functions in PDG^-.

PROPOSITION 6.56. *Suppose $u \in PDG^-(\Omega)$ is nonnegative in $B(r) \times [0,\alpha_1]$ for some positive α_1 such that $B(r) \times [0,\alpha_1] \subset \Omega$ and that $\bar{u} \geq A$ on $B(\delta r) \times \{s\}$ for some $A > 0$ and $\delta \in (0,1/2)$. Then for any $\alpha_0 \in (0,\alpha_1)$, there are positive constants C_1 and κ determined only by n, α, α_0, α_1, and ε such that*

$$\bar{u} \geq C_1\delta^\kappa A \text{ on } B(r/2) \times [s+\alpha_0 r^2, s+\alpha_1 r^2]. \tag{6.77}$$

PROOF. Let k be the positive integer such that $2^{-k} < \delta \leq 2^{1-k}$ and choose $\eta \leq \theta$, the constant from Lemma 6.51, so that $\tau = 16(1-4^{-k})\eta/3 \leq \alpha_0$. Then set $t_0 = s$ and $r_0 = 2^{-k}r$ and define r_i, t_i, and A_i inductively by the expressions

$$r_i = 2r_{i-1},\ t_i = t_{i-1} + \eta r^2,\ A_i = \inf_{B(r_i)\times\{t_i\}} \bar{u}.$$

From Proposition 6.55, we have that $A_i \geq \delta_0 A_{i-1}$ and hence, since $r_k = r/2$ and $t_k = s + \tau r^2$, it follows that

$$\inf_{B(r/2)\times\{s+\tau r^2\}} \bar{u} = A_k \geq \delta_0^k A \geq \delta^\kappa A$$

for $\kappa = \ln(\delta_0)/\ln(1/2)$.

Next Proposition 6.55 implies that $\bar{u} \geq \delta^\kappa A$ in $B(r/2) \times [s+\tau r^2/2, s+\tau r^2]$. A simple induction argument shows that $\bar{u} \geq \delta_0^m \delta^\kappa A$ in $B(r/2) \times [s+m\tau r^2/2, s+\min\{\alpha_1,(m+1)\tau/2\}r^2]$ for any positive integer m. The proof is completed by choosing m so that $m\tau/2 < \alpha_1 \leq (m+1)\tau/2$ and setting $C_1 = \delta_0^m$. □

As we shall see in Chapter VII, Lemma 6.53 and Proposition 6.56 imply that u satisfies the weak Harnack inequality,

$$\left(R^{-n-2}\int_{\Theta(R)} u^\sigma\, dX\right)^{1/\sigma} \leq C[\inf_{Q(R)} u + \chi R^{(n+2)\varepsilon/2}]$$

for some positive σ, provided u is nonnegative in $Q(4R)$.

Notes

The theory of weak solutions of parabolic equations originates in the Hilbert space approach pioneered by Hilbert [111] and Lebesgue [195] for Laplace's equation in the early 1900's. The theory was more fully developed by many authors, especially Friedrichs [90] and Gårding [91]. An important distinction between the elliptic and parabolic theories is that, although the space V_0 in which solutions lie is a Hilbert space, additional structure of the problem is used because the Riesz representation theorem is not applicable. (But see [302, Chapter IV] for a Hilbert space approach in cylindrical domains.) Our approach to the Cauchy-Dirichlet problem (existence and uniqueness of solutions and higher regularity) is taken from Ladyzhenskaya and Ural′tseva's work [183]. In particular, Theorem 6.1 and the discussion in Section VI.2 are taken from [183, Chapter III]. The proof of Theorem 6.2 is based on our existence result from Chapter V, but otherwise it is close to that of the corresponding theorem in [183]. Lemma 6.3 and Theorem 6.4 come from the work of Brown, Hu, and Lieberman [25] which extended Theorem 6.1 to H_1 noncylindrical domains under the same hypotheses on the coefficients. Alternative approaches to weak solutions in noncylindrical domains can be found in [42] or [344]. Note also that Theorem 6.1 is true for arbitrary domains if $u \in W^{1,2}$ or if u is continuous up to $\mathscr{P}\Omega$. In fact, all the results in this chapter are true under these somewhat different hypotheses. It is in this context that the hypothesis of condition (A) is most useful. References to the corresponding elliptic results, which are proved by very similar methods, can be found in [101]. In fact, our hypotheses can be weakened somewhat further. Petrushko [277] showed that it is possible to define weak solutions in cylindrical domains with $H_{1+\alpha}$ boundary if the boundary data are only in L^p for some $p > 1$, and Lewis and Murray [198] develop a theory of weak solutions (with L^2 boundary data) in domains which satisfy a condition slightly weaker that $\mathscr{P}\Omega \in H_1$. In both cases, more delicate arguments are needed.

Our proof of the Sobolev imbedding theorem, Theorem 6.7, is also taken from [183, pp. 63–65]. A more detailed study of this imbedding theorem, including imbeddings of $W^{k,p}(\Omega)$ in $L^q(\Sigma)$ for lower-dimensional subsets Σ of Ω, can be found in the books [2,78,258,260,348]. Theorem 6.8 is due to Morrey [259], but is usually derived from a potential representation. The proof given here is based on the one in [311]. Theorem 6.9 is also from [183] (it's equation (3.2) of Chapter III in the special case $q = r = 2(n+2)/n$), and is a simple consequence of the Gagliardo-Nirenberg multiplicative Sobolev-type inequalities from [93]. Corollary 6.10 comes from [232, Lemma 1.3].

The discussion of Poincaré's inequality is based on the basic estimate Lemma 6.12, which was proved in [262, Lemma 3] only for $p = 2$, as a preliminary step for the Harnack inequality. As we see here, the same proof works for any p. The usual proofs of Poincaré's inequality proceed either via a potential representation (see, e.g., [101, (7.45)]) or by a contradiction argument using the compactness of the imbedding of $W^{1,p}$ in L^p (see, e.g., [348, Section 4.2]). Our Lemma 6.13 is based on [19, Lemma 2].

Our discussion of boundedness of solutions (Theorems 6.15 and 6.17) is based on [304, Theorem 1.3], which uses a slightly different test function similar to the one used in Lemma 6.19. This work, in turn, is based on the Moser iteration scheme [261]. These pointwise bounds on solutions were first proved by a different iteration scheme ([60, 183]), which we discuss more thoroughly in connection with Section VI.13. The weak Harnack inequality Theorem 6.18 is also due to Moser [262] (as a simple consequence of Equation (3.10) and Theorem 4.1 in that paper) for linear parabolic equations and Trudinger [304, Theorem 1.2] for quasilinear equations. It should be noted, however, that Trudinger was the first to recognize the significance of the weak Harnack inequality even though it was an easy consequence of previously known results. The proof presented here differs from those cited as it is based on an alternative method developed by Trudinger [306, Section 4] to deal with more general classes of degenerate elliptic equations. Although that reference promised a forthcoming discussion of the parabolic case, this book marks the first published application of the technique to parabolic equations.

The weak maximum principle, Corollary 6.26, is well-known, but the present form of the strong maximum principle, Theorem 6.25 seems to be new. The Hölder continuity of solutions, Theorem 6.28, is originally due to Nash [264], who assumed that only the coefficients a^{ij} were non-zero. An alternative proof of Hölder continuity was given by DeGiorgi [60] for elliptic equations, and this proof was generalized to parabolic equations by Ladyzhenskaya and Ural′tseva [184, Section 1-4]; they allow lower order terms. Several other proofs of this important result are known. It has been noted many times that the lower terms may be in certain Morrey spaces. The first such observation is in [304, Theorem 5.1], and a fairly thorough development of the Morrey space theory is given in [343]. Our interest in such weak regularity hypotheses comes from a simplified version of regularity for weak solutions of quasilinear equations (see [223] as a response to [64], and [231] as a response to [280]). A weak Harnack inequality was proved in [231] for quasilinear elliptic equations with coefficients in suitable Morrey spaces (analogous to the ones given here) based on the Sobolev-type imbedding theorem of Adams [1]; the parabolic version of this Sobolev-type theorem is not yet known. Still, it has been shown that weak solutions of linear parabolic *equations* are bounded and satisfy a weak Harnack inequality under even weaker conditions on the coefficients [300, 346].

The boundary estimates of Section VI.9 are contained in [304, Section 4] and the Hölder estimate is [183, Theorem III.10.1]. The results in Section VI.10 are simple variants of the corresponding results for the Cauchy-Dirichlet problem. Weaker conditions on $\mathscr{P}\Omega$ which imply continuity of the solution (for the Cauchy-Dirichlet problem) up to the parabolic boundary are discussed in, e.g., [94].

The existence and regularity in Hölder spaces from Section VI.12 is a straightforward extension of the results for elliptic equations given in [101, Section 8.11] and improves the results in Section III.11 of [183] slightly.

Section VI.13 is a modern version of ideas going back to DeGiorgi [60]. In that work, he showed that weak solutions of linear elliptic equations in divergence form, without lower order terms, were bounded and Hölder continuous. His ideas were extended to linear parabolic equations with lower order terms and to quasilinear parabolic equations by Ladyzhenskaya and Ural′tseva in [184]. They also pointed out that the maximum estimates were valid for functions in either class PDG^+ or PDG^-, which they called $\mathfrak{A}$ and $\mathfrak{B}$, respectively, and that the Hölder estimate holds if $\pm u$ are both in PDG^-. DiBenedetto and Trudinger [65] showed that nonnegative functions in the elliptic DeGiorgi class, which corresponds to supersolutions of elliptic equations, satisfy a weak Harnack inequality. In addition to some basic test function arguments, the key to their result is a measure theoretic lemma of Krylov and Safonov [180], which we shall revisit in Chapter VII. The results in Section VI.13 are taken from [328], in which Wang proves a weak Harnack inequality for functions in the space PDG^-. The intermediate steps which we have reproduced here, in slightly different form, follow fairly closely some of the steps used in [64] to prove Hölder continuity of solutions to degenerate parabolic equations. An important difference is that DiBenedetto and Friedman prove an extra inequality which implies that $\theta = 1$ in Lemma 6.51. Results on functions in the DeGiorgi classes are applicable to problems other than parabolic equations. The most important of these is the parabolic Q-minima introduced by Wieser [342]. (See also [331].)

Exercises

6.1 Prove Hardy's inequality: If $f \in L^1(\mathbb{R})$ and F is defined by

$$F(x) = \frac{1}{x}\int_0^x f(t)\,dt,$$

then

$$\|F\|_p \leq C(p)\,\|f\|_p$$

for $p > 1$.

6.2 Define $L^{p,q}(\Omega)$ to be the set of all functions $u \in L^1(\Omega)$ such that

$$\|u\|_{p,q} = \sup_{r \leq \operatorname{diam}\Omega} \left(\int_{I(\Omega)} \left(\int_{\Omega(t)} |u(X)|^p\,dx \right)^{q/p} \right)^{1/q}$$

is finite.

(a) Show that

$$\|u\|_{p,q} \leq C(n,p,q)\,\|u\|_V$$

for all $u \in V_0$ if $\frac{1}{q} + \frac{n}{2p} \geq \frac{n}{4}$.

(b) Show that the results in Section VI.1 hold under the weaker coefficient hypotheses

$$|b|^2, c^0 \text{ in } L^{p,q},\ (c^i)^2 \in L^{P,Q}$$

with $\frac{1}{q} + \frac{n}{2p} < 1$, $q > n/2$, $\frac{1}{Q} + \frac{n}{2P} = 1$, and $Q \geq 1$.

6.3 For a domain $\Omega \subset \mathbb{R}^n$, we say that $u \in BMO$ if the quantity

$$R^{-n}\int_{B(R)} |u - \{u\}_R| \, dX$$

is uniformly bounded for all $R > 0$ such that $B(R) \subset \Omega$, and the supremum of this quantity is called the BMO seminorm of u. Show that $W^{1,n}(\Omega)$ is continuously imbedded in BMO.

6.4 Let $(X, \mathcal{M}, \mu)$ be an arbitrary measure space, and for $p \in \mathbb{R}\setminus\{0\}$, define

$$\|u\|_p = (\int_X |u|^p \, d\mu)^{1/p}.$$

(a) Show that if $u \in L^p(X)$ for all $p \geq p_0$, then $\lim_{p\to\infty} \|u\|_{p,X}$ exists if and only if $u \in L^\infty(X)$, in which case $\|u\|_\infty = \lim_{p\to\infty} \|u\|_p$.
(b) Show that if u is μ-measurable and positive in X and if $\|u\|_p$ is finite for $p < -p_0$, then $\lim_{p\to-\infty} \|u\|_p$ exists if and only if $\inf|u| > 0$, in which case the limit equals the infimum.
(c) Show that if $\mu(X) = 1$, if $u \neq 0$ almost everywhere in X, and if $\lim_{p\to 0} \|u\|_p$ exists, then this limit is $\exp(\int_X \ln|u| \, d\mu)$.

6.5 Show that a result like Theorem 6.17 remains valid if the hypotheses on b^i, c^i, and c^0 are relaxed to those in Exercise 6.2(c) and $(f^i)^2$ and g are in $L^{p,q}$. Prove a corresponding result for Theorem 6.18.

6.6 Show that if $\beta < 1/e$, then there is a constant $C(\beta)$ such that

$$\sum_{m=0}^{\infty} \frac{\beta^m (w+m)^m}{m!} \leq Ce^w$$

for all $w \geq 0$. (Hint: Use the ratio test to show that the series converges for all w, and then consider the formal derivative of the series.)

6.7 Give an example of a nonnegative solution u of the heat equation in $Q(4R)$ such that $\inf_{Q(R)} u = 0$ and $\sup_{Q(R)} u > 0$.

CHAPTER VII

STRONG SOLUTIONS

Introduction

So far, we have considered two kinds of solutions of parabolic equations: A weak solution, which is only once weakly differentiable with respect to x, and a classical solution, which is twice continuously differentiable with respect to x and once continuously differentiable with respect to t. In addition, weak solutions are only defined if the equation can be written in divergence form. In this chapter, we consider an intermediate regularity situation in which solutions have weak derivatives Du, D^2u and u_t. When these derivatives are in L^p for some $p \geq 1$, we say that $u \in W_p^{2,1}$. Such functions are then possible solutions for equations in nondivergence form. In other words, we define the operator L on $W_1^{2,1}$ by

$$Lu = -u_t + a^{ij}D_{ij}u + b^iD_iu + cu. \tag{7.1}$$

Assuming suitable regularity of the coefficients of L, for example, their membership in L^∞, we say that u is a strong solution of $Lu = f$ if $u \in W_1^{2,1}$ and the equation is satisfied almost everywhere. (As mentioned in the introduction, we shall not consider other situations with still weaker regularity hypotheses on the solution, such as viscosity solutions which are merely continuous, or mild solutions, which are in L^1.)

Two strands run through this chapter. First is a theory of strong solutions in $W_p^{2,1}$ similar to the Schauder theory of Chapters IV and V, which is the focus of Sections VII.4–VII.8. The second is the development of maximum principles (in Section VII.1) and local properties of strong solutions (in Sections VII.9–VII.11) analogous to those in Chapters II, III, and VI.

1. Maximum principles

We start with an extension of the classical maximum principle to strong solutions. To state and prove this extension, we introduce some useful notation. Writing $\mathscr{D} = \det(a^{ij})$, we note that $\mathscr{D}^* = \mathscr{D}^{1/(n+1)}$ is well-defined. We also define the *upper contact set* $E(u)$ of u to be the set of all $X \in \overline{\Omega} \setminus \mathscr{P}\Omega$ such that there is $\xi \in \mathbb{R}^n$ with

$$u(X) + \xi \cdot (y - x) \geq u(Y)$$

for all $Y \in \Omega$ with $s \le t$. If $|x| \le R$ in Ω and $\beta_0 \ge 0$, we write $E^+(u)$ for the subset of $E(u)$ on which $u > 0$ and $(R+\beta_0)|\xi| < u(X) - \xi \cdot x < \sup u^+/2$. To deal directly with oblique derivative problems and Dirichlet boundary data, we suppose also that β is an inward pointing vector field on $\mathscr{P}\Omega$ with $|\beta| \le \beta_0$ and $\beta^{n+1} = 0$, and we define $\mathscr{M}$ by $\mathscr{M}u = -u + \beta \cdot Du$. We then have the following maximum principle which extends Theorem 2.10.

THEOREM 7.1. *Let $Lu \ge f$ in Ω for some function $u \in W^{2,1}_{n+1,loc}(\Omega) \cap C^0(\overline{\Omega})$. Suppose that $|x| \le R$ and $0 < t < T$ for $(x,t) \in \Omega$, and define*

$$B_0 = \|b/\mathscr{D}^*\|^{n+1}_{n+1,E^+(u)} + R + \beta_0. \tag{7.2a}$$

If $\beta = 0$ on $B\Omega$ and if $c \le k$ in Ω for some nonnegative constant k, then there is a constant $c_1(n)$ such that

$$\sup_\Omega u \le e^{kT}(\sup_{\mathscr{P}\Omega}(\mathscr{M}u)^- + c_1 B_0^{n/(n+1)} \|f/\mathscr{D}^*\|_{n+1;E^+(u)}). \tag{7.2b}$$

The proof of Theorem 7.1 is accomplished by first proving some related results under stronger hypotheses on u. We also write $E^+(u;\theta)$ (with $\theta \in (0,1]$) for the subset of $E(u)$ on which $u > 0$ and

$$\frac{R+\beta_0}{\theta}|\xi| < u(X) - \xi \cdot x < \frac{1}{2}\sup u^+.$$

Note that $E^+(u;\theta) \subset E^+(u;1) = E^+(u)$.

LEMMA 7.2. *Suppose Ω, R and T are as in Theorem 7.1 and let $\theta \in (0,1]$. If $u \in C^2(\Omega) \cap C^0(\overline{\Omega})$ and if $\mathscr{M}u \ge 0$ on $\mathscr{P}\Omega$, then*

$$\sup_\Omega u \le C(n)\left(\frac{R+\beta_0}{\theta}\right)^{n/(n+1)} \left(\int_{E^+(u;\theta)} |u_t \det D^2 u| \, dX\right)^{1/(n+1)}. \tag{7.3}$$

PROOF. Define the function Φ: $\Omega \to \mathbb{R}^{n+1}$ by $\Phi(X) = (Du, u - x \cdot Du)$. Then

$$D_i\Phi^j = \begin{cases} D_{ij}u & \text{if } i,j \le n \\ D_i u_t & \text{if } i \le n, j = n+1 \\ -x^k D_{jk}u & \text{if } i = n+1, j \le n \\ u_t - x^k D_k u_t & \text{if } i = j = n+1. \end{cases}$$

By simple row operations (adding x^k times the k-th row to the last row for $k = 1,\dots,n$), we find that the determinant of $D\Phi$ is the same as the determinant of the matrix F with entries given by

$$F_{ij} = \begin{cases} D_{ij}u & \text{if } i,j \le n \\ D_i u_t & \text{if } i \le n, j = n+1 \\ 0 & \text{if } i = n+1, j \le n \\ u_t & \text{if } i = j = n+1, \end{cases}$$

and the determinant of this matrix is easily computed to be $u_t \det D^2 u$. It follows that

$$\int_{E^+(u;\theta)} \left|u_t \det D^2 u\right| dX \geq \int_{\Phi(E^+(u;\theta))} 1\, d\Xi = \left|\Phi(E^+(u;\theta))\right|. \qquad (7.4)$$

Now let $M = \sup u$ and suppose that $u(X_0) = M$. Then write Σ for the set of all $\Xi = (\xi, h)$ in $\mathbb{R}^{n+1}$ such that

$$\frac{R+\beta_0}{\theta}\,|\xi| < h < \frac{M}{2},$$

and, for $\Xi \in \Sigma$, define $p(Y) = \xi \cdot y + h$. By hypothesis, $p > 0$ in $\overline{\Omega}$ and hence $p > u$ on $B\Omega$. In addition, $p(X_0) < u(X_0)$, so there is a point $Y_1 \in \overline{\Omega}$ with $s_1 \leq t_0$ such that $p(Y_1) = u(Y_1)$ and $p(Y) \geq u(Y)$ if $s \leq s_1$. If $Y_1 \in \mathscr{P}\Omega$, then $\beta \cdot D(u-p)(Y_1) \leq 0$, so

$$p(Y_1) = u(Y_1) \leq \beta \cdot Du(Y_1) \leq \beta \cdot Dp(Y_1) \leq \beta_0\,|\xi|.$$

Since $p(Y_1) = \xi \cdot y_1 + h \geq h - R|\xi|$, it follows that $h \leq (R+\beta_0)\,|\xi|$, which contradicts the assumption $\Xi \in \Sigma$, so $Y_1 \in \overline{\Omega} \setminus \mathscr{P}\Omega$. Therefore, if $Y \in \Omega$ and $s \leq s_1$, we have

$$u(Y) \leq \xi \cdot y + h = \xi \cdot (y - y_1) + p(Y_1) = \xi \cdot (y - y_1) + u(Y_1),$$

and hence $Y_1 \in E^+(u)$. Moreover, it is easy to check that $\Phi(Y_1) = \Xi$ and hence $\Sigma \subset \Phi(E^+(u))$. It follows that

$$\left|\Phi(E^+(u))\right| \geq |\Sigma| = \frac{\omega_n M^{n+1}\theta^n}{n(n+1)2^{n+1}(R+\beta_0)^n}.$$

We then infer (7.3) from this inequality and (7.4). □

The next step is to show that Lemma 7.2 remains valid under the regularity hypotheses of Theorem 7.1.

PROPOSITION 7.3. *Lemma 7.2 is true if the hypothesis $u \in C^2(\Omega) \cap C^0(\overline{\Omega})$ is relaxed to $u \in W^{2,1}_{n+1,loc} \cap C^0(\overline{\Omega})$.*

PROOF. Suppose first that $\mathscr{M}u > 0$ on $\mathscr{P}\Omega$. If $X \in \overline{E^+(u;\theta)}$ and $X \in \mathscr{P}\Omega$, then there is ξ with $|\xi| \leq u(X) - \xi \cdot x$ such that $u(X) + \xi \cdot (x - y) \geq u(Y)$ for $y = h\beta + x$ and $s = t$ provided $h \geq 0$ is sufficiently small. It then follows that

$$\beta \cdot Du(X) \leq \xi \cdot \beta \leq \beta_0\,|\xi|.$$

From the boundary condition, $u(X) < \beta_0\,|\xi|$, but $X \in \overline{E^+(u;\theta)}$ implies that

$$u(X) \geq (R+\beta_0)\,|\xi| + \xi \cdot x \geq \beta_0\,|\xi|.$$

From this contradiction, we infer that there is a positive constant ε such that $d^*(X) = \operatorname{dist}(X, \mathscr{P}\Omega) > \varepsilon$ if $X \in E^+(u;\theta)$.

Now, let $(\tilde{u}_k)$ be a sequence of C^2 functions converging to u in $W^{2,1}_{n+1}(\Omega \cap \{d^* > \varepsilon/2\})$ and let η be a nonnegative $C^2(\overline{\Omega})$ function such that $\eta(X) = 0$ if

$d^*(X) < \varepsilon/2$ and $\eta(X) = 1$ if $d^*(X) > \varepsilon$. Finally, set $u_k = \eta\tilde{u}_k + (1-\eta)u$. By invoking (7.4) for each u_k, we reduce the proof to showing that

$$\int_{E^+(u;\theta)} \left|u_t \det D^2u\right| dX \ge \limsup_k \int_{E^+(u_k;\theta)} \left|u_{k,t} \det D^2u_k\right| dX. \tag{7.5}$$

Now we let G be an arbitrary open set in $\mathbb{R}^{n+1}$ with $E^+(u;\theta) \subset G$ and $d^* > \varepsilon$ in G. We claim that $E^+(u_k;\theta) \subset G$ for all sufficiently large k. Otherwise, there is a subsequence $(u_{k(j)})$ and a sequence of points $(X_j) \in E^+(u_{k(j)};\theta) \setminus G$. Since $\overline{\Omega}$ is compact, we may assume that $X_j \to X \in \overline{\Omega}$ and clearly $X \notin \mathscr{P}\Omega$. Now let ξ_j be a vector from the definition of $X_j \in E^+(u_{k(j)};\theta)$ and note that $|\xi_j| \le \sup u_{k(j)}/(2R)$ so that the sequence (ξ_j) is also uniformly bounded, and, again, we may assume that this sequence converges to some vector ξ. If $Y \in \Omega$ and $s \le t$, then

$$u_{k(j)}(Y) \le u_{k(j)}(X_j) + \xi_j \cdot (y - x_j).$$

Taking the limit as $j \to \infty$ gives that $X \in E^+(u;\theta)$, which contradicts the assumption that each X_j is not in G and hence $X \notin G$.

Therefore, we have

$$\int_G \left|u_t \det D^2u\right| dX \ge \limsup_k \int_G \left|u_{k,t} \det D^2u_k\right| dX.$$

Since G is an arbitrary open set with $E^+(u;\theta) \subset G$ and $E^+(u_k;\theta) \subset G$, (7.5) follows immediately, and hence so does the conclusion of this proposition if $\mathscr{M}u > 0$ on $\mathscr{P}\Omega$. To extend to the case that $\mathscr{M}u \ge 0$ on $\mathscr{P}\Omega$, we apply the proposition to $u - 1/k$ and note that $E^+(u-1/k;\theta) \subset E^+(u;\theta)$. □

We are now ready to prove Theorem 7.1 by making a suitable change of dependent variable and a good choice for g.

PROOF OF THEOREM 7.1. We set

$$w = e^{-kt}u - \sup_{\mathscr{P}\Omega} e^{-kt}u^+,$$

and $E = E^+(w;\theta)$ for some $\theta \in (0,1]$ to be determined. Since $0 < e^{-kt} \le 1$, we have $-w_t + a^{ij}D_{ij}w \ge -f^- - |b|\,|Dw|$ on E and $w \le 0$ on $\mathscr{P}\Omega$. From the differential equation, we find that

$$\frac{w_t - a^{ij}D_{ij}w}{(n+1)\mathscr{D}^*} \le \frac{f^- + |b|\,|Dw|}{(n+1)\mathscr{D}^*}$$

on E.

Now we note that $w(x,t) \ge w(x,s)$ for all $X \in E$ and $s \le t$ and hence $w_t \ge 0$ a.e. on E. Moreover,

$$w(x,t) + \xi \cdot h \ge w(x+h,t) \text{ and } w(x,t) - \xi \cdot h \ge w(x-h,t)$$

for all $X \in E$ and all sufficiently small vectors h, so

$$2w(x,t) \ge w(x+h,t) + w(x-h,t).$$

Now let us choose $h = \eta H$ for some unit vector H. Since

$$\frac{w(x+h,t)+w(x-h,t)-2w(x,t)}{\eta^2}$$

converges weakly to the linear combination of derivatives $D_{ij}wH^iH^j$, it follows that $D^2w \le 0$ a.e. in E. Since $(\det A \det B)^{1/(n+1)} \le (\operatorname{trace} AB)/(n+1)$ for any $(n+1) \times (n+1)$ positive semidefinite matrices A and B, it follows that $w_t - a^{ij}D_{ij}w \ge (n+1)\mathscr{D}^* \left|w_t \det D^2w\right|^{1/(n+1)}$ on E. Furthermore, $|Dw| \le M\theta/(R+\beta_0)$ on E, so Proposition 7.3 implies that

$$M \le C(n)\left(\frac{R+\beta_0}{\theta}\right)^{n/(n+1)} \|f/\mathscr{D}^*\|_{n+1;E^+(u)}) + C(n)\left(\frac{\theta B_0}{R+\beta_0}\right)^{1/(n+1)} M$$

because $E^+(w;\theta) \subset E^+(w) \subset E^+(u)$. We now take

$$\theta = \frac{R+\beta_0}{B_0}\min\{1,(2C(n))^{-n-1}\}$$

to infer (7.2b). □

An immediate consequence of Theorem 7.1 is a general comparison theorem.

COROLLARY 7.4. *Suppose $b/\mathscr{D}^* \in L^{n+1}_{loc}$ and $c \le k$ in a bounded domain Ω. Suppose also that β is an inward pointing vector field which is bounded on $\mathscr{P}\Omega$. If $u \in W^{2,1}_{n+1,loc} \cap C^0(\overline{\Omega})$ satisfies the inequalities $Lu \ge 0$ in Ω, $\mathscr{M}u \ge 0$ on $\mathscr{P}\Omega$, then $u \le 0$ in Ω.*

PROOF. For $\delta > 0$, set $u_\delta = u - e^{kt}\delta$. Then

$$Lu_\delta = e^{kt}Lu - \delta c e^{kt} + k\delta e^{kt} \ge 0$$

in Ω and $\mathscr{M}u_\delta = \mathscr{M}u + \delta e^{kt} \ge \delta$ on $\mathscr{P}\Omega$, assuming without loss of generality that $0 < t$ in Ω. From the proof of Theorem 7.1, we see that $b/\mathscr{D}^* \in L^{n+1}(E^+(u_\delta))$ and hence Theorem 7.1 implies that $u_\delta \le 0$ in $\overline{\Omega}$. The proof is completed by sending $\delta \to 0$. □

2. Basic results from harmonic analysis

In this section, we prove some results from harmonic analysis that will be useful in our proof of the L^p estimates and for studying local properties of strong solutions as well. Our first step is to examine a parabolic version of the Calderón-Zygmund decomposition in a cube.

For this purpose, we consider a cube $K_0 = \{|x^i - x^i_0| < R, |t - t_0| < T\}$ for some point X_0 and positive constants R and T, and we suppose that $f \in L^1(K_0)$ is nonnegative and that τ is a positive constant with

$$\tau \ge \frac{1}{|K_0|}\int_{K_0} f(X)\,dX.$$

We divide K_0 into 2^{n+2} congruent subcubes with disjoint interiors

$$K_{1,1},\dots,K_{1,2n+2}$$

so that

$$K_{1,j} = \{|x^i - x^i_{1,j}| < R/2, |t - t_{1,j}| < T/4\}$$

for some $X_{1,j} \in K_0$. Each subcube K such that

$$\int_K f(X)\,dX \le \tau|K|$$

is subdivided the same way and this process is repeated indefinitely, creating at the m-th step a finite set of cubes $K_{m,1},\dots,K_{m,k(m)}$. We write $\mathscr{S}$ for the set of all subcubes K obtained by this division such that

$$\int_K f(X)\,dX > \tau,$$

and for each $K \in \mathscr{S}$, we write $\tilde{K}$ for the subcube whose subdivision gives K. Since $|\tilde{K}|/|K| = 2^{n+2}$, it follows that

$$\tau < \frac{1}{|K|}\int_K f(X)\,dX \le 2^{n+2}\tau$$

for any $K \in \mathscr{S}$. We now define $B = \cup_{K\in\mathscr{S}}K$, and $G = K_0 \setminus B$. The following variant of the Lebesgue differentiation theorem is now used.

LEMMA 7.5. *Let $f \in L^1(K_0)$ be nonnegative, and for each $X \in K_0$, let $K_m(X)$ be the cube obtained at the m-th step of the Calderón-Zygmund decomposition such that $X \in K_m(X)$, and set*

$$f_m(X) = \frac{1}{|K_m(X)|}\int_{K_m(X)} f(Y)\,dY. \tag{7.6}$$

Then $f_m \to f$ in $L^1(K_0)$.

PROOF. Let $\rho_m(X)$ denote the center of $K_m(X)$ and set $\sigma_m(X) = X + \sigma_m(X)$. Then

$$\int_{K_0} |f(X) - f_m(X)|\,dX = \int_{K_0}\left|f(X) - \frac{1}{|K_m|}\int_{K_m} f(Y+\sigma_m(X))\,dY\right|dX$$
$$\le \frac{1}{|K_m|}\int_{K_m}\int_{K_0}|f(X) - f(Y+\sigma_m(X))|\,dX\,dY,$$

where K_m is the cube with the same dimensions as $K_m(X)$ centered at the origin.

Now fix $\varepsilon > 0$ and choose $g \in C(\overline{K_0})$ such that $\|f-g\|_1 < \varepsilon$. Since $X + Y + \rho_m(X)$ and X are in the same cube $K_m(X)$, it follows by continuity that $|g(X) - g(X+Y+\rho_m(X))| < \varepsilon/3$ for all sufficiently large m, independent of the choice of X. For such an m, we easily see that $\|f - f_m\|_1 < \varepsilon$, and hence $f_m \to f$ in L^1. □

From this lemma, we see that $f \le \tau$ a.e. in G. In addition, if we define $\tilde{B} = \cup_{K\in\mathscr{S}} \tilde{K}$, it follows that

$$\int_{\tilde{B}} f(X)\,dX \le \tau\left|\tilde{B}\right|.$$

In particular, if f is the characteristic function of some set Γ, then

$$|\Gamma| = \left|\Gamma \cap \tilde{B}\right| \le \tau\left|\tilde{B}\right|. \tag{7.7}$$

Our next preliminary step is the following lemma on the distribution function $\mu(h,f)$ of a function f defined by

$$\mu(h,f) = |\{X : f(X) > h\}|.$$

When the function f is understood, we write $\mu(h)$ instead of $\mu(h,f)$.

LEMMA 7.6. *If $f \in L^p(\Omega)$ is nonnegative for some open set Ω and some $p \ge 1$, then*

$$\mu(h) \le h^{-p} \int_\Omega f^p\,dX, \tag{7.8}$$

and

$$\int_\Omega f^p\,dX = p\int_0^\infty h^{p-1}\mu(h)\,dh. \tag{7.9}$$

PROOF. To prove (7.8), we note that

$$\mu(h)h^p \le \int_{\{f>h\}} f^p\,dX \le \int_\Omega f^p\,dX.$$

For (7.9), we use Fubini's theorem to infer that

$$\begin{aligned}\int_\Omega f^p\,dX &= \int_\Omega \int_0^{f(X)^p} dh\,dX = \int_\Omega \int_0^{f(X)} ps^{p-1}\,ds\,dX \\ &= \int_0^\infty ps^{p-1} \int_{\{f>s\}} dX\,ds = p\int_0^\infty s^{p-1}\mu(s)\,ds.\end{aligned}$$

A change of dummy variable gives (7.9). □

From these preliminaries (some of which we shall use directly in our study of strong solutions), we now derive a special case of the well-known Marcinkiewicz interpolation theorem.

THEOREM 7.7. *Let $1 \le q < \infty$ and let J be a linear operator which maps functions in $L^q(\Omega) + L^\infty(\Omega)$ to measurable functions for some open set Ω. If there are constants J_q and J_∞ such that*

$$\mu(h, Jf) \le \left(\frac{J_q\,\|f\|_q}{h}\right)^q, \tag{7.10a}$$

$$\|Jf\|_\infty \le J_\infty\,\|f\|_\infty, \tag{7.10b}$$

then J is a linear operator from L^p to itself for all $p \in (q,\infty)$ and

$$\|Jf\|_p \le C(p,q,)J_q^{q/p}J_\infty^{(p-q)/q}\|f\|_p. \tag{7.11}$$

PROOF. For each $h > 0$, we take $s = h/(3J_\infty)$, and define

$$f_1(X) = \begin{cases} f(X) & \text{if } |f(X)| > s, \\ 0 & \text{otherwise,} \end{cases}$$

and $f_2 = f - f_1$. Then $\|Jf_2\|_\infty \le h/3$ and therefore

$$\mu(h,Jf) \le \mu(h/2,Jf_1) \le \left(\frac{2J_q\|f_1\|_q}{h}\right)^q.$$

Hence

$$\begin{aligned}\int_\Omega |Jf|^p\,dX &\le p(2J_q)^q \int_0^\infty h^{p-1-q}\Big(\int_{\{|f|>s\}} |f|^q\,dX\Big)\,dh \\ &= p(2J_q)^q(3J_\infty)^{p-q}\int_\Omega |f|^p\,dX,\end{aligned}$$

which is (7.11). □

We now prove a parabolic variant of the Vitali covering lemma. It will be useful to introduce the centered cylinders:

$$C(X_0,R) = \{X \in \mathbb{R}^n : |x - x_0| < R,\ |t - t_0| < R/2\}.$$

We call R the *radius* of $C(X_0,R)$.

LEMMA 7.8. *Let A be a bounded subset of $\mathbb{R}^{n+1}$ and let $\mathscr{C}$ be a covering of A by centered cylinders with bounded radius. Then there is a sequence $(C(X_i,R_i))$ of disjoint centered cylinders from $\mathscr{C}$ such that*

$$A \subset \bigcup_i C(X_i,5R_i). \tag{7.12}$$

PROOF. Choose $C_1 = C(X_1,R_1)$ in $\mathscr{C}$ so that

$$R_1 \ge \frac{1}{2}\sup\{R : R \text{ is the radius of some } C \in \mathscr{C}\}.$$

Then we define $C_k = C(X_k,R_k)$ inductively so that $C_k \cap C_i = \emptyset$ for $i < k$ and

$$R_k \ge \frac{1}{2}\sup\{R : R \text{ is the radius of some } C \in \mathscr{C}_k\},$$

where $\mathscr{C}_k$ denote the set of all $C \in \mathscr{C}$ which are disjoint from $C_1,\dots,C_{k-1}$.

Next, we fix a point $X \in A$ and let $C(Y,r)$ be an element of $\mathscr{C}$ with $X \in C(Y,r)$. If the sequence (C_k) is infinite, we note that the union of the C_k has finite measure. It follows that $\sum|C_k| < \infty$ and hence $R_k \to 0$ as $k \to \infty$. Therefore there is a last positive integer k such that $R_k \ge r/2$. If the sequence is finite, then this statement is immediate. If $C(Y,r)\cap C(X_i,R_i) = \emptyset$ for all $i < k$, then we contradict our definition of R_k and hence $C(Y,r) \cap C(X_i,R_i)$ is nonempty for some $i < k$. With this i fixed and $Y_1 \in C(Y,r) \cap C(X_i,R_i)$, we have

$$|x-x_i| \leq |x-y| + |y-y_1| + |y_1-x_i| < r+r+R_i \leq 5R_i$$

and, similarly $|t-t_i| < (5/2)R_i$. Therefore, $X \in C(X_i, 5R_i)$, which immediately implies (7.12). □

Next, we introduce the parabolic version of the Hardy-Littlewood maximal function. If $v \in L^1(\mathbb{R}^{n+1})$, we set

$$\mathcal{M}v(X) = \sup_{r>0} \frac{1}{|C(X,r)|} \int_{C(X,r)} |v(Y)|\,dY.$$

If $v \in L^1(\Omega)$ for some measurable subset Ω of $\mathbb{R}^{n+1}$, we define $\tilde{v}$ by

$$\tilde{v}(X) = \begin{cases} v(X) & \text{if } X \in \Omega, \\ 0 & \text{if } X \notin \Omega, \end{cases}$$

and we define $\mathcal{M}_\Omega v = \mathcal{M}(\tilde{v})$.

LEMMA 7.9. *If $v \in L^1(\mathbb{R}^{n+1})$, then for any $\alpha > 0$, we have*

$$|\{X : \mathcal{M}v(X) > \alpha\}| \leq \frac{5^{n+2}}{\alpha} \|v\|_1. \tag{7.13}$$

If $v \in L^p$ for some $p > 1$, then

$$\|\mathcal{M}v\|_p \leq C(n,p)\|v\|_p. \tag{7.14}$$

PROOF. For $\alpha > 0$, set

$$E_\alpha = \{X : \mathcal{M}v(X) > \alpha\}.$$

Then for every $X \in E_\alpha$, there is a positive $r(X)$ such that

$$\int_{C(X,r(X))} |v|\,dX \geq \alpha |C(X,r(X))|.$$

Lemma 7.8 now gives a sequence (X_i) in E_α such that $C(X_i,r(X_i)) \cap C(X_j,r(X_j)) = \emptyset$ if $i \neq j$ and

$$\sum_i |C(X_i,r(X_i))| \geq 5^{-n-2}|E_\alpha|.$$

If we set $C_i = C(X_i,r(X_i))$ as before, we have

$$\begin{aligned} \|v\|_1 &\geq \int_{\cup C_i} |v|\,dX \geq \alpha |\cup C_i| \\ &= \alpha \sum |C_i| \geq 5^{-n-2}\alpha |E_\alpha|. \end{aligned}$$

Rearranging this inequality gives (7.13).

Next, we observe that (7.14) is immediate with $C = 1$ when $p = \infty$. An application of Theorem 7.7 completes the proof. □

From this lemma, we readily infer the parabolic analog of the Lebesgue differentiation theorem.

COROLLARY 7.10. *If* $v \in L^1(\mathbb{R}^{n+1})$, *then*

$$v(X) = \lim_{r\to 0} \frac{1}{|C(X,r)|} \int_{C(X,r)} v(Y)\,dY \tag{7.15}$$

for almost all X. *Therefore* $|v| \le \mathcal{M}v$ *a.e.*

PROOF. Set

$$V(X,r) = \frac{1}{|C(X,r)|} \int_{C(X,r)} v(Y)\,dY.$$

Then the argument of Lemma 7.5 shows that $\lim_{r\to 0} \|V(\cdot,r) - v\|_1 = 0$ and hence there is a sequence (r_k) such that $r_k \to 0$ and $V(\cdot,r_k) \to v$ almost everywhere. Therefore, it suffices to show that $\lim_{r\to 0} V(\cdot,r)$ exists almost everywhere. To show the existence of this limit, we define $\overline{v}$ and $\underline{v}$ by

$$\overline{v} = \limsup_{r\to 0} V(\cdot,r), \quad \underline{v} = \liminf_{r\to 0} V(\cdot,r).$$

Now, let $\varepsilon > 0$ and $\eta > 0$ be arbitrary and let h be a uniformly continuous function such that $\|v-h\|_1 \le \eta\varepsilon/(2\cdot 5^{n+1})$. Then $\overline{h} = \underline{h}$ and hence for $w = v - h$, we have $0 \le \overline{v} - \underline{v} = \overline{w} - \underline{w}$. Lemma 7.9 implies that

$$|\{\overline{w} - \underline{w} > \varepsilon\}| \le 2\frac{5^{n+2}}{\varepsilon}\|w\|_1 < \eta,$$

and hence

$$|\{\overline{v} - \underline{v} > \varepsilon\}| < \eta.$$

Since η is arbitrary, it follows that $|\{\overline{v} - \underline{v} > \varepsilon\}| = 0$ for all $\varepsilon > 0$ and hence $\overline{v} = \underline{v}$ except on a set of measure zero. □

We are now ready to prove a further variant of the Vitali lemma.

LEMMA 7.11. *Let* $0 < \varepsilon < 1$, *let* $\rho > 0$, *let* $X_0 \in R^{n+1}$, *and let* A *and* B *be subsets of* $Q(X_0,\rho)$ *such that* $A \subset B$ *and* $|A| < \varepsilon|Q(X_0,\rho)|$. *Suppose also that, for every* $X \in Q(0,\rho)$ *and every* $r \in (0,\rho)$ *such that* $|A \cap C(X,r)| \ge \varepsilon|C(X,r)|$, *we have* $C(X,r) \cap Q(X_0,\rho) \subset B$. *Then*

$$|A| \le 20^{n+2}\varepsilon|B|. \tag{7.16}$$

PROOF. First, by applying Corollary 7.10 to χ_A, we see that, for almost all $X \in A$, there is a number $r(X) \in (0,2\rho)$ such that

$$|A \cap C(X,r(X))| = \varepsilon|C(X,r(X))|$$

and

$$|A \cap C(X,r)| < \varepsilon|C(X,r)|$$

for all $r > r(X)$. Then Lemma 7.8 gives a sequence (X_i) (possibly finite) such that $C(X_i, r(X_i)) \cap C(X_j, r(X_j)) = \emptyset$ if $i \neq j$ and

$$A \subset Q(X_0, \rho) \cap \bigcup_i C(X_k, 5r(X_i)).$$

For simplicity, we now set $C_i = C(X_i, r(X_i))$ and $5C_i = C(X_i, 5r(X_i))$. Then

$$|A \cap 5C_i| < \varepsilon |5C_i| = 5^{n+2} \varepsilon |C_i|.$$

In addition, because $X_i \in Q(X_0, \rho)$ and $r(X_i) < 2\rho$, we have

$$|C_i| \leq 4^{n+2} |C_i \cap Q(X_0, \rho)|.$$

It follows that

$$\begin{aligned} |A| &= \left| \bigcup_i (A \cap 5C_i) \right| \leq \sum_i |A \cap 5C_i| \\ &\leq 5^{n+2} \varepsilon \sum_i |C_i| \leq 20^{n+2} \varepsilon \sum |C_i \cap Q(X_0, \rho)| \\ &= 20^{n+2} \varepsilon \left| \bigcup_i (C_i \cap Q(X_0, \rho)) \right| \\ &\leq 20^{n+2} \varepsilon |B|. \end{aligned}$$

□

3. L^p estimates for constant coefficient divergence structure equations

As a preliminary step, we prove L^p estimates for first derivatives of solutions of divergence structure equations in analogy with the situation in Section IV.3. Our first step is a measure-theoretic estimate for the maximal functions of these derivatives. (Note that, strictly speaking, the maximal function in Proposition 7.12 should be replaced by $\mathcal{M}_\Omega$ and similarly throughout the section, but we leave it to the reader to check that this change is immaterial to the proofs.)

PROPOSITION 7.12. *Let λ and Λ be positive constants and let (A^{ij}) be a constant matrix such that*

$$\lambda |\xi|^2 \leq A^{ij} \xi_i \xi_j \leq \Lambda |\xi|^2 \tag{7.17}$$

for all $\xi \in \mathbb{R}^n$. Let $X_0 \in \mathbb{R}^{n+1}$, let $R > 0$ and set $\Omega = Q(X_0, 6R)$. Suppose u is a weak solution of the equation

$$-u_t + D_i(A^{ij} D_j u) = D_i f^i \text{ in } \Omega \tag{7.18}$$

for some $f \in L^2(\Omega)$. Suppose also that $X_1 \in \Omega$ and $r \in (0, R)$ are chosen so that $Q(X_1, 6r) \subset \Omega$. Then there are positive constants $M(\lambda, \Lambda, n)$ and $\delta(\varepsilon, \lambda, \Lambda, n)$ such that, if

$$\{Y \in Q(X_1, r) : \mathcal{M}|f|^2(Y) \leq \delta^2 \theta^2, \ \mathcal{M}(|Du|^2)(Y) \leq \theta^2\} \neq \emptyset \tag{7.19}$$

for some $\theta > 0$, then

$$\left|\{X \in Q(X_1,r) : \mathcal{M}(|Du|^2) > \theta^2 M^2\}\right| \le \varepsilon |Q(X_1,r)|. \tag{7.20}$$

PROOF. First, let Y_0 be a point in $Q(X_1,r)$ so that $\mathcal{M}|f|^2(Y_0) \le \delta^2\theta^2$ and $\mathcal{M}(|Du|^2)(Y_0) \le \theta^2$. Then

$$\frac{1}{\omega_n(6r)^{n+2}} \int_{C(Y_0,6r)\cap\Omega} |Du|^2\, dX \le \theta^2$$

and

$$\frac{1}{\omega_n(6r)^{n+2}} \int_{C(Y_0,6r)\cap\Omega} |f|^2\, dX \le \delta^2\theta^2.$$

Next, we set

$$t^* = \min\{t_1 + 2r, t_0\}, \quad X^* = (x_1, t^*),$$

and we write Q_k $(k > 0)$ for $Q(X^*, kr)$. It follows that $Q_4 \subset C(Y_0, 6r) \cap \Omega$. Since $|Q_4| = \omega_n(4r)^{n+2}$, we therefore have

$$\int_{Q_4} |Du|^2\, dX \le 2^{n+2}\theta^2 |Q_4|.$$

Now let w solve

$$-w_t + D_i(A^{ij}D_j w) = D_i f^i \text{ in } Q_4, \quad w = 0 \text{ on } \mathcal{P}Q_4, \tag{7.21}$$

and set $h = u - w$. It follows from the comments following Theorem 6.1 that there is a constant c_1 (determined only by λ, Λ, and n) such that

$$\int_{Q_4} |Dw|^2\, dX \le c_1 \int_{Q_4} |f|^2\, dX \le 2^{n+2} c_1 \delta^2\theta^2 |Q_4|,$$

and hence

$$\int_{Q_4} |Dh|^2\, dX \le 2\int_{Q_4} [|Dw|^2 + |Du|^2]\, dX \le 2^{n+3}(1 + c_1)\theta^2 |Q_4|.$$

Since $Lh = 0$ in Q_4, it follows (from Theorem 6.6 and Lemma 3.18) that $D_i h$ is a weak solution of the equation $L(D_i h) = 0$ for $i = 1, \ldots, n$. Hence we can apply the local maximum principle (Theorem 6.17) in $Q(X,r)$ for each $X \in Q_3$ to infer that

$$\sup_{Q_3} |Dh|^2 \le M_0\theta^2 \tag{7.22}$$

for some M_0, determined only by λ, Λ, and n.

Now we apply Lemma 7.9 (specifically (7.13)) to conclude that

$$\begin{aligned}
5^{-n-2}\alpha |\{X \in Q(X_1,r) : \mathcal{M}_{Q_3}|Dw|^2 > \alpha\}| &\le \int_{Q_3} |Dw|^2\, dX \\
&\le 2^{n+2} c_1 \delta^2\theta^2 |Q_4| \\
&= 8^{n+2} c_1 \delta^2\theta^2 |Q(X_1,r)|
\end{aligned}$$

for any $\alpha > 0$. In particular, for $\alpha = \theta^2$, we have

$$|\{X \in Q(X_1,r) : \mathcal{M}_{Q_3}|Dw|^2 > \theta^2\}| \le 40^{n+2}c_1\delta^2|Q(X_1,r)|.$$

Next, we note that $|Du|^2 \le 2|Dw|^2 + 2M_0\theta^2$ in Q_3. Suppose that $Y \in Q(X_1,r)$ is a point such that $\mathcal{M}_{Q_3}|Dw|^2(Y) \le \theta^2$. For $\rho \le 2r$, we have $C(Y,\rho) \cap \Omega \subset Q_3$ and hence

$$\frac{1}{|C(Y,\rho)|}\int_{C(Y,\rho)\cap\Omega} |Du|^2\,dX \le 2\mathcal{M}_{Q_3}|Dw|^2(Y) + 2M_0\theta^2 \le (2M_0+2)\theta^2.$$

For $\rho > 2r$, we have $C(Y,\rho) \subset C(Y_0,2\rho)$ and hence

$$\frac{1}{|C(Y,\rho)|}\int_{C(Y,\rho)\cap\Omega} |Du|^2\,dX \le \frac{1}{|C(Y,\rho)|}\int_{C(Y_0,2\rho)\cap\Omega} |Du|^2\,dX \le 2^{n+2}\theta^2.$$

For $M = \max\{2^{(n+2)/2}, (2M_0+2)^{1/2}\}$, it follows that $\mathcal{M}_{Q_3}|Dw|^2(Y) \le \theta^2$ implies $\mathcal{M}|Du|^2(Y) \le M^2\theta^2$. Therefore

$$\{X \in Q(X_1,r) : \mathcal{M}|Du|^2 > M^2\theta^2\} \subset \{X \in Q(X_1,r) : \mathcal{M}_{Q_3}|Dw|^2 > \theta^2\},$$

and hence

$$|\{X \in Q(X_1,r) : \mathcal{M}|Du|^2 > M^2\theta^2\}| \le 40^{n+2}M_0\delta^2|Q(X_1,r)|.$$

The estimate (7.20) now follows by taking $\delta = \varepsilon/(1+40^{n+2}M_0)^{1/2}$. □

Next, we draw some useful conclusions about the measure of various sets.

COROLLARY 7.13. *Let λ, Λ, (A^{ij}), X_0, R, u and f be as in Proposition 7.12. Let $\varepsilon > 0$, let M and δ be the constants from Proposition 7.12, and set $\varepsilon_1 = 20^{n+2}\varepsilon$. If θ is a positive constant such that*

$$|\{\mathcal{M}|Du|^2 > \theta^2M^2\}| \le \varepsilon|Q(X_0,R)|, \tag{7.23}$$

then

$$\begin{aligned}|\{\mathcal{M}|Du|^2 > \theta^2M^2\}| \le{}& \varepsilon_1|\{\mathcal{M}|Du|^2 > \theta^2\}| \\ &+ \varepsilon_1|\{\mathcal{M}|f|^2 > \delta^2\theta^2\}|,\end{aligned} \tag{7.24a}$$

$$\begin{aligned}|\{\mathcal{M}|Du|^2 > \alpha^2\theta^2M^2\}| \le{}& \varepsilon_1|\{\mathcal{M}|Du|^2 > \alpha^2\theta^2\}| \\ &+ \varepsilon_1|\{\mathcal{M}|f|^2 > \alpha^2\theta^2\delta^2\}|\end{aligned} \tag{7.24b}$$

for any $\alpha \ge 1$, and

$$\begin{aligned}|\{\mathcal{M}|Du|^2 > \theta^2M^{2j}\}| \le{}& \varepsilon_1^j|\{\mathcal{M}|Du|^2 > \theta^2\}| \\ &+ \sum_{i=1}^{j} \varepsilon_1^i|\{\mathcal{M}|f|^2 > \theta^2M^{2(j-i)}\delta^2\}|\end{aligned} \tag{7.24c}$$

for any positive integer j, where $\{\mathcal{M}g > h\}$ denotes the subset of $Q(X_0,R)$ on which $\mathcal{M}g > h$ for $g = |Du|^2$ or $g = |f|^2$.

PROOF. First, we set $A = \{\mathcal{M}|Du|^2 > \theta^2 M^2\}$ and $B = \{\mathcal{M}|Du|^2 > \theta^2\} \cup \{\mathcal{M}|f|^2 > \theta^2\delta^2\}$. It's easy to check that the hypotheses of Lemma 7.11 are satisfied by using the contrapositive of Proposition 7.12 to check that $|A \cap C(X,r)| \geq \varepsilon|C(X,r)|$ implies that $C(X,r) \cap Q(X_0,R) \subset B$. Then (7.24a) is just (7.16).

Next, we apply (7.24a) with $\alpha\theta$ replacing θ to infer (7.24b), and then (7.24c) follows from (7.24b) by induction, using $\alpha = 1, M, \ldots, M^j$. □

We are now in a position to prove the basic L^p estimate for $p \geq 2$.

PROPOSITION 7.14. *Let $p \geq 2$, let λ and Λ be positive constants, let (A^{ij}) be a constant matrix satisfying* (7.17) *for all $\xi \in \mathbb{R}^n$, and let $u \in W_2^{2,1}(Q(X_0,36R))$ satisfy* (7.18) *with some $f \in L^p(Q(X_0,36R))$ for some $X_0 \in \Omega$ and $R > 0$. Then $u \in W_p^{2,1}(Q(X_0,R))$ and there is a constant C, determined only by λ, Λ, and n such that*

$$\|Du\|_{p,Q(X_0,R)} \leq C\left(\|f\|_{p,Q(X_0,36R)} + \|u\|_{p,Q(X_0,36R)}\right). \tag{7.25}$$

PROOF. For M the constant from Proposition 7.12, set

$$\varepsilon = \frac{1}{2}\min\left\{1, \frac{1}{20^{n+2}M^p}\right\}$$

and let δ be the corresponding constant from Proposition 7.12. We also set

$$F = \|f\|_{p,Q(X_0,36R)} + R^{-1}\|u\|_{p,Q(X_0,36R)}$$

and note that

$$\|f\|_{2,Q(X_0,36R)} + R^{-1}\|u\|_{2,Q(X_0,36R)} \leq C(n)FR^{(p-2)/(2p)}.$$

We now take $\eta \in C^1(\overline{Q(X_0,36R)})$ a function which vanishes on $\mathcal{P}Q(X_0,36R)$ and is identically one on $Q(X_0,6R)$ such that $|D\eta| \leq 2/R$ and $|\eta_t| \leq 4/R^2$ and we use $\eta^2\bar{u}$ as test function in the weak form of the differential equation for u with and $\bar{u}$ a suitable Steklov average of u. As in the proof of Theorem 6.1, we infer that

$$\|Du\|_{2,Q(X_0,R)} \leq c_1 F|Q(X_0,R)|^{(p-2)/(2p)}$$

for some c_1 determined only by λ, Λ, and n. In addition, we see from (7.13) that there is a constant $c_2(\lambda,\Lambda,n)$ such that

$$\left|\{X \in Q(X_0,R) : \mathcal{M}|Du|^2 > \theta^2M^2\}\right| \leq \frac{c_2R^{-2(n+2)/p}}{\theta^2M^2}F^2|Q(X_0,R)|$$

for any $\theta > 0$. In particular, if we choose

$$\theta = \frac{c_2F}{MR^{(n+2)/p}}\varepsilon^{-1/2},$$

we obtain (7.23).

Next, we set $I = \|\mathcal{M}|Du|^2\|_{p/2,Q(X_0,R)}$, so

$$I = p\int_0^\infty \sigma^{p-1}|\{\mathcal{M}|Du|^2 > \sigma^2\}|\,d\sigma$$
$$\leq \theta^p M^p|Q(X_0,R)| + p\int_{\theta M}^\infty \sigma^{p-1}|\{\mathcal{M}|Du|^2 > \sigma^2\}|\,d\sigma.$$

Now

$$p\int_{\theta M}^\infty \sigma^{p-1}|\{\mathcal{M}|Du|^2 > \sigma^2\}|\,d\sigma = \sum_{j=1}^\infty p\int_{\theta M^j}^{\theta M^{j+1}} \sigma^{p-1}|\{\mathcal{M}|Du|^2 > \sigma^2\}|\,d\sigma,$$

and

$$p\int_{\theta M^j}^{\theta M^{j+1}} \sigma^{p-1}|\{\mathcal{M}|Du|^2 > \sigma^2\}|\,d\sigma \leq \theta^p M^{(j+1)p}|\{\mathcal{M}|Du|^2 > \theta^2 M^{2j}\}|.$$

Applying Corollary 7.13 yields

$$I \leq \theta^p M^p(|Q(X_0,R)| + I_1 + I_2)$$

for

$$I_1 = \sum_{j=1}^\infty M^{jp}\varepsilon_1^j|\{\mathcal{M}|Du|^2 > \theta^2 M^2\}|$$

and

$$I_2 = \sum_{j=1}^\infty M^{jp}\sum_{i=1}^j \varepsilon_1^i|\{\mathcal{M}f^2 > M^{2(j-i)}\delta^2\theta^2\}|.$$

Our choice for ε immediately implies that

$$I_1 \leq |Q(X_0,R)|$$

while we can rewrite

$$I_2 = \sum_{i=1}^\infty \varepsilon_1^i M^{ip}\sum_{j=i}^\infty M^{(j-i)p}|\{\mathcal{M}f^2 > M^{2(j-i)}\delta^2\theta^2\}|$$
$$= \sum_{i=1}^\infty \varepsilon_1^i M_{ip}\sum_{j=0}^\infty M^{jp}|\{\mathcal{M}f^2 > M^{2j}\delta^2\theta^2\}|$$
$$\leq \sum_{j=0}^\infty M^{jp}|\{\mathcal{M}f^2 > M^{2j}\delta^2\theta^2\}|$$
$$\leq |\{\mathcal{M}f^2 > \delta^2\theta^2\}| + M^p\sum_{j=1}^\infty M^{(j-1)p}|\{\mathcal{M}f^2 > M^{2j}\delta^2\theta^2\}|.$$

From the integral test for convergence, we infer that

$$\sum_{j=1}^{\infty} M^{(j-1)p}|\{\mathcal{M}f^2 > M^{2j}\delta^2\theta^2\}| \le M^p \int_0^\infty \sigma^{p-1}|\{\mathcal{M}f^2 > \sigma^2\delta^2\theta^2\}|\,d\sigma$$
$$= M^p \int_{Q(X_0,6R)} \frac{\mathcal{M}f^p}{(\delta\theta)^p}\,dX$$
$$\le c(n,p)M^p \int_{Q(X_0,6R)} \frac{f^p}{(\delta\theta)^p}\,dX.$$

Combining all these inequalities and using Corollary 7.10 shows that

$$\int_{Q(X_0,R)} |Du|^p\,dX \le C(\lambda,\Lambda,n,p)F$$

as desired. □

In order to manage the case $p < 2$, we want to prove our first variant of Proposition 7.12. As a preliminary step, we prove an estimate which generalizes Lemma 4.12.

LEMMA 7.15. *Let λ and Λ be positive constants and let (A^{ij}) be a constant matrix satisfying* (7.17) *for all $\xi \in \mathbb{R}^n$. Let $X_0 \in \mathbb{R}^{n+1}$ with $x_0^n = 0$, let $R > 0$, and set $\Omega = Q(X_0,R)$. Let $r \in (0,R)$, let $X_1 \in \mathcal{P}\Omega$, and suppose that h satisfies*

$$Lh = 0 \text{ in } \Omega[X_1,r], \quad h = 0 \text{ on } \mathcal{P}\Omega[X_1,r], \tag{7.26}$$

and $|h| \le M$ in $\Omega[X_1,r]$ for some nonnegative constant M. Then there is a constant C, determined only by n, λ, and Λ such that

$$\operatorname*{osc}_{\Omega[X_1,r]} h \le CM\frac{\rho}{r} \tag{7.27a}$$

for all $\rho \in (0,r)$ and

$$\sup_{\Omega[X_1,r/2]} |Dh| \le CM/R. \tag{7.27b}$$

PROOF. Note first that we can extend h to be zero for $t < t_0 - R^2$. Now rotate axes so that the positive x^n-axis is parallel to $x_0 - x_1$. Then the argument in Lemma 4.12 yields (7.27a). In conjunction with the local gradient estimate (4.12), we then obtain (7.27b). □

We are now ready to modify Proposition 7.12.

PROPOSITION 7.16. *Let λ and Λ be positive constants and let (A^{ij}) be a constant matrix satisfying* (7.17) *for all $\xi \in \mathbb{R}^n$. Let $X_0 \in \mathbb{R}^{n+1}$, let $R > 0$ and set $\Omega = Q(X_0,6R)$. Suppose u is a weak solution of*

$$-u_t + D_i(A^{ij}D_ju) = D_if^i \text{ in } \Omega,\ u = 0 \text{ on } \mathcal{P}\Omega. \tag{7.28}$$

Finally, let $X_1 \in \Omega$ and $r \in (0,R)$. Then there are positive constants $M(\lambda,\Lambda,n)$ and $\delta(\varepsilon,\lambda,\Lambda,n)$ such that, if

$$\{Y \in Q(X_1,r)\cap\Omega : \mathcal{M}|f|^2(Y) \leq \delta^2\theta^2,\ \mathcal{M}(|Du|^2)(Y) \leq \theta^2\} \neq \emptyset \tag{7.29}$$

for some $\theta > 0$, then

$$\left|\{X \in Q(X_1,r)\cap\Omega : \mathcal{M}(|Du|^2) > \theta^2 M^2\}\right| \leq \varepsilon |Q(X_1,r)|. \tag{7.30}$$

PROOF. If $d(X_1) \geq 4r$, then the argument of Proposition 7.12 can be used without change.

If $d(X_1) < 4r$, we first note that there is a constant $c_1(n)$ such that there are points $Y_1,\dots,Y_k$ in $\mathcal{P}\Omega\cap Q(X_1,6r)$, $Y_{k+1},\dots,Y_m$ in $\Omega\cap Q(X_1,6r)$ with $m \leq c_1(n)$ such that

$$\mathcal{P}\Omega\cap Q(X_1,r) \subset \bigcup_{j=1}^{k} Q(Y_i,\frac{r}{6}),$$

$$\Omega\cap Q(X_1,r) \subset \bigcup_{j=1}^{k} Q(Y_i,\frac{r}{6}) \cup \bigcup_{j=k+1}^{m} Q(Y_i,\frac{r}{c_1(n)}),$$

$$d(Y_j) \geq \frac{4r}{c_1(n)} \text{ for } j > k.$$

A straightforward modification of the proof of Proposition 7.12 shows that there are constants M_1 and c_2, determined only by λ, Λ, and n such that

$$|\{X \in Q(Y_j,\frac{r}{c_1(n)} : \mathcal{M}|Du|^2 > M_1^2\theta^2\}| \leq c_2\delta^2 r^{n+2} \tag{7.31}$$

for $j > k$.

If $i \leq k$, then we define

$$t^* = \min\{t_1 + \frac{r}{3}, t_0\}, \quad X^* = (y_i,t^*),$$

and we write Q_j for $Q(X^*, jr/6)\cap\Omega$. We now note that there a constant c_3, determined only by n such that

$$\frac{r^{n+2}}{c_3} \leq |Q_j| \leq c_3 r^{n+2}$$

for $j \in [1,6]$. It follows that

$$\int_{Q_4} |Du|^2\, dX \leq c(n)\theta^2 r^{n+2}.$$

With w solving (7.21) and $h = u - w$, we infer as before that

$$\int_{Q_4} |Dw|^2\, dX \leq c\theta^2 r^{n+2}$$

for some c determined only by λ, Λ, and n. The proof of Lemma 3.10 shows that

$$\int_{Q_4} h^2\,dX \le cr^{n+4}$$

and then Theorem 6.17 gives

$$\sup_{Q_{7/2}} |h| \le cr.$$

Lemma 7.15 then implies (7.22). Hence there are constants M_2 and c_4 (determined by λ, Λ, and n) such that

$$|\{X \in Q(Y_i, \frac{r}{6}) \cap \Omega : \mathcal{M}|Du|^2 > M_2^2\theta^2\}| \le c_4\delta^2 r^{n+2} \tag{7.32}$$

for $i \le k$. Summing (7.31) and (7.32) over i and j gives

$$|\{X \in Q(X_1,r) \cap \Omega : |Du|^2 > M^2\theta^2\}| \le c_5\delta^2 |Q(X_1,r)|$$

for $M = \max\{M_1, M_2\}$ and the proof is completed as before. □

We then obtain the following estimate for weak solutions of the Cauchy-Dirichlet problem in a cylinder. Because our theory of weak solutions is not applicable to (7.28) if $p < 2$, we work only with solutions that are known to have more smoothness. As we shall see, this form of the estimate is adequate for our purposes.

PROPOSITION 7.17. *Let λ and Λ be positive constants and let (A^{ij}) be a constant matrix satisfying* (7.17) *for all $\xi \in \mathbb{R}^n$. Let $X_0 \in \mathbb{R}^{n+1}$, let $R > 0$ and set $\Omega = Q(X_0, 6R)$. Let u be a weak solution of* (7.28) *for some $f \in L^\infty$. Then for any $p > 1$, there is a constant C, determined only by n, λ, Λ, and p such that*

$$\|Du\|_p \le C\|f\|_p. \tag{7.33}$$

PROOF. If $p \ge 2$, we follow the proof of Proposition 7.14 with Proposition 7.16 in place of Proposition 7.12.

If $p < 2$, we set $g = |Du|^{p-2}Du$ and note that $g \in L^\infty$ with $\|g\|_q = \|Du\|_p$. Then we take v to be the solution of the adjoint problem:

$$v_t + D_j(A^{ij}D_iv) = D_jg^j \text{ in } \Omega,\ v = 0 \text{ on } S\Omega \cup (B(x_0,6R) \times \{t_0\}).$$

By making the transformation $\tau = t_0 - t$, we see that this is equivalent to a Cauchy-Dirichlet problem and hence $\|Dv\|_q \le C\|g\|_q$ by the case $p \ge 2$. Now extend v and g to be zero in $B(x_0,6R) \times (0,1)$ and let v_h and g_h be their Steklov averages for $h \in (0,1)$. It follows that v_h is a $W^{1,2}$ solution of

$$v_{h,t} + D_j(A^{ij}D_iv_h) = D_jg_h^j \text{ in } \Omega, v_h = 0 \text{ on } S\Omega \cup (B(x_0,6R) \times \{t_0\})$$

and therefore

$$\int D_iug_h^i\,dX = \int uv_{h,t} - D_juA^{ij}D_iv_h\,dX.$$

But using v_h as a test function in the weak version of (7.28) gives

$$\int uv_{h,t} - D_i v_h A^{ij} D_j u\,dX = \int D_i v_h f^i\,dX.$$

It follows that

$$\int D_i u g_h^i\,dX = \int D_i v_h f^i\,dX.$$

Since $Dv_h \to Dv$ and $g_h \to g$ in L^q, it follows that

$$\int |Du|^p\,dX = \int D_i u g^i\,dX = \int D_i v f^i\,dX \le ||Dv||_q ||f||_p \le C||Du||_p^{p/q} ||f||_p.$$

Simple rearrangement completes the proof. □

4. Interior L^p estimates for solutions of nondivergence form constant coefficient equations

We now prove the basic L^p estimates for solutions of simple equations with constant coefficients. These estimates are parabolic analogs of the Calderón-Zygmund estimates for elliptic equations. We start with an L^p version of Theorem 4.9 in which we abbreviate

$$||g||_{p,R} = ||g||_{p,Q(X_0,R)}.$$

PROPOSITION 7.18. *Let $R > 0$, let $X_0 \in \mathbb{R}^{n+1}$, let $p > 1$, and let (A^{ij}) be a constant matrix satisfying* (7.17). *Let $u \in W_p^{2,1}(Q(X_0,2R))$ and set $f = Lu$. Then there is a constant $C(n,p,\lambda,\Lambda)$ such that*

$$\left\|D^2u\right\|_{p,R} + ||u_t||_{p,R} \le C\left(||f||_{p,2R} + R^{-1}||Du||_{p,2R} + R^{-2}||u||_{p,2R}\right). \tag{7.34}$$

PROOF. Let (u_m) be a sequence of $C^{2,1}$ functions which converge to u in $W_p^{2,1}$, and let η be a nonnegative $C^2(\overline{Q(X_0,2R)})$ function which vanishes, along with its spatial gradient, on $\mathscr{P}Q(X_0,2R)$. We may take η so that $|\eta_t| + |D^2\eta| + R^{-1}|D\eta| \le C(n)R^{-2}$. Then we set $v_m = \eta u_m$. It follows that $v_m = 0$ and $Dv_m = 0$ on $\mathscr{P}Q(X_0,2R)$ and that $Lv_m = \tilde{f}_m$ for

$$\tilde{f}_m = \eta Lu_m - u_m L\eta + A^{ij}D_i u_m D_j\eta + A^{ij}D_i\eta D_j u_m.$$

Then each component $D_k v_m$ is a weak solution of

$$-w_t + D_i(A^{ij}D_j w) = D_i f^i \text{ in } Q(X_0,36R),\ w = 0 \text{ on } \mathscr{P}Q(X_0,36R)$$

for $f^i = \delta^{ik}\tilde{f}_m$. It follows from Proposition 7.17 that

$$||Dv_m||_p \le C||\tilde{f}_m||_p$$

and the differential equation implies that

$$||v_{m,t}||_p \le C||\tilde{f}_m||_p.$$

The properties of η now imply (7.34) with u_m in place of u and Lu_m in place of f. Since $u_m \to u$ in $W_p^{2,1}$, it follows that the corresponding norms converge and that $Lu_m \to f$ in L^p. □

5. An interpolation inequality

In order to study equations with nonconstant coefficients, we use an L^p analog of the interpolation inequality Proposition 4.2. As a first step, we prove it in $\mathbb{R}^n$.

LEMMA 7.19. *Let $u \in W^{2,p}(\mathbb{R}^n)$ for some $p > 1$. Then, for any $\varepsilon > 0$, we have*

$$\|Du\|_p \le \varepsilon \left\|D^2 u\right\|_p + 36\frac{n}{\varepsilon}\|u\|_p. \tag{7.35}$$

PROOF. We start by proving (7.35) in the one dimensional case. Let $v \in C_0^2(\mathbb{R})$, fix $a \in \mathbb{R}$, set $b = a + \varepsilon/2$, and choose $x_1 \in (a, a+\varepsilon/6)$ and $x_2 \in (b - \varepsilon/6, b)$. Then the mean value theorem implies that there is x_3 with $x_1 < x_3 < x_2$ and

$$|v'(x_3)| = \left|\frac{v(x_1) - v(x_2)}{x_1 - x_2}\right| \le \frac{6}{\varepsilon}(|v(x_1)| + |v(x_2)|),$$

and therefore

$$|v'(x)| \le \frac{6}{\varepsilon}(|v(x_1)| + |v(x_2)|) + \int_a^b |v''(y)|\, dy$$

for any $x \in (a,b)$. Integrating this inequality with respect to x_1 and x_2 now gives

$$|v'(x)| \le \int_a^b |v''(y)|\, dy + \frac{36}{\varepsilon^2}\int_a^b |v(y)|\, dy.$$

From Hölder's inequality, we infer that

$$|v'(x)|^p \le 2^p \left[\left(\frac{\varepsilon}{2}\right)^{p-1}\int_a^b |v''(y)|^p\, dy + \left(\frac{18}{\varepsilon}\right)^p \frac{2}{\varepsilon}\int_a^b |v(y)|^p\, dy\right].$$

Integration with respect to x then yields

$$\int_a^b |v'(x)|^p\, dy \le \varepsilon^p \int_a^b |v''(y)|^p\, dy + \left(\frac{36}{\varepsilon}\right)^p \int_a^b |v(y)|^p\, dy.$$

If we then subdivide $\mathbb{R}$ into intervals of length $\varepsilon/2$ with disjoint interiors and add the resultant inequalities, we find that

$$\int |v'(x)|^p\, dx \le \varepsilon^p \int |v''(y)|^p\, dy + \left(\frac{36}{\varepsilon}\right)^p \int |v(y)|^p\, dy.$$

For the higher dimensional case, we apply this inequality to u considered as a function of the single variable x^i and then integrate over the other coordinates. In this way, we obtain

$$\int |D_i u|^p\, dx \le \varepsilon^p \int |D_{ii} u|^p\, dy + \left(\frac{36}{\varepsilon}\right)^p \int |u|^p\, dy$$

for any $u \in C_0^2$. By approximation, this inequality is also valid for $u \in W^{2,p}$. We then obtain (7.35) by adding with respect to i and applying Minkowski's inequality. □

In our applications, we need a version of this inequality in bounded domains.

LEMMA 7.20. *Let $u \in W^{2,p}(\omega)$ for some $p > 1$ and some $\omega \subset \mathbb{R}^n$ with $\partial\Omega \in C^{1,1}$. Then, for any $\varepsilon > 0$, we have*

$$\|Du\|_p \leq \varepsilon \left\|D^2 u\right\|_p + \frac{C(n,\Omega)}{\varepsilon} \|u\|_p. \tag{7.36}$$

PROOF. It suffices to extend u to a function $\hat{u} \in W^{2,p}(\mathbb{R}^n)$ with compact support so that

$$\left\|D^2 \hat{u}\right\|_p \leq C[\left\|D^2 u\right\|_p + \|Du\|_p + \|u\|_p], \; \|\hat{u}\|_p \leq C\|u\|_p.$$

As in Theorem 6.35, we may assume that $u \in W^{2,p}(B^+(R))$ with $u(x) = 0$ and $Du = 0$ if $|x| = R$. It's straightforward to check (as in Lemma 6.34) that

$$\hat{u}(x) = \begin{cases} u(x) & \text{if } x^n \geq 0, \\ -3u(x', -x^n) + 4u(x', -\frac{1}{2}x^n) & \text{if } x^n \geq 0 \end{cases}$$

has all these properties. □

COROLLARY 7.21. *Suppose $\omega = B(R)$ or $B^+(R)$ for some $R > 0$. If $u \in W^{2,p}(\omega)$ for some $p > 1$, then*

$$\|Du\|_p \leq \varepsilon \left\|D^2 u\right\|_p + \frac{C(n)}{\varepsilon} \|u\|_p. \tag{7.37}$$

PROOF. If $\omega = B(R)$, then we use a scaling argument to infer that the constant in (7.37) is independent of R. If $\omega = B^+(R)$, we use the extension argument of Lemma 7.20 to extend u to a $W^{2,p}(B(R))$ function $\hat{u}$. and then apply the inequality to $\hat{u}$. □

6. Interior L^p estimates

We now derive interior L^p estimates for the derivatives u_t and D^2u when u is a strong solution of a second order parabolic equation. The basic idea is to use a perturbation argument like that in Chapter IV.

THEOREM 7.22. *Let $\Omega \subset \mathbb{R}^{n+1}$ be a domain, let $u \in W^{2,p}_{loc}(\Omega) \cap L^p(\Omega)$ for some $p \geq 2$. Define the operator L by (7.1) and suppose that the coefficients* *satisfy*

$$\lambda|\xi|^2 \le a^{ij}\xi_i\xi_j \le \Lambda|\xi|^2, \tag{7.38a}$$

$$|b^i(X)| \le B/d(X),\ |c(X)| \le c_1/d(X)^2, \tag{7.38b}$$

$$|a^{ij}(X) - a^{ij}(Y)| \le \omega(\frac{|X-Y|}{d(X,Y)}) \tag{7.38c}$$

for some positive constants λ, Λ, B, *and* c_1 *and a positive, continuous, increasing function* ω *with* $\omega(0) = 0$. *Then, for any subdomain* Ω' *with* $d(\Omega', \mathscr{P}\Omega) > \varepsilon$, *there is a constant* C *determined only by* λ, Λ, B, c_1, ε, *and* ω *such that*

$$\|D^2u\|_{p,\Omega'} + \|u_t\|_{p,\Omega'} \le C(\|Lu\|_{p,\Omega} + \|u\|_{p,\Omega}). \tag{7.39}$$

PROOF. Fix $X_0 \in \Omega'$ and define $L_0 = a^{ij}(X_0)D_{ij} - D_t$ and $f = Lu$. For $\delta \in (0,1/2)$ to be chosen, set $R = \delta d(X_0)$, let η be a cut-off function supported in $Q(R,X_0)$, and let $v = \eta u$. Then, by direct calculation,

$$L_0v = (a^{ij}(X_0) - a^{ij})D_{ij}v + (L\eta)u + \eta f + 2a^{ij}D_iuD_j\eta - \eta[b^iD_iu + cu].$$

Then Proposition 7.18 (applied to v) implies that

$$\begin{aligned}\|D^2v\|_p \le C_1\omega(\delta)\|D^2v\|_p &+ C\|f\|_p \sup\eta \\ &+ C\|Du\|_{p,\operatorname{supp}\eta}[\frac{\sup\eta}{R} + \sup|D\eta|] \\ &+ C\|u\|_p[\frac{\sup\eta}{R^2} + \sup|D^2\eta| + \sup|\eta_t|].\end{aligned}$$

Now we take δ so that $C_1\omega(\delta) \le \frac{1}{2}$. For $\sigma \in (\frac{1}{2}, 1)$, we take η so that

$$0 \le \eta \le 1 \text{ in } Q(R),$$

$$\eta = 1 \text{ in } Q(\sigma R), \eta = 0 \text{ in } Q(R) \setminus Q(\frac{1+\sigma}{2}R),$$

$$|D\eta| \le \frac{4R}{1-\sigma}, |D^2\eta| + |\eta_t| \le \frac{C(n)R^2}{(1-\sigma)^2}.$$

With this choice for η, we have

$$\|D^2u\|_{p,\sigma R} \le C(\|f\|_{p,R} + \frac{1}{(1-\sigma)R}\|Du\|_{p,(1+\sigma)R/2} + \frac{1}{(1-\sigma)^2R^2}\|u\|_{p,R}),$$

where we use the subscript p,r to indicate the $L^p(Q(r))$ norm.

For any nonnegative integer k, we now define the seminorm

$$[u]_k = \sup_{0<\sigma<1}(1-\sigma)^kR^k\left\|D^ku\right\|_{p,\sigma R}.$$

With this definition, our estimate can be written as

$$[u]_2 \le C_2(R^2\|f\|_p + [u]_1 + [u]_0).$$

To eliminate the term $[u]_1$ from the right side of this inequality, we note that, for any $\theta > 0$, there is a $\sigma(\theta)$ such that

$$[u]_1 \leq (1-\sigma(\theta))R\|Du\|_{p,\sigma(\theta)R} + \theta.$$

Hence

$$\begin{aligned}[u]_1 &\leq \frac{1}{2C_2}(1-\sigma(\theta))^2R^2\left\|D^2u\right\|_{p,\sigma(\theta)R} + C\|u\|_{p,\sigma(\theta)} + \theta \\ &\leq \frac{1}{2C_2}[u]_2 + C[u]_0 + \theta\end{aligned}$$

by Lemma 7.19 and the definition of the seminorms. Therefore, after sending θ to zero, we find that

$$[u]_1 + [u]_2 \leq C(R^2\|f\|_p + [u]_0).$$

It follows that

$$R^2\left\|D^2u\right\|_{p,R/2} + R\|Du\|_{p,R/2} \leq C(\|f\|_p + \|u\|_p).$$

If we now cover Ω' by a finite number of cylinders of the form $Q(\delta d(X_0), X_0)$, we obtain the estimate on D^2u. The estimate on u_t follows from this estimate by taking the differential equation into account. □

7. Boundary and global estimates

In order to derive estimates near the parabolic boundary of a domain, we first consider the case of a flat boundary portion. The key new step is a boundary version of Proposition 7.12. To this end, we recall that $Q^+(X_0,R)$ is the set of all $x \in Q(X_0,R)$ with $x^n > 0$.

PROPOSITION 7.23. *Let λ and Λ be positive constants and let (A^{ij}) be a constant matrix satisfying* (7.17) *for all $\xi \in \mathbb{R}^n$. Let $X_0 \in \mathbb{R}^{n+1}$ with $x_0^n = 0$, let $R > 0$, and set $\Omega = Q^+(X_0,R)$. Suppose u is a weak solution of* (7.28). *Finally, let $X_1 \in \Omega$, let $r \in (0,R)$, and let $\varepsilon > 0$. Then there are constants $M(\lambda,\Lambda,n)$ and $\delta(\varepsilon,\lambda,\Lambda,n)$ such that* (7.29) *for some $\theta > 0$ implies* (7.30).

PROOF. Replace Lemma 3.10 by Corollary 3.11 in the proof of Proposition 7.16, and note that Lemma 7.15 remains true if we replace Q by Q^+. □

PROPOSITION 7.24. *Proposition 7.17 is valid for $\Omega = Q^+(X_0, 6R)$.*

PROOF. The result follows from Proposition 7.23 using the arguments in Proposition 7.17. □

As in Theorem 4.22, in order to obtain boundary $W_p^{2,1}$ estimates, we also need a result on the oblique derivative problem. In this case, the geometry is somewhat more complicated because we only have existence results in domains of the type Σ^+ from Section IV.8.

We start by recalling some notation. First, we assume that β is a constant vector and that there is a nonnegative constant μ such that

$$|\beta'| \le \mu\beta^n. \tag{7.40}$$

We then fix the constant $\bar{\mu} \in (\mu/(1+\mu^2)^{1/2}, 1)$, and, for $R > 0$ and $X_0' \in \mathbb{R}^{n+1}$ with $x_0^n = 0$, we set

$$x_0 = (x_0^1, \dots, x_0^{n-1}, -\bar{\mu}R, t_0)$$

and we write $\Sigma^+(X_0,R)$ and $\Sigma^0(X_0,R)$ for the subsets of $Q(X_0,R)$ on which $x^n > 0$ and $x^n = 0$, respectively. We also define $\sigma(X_0,R) = \mathscr{P}\Sigma^+(X_0,R) \setminus \Sigma^0(X_0,R)$. Moreover, with this X_0' fixed and $X_1' \in \mathbb{R}^{n+1}$ with $|x_1' - x_0| = R$ and $x_1^n = 0$, we define

$$X_2 = X_1 + \left(1 - \frac{r}{R}\right)(X_1 - X_0),$$

and

$$\begin{aligned} \Sigma^+(X_1;R,r) &= \Sigma^+(X_0,R) \cap \Sigma^+(X_2,R), \\ \Sigma^0(X_1;R,r) &= \Sigma^0(X_0,R) \cap \Sigma^0(X_2,R), \\ \sigma(X_1;R,r) &= \mathscr{P}\Sigma^+(X_0;R,r) \setminus \Sigma^0(X_1;R,r). \end{aligned}$$

We then have our final version of Proposition 7.12,

PROPOSITION 7.25. *Let λ and Λ be positive constants and let (A^{ij}) be a constant matrix satisfying* (7.17) *for all $\xi \in \mathbb{R}^n$. Let $X_0 \in \mathbb{R}^{n+1}$, let $R > 0$ and set $\Omega = \Sigma^+(X_0,6R)$. Suppose u is a weak solution of*

$$Lu = D_i f^i \text{ in } \Omega, \quad u = 0 \text{ on } \sigma(X_0,6R), \quad \beta \cdot Du = g \text{ on } \Sigma^0(X_0,6R) \tag{7.41}$$

for $g = \beta^n f^n / A^{nn}$, and let $X_1 \in \overline{\Omega}$ and $r \in (0,R)$. Then there are positive constants $M(\lambda,\Lambda,n,\mu,\bar{\mu})$ and $\delta(\varepsilon,\lambda,\Lambda,n,\mu,\bar{\mu})$ such that (7.29) *for some $\theta > 0$ implies* (7.30).

PROOF. This time, we cover $Q(X_1,r)$ by sets of four types: $\Sigma^+(Y_i;R,r_i)$ with $Y_i \in \overline{\Sigma^0(X_0,R)} \cap \sigma(X_0,R)$, $\Sigma^+(Y_i,r_i)$ with $Y_i \in \overline{\Sigma^0(X_0,R)}$, $Q(Y_i,r_i) \cap \Omega$ with $Y_i \in \sigma(X_0,R)$, and $Q(Y_i,r_i)$. The important geometric facts here are that sets of the second type do not intersect $\sigma(X_0,R)$, sets of the third type do not intersect $\Sigma^0(X_0,R)$, and sets of the fourth type do not intersect $S\Omega$.

We observe that u satisfies

$$-u_t + D_i(\tilde{A}^{ij} D_j u) = D_i f^i \text{ in } \Omega, \ \tilde{A}^{nj} D_j u = f^n \text{ on } \Sigma^0(X_0,R)$$

for the matrix $(\tilde{A}^{ij})$ given by

$$\tilde{A}^{ij} = \begin{cases} A^{ij} & \text{for } i,j < n, \\ A^{nn}\beta^j/\beta^n & \text{for } i = n, j < n, \\ A^{in} + A^{ni} - A^{nn}\beta^i/\beta^n & \text{for } i < n, j = n. \end{cases}$$

(See the proof of Lemma 4.17.) Moreover, this matrix satisfies the inequalities (7.17) with λ and Λ replaced by suitable constants determined by λ, Λ, and μ.

For sets of the first or second type, when we estimate Dh, we use the proof of Lemma 4.20 to obtain (4.13) in our current version of Lemma 4.12. For the L^2 estimates in sets of the first type, we invoke Theorems 6.45, 6.46, 6.47 in place of Theorems 6.1, 6.2, and 6.17, respectively. For sets of the second type, we use Theorems 6.45, 6.46, and 6.41.

Finally, estimates in sets of the third or fourth type follow the exact reasoning of Proposition 7.16. □

We then obtain an L^p estimate for Du if u solves the appropriate mixed boundary value problem.

PROPOSITION 7.26. *Let λ and Λ be positive constants and let (A^{ij}) be a constant matrix satisfying* (7.17) *for all $\xi \in \mathbb{R}^n$. Let $X_0 \in \mathbb{R}^{n+1}$, let $R > 0$ and set $\Omega = \Sigma^+(X_0, 6R)$. Then, for any $f \in L^\infty(\Omega)$, there is a unique weak solution u of* (7.41). *Moreover, for any $p > 1$,*

$$||Du||_p \leq C(n, \lambda, \Lambda, p, \mu)||f||_p. \tag{7.42}$$

PROOF. For $p \geq 2$, we argue as before with Proposition 7.25 in place of Proposition 7.12. For $p < 2$, we modify the duality argument only a little. With $(\tilde{A}^{ij})$ as in Proposition 7.25, we take v to be the solution of

$$-v_t + D_j(\tilde{A}^{ij} D_i v) = D_j g^j \text{ in } \Omega,\ v = 0 \text{ on } \sigma,\ \tilde{A}^{jn} D_j v = 0 \text{ on } \Sigma^0.$$

□

In order to apply our results, we also need a special cut-off function.

LEMMA 7.27. *Let μ be a nonnegative constant and let β be a constant vector such that $|\beta'| \leq \mu\beta^n$. Then, for any $R > 0$, there is a nonnegative function $\zeta \in C^{2,1}(\overline{Q(2R)})$ with support in $\overline{Q(3R/2)}$ such that $\zeta \leq 1$ and $R|D\zeta| + R^2|D^2\zeta| + R^2|\zeta_t| \leq C(n, \mu)$ in $Q(2R)$, $\zeta \equiv 1$ in $Q(R)$, and $\beta \cdot D\zeta \equiv 0$ on $Q^0(2R)$.*

PROOF. Let $g \in C^2([0,1])$ with $g(0) = 0$, $g'(0) = 1$, and $0 \leq 18\mu g \leq 1$ on $[0,1]$. Also, let $h \in C^2([0,\infty)$ be decreasing with $h \equiv 1$ in $[0, 1/3)$ and $h \equiv 0$ in $[1/2, \infty)$. We then define G on $B(2R)$ by

$$G(x) = \frac{|x|^2}{4R^2} + 9\mu^2 \left[g\left(\frac{x^n}{2R}\right) \right]^2 - g\left(\frac{x^n}{2R}\right) \frac{\beta' \cdot x}{\beta^n R},$$

and we take

$$\zeta(X) = h \circ G(x) h\left(-\frac{3t}{2R^2}\right).$$

From Cauchy's inequality, we have

$$\frac{2}{9}\frac{|x|^2}{R^2} \leq G(x) \leq \frac{5}{18}\frac{|x|^2}{R^2} + 18\mu^2 \left[g\left(\frac{x^n}{2R}\right) \right]^2,$$

which implies the statement about the support of ζ, the estimate $\zeta \le 1$, the estimates on the derivatives of ζ, and the fact that $\zeta \equiv 1$ in $Q(R)$. A simple calculation (using $g(0) = 0$ and $g'(0) = 1$) leads to $\beta \cdot D\zeta \equiv 0$ on Q^0. □

From this result, we can infer a local boundary $W_p^{2,1}$ estimate.

PROPOSITION 7.28. *Let $R > 0$, let $X_0 \in \mathbb{R}^{n+1}$ with $x^n = 0$, and let $u \in W_p^{2,1}(Q^+(X_0,2R))$ for some $p > 1$ with $u = 0$ on $Q^0(X_0,R)$. Let (A^{ij}) be a constant matrix satisfying* (7.17) *and set $f = Lu$. Then* (7.34) *is valid.*

PROOF. First, let ζ be the function from Lemma 7.27 corresponding to β defined by

$$\beta^k = \begin{cases} A^{nk} + A^{kn} & \text{if } k < n, \\ A^{nn} & \text{if } k = n. \end{cases}$$

We argue as in Proposition 7.18 (but with Proposition 7.24 in place of Proposition 7.17) to obtain L^p estimates for $D_{ij}u$ with $i + j < 2n$. Then we observe that $w = D_n(\zeta u)$ satisfies $Lw = D_i f^i$ in $Q^+(X_0,2R)$ and $\beta \cdot Dw = f^n$ on $Q^0(X_0,2R)$ with

$$f^i = \delta^{in}[\zeta f + uL\zeta + \beta^n D_n u D_n \zeta].$$

Working in $\Sigma^+(X_0,c_1R)$ for c_1 large enough that $Q^+(X_0,2R) \subset \Sigma^+(X_0,c_1R)$ gives the L^p estimate for $D_{in}u$ for $i \le n$. As in Proposition 7.18 (with Proposition 7.26 in place of Proposition 7.17), the L^p estimate for all the second derivatives implies the same estimate for u_t. □

Regularity near $B\Omega$ is also straightforward using the notation

$$Q_T(X_0,r) = B(X_0,r) \times (0,T), \|g\|_{p,R,T} = \|g\|_{p,Q_T(X_0,R)}.$$

PROPOSITION 7.29. *Let $R > 0$, let $T > 0$, let $X_0 \in \mathbb{R}^{n+1}$ with $x^n = 0$, let $p > 1$, and let $u \in W_p^{2,1}(B^+(X_0,2R) \times (0,T))$ with $u = 0$ on $B^0(X_0,2R) \times (0,T)$. Let (A^{ij}) be a constant matrix satisfying* (7.17) *and set $f = Lu$. Then there is a constant C, determined only by n, p, λ, Λ, n and T/R^2 such that*

$$\left\|D^2u\right\|_{p,R,T} + \|u_t\|_{p,R,T} \le C\left(\|f\|_{p,R,T)} + R^{-1}\|Du\|_{p,R,T}\right). \tag{7.43}$$

PROOF. We first assume that $T < R^2$. Then, if we extend u and f to be zero for $t < 0$, it follows that $u \in W_p^{2,1}(Q(X_0,2R))$ and hence we obtain the estimates for D^2u and u_t in $L^p(Q(X_0,R))$. But these norms are the same as the $L^p(Q_T(X_0,R)$ norms. If $R^2 \le T < 2R^2$, we note that $Q_T(X_0,R)$ is the union of $Q(X_0,R)$ and $Q_{T-R^2}(X_0,R)$ and the extension argument gives the L^p estimates in each of these subcylinders. A similar decomposition (but also invoking the estimates from Proposition 7.28 directly) gives the general result. □

The arguments of Proposition 7.18 are easily modified to obtain an estimate when a homogeneous Dirichlet boundary condition is prescribed on part of $S\Omega$.

The only complication is that the change of variables used to transform a portion of $S\Omega$ into a flat boundary must preserve the continuity hypothesis for a^{ij}. For this preservation, we use $H_{1+\alpha}$ domains for suitable α. Thus, we have the following result.

THEOREM 7.30. *Let $p > 1$ and $\alpha \in (0,1)$ such that $p(1-\alpha) < 1$. Let $\Omega \subset \mathbb{R}^{n+1}$, with an $H_{1+\alpha}$ boundary portion $\Delta \subset S\Omega$ and suppose L is defined by* (7.1) *with coefficients satisfying* (7.38a) *and*

$$|b^i| \le B, |c| \le c_1, \tag{7.44a}$$

$$|a^{ij}(X) - a^{ij}(Y)| \le \omega(|X-Y|). \tag{7.44b}$$

If $u \in W^{2,1}_p$ and $u = 0$ on Δ, then, for any $\varepsilon > 0$ and any subdomain Ω' with $\operatorname{dist}(\Omega', \mathscr{P}\Omega \setminus \Delta) > \varepsilon$, *there is a constant C such that* (7.39) *holds.*

PROOF. The only thing that needs to be checked is the behavior of the terms associated with the change of variables which transforms Δ to a subset of $\mathbb{R}^{n+1}_0$. In particular (referring back to Section IV.7), we note that the troublesome terms are of the form $V^k D_k u$ with $|V| \le c(x^n)^{\alpha-1}$. Note that we may assume that u is supported in $Q(R)$ for some $R > 0$.

To estimate

$$\int (x^n)^{p(\alpha-1)} |Du|^p \, dX,$$

we first perform the integration with respect to x^n, suppressing the other variables from the notation, and noting that $Du(R) = 0$:

$$\begin{aligned}\int_0^R (x^n)^{p(\alpha-1)} |Du|^p \, dx^n &\le \int_0^R (x^n)^{p(\alpha-1)} \, dx^n \sup_{0<r<R} |Du(r)|^p \\ &= C(p) R^{p\alpha} R^{1-p} \sup_{0<r<R} |Du(r)|^p \\ &\le C R^{p\alpha} R^{1-p} \Big(\int_0^R |D_n Du(x^n)| \, dx^n\Big)^p \\ &\le C R^{p\alpha} \int_0^R |D_n Du(x^n)|^p \, dx^n\end{aligned}$$

since $Du(R) = 0$. It follows that

$$\int (x^n)^{p(\alpha-1)} |Du|^p \, dX \le C R^{p\alpha} \int |D^2 u|^p \, dX.$$

□

When the boundary portion Δ is all of $S\Omega$ and we also assume that $u = 0$ on $B\Omega$, we can prove a better estimate.

COROLLARY 7.31. *Let α and p be as in Theorem 7.30. If $\mathscr{P}\Omega \in H_{1+\alpha}$, and $u \in W_p^{2,1}$ is a solution of $Lu = f$ in Ω, $u = 0$ on $\mathscr{P}\Omega$, then*

$$\|D^2u\|_p + \|u_t\|_p \le C\|f\|_p. \tag{7.45}$$

PROOF. From Theorem 7.30, it suffices to show that we can estimate $\|u\|_p$ in terms of $\|f\|_p$. For this estimate, we note that the function v defined by

$$v(X) = \begin{cases} u(X) & \text{if } X \in \Omega \\ 0 & \text{if } X \notin \Omega \end{cases}$$

is in $W^{1,p}$. It is easy to see that

$$\|u\|_p = \|v\|_p, \|u_t\|_p = \|v_t\|_p.$$

Moreover,

$$\begin{aligned} \int_{\mathbb{R}^n} |v(x,\tau)|^p\, dx &= \int_0^\tau \int_{\mathbb{R}^n} \frac{d}{dt} |u(X)|^p\, dx dt \\ &= \int_0^\tau \int_{\mathbb{R}^n} p|u(X)|^{p-1} |u_t(X)|\, dX. \end{aligned}$$

We now estimate this last integral using Hölder's inequality and then note that

$$\int_0^\tau \int_{\mathbb{R}^n} |u_t(X)|^p\, dX \le C(\int_0^\tau \int_{\mathbb{R}^n} |u(X)|^p\, dX + \|f\|_p^p).$$

Hence $V(\tau) = \int_{\mathbb{R}^n} |v(x,\tau)|^p\, dx$ satisfies the inequality

$$V(\tau) \le C(\int_0^\tau V(t)\, dt + \|f\|_p^p).$$

Hence Gronwall's inequality gives a bound for V, and therefore

$$\|u\|_p = (\int_0^T V(t)\, dt)^{1/p} \le C\|f\|_p.$$

From this inequality and (7.34), we infer (7.45). □

From this estimate, we can infer an existence and uniqueness theorem analogous to Theorem 5.14.

THEOREM 7.32. *Suppose $\Omega \subset \mathbb{R}^{n+1}$ with $\mathscr{P}\Omega \in H_2$, and let L, p, and α be as in Theorem 7.30. Then for any $\varphi \in W_p^{2,1}$ and any $f \in L^p(\Omega)$, there is a unique solution of $Lu = f$ in Ω, $u = \varphi$ on $\mathscr{P}\Omega$. Moreover, u satisfies the estimate*

$$\|u\|_p + \|Du\|_p + \|D^2u\|_p + \|u_t\|_p \le C(\|f\|_p + \|\varphi\|_p + \|D\varphi\|_p + \|D^2\varphi\|_p + \|\varphi_t\|_p). \tag{7.46}$$

PROOF. If u is a solution of $Lu = f$ in Ω, $u = \varphi$ on $\mathscr{P}\Omega$, then $v = u - \varphi$ is a solution of $Lv = g$ in Ω, $v = 0$ on $\mathscr{P}\Omega$ for $g = f - L\varphi$ and conversely, given v a solution of this problem, $u = v + \varphi$ is a solution of the original one. To solve the problem for v, first mollify the coefficients of L and mollify g. This mollified problem has a unique solution by Theorem 5.15 for $b < 2 - 1/p$, and then Corollary 7.31 gives uniform estimates on the approximating problems. Extracting a suitable subsequence gives a solution and then Corollary 7.31 shows that this solution is unique. □

8. $W_p^{2,1}$ estimates for the oblique derivative problem

When the Dirichlet condition on $S\Omega$ is replaced by an oblique derivative condition, the considerations of the previous section need only a minor modification. The only serious technical difficulty is connected with the extension of nonzero boundary data. For zero data and operators with constant coefficients, we have a simple variant of Proposition 7.28.

PROPOSITION 7.33. *Let $R > 0$, let $X_0 \in \mathbb{R}^n$ with $x^n = 0$, let β be a constant vector with $\beta^n > 0$, and let $u \in W_p^{2,1}(Q^+(X_0,R))$ satisfy $\beta \cdot Du = 0$ on $Q^0(X_0,R)$. Let A^{ij} be a constant matrix satisfying* (7.17) *and set $f = Lu$. If μ is a constant such that $|\beta'| \le \mu$, then there is a constant C, determined only by n, λ, Λ, μ such that* (7.34) *holds.*

PROOF. With ζ from Lemma 7.27, we first apply Proposition 7.26 to $w = D_j(\zeta u)$ if $j < n$ and note that $Lw = D_i f^i$ with $f^n = 0$ on Q^0, and then we apply Proposition 7.24 to $\beta \cdot D(\zeta u)$. □

Our next step is an extension lemma for functions defined on $\mathbb{R}^n_+ \times (0,T)$.

LEMMA 7.34. *Let B^i, $i = 0, \dots, n$ be functions in $H_\alpha(\mathbb{R}^n_0 \times (0,T))$ and let $u \in W_p^{2,1}(\mathbb{R}^n_+ \times (0,T))$ with $u = 0$ on $\mathbb{R}^n_+ \times \{0\}$ and support in $Q(R,X_0)$ for some positive R and some $X_0 \in \mathbb{R}^n_0 \times (0,T)$. If $p(1-\alpha) < 1$ and if $B^i(X_0) = 0$ for all $i > 0$, then there are functions z_1 and z_2 in $W_p^{2,1}$ such that*

$$z_1 = z_2 = 0 \text{ on } \mathbb{R}^n_0 \times (0,T), \tag{7.47a}$$

$$D_n z_1 = \sum_{i>0} B^i D_i u \text{ and } D_n z_2 = B^0 u \text{ on } \mathbb{R}^n_0 \times (0,T), \tag{7.47b}$$

$$\|z_{1,t}\|_p + \|D^2 z_1\|_p \le C(n,p,\alpha,[B^i]_\alpha) R^\alpha (\|D^2 u\|_p + \|u_t\|_p). \tag{7.47c}$$

$$\|z_{2,t}\|_p + \|D^2 z_2\|_p \le C(n,p,\alpha,|B^0|_\alpha)(1+R^\alpha)\|Du\|_p. \tag{7.47d}$$

PROOF. Extend each B^i as an $H_2^{(-\alpha)}$ function. With φ and η as in Lemma 4.3, we then define z_1 by

$$z_1(X) = x^n B^m \int_{\mathbb{R}^{n+1}_+} D_m u(x - x^n y/2, t - (x^n)^2 s)\varphi(y)\eta(s)\,dY,$$

with the index m running from 1 to n. It's easy to check that $z_1 = 0$ and $D_n z_1 = B^m D_m u$ on $\mathbb{R}^n_0 \times (0,T)$, so we only need to estimate the higher derivatives. Suppressing the argument of the functions involved and writing ∂_i for $\partial/\partial y^i$, we have

$$\begin{aligned} D_{ij}z_1 &= (D_{ij}B^m x^n + D_j B^m \delta_{in} + \delta_{jn} D_i B^m) \int D_m u \varphi \eta \, dY \\ &\quad + D_i B^m \int D_m u[\partial_j \varphi \eta + \delta_{jn} \partial_k (y^k \varphi) - 2\delta_{jn} \varphi \eta'] \, dY \\ &\quad + B^m \int D_{mi} u[\partial_j \varphi \eta + \delta_{jn} \partial_k (y^k \varphi) - 2\delta_{jn} \varphi \eta'] \, dY. \end{aligned}$$

Using our estimates on B and its derivatives along with Tonelli's theorem, we find that

$$\int |D^2 z_1|^p \, dX \le C R^{p\alpha} \int |D^2 u|^p \, dX + C \int (x^n)^{p(\alpha-1)} |Du|^p \, dX.$$

To illustrate the derivation, we look at

$$I = \int_{\mathbb{R}^{n+1}_+} \left| D_i B_m \int_{Q(1)} D_m u \partial_j \varphi \eta \, dY \right|^p \, dX.$$

From our estimates on B and its derivatives and the assumed bounds on $\partial\varphi$ and η, we have

$$\begin{aligned} I &\le C \int_{\mathbb{R}^{n+1}_+} (x^n)^{p(\alpha-1)} \left(\int_{Q(1)} |Du| \, dY \right)^p \, dX \\ &\le C \int_{\mathbb{R}^{n+1}_+} (x^n)^{p(\alpha-1)} \int_{Q(1)} |Du|^p \, dY \, dX \end{aligned}$$

(by Hölder's inequality)

$$= C \int_{Q(1)} \int_{\mathbb{R}^{n+1}_+} (x^n)^{p(\alpha-1)} |Du(x - \frac{x^n y}{2}, t - (x^n)^2 s)|^p \, dX \, dY.$$

Next we note that $\frac{1}{2}x^n \le x^n - \frac{1}{2}x^n y^n \le \frac{3}{2}x^n$, so

$$\begin{aligned} &\int_{\mathbb{R}^{n+1}_+} (x^n)^{p(\alpha-1)} |Du(x - \frac{x^n y}{2}, t - (x^n)^2 s)|^p \, dX \\ &\quad \le C \int_{\mathbb{R}^{n+1}_+} (x^n - \frac{x^n y^n}{2})^{p(\alpha-1)} |Du(x - \frac{x^n y}{2}, t - (x^n)^2 s)|^p \, dX \\ &\quad = C \int_{\mathbb{R}^{n+1}_+} (x^n)^{p(\alpha-1)} |Du(x,t)|^p \, dX. \end{aligned}$$

Since $|Q(1)| = C(n)$, it follows that

$$I \le C \int (x^n)^{p(\alpha-1)} |Du|^p \, dX,$$

and the other terms in this representation of z_1 are estimated similarly.

Recalling the estimate in the proof of Theorem 7.30, we conclude that

$$\|D^2 z_1\|_p \le CR^\alpha \|D^2 u\|_p .$$

To estimate $z_{1,t}$, we note that

$$z_{1,t} = x^n B_t^m \int D_m u \varphi \eta \, dY \\ + B^m \int u_t [2\partial_m \varphi \eta + \delta_{mn}(\partial_j (y^j \varphi) - 2\varphi (s\eta)'] \, dY,$$

so

$$\|z_{1,t}\|_p \le CR^\alpha \left(\|D^2 u\|_p + \|u_t\|_p \right).$$

The same arguments work for the derivatives of z_2 except that we take $g = B^0 u$ and note that $|B^0| \le C$. □

From these preliminaries, we can prove the analog of Theorem 7.30 for the oblique derivative problem.

THEOREM 7.35. *Suppose Ω, Δ, p, α, and L are as in Theorem 7.30. Let β be a vector field defined on Δ such that $\beta \cdot \gamma > 0$ on Δ, and let β^0 be a scalar function defined on Δ. If β and β^0 are in $H_\alpha(\Delta)$, then for any $f \in L^p$ and $u \in W_p^{2,1}$ such that $Lu = f$ in Ω and $\beta \cdot Du + \beta^0 u = 0$ on Δ, there is a constant C, determined also by α, $|\beta|_\alpha$, $|\beta_1|_\alpha$ and $\min |\beta'| / \beta \cdot \gamma$, such that (7.39) is valid. If $\Delta = S\Omega$, then (7.45) holds.*

PROOF. Fix $X_0 \in \Delta$, choose R as before, and transform Δ locally to a flat boundary portion. In the transformed variables, apply Lemma 7.34 to $B^i = \beta^i - \beta^i(X_0)$ for $i > 0$, $B^0 = \beta^0$, and u replaced by ηu. Then $w = \eta u - (z_1 + z_2)$ is a solution of $-w_t + A^{ij} D_{ij} w = F$ in $\mathbb{R}_+^n \times (0,T)$, $\beta(X_0) \cdot Dw = 0$ on $\mathbb{R}_0^n \times (0,T)$, $w = 0$ on $\mathbb{R}_+^n \times \{0\}$ for a suitable F and w has support in the cylinder $Q(R, X_0)$. Hence we infer an estimate on w and its derivatives from Proposition 7.33. The perturbation argument of Theorem 7.22 then yields the desired estimate for u. □

Note that we can also study the nonhomogeneous problem

$$Lu = f \text{ in } \Omega, \ Mu = \psi \text{ on } S\Omega, \ u = \varphi \text{ on } B\Omega$$

provided there is a function $\varphi_0 \in W_p^{2,1}$ such that $M\varphi_0 = \psi$ on $S\Omega$ and $\varphi_0 = \varphi$ on $B\Omega$. See the notes for a discussion of this compatibility condition.

9. The local maximum principle

We now return to the study of strong solutions without any continuity hypotheses on the highest order coefficients. We shall prove analogs of the results in Chapter VI. In this section, we show that a local maximum estimate for a subsolution can be obtained in terms of an L^p estimate. For this estimate, we assume that there are constants B, c_0, λ_0, λ_1, and μ such that

$$\lambda_1 \le \lambda \le \lambda_0,\ \Lambda \le \mu\lambda, \tag{7.48a}$$

$$|b| \le B\lambda,\ c \le c_0\lambda. \tag{7.48b}$$

Our estimate then takes the following form.

THEOREM 7.36. *If $u \in W^{2,1}_{n+1}$ satisfies $Lu \ge f$ in $Q(R)$ with the coefficients of L satisfying* (7.48), *then for any $p > 0$ and $\rho \in (0,1)$, there is a constant C determined only by p, λ_0, λ_1, μ, ρ, and $BR + c_0R^2$ such that*

$$\sup_{Q(\rho R)} u \le C\left[(R^{-n-2}\int_{Q(R)} (u^+)^p\,dX)^{1/p} + R^{n/(n+1)}\|f\|_{n+1}\right]. \tag{7.49}$$

PROOF. Let $\zeta = (1-|x|^2/R^2)^+(1+t/R^2)^+$, and for $q \ge 2$ to be chosen, set $\eta = \zeta^q$. We define P by $Pv = -v_t + a^{ij}D_{ij}v$ and set $v = \eta u$. A direct calculation yields

$$Pv = \eta f - \eta(b^iD_iu + cu) + uP\eta + 2a^{ij}D_iuD_j\eta.$$

Since $|Du| \le |Dv|/\eta + |v||D\eta|/\eta$ and $|Dv| \le v/(R-|x|)$ on $E^+(v)$, we have $|Du| \le 2(1+q)v/(R\eta^{1+1/q})$ on $E^+(v)$ and hence

$$\begin{aligned} Pv &\ge \eta f - v\zeta^{-2}[\frac{4(1+q)B\lambda\zeta}{R} + \frac{q(2n+1)\Lambda/\zeta^2}{R^2} + c_0\zeta^2 + \frac{q}{4R^2}] \\ &\ge f - Cv\zeta^{-2}R^{-2}\mathscr{D}^*. \end{aligned}$$

Therefore, Theorem 7.1 implies that

$$\sup v \le C[R^{n/(n+1)}\|f\|_{n+1} + R^{n/(n+1)-2}\|v\zeta^{-2}\|_{n+1}].$$

Moreover, $v\zeta^{-2} = u^{2/q}v^{1-2/q}$ and $n/(n+1) - 2 = -(n+2)/(n+1)$, so

$$\sup v \le C\left[F + (\sup v)^{1-2/q}\left(R^{-n-2}\int_{Q(R)} (u^+)^{2(n+1)/q}\,dX\right)^{1/(n+1)}\right]$$

for $F = R^{n/(n+1)}\|f\|_{n+1}$. If $p \le n+1$, we take $q = 2(n+1)/p$ and then (7.49) follows from Young's inequality. For $p > n+1$, we take $q = 2$ and then use Hölder's inequality. □

10. The weak Harnack inequality

For our analog of Theorem 6.18, we recall the definition of $\Theta(R)$. For $Y \in \Omega$ and $R > 0$ so that $Q(Y,R) \subset \Omega$, we set $\Theta(R) = Q((y,s-4R^2),R)$. With this definition, we have the following weak Harnack inequality. We also recall the definition of the Morrey space $M^{p,q}$ as the set of all functions u in L^p with finite norm

$$\|u\|_{p,q} = \sup_{r\leq \operatorname{diam}\Omega} \left(r^{-q}\int_{Q(r)} |u|^p\,dX\right)^{1/p}.$$

THEOREM 7.37. *If $u \in W^{2,1}_{n+1}$ is nonnegative in $Q(4R)$ and satisfies the inequality $Lu \leq f$ in $Q(4R)$ for some L with coefficients satisfying* (7.48a),

$$\|b/\lambda\|_{n+1,1} \leq B, \tag{7.48b$'$}$$

and $c = 0$, then there are positive constants C, B_1, and p determined only by B, n, λ_0, λ_1, and μ such that $B \leq B_1$ implies

$$\left(R^{-n-2}\int_{\Theta(R)} u^p\,dX\right)^{1/p} \leq C(\inf_{Q(R)} u + R^{n/(n+1)}\|f\|_{n+1}). \tag{7.50}$$

As with the proof of Theorem 6.18, we proceed in several stages. Unlike the situation in that previous theorem, the function $w = -\log(u+k)$ is not needed here, and we do not estimate the analogs of L^p norms with p running from $-\infty$ to some positive number. Instead, we use some pointwise estimates for u based on the size of the set on which u is large. Our first lemma shows that if u is large on most of Q, then u is large on all of a smaller cylinder. To simply notation, we set

$$k = R^{n/(n+1)}\|f\|_{n+1}, \bar{u} = u + k.$$

LEMMA 7.38. *There are constants β_0 and ζ determined only by n, λ_0, and λ_1 such that if $B \leq \beta_0$, $r \leq R$, and*

$$|\{X \in Q(X_0,r) : \bar{u} > h\}| < \zeta\,|Q(r)| \tag{7.51}$$

for some $h \geq k$, then

$$\inf_{Q(r/2)} \bar{u} \geq C_1(n,\lambda_0,\lambda_1)h. \tag{7.52}$$

PROOF. Set

$$v = h\left[1 - \frac{|X-X_0|^2}{r^2}\right] - \bar{u}, Q^* = \{X \in Q(X_0,r) : \bar{u} > h\}$$

and compute

$$Lv = -Lu - hr^{-2}[\mathscr{T} + 1 + b^i(x^i - x_0^i)]$$

so that

$$r^{n/(n+1)}\|Lv/\mathscr{D}^*\|_{n+1,Q^*} \leq Ck + C[\zeta^{1/(n+1)} + \beta_0]h$$

for some $C(n,\lambda_0,\lambda_1)$ provided $\beta_0 \leq 1$. Since $v \leq 0$ on $\mathscr{P}Q^*$, it follows from Theorem 7.1 that

$$\sup_{Q^*} v \leq Ck + C[\zeta^{1/(n+1)} + \beta_0]h.$$

Now take $\zeta^{1/(n+1)} + \beta_0 \le 1/(4C)$ to see that

$$Ck + \frac{1}{4}h \ge v \ge \frac{1}{2}h - \bar{u}$$

on $Q(r/2)$. A simple rearrangement gives (7.52) with $C_1 = 1/(2+2C)$. □

The next step is a comparison theorem, which states that if u is large on a ball for some fixed time, then there is a lower bound for u at a given later time on any concentric ball.

LEMMA 7.39. *Let α_0, α_1, and ε be positive constants with $\alpha_0 < \alpha_1$ and $\varepsilon \le 1/2$, and suppose that $Lu \ge f$ in $Q(R)$. Let r be so small that $B(r) \times (-r^2, (\alpha_1 - 1)r^2) \subset Q(R)$. Then there are positive constants $C_2(n, \lambda_0, \mu)$, $\beta_1(n, \lambda_0, \lambda_1, \varepsilon)$, and $\kappa(n, \alpha_0, \alpha_1, \lambda_0, \lambda_1, \mu)$ such that if $Br^{n/(n+1)} \le \beta_1$ and*

$$\bar{u} \ge A \text{ on } \{|x| < \varepsilon r, t = -r^2\} \tag{7.53}$$

for some positive constant A, then

$$\bar{u} \ge C_2 \varepsilon^{\kappa} A \text{ on } \{|x| < r/2, t = (\alpha - 1)r^2\} \tag{7.54}$$

for any $\alpha \in [\alpha_0, \alpha_1]$.

PROOF. Fix $\alpha \in [\alpha_0, \alpha_1]$ and set

$$\psi_0 = \frac{(1-\varepsilon^2)(t+r^2)}{\alpha} + \varepsilon^2 r^2, \psi_1 = (\psi_0 - |x|^2)^+,$$

and $\psi = \psi_1^2 \psi_0^{-q}$ for some $q \ge 2$ to be chosen. As in Lemma 2.6, we have $-\psi_t + a^{ij} D_{ij} \psi \ge 0$ in Q if q is sufficiently large (determined only by n, α_0, α_1, λ_0, λ_1, and μ). With this choice of q, we also have

$$L\psi \ge b^i D_i \psi = -4\psi_1 \psi_0^{-q} b \cdot x \ge -4|b| r(\varepsilon r)^{2-2q}.$$

In addition, the proof of Lemma 2.6 gives

$$\psi(x, -r^2) \le (\varepsilon r)^{-2q+4} \tag{7.55a}$$

if $|x| \le r$ and

$$\psi(x, (\alpha-1)r^2) \ge \frac{9r^{-2q+4}}{16} \tag{7.55b}$$

if $|x| \le r/2$. Now set $v = \bar{u} - A(\varepsilon r)^{2q-4}\psi$ and $Q = B(r) \times (-r^2, (\alpha-1)r^2)$. Since $\psi(X) = 0$ for $|x| = r$, we infer from (7.53) and (7.55a) that $v \ge 0$ on $\mathscr{P}Q$. Applying Theorem 7.1 to v then yields $v \ge -Ck - C\beta_1 A\varepsilon^{-2}$ in Q and hence

$$\bar{u} \ge A[(\varepsilon r)^{2q-4}\psi - C\beta_1 \varepsilon^{-2}] - Ck \ge A\frac{\varepsilon^{2q-4}}{2} - Ck$$

for $|x| \le r/2$ and $t = (\alpha-1)r^2$ by (7.55b) provided β_1 is small enough. For $\kappa = 2q-4$ and $C_2 = 1/(2+2C)$, this inequality is just (7.54). □

Our next step is a measure-theoretic lemma of Krylov and Safonov. To state this lemma, we first give some definitions. For any $X_0 \in \mathbb{R}^{n+1}$ and $R > 0$, we write $K(X_0,R)$ for the cube $\{X : \max_i |x^i - x_0^i| < R, 0 < t - t_0 < R^2\}$. With X_0 and R fixed, we abbreviate $K_0 = K(X_0,R)$. For $\eta \in (0,3/4)$ to be further specified and any cube $K(X_1,r) \subset K_0$, we then define

$$K_1(X_1,r) = K_0 \cap \{\max_i |x^i - x_1^i| < 3r, -3r^2 < t - t_1 < 4r^2\}$$

and

$$K_2(X_1,r) = \{\max_i |x^i - x_1^i| < 3r, \max_i |x^i - x_0^i| < r,$$
$$r^2 < t - t_1 < (1 + \frac{4}{\eta})r^2\}.$$

Next, for $\xi \in (0,1)$ and $\Gamma \subset K_0$, we define $G^* = G^*(\Gamma,\xi)$ to be the family of all subcubes K of K_0 such that $|K \cap \Gamma| \geq \xi |K|$ and set $Y_i = \cup_{K \in G^*} K_i$ for $i = 1,2$.

LEMMA 7.40. *Let $\Gamma \subset K_0$. Then*

$$|\Gamma \setminus Y_1| = 0, \tag{7.56a}$$
$$|\Gamma| \leq \xi |K_0| \text{ implies } |\Gamma| \leq \xi |Y_1|, \tag{7.56b}$$
$$|Y_1| \leq (1+\eta) |Y_2|. \tag{7.56c}$$

PROOF. If $|\Gamma| > \xi |K_0|$, then $Y_1 = K_0$ (by using $K = K_0$), which gives (7.56a) in this case. Of course, (7.56b) is vacuously true.

If $|\Gamma| \leq \xi |K_0|$, then we use the Calderón-Zygmund decomposition from Section VII.2 with $\tau = \xi$. If $K(X_1,r) \in \mathscr{S}$, then $\tilde{K} \subset K_1(X_1,r)$, so $\tilde{B} \subset Y_1$ and hence (7.56a,b) follow from (7.7).

To prove (7.56c), we first fix an x such that the set $I(x) = \{t : (x,t) \in Y_2\}$ is nonempty. Since $I(x)$ is an open subset of $\mathbb{R}$, we can write it as a union of disjoint open intervals $I_m = (t_m, \tau_m)$ and each I_m is a union of (not necessarily disjoint) open intervals

$$I_{m,k} = (t_{m,k} + r_{m,k}^2, t_{m,k} + (1 + \frac{4}{\eta})r_{m,k}^2),$$

each one corresponding to some K_2. If we set

$$J_{m,k}^* = (t_{m,k} - 3r_{m,k}^2, t_{m,k} + 4r_{m,k}^2), I_m^* = \cup J_{m,k}^*,$$

with the union taken over all k for which there is an $I_{m,k}$, then

$$\{t : (x,t) \in Y_1 \cup Y_2\} \subset \cup (I_m^* \cup I_m).$$

By construction, we have

$$t_{m,k} + r_{m,k}^2 \geq t_m \text{ and } t_{m,k} + (1 + \frac{4}{\eta})r_{m,k}^2 \leq \tau_m,$$

and therefore $J^*_{m,k} \subset (t_m - \eta(\tau_m - t_m), \tau_m)$ because $\eta \le 3/4$. It follows that $I_m \cup I^*_m \subset (t_m - \eta(\tau_m - t_m), \tau_m)$, and hence

$$|\{t : (x,t) \in Y_1 \cup Y_2\}| \le \sum (1+\eta)(\tau_m - t_m) = (1+\eta)\,|I(x)|\,.$$

If we now write χ_1 and χ_2 to denote the characteristic functions of Y_1 and Y_2, respectively, we have that

$$\int_{\mathbb{R}} \chi_1(x,t)\,dt \le (1+\eta) \int_{\mathbb{R}} \chi_2(x,t)\,dt$$

for any x. (If $I(x)$ is nonempty, this is what we have just proved since $Y_1 \subset Y_1 \cup Y_2$. If $I(x)$ is empty then $\chi_1(x,t) = 0$ for all t.) Integrating this inequality with respect to x and using Fubini's theorem give (7.56c). □

We are now ready to prove Theorem 7.37.

PROOF OF THEOREM 7.37. For simplicity, let $X_0 = 0$ and set

$$B_1 = \min\{\beta_0, \beta_1(n, \lambda_0, \lambda_1, 1/12)\}.$$

We then take $\xi = \zeta$, the constant from Lemma 7.38, $\eta = \frac{1}{2}\min\{\frac{3}{4}, \xi^{-1/2} - 1\}$ and, for a fixed $h > k$, we set

$$\Gamma = \Gamma(h) = \{X \in K_0 : \bar{u} > h\}.$$

Now write $G^*(h)$ for $G^*(\Gamma(h), \xi)$ and fix X_1 and r such that $K(X_1, r) \in G^*(h)$. Then, for

$$K'(X_1, r) = \{\max_i |x^i - x^i_1| < r/2, t_1 + \frac{3}{4}r^2 < t < t_1 + r^2\},$$

we have $\inf_{K'} \bar{u} \ge C_1 h$ by Lemma 7.38. With $\varepsilon = \frac{1}{12}$, $\alpha_0 = \frac{1}{4}$, $\alpha_1 = 1 + 4/\eta$, and $A = C_1 h$ in Lemma 7.39, we infer that $\bar{u} \ge C_3 h$ in K_2 for C_3 depending on C_1, C_2 and κ.

Suppose that $|\Gamma| \le \xi\,|K_0|$, and set $\gamma = C_3/2$. Then $|Y_1| \ge |\Gamma|/\xi$ from Lemma 7.40 and we have just shown that $\bar{u} \ge \gamma h$ on Y_2. Therefore

$$|\Gamma(\gamma h)| \le \xi^{-1/4}|\Gamma(h)|$$

implies that

$$|Y_2 \cap \Theta| \le \xi^{-1/4}|\Gamma(h)|,$$

and hence

$$|Y_2 \setminus \Theta| \ge (1 - \xi^{1/4})\xi^{-1/2}|\Gamma(h)|,$$

in which case there is a cube $K(X_1, r) \in G^*(h)$ such that the intersection of K_2 with

$$\{t \ge (1 - \xi^{1/4})\xi^{-1/2}|\Gamma(h)|\,r^{-n} - 3R^2\}$$

has positive measure. It follows that

$$t_1 + r^2 + \frac{4}{\eta}r^2 \ge (1 - \xi^{1/4})\xi^{-1/2}|\Gamma(h)|\,r^{-n} - 3R^2,$$

and, because $K \in G^*$, $t_1 + r^2 \leq -3R^2$, so

$$r^2 \geq \frac{\eta}{4}(1-\xi^{1/2})\frac{|\Gamma(h)|}{\xi^{1/4}}r^{-n}.$$

Since $\bar{u} \geq \gamma h$ on K_2, we have that $\bar{u} \geq \gamma h$ on $B(x_1, r) \times \{-3R^2/4\}$ and then Lemma 7.39 with $r = R$ and $\varepsilon = r/(4R)$ gives

$$\bar{u} \geq C_2 \left(\frac{r}{4R}\right)^{\kappa} \gamma h$$

on $Q(R)$. Hence

$$\inf_{Q(R)} \bar{u} \geq C_4 h \left(\frac{|\Gamma(h)|}{R^{n+2}}\right)^{\kappa/(n+2)}$$

under all these hypotheses.

We summarize the three possibilities for any value of h.

(1) $|\Gamma(h/\gamma)| \leq \xi^{-1/4}|\Gamma(h)|$,
(2) $|\Gamma(h)| \leq C_5(\inf_Q \bar{u}/h)^q R^{n+2}$,
(3) $|\Gamma(h)| > \xi\,|K_0|$,

where $q \leq (n+2)/\kappa$ is chosen so that $\gamma^q \geq \xi^{1/4}$. Using these three possibilities, we now obtain an estimate for $|\Gamma(h)|$ for any $h > 0$. To this end, we choose $h_0 > 0$ such that $|\Gamma(h_0)| \leq \xi\,|K_0|$ and $|\Gamma(\gamma h_0)| > \xi\,|K_0|$. Note that $\inf_Q \bar{u} \geq C_1 h_0$ by Lemma 7.38. Next, we set $h_1 = \inf_Q \bar{u}/C_1$ and, for any positive integer m, define

$$\alpha = C_5 R^{n+2}, \beta_m = |\Gamma(h_1/\gamma^m)|, b_1 = \max\{\beta_1, \alpha\}.$$

Note that $b_1 \leq CR^{n+2}$, and an easy induction shows that $\beta_m \leq C\gamma^{mq}R^{n+2}$. Now we fix $h > h_0$, and we write m for the positive integer such that $\gamma^{mq}h_1 \geq h > \gamma^{(m-1)q}h_1$. It follows that

$$|\Gamma(h)| \leq C\gamma^{q(m-1)}R^{n+2} \leq C\left(\frac{h_1}{h}\right)^q R^{n+2}.$$

For $p \in (0, q)$, we infer that

$$\int_{h_1}^{\infty} h^{p-1}\,|\Gamma(h)|\,dh \leq CR^{n+2}h_1^q \int_{h_1}^{\infty} h^{p-1-q}\,dh = CR^{n+2}h_1^p,$$

and so we have

$$\int_{\Theta} \bar{u}^p\,dX \leq CR^{n+2}h_1^p.$$

This inequality implies (7.50) because of the definition of h_1. □

From Theorem 7.37, we infer a Hölder estimate by the same argument as in Theorem 6.28.

COROLLARY 7.41. *If $Lu = f$ and the coefficients of L satisfy* (7.48a) *and* (7.48)′, *then*

$$\operatorname*{osc}_{Q(R)} u \le C\left(\frac{R}{R_0}\right)^{\alpha}\left(\operatorname*{osc}_{Q(R_0)} u + \|f - cu\|_{n+1;Q(R_0)} R_0^{n/(n+1)}\right). \tag{7.57}$$

Furthermore, Theorems 7.36 and 7.37 lead directly to a parabolic Harnack inequality.

COROLLARY 7.42. *If $Lu = 0$ and $u \ge 0$, then*

$$\sup_{\Theta(R/2)} u \le C \inf_{Q(R)} u. \tag{7.58}$$

11. Boundary estimates

The local maximum principle Theorem 7.36 is easily extended to the case when the cylinder intersects the parabolic boundary of the domain. To state the results more easily, for a fixed point X_0, and any number r, we define $\Omega[r] = Q(r) \cap \Omega$, with a similar definition for $\mathcal{P}\Omega[r]$.

THEOREM 7.43. *Let $u \in W^{2,1}_{n+1}$ satisfy $Lu \ge f$ in Ω and $u \le 0$ on $\mathcal{P}\Omega[2R]$. If the coefficients of L satisfy* (7.48), *then, for any $p > 0$, there is a constant $C(BR + c_0R^2, n, \lambda_0, \mu)$ such that*

$$\sup_{\Omega[R]} u \le C\left[(R^{-n-2}\int_{\Omega[2R]} (u^+)^p\, dX)^{1/p} + R^{n/(n+1)}\|f\|_{n+1}\right]. \tag{7.59}$$

PROOF. With η and v as in the proof of Theorem 7.36, we have $v \le 0$ on $\mathcal{P}\Omega[2R]$, so we can imitate the proof of that theorem provided we take $E^+(v)$ with respect to $\Omega[2R]$. □

The weak Harnack inequality at the boundary is slightly more complicated.

THEOREM 7.44. *Let $Lu \le f$ in Ω, $u \ge 0$ in $\Omega[4R]$ and set $m = \inf_{\mathcal{P}\Omega[4R]} u$ and*

$$u_m = \begin{cases} \inf\{u, m\} & \text{in } \Omega \\ m & \text{outside } \Omega. \end{cases} \tag{7.60}$$

If the coefficients of L satisfy (7.48a) *and* (7.48b)′ *with $c = 0$, then there are positive constants C, B_1, and p determined only by n, λ_0, λ_1, and μ such that $B \le B_1$ implies*

$$(R^{-n-2}\int_{\Theta(R)} u_m^p\, dX)^{1/p} \le C(\inf_{Q(R)} u_m + R^{n/(n+1)}\|f\|_{n+1}). \tag{7.61}$$

PROOF. Replace u by u_m in the proof of Theorem 7.37. □

Boundary modulus of continuity estimates then follow as in Chapter VI.

COROLLARY 7.45. *Let $Lu = f$ in Ω and suppose that Ω satisfies condition (A) at $X_0 \in \mathscr{P}\Omega$. Then there are constants C and α determined only by A, n, λ_0, Λ_0, μ, and B such that if σ is an increasing function with $\tau^{-\delta}\sigma(\tau)$ a decreasing function of τ for some $\delta < \alpha$ and* $\operatorname{osc}_{\mathscr{P}\Omega[r]} u \le \sigma(r)$, *then*

$$\operatorname*{osc}_{\Omega[r]} u \le C\left(\left(\frac{r}{R}\right)^{\alpha} \operatorname*{osc}_{\Omega[R]} u + \|f - cu\|_{n+1} r^{n/(n+1)} + \sigma(r)\right) \tag{7.62}$$

for $0 < r < R$.

Another useful boundary estimate is one for the oscillation of the normal derivative proved by Krylov. To state this estimate, we introduce the following sets:

$$G(\rho,R) = \{|x'| < R, 0 < x^n < \rho R, -R^2 < t < 0\}$$
$$G'(\rho,R) = \{|x'| < R, \rho R < x^n < 2\rho R, -R^2 < t < 0\}$$

for positive constants ρ and R. We also assume for the moment that L has no lower order terms. In other words,

$$Lu = -u_t + a^{ij}D_{ij}u, \tag{7.63a}$$

and there are positive constants λ and Λ such that

$$\max a^{ij} \le \Lambda, a^{ij}\xi_i\xi_j \ge \lambda|\xi|^2. \tag{7.63b}$$

LEMMA 7.46. *Suppose that L has the form (7.63) in Q^+ with $u \ge 0$ in Q^+. Suppose also that $Lu \le \lambda F_1 (x^n)^{\delta-1}$ for some nonnegative constant F_1 and $\delta \in (0,1)$. If $\rho = \lambda/(2+(2n+4)\Lambda)$ and $10r^2 \le R^2$, then*

$$\inf_{G'(\rho,2r)} \frac{u}{x^n} \le 4 \inf_{G(\rho,r)} \frac{u}{x^n} + \frac{8}{\delta}(\rho r)^{\delta} F_1. \tag{7.64}$$

PROOF. Set $A = \inf_{G'(\rho,2r)} u/x^n$ and

$$w_1 = \left(1 - \frac{x^n}{2\rho r} + \frac{|x'|^2}{r^2} - \frac{t}{r^2}\right)x^n,$$

$$w_2 = \frac{1}{\delta}[(2\rho r)^{\delta} - (x^n)^{\delta}]x^n,$$

$$w = u - Ax^n + \frac{A}{4}w_1 + F_1 w_2.$$

A simple calculation shows that $Lw \le 0$ in $G(\rho,2r)$. Moreover, because $w = u$ on $\{x^n = 0\}$, $w \ge -Ax^n + Aw_1/4$ on $\{t = -4r^2\}$, $w \ge u - Ax^n + Aw_1/4$ on $\{|x'| = 2r\}$, and $w \ge u - Ax^n$ on $\{x^n = 2\rho r\}$, we infer that $w \ge 0$ on $\mathscr{P}G(\rho,2r)$. The maximum principle implies that $w \ge 0$ in $G(\rho,2r)$ and hence in $G(\rho,r)$. Since $w_1 \le 3x^n$ in $G(\rho,r)$, it follows that $u \ge [A/4 - 2(F_1/\delta)(\rho r)^{\delta}]x^n$ in $G(\rho,r)$. Dividing this inequality by x^n gives (7.64). □

From this estimate and the weak Harnack inequality, we infer an oscillation estimate for u/x^n.

LEMMA 7.47. *Suppose that there are positive constants F_1, K, and $\delta \in (0,1)$ such that*

$$|Lu| \le \lambda F_1 (x^n)^{\delta-1} \text{ in } Q^+, \tag{7.65a}$$

$$|u| \le Kx^n \text{ in } Q^+. \tag{7.65b}$$

If ρ is the same as in Lemma 7.46, then there are positive constants θ and C determined only by n, δ, λ, and Λ such that

$$\operatorname*{osc}_{G(\rho,r)} \frac{u}{x^n} \le C\left(\frac{r}{R}\right)^{\theta} [\operatorname*{osc}_{G(\rho,R)} \frac{u}{x^n} + F_1 R^{\delta}] \tag{7.66}$$

for all $r \in (0,R)$.

PROOF. Without loss of generality, we may assume that $10r^2 \le R^2$. We define

$$m_i = \inf_{G(\rho,ir)} \frac{u}{x^n}, M_i = \sup_{G(\rho,ir)} \frac{u}{x^n}$$

for $i = 1,4$ and then set

$$\Sigma = \{|x'| < r, \rho r < x^n < \frac{3}{4}\rho r, -4r^2 < t < -2r^2\},$$
$$\Sigma' = G'(\rho, 2r),$$
$$\Sigma'' = \{|x'| < 2r, \frac{1}{2}\rho r < x^n < \rho r, -r^2 < t < 0\}.$$

If we apply the weak Harnack inequality to $u - m_2 x^n$ in Σ' and note that x^n/r is trapped between ρ and 2ρ in Σ', we obtain

$$\left(\frac{1}{|\Sigma|}\int_\Sigma (u - m_2 x^n)^p \, dX\right)^{1/p} \le C(\inf_{\Sigma'}(u - m_2 x^n) + r^{1+\delta} F_1)$$
$$\le Cr(\inf_{\Sigma'}(\frac{u}{x^n} - m_2) + r^{\delta} F_1)$$

and then Lemma 7.46 gives

$$\left(\frac{1}{|\Sigma|}\int_\Sigma (u - m_2 x^n)^p \, dX\right)^{1/p} \le Cr[m_1 - m_2 + r^{\delta} F_1].$$

Similar reasoning with $M_2 x^n - u$ yields

$$\left(\frac{1}{|\Sigma|}\int_\Sigma (M_2 x^n - u)^p \, dX\right)^{1/p} \le Cr[M_2 - M_1 + r^{\delta} F_1].$$

Adding these inequalities gives

$$(M_2 - m_2) r \le Cr[m_1 - m_2 + M_2 - M_1 + rF_1]$$

because

$$(\int_\Sigma (w+v)^p\,dX)^{1/p} \le C(p)[(\int_\Sigma w^p\,dX)^{1/p} + (\int_\Sigma v^p\,dX)^{1/p}]$$

for any nonnegative functions w and v. The desired estimate now follows after some elementary algebra by virtue of Lemma 4.5. □

Since u/x^n tends to $D_n u$ as x^n tends to zero, this lemma gives an estimate on the oscillation of the normal derivative on Q^0 in terms of known quantities. We shall leave the details of this estimate to a later chapter, in the context of quasilinear equations.

We can also prove versions of the local maximum principle and the weak Harnack inequality near the boundary when the Dirichlet boundary condition is replaced by an oblique derivative condition.

THEOREM 7.48. *Suppose $u \in W^{2,1}_{n+1}(Q^+)$ satisfies $Lu \ge f$ in Q^+, and $\mathscr{M}u \ge -\Psi\beta^n$ on Q^0. If L satisfies (7.48) and if there are constants μ and ν such that*

$$|\beta'| \le \mu\beta^n,\ \beta^0 \le \nu\beta^n, \tag{7.67}$$

then, for any $p > 0$, there is a constant C determined by $BR + c_0R^2$, n, p, λ_0, μ, and νR such that

$$\sup_{Q(R/2)} u \le C[\left(R^{-n-2}\int_Q (u^+)^p\,dX\right)^{1/p} + R^{n/(n+1)}\|f/\mathscr{D}^*\|_{n+1} + \Psi R]. \tag{7.68}$$

PROOF. Define

$$\eta(X) = (\frac{x^n}{R} + 1 - \frac{|x|^2}{9\mu^2R^2})^+,\ \zeta(X) = \eta(X)(1 - \frac{t-s}{4R^2})^+,$$

and note that, on the set $\{x^n = 0, \eta(X) > 0\}$, we have $|x'| < 3\tau R$ and hence

$$\beta\cdot D\eta(X) = \frac{\beta^n}{R} - \frac{2}{9\mu R^2}\beta'\cdot x' \ge \frac{\beta^n}{3r}.$$

Now set $v = \zeta^q u$ for $q \ge 3\nu R + 3$ to be chosen and define $\mathscr{M}_0$ by

$$\mathscr{M}_0 w = \beta\cdot Dw + (\beta^0 - \frac{q\beta^n}{3R})w.$$

Then

$$\beta^0 - \frac{q\beta^n}{3R} \le -\frac{\beta^n}{R}$$

and $\mathscr{M}_0 v \ge -\Psi\beta^n$ on Q^0. As in Theorem 7.36, we also have $Pv \ge \eta f - \frac{Cv}{\zeta^2R^2}\mathscr{D}^*$ in Q^+ provided q is sufficiently large, depending only on νR and n. Then the maximum principle, Theorem 7.1, gives

$$\sup v \le C[F + R\Psi + (\sup v)^{1-2/q}(\frac{1}{|Q|}\int_Q (u^+)^{2(n+1)/q}\,dX)^{1/(n+1)}]$$

for $F = R^{n/(n+1)}\|f\|_{n+1}$, and (7.68) follows from this inequality as before. □

From an H_2 change of variables, we obtain the same result in smooth domains.

COROLLARY 7.49. *Suppose that* $\mathscr{P}\Omega \in H_2$ *and that* $u \in W^{2,1}_{n+1}$ *satisfies* $Lu \geq f$ *in* $\Omega[2R]$, $\mathscr{M}u \geq -\Psi\beta^n$ *on* $S\Omega[2R]$, *and* $u \leq 0$ *on* $B\Omega[2R]$. *If L has the form* (7.63) *and if there are constants* μ *and* ν *such that*

$$|\beta - (\beta\cdot\gamma)\gamma| \leq \mu\beta\cdot\gamma, \beta^0 \leq \nu\beta\cdot\gamma, \tag{7.67$'$}$$

then, for any $p > 0$, *there is a constant C determined by* $BR + c_0R^2$, n, p, λ_0, μ, νR, *and* Ω *such that*

$$\sup_{\Omega[R/2]} u \leq C[\left(R^{-n-2}\int_{\Omega[R]}(u^+)^p\,dX\right)^{1/p} + R^{n/(n+1)}\|f\|_{n+1} + \Psi R]. \tag{7.68$'$}$$

A weak Harnack inequality completely analogous to the one in Theorem 7.37 can be proved in H_2 domains. For our applications to quasilinear problems, we present here a slightly different version for $H_{1+\delta}$ domains.

PROPOSITION 7.50. *Let* $\mathscr{P}\Omega \in H_{1+\delta}$ *for some* $\delta \in (0,1)$, *and suppose that* $u \in W^{2,1}_{n+1,loc} \cap C(\overline{\Omega})$ *satisfies* $Lu \leq f$ *in* Ω, $\mathscr{M}u \leq \Psi\beta\cdot\gamma$ *on* $S\Omega$. *Suppose also that* (a^{ij}) *satisfies* (7.63b), *that there are nonnegative constants B and F such that* $|b| \leq \lambda B d^{\delta-1}$ *and* $|f| \leq \lambda F d^{\delta-1}$ *in* Ω, *where d is the distance to* $S\Omega$, *and that* $c = 0$ *and* $\beta^0 = 0$. *If* $u \geq 0$ *in* $\Omega[4R]$ *for some R such that* $B\Omega[4R]$ *is empty, and if R is sufficiently small (determined only by B,* δ, *and* Ω), *then there are constants C and p determined only by B, n, p,* λ_0, Λ_0, μ, τ, *and* νR *such that*

$$\left(R^{-n-2}\int_{\Theta(X^*,R)} u^p\,dX\right)^{1/p} \leq C(\inf_{\Omega[R]} u + FR^{1+\delta} + \Psi R^\delta), \tag{7.69}$$

where $d(X^*) \geq R$ *and* $X^* \in \Omega[2R]$.

PROOF. We note that (7.69) follows, via an $H_2^{(-1-\delta)}$ change of variables and a covering argument, from the estimate

$$\left(\frac{1}{|\Sigma|}\int_\Sigma u^p\,dX\right)^{1/p} \leq C(\inf_{G(\rho,R/2)} u + FR^{1+\delta} + \Psi R), \tag{7.69$'$}$$

with ρ suitably chosen and

$$\Sigma = \{|x'| < R, \rho R < x^n < 2\rho R, -4R^2 < t < -3R^2\}.$$

The weak Harnack inequality Theorem 7.37 implies that

$$\left(\frac{1}{|\Sigma|}\int_\Sigma u^p\,dX\right)^{1/p} \leq C(\inf_{G'(\rho,R)} u + R^{n/(n+1)}\|f/\mathscr{D}^*\|_{n+1} + R\Psi) \tag{7.70}$$

for $\rho \le \lambda/(2+(2n+4)\Lambda)$, so we are left with estimating the infimum in this inequality. To this end, we set

$$A = \inf_{G'(\rho,R)} u,$$
$$w_1 = 2 - \left(\frac{x^n}{\rho R}\right)^2 - \frac{x^n}{\rho R} + \frac{|x'|^2}{R^2},$$
$$w_2 = \rho R - x^n,$$
$$w_3 = \frac{1}{\delta(1+\delta)}[(2\rho R)^{1+\delta} - (x^n)^{1+\delta}],$$
$$w = u - A + \frac{A}{4}w_1 + \Psi w_2 + F_0 w_3$$

with $F_0 = 2F + 2\Psi + 2ABR^{-1}(1+1/\rho)$. Then we have

$$Lw \le F_0[\frac{BR^\delta}{\delta} - \frac{1}{2}] \le 0$$

in $G(\rho,R)$ provided $BR^\delta \le \delta/2$, $\mathscr{M}w \le 0$ on $G'(\rho,R)$ provided $\rho \le 1/2\mu$, and $w \ge 0$ on $\mathscr{P}G(\rho,R) \setminus G'(\rho,R)$, so Corollary 7.4 implies that $w \ge 0$ in $G(\rho,R)$ and hence in $G(\rho,R/2)$. Since $\rho \le \frac{1}{4}$, it follows that

$$u \ge [\frac{1}{4} - 5\frac{BR^\delta}{\delta}]A - 4R\Psi - \frac{2}{\delta}FR^{1+\delta}$$

in $G(\rho,R)$ and (7.69)′ follows by combining this inequality with (7.70) if $BR^\delta \le \delta/40$. □

The usual Hölder estimate follows immediately from Proposition 7.50.

COROLLARY 7.51. *Let $u \in W^{2,1}_{n+1,loc}(\Omega) \cap C(\overline{\Omega})$ satisfy $Lu = f$ in Ω, $|\mathscr{M}u| \le \Psi\beta \cdot \gamma$ on $S\Omega$. If L and M are as in Proposition 7.50, then there are constants C and α determined only by B, n, λ_0, Λ_0, μ, τ, and Ω such that*

$$\operatorname*{osc}_{\Omega[r]} u \le C\left(\frac{r}{R}\right)^\alpha (\operatorname*{osc}_{\Omega[R]} u + R\Psi + R^{1+\delta}F) \tag{7.71}$$

for any $R > 0$ such that $B\Omega[4R]$ is empty, and $r \le R$.

We leave to the reader the easy variation of this corollary in which $B\Omega[4R]$ is nonempty and u is Hölder continuous on $B\Omega$.

Notes

The maximum principle Theorem 7.1 (in the special case $\beta = 0$) is due to Krylov [167] (see also [168] and note that the coefficient b is required to be bounded in these works; Tso [316] showed that $b \in L^{n+1}$ is allowable) and is based on the corresponding elliptic results of Aleksandrov, Bakel′man and Pucci

[3, 12, 279]. Our proof is modelled on a different method, used by Tso [316], and the modifications for oblique boundary conditions come from work of Lions, Trudinger, and Urbas [240, Theorem 2.1] and the author [218, Lemma 1.1]. The idea of introducing the parameter θ comes from [234] (see Lemmata 3.1 and 5.1 there). An alternative approach, which is much more geometric, is given in [266]. A slightly different version of Theorem 7.1 with general β appears in [267, Theorem 1]. Further variations are given in [7, Theorem 2.1; 28; 234]. In another direction, Crandall, Fok, Kocan, and Święch [56, Theorem 0.2] showed that u can be estimated in terms of $||f||_p$ for some $p < n+1$ if b is bounded and the estimate and the number p are allowed to depend on the maximum and minimum eigenvalues of (a^{ij}). In addition, Escauriaza [75] showed that, if there is a function f^* depending only on x such that $|f(x,t)| \leq f^*(x)$ for almost all X and if Ω is a cylinder, then the maximum of u can be estimated in terms of $||f^*||_n$.

The Calderón-Zygmund decomposition was first used by these two authors to study singular integral operators [37]. The parabolic version given here is a simple modification of the description from [101, Section 9.2]. Our proof of the Marcinkiewicz interpolation theorem [249] is taken from [101, Theorem 9.8], which comes, in turn, from [299, Section I.4]. Our presentation of the Vitali type lemma (Lemma 7.8) also comes from [299] in which it is pointed out that Lemma 7.8 is only similar to the actual Vitali covering lemma. The second Vitali type lemma (Lemma 7.11) is [335, Theorem 3].

Solonnikov [296] first proved the L^p estimates for parabolic equations and a large class of higher order systems as well. By invoking a suitable integral representation for solutions of constant coefficients equations, he was able to avoid the interpolation theorems altogether. He also proved Lemma 7.34 via a different representation for z_1 and z_2. In addition, his more careful study of trace theorems (further developed by Weidemaier [340]) leads to better results concerning nonhomogeneous boundary data; in particular, he states exact compatibility conditions for these problems. Other integral representations can be used, leading to a weak L^1 estimate, in which case only the Marcinkiewicz interpolation theorem is used. The details for elliptic equations are in [101, Theorem 9.9]. Our proof for constant coefficient parabolic equations is based on Wang's geometric approach [335] although our extension to variable coefficients follows the elliptic case in [101, Theorem 9.11]. In addition, Wang only considered interior estimates in [335], so this proof for the boundary estimates appears in print for the first time. A different but related method, which gives the interior estimate if p is sufficiently large, was developed for elliptic equations by Caffarelli [30, Theorem 1], and extended to parabolic equations by Wang [333, Section 5]. This method applies to a large class of nonlinear equations as well, and it also shows that the hypothesis of continuity for the coefficient matrix (a^{ij}) can be weakened. We refer also to

[8, 23, 237, 253, 275, 294, 295, 341] for more detail on the L^p estimates and Morrey space estimates with discontinuous a^{ij}. Further variations on the L^p estimates which involve anisotropic norms can be found in [177–179].

We also point out that our boundary assumption (that $\mathscr{P}\Omega \in H_{1+\alpha}$) can be modified in various ways. In [183, Section IV.9], the coefficients b^i and c are allowed to lie in L^q and $L^{q'}$ with

$$q > \min\{p, n+2\}, \quad q \geq \max\{p, n+2\},$$
$$q' > \min\{p, \frac{n+2}{2}\}, \quad q' \geq \max\{p, \frac{n+2}{2}\}.$$

Hence, we may assume that $\partial\Omega \in W_q^{2,1}$. Further weakening of the boundary regularity appears in [27].

Krylov and Safonov [180] proved the Harnack and Hölder estimates in Section VII.10; some of their ideas were inspired by earlier work of Landis [192] on elliptic equations with special structure; the argument of Landis was carried out for parabolic equations by Glagoleva [104]. The weak Harnack inequality and local maximum principle were proved by Gruber [105, Theorems 2.1 and 3.1] using probability methods as well as ideas of Trudinger [308] which were inspired by [180]. Our proof of the local maximum principle and weak Harnack inequality is taken from Reye's unpublished thesis [281], which uses many of the ideas in [308]. Reye's main contribution is our Lemma 7.39 which simplifies a number of comparisons. In fact, the details of our proof relating to the Morrey space hypothesis on b come from [190], which has a version of our Lemma 7.39 as well. Dong [68, Theorem X.7] made further modifications in the proof which are useful for considering certain nonuniformly parabolic equations. We also note that Wang and Liu [330] proved the boundary estimates Theorems 7.28 and 7.29, along with Corollary 7.30.

The normal derivative estimates in Lemmata 7.46 and 7.47 were originally proved by Krylov [170, Theorem 4.2] using different means. He studied the ratio u/x^n by introducing four new variables $y^1, \ldots, y^4$ and defining a new function

$$v(x^1, \ldots, x^{n-1}, y^1, \ldots, y^4, t) = \frac{u(x^1, \ldots, x^{n-1}, \Sigma_{i=1}^4 (y^i)^2)}{\Sigma (y^i)^2}.$$

It can be shown that v satisfies a degenerate parabolic equation and that the boundary $\{x^n = 0\}$ is converted to an interior singular set for the equation. Moreover, v satisfies a certain Harnack inequality. Caffarelli suggested a simplification of this approach, namely to estimate the quantity $u - Ax^n$ rather than u/x^n. The details of Caffarelli's idea were first published by Kazdan [157, Theorem 4.28] (and attributed there to Caffarelli); in particular, a two-sided Harnack inequality was used and a somewhat different comparison function. Our use of the weak Harnack inequality is closer in spirit to Krylov's original proof and was first used by the present author [210, Lemmata 4.2 and 4.3]. (This works was written at the same time as [157], and Caffarelli's suggestion (but not the details) is the basis for the boundary Hölder gradient estimates there. However, I was unaware of [157] until quite recently.) In [191, Lemma 5.5], Ladyzhenskaya and Ural′tseva prove a version of Lemma 7.47 under slightly different hypotheses: L has the form (7.1) with $b \in L^q$ for $q > n+2$ and $c \equiv 0$, and $f \in L^q$. More recently, Apushkinskaya and Nazarov [7, Lemma 4.5] have proved such results with $b = b_1 + b_2$ and $f = f_1 + f_2$, where b_1 and f_1 are in L^q and $|b_2|$ and $|f_2|$ are bounded by $Cd^{\delta-1}$.

Our proofs of the local maximum principle and weak Harnack inequality for oblique derivative conditions are simple variants of the corresponding elliptic results given in [218; 238, Section 2]. Hölder estimates for parabolic oblique derivative problems have also been stated and proved by Dong [67] and Nazarov [265]. An alternative approach, which gives the local maximum principle and weak Harnack inequality for parabolic oblique derivative problems in Lipschitz domains, is given in [235, Section 7].

Exercises

7.1 Show that Theorem 7.1 is valid for the operator $Lu = -ru_t + a^{ij}D_{ij}u + b^iD_iu + cu$ if we define $\mathscr{D}^* = [\det(a^{ij})r]^{1/(n+1)}$.

7.2 Show that Theorem 7.1 is valid for the boundary operator

$$\mathscr{M}u = \beta \cdot Du + \beta^{n+1}u_t + \beta^0 u$$

provided (β, β^{n+1}) is inward pointing and $\beta^{n+1} \le 0$.

7.3 Prove that the Marcinkiewicz interpolation theorem can be generalized to allow differing exponents and measure spaces for the domain and range of J. Specifically, let $1 \le q < r < \infty$, $1 \le \bar{q} < \bar{r} < \infty$, and let Ω and Ω' be arbitrary measurable sets, let J be a linear operator from $L^q \cap L^r(\Omega)$ to $L^{\bar{q}} \cap L^{\bar{r}}(\Omega')$, and suppose that there are constants J_1 and J_2 such that

$$\mu(h, Jf) \le \left(\frac{J_1 \|f\|_q}{h}\right)^{\bar{q}}, \qquad \mu(h, Jf) \le \left(\frac{J_1 \|f\|_r}{h}\right)^{\bar{r}}$$

for all $f \in L^q \cap L^r(\Omega)$. Show that J extends as a bounded linear from $L^p(\Omega)$ to $L^{\bar{p}}(\Omega')$ for any p and $\bar{p}$ such that

$$\frac{1}{p} = \frac{\alpha}{q} + \frac{1-\alpha}{r}, \qquad \frac{1}{\bar{p}} = \frac{\alpha}{\bar{q}} + \frac{1-\alpha}{\bar{r}}$$

for some $\alpha \in (0,1)$.

Also give the bound for the norm of J in terms of J_1 and J_2, and examine the case when r or $\bar{r}$ is infinite.

7.4 Show that Theorems 7.22, 7.30, 7.32, and 7.35 hold if a^{ij} is only uniformly continuous with respect to x and measurable with respect to t.

(Hint: First show that Propositions 7.28 and 7.33 hold if A^{ij} depends only on t.)

7.5 Use Lemma 7.47 to show that the hypothesis of continuity (4.57) on (a^{ij}) in Proposition 4.25 can be replaced by a smallness condition on α.

7.6 Show that the $W_p^{2,1}$ estimates hold if b^i and c are in suitable $L^{q,q'}$ spaces. How are n, p, q and q' related?

CHAPTER VIII

FIXED POINT THEOREMS AND THEIR APPLICATIONS

Introduction

In previous chapters, we saw that the existence, uniqueness, and regularity of solutions to linear initial-boundary value problems could be derived from certain *a priori* estimates. This chapter lays the groundwork for a corresponding study of nonlinear problems. Primarily, we consider the functional analytic aspect of these equations, and we prove some general solvability results, which reduce the question of existence of solutions to the establishment of *a priori* estimates. While the method of continuity suffices for linear problems, its application to nonlinear problems is not without difficulties. The most important for our investigation has to do with regularity hypotheses. As we saw in Exercise 5.4, this method assumes that the coefficients of the differential equation have Hölder continuous derivatives with respect to the u and Du variables to prove the solvability of the Cauchy-Dirichlet problem in $H_{2+\alpha}$. On the other hand, the linear theory then implies that this solution is in $H_{3+\beta,\mathrm{loc}}$ for some β. Although this higher regularity hypothesis can be removed by approximation arguments, it is very convenient to have a method which can deal with this regularity issue directly. For the oblique derivative problem, it does not seem possible to handle this regularity issue directly, but the technique we develop for such problem has other important applications.

In comparison with the existence program developed in [183], the present one emphasizes the connection of solutions on the entire domain in question (called *global solutions*) with solutions only on some small subset of the domain (called *local solutions*). A motivating example is the well-studied, semilinear problem:

$$-u_t + \Delta u + u^p = 0 \text{ in } \Omega,$$
$$u = 0 \text{ on } S\Omega, u = u_0 \text{ on } B\Omega,$$

where $p > 1$. If $\Omega = \omega \times (0,T)$ and if u_0 is a suitable nonnegative function, it is well-known [197] that there is a positive constant T_∞ (determined by u_0 and ω) such that this problem has a solution only for $T < T_\infty$ and that u is unbounded as $T \to T_\infty$. In addition, for many of the problems considered in this book, the solution can fail to exist for any prescribed positive T.

We study two types of fixed point theorems in this chapter. The first is an infinite-dimensional version of the Brouwer fixed point theorem, which states that any continuous map of a closed bounded convex set in $\mathbb{R}^n$ to itself has a fixed point. The second is a variant of a theorem due to Cacciopoli. Alternative fixed point theorems which have been used for studying nonlinear parabolic equations are discussed in the Notes to this chapter.

In Chapters VIII through XIII, we are concerned with problems with quasilinear equations, that is, equations of the form $Pu = 0$, where the operator P is defined on $C^{2,1}(\Omega)$ for some domain $\Omega \subset \mathbb{R}^{n+1}$ by

$$Pu = -u_t + a^{ij}(X,u,Du)D_{ij}u + a(X,u,Du). \tag{8.1}$$

The coefficients a^{ij} and a are assumed to defined for all values of their arguments. Specifically, $a^{ij}(X,z,p)$ and $a(X,z,p)$ are defined for all $(X,z,p) \in \Omega \times \mathbb{R} \times \mathbb{R}^n$.

Before discussing the functional analysis, we present here some definitions which will be used in the next six chapters. We say that P is *parabolic* in a subset $\mathcal{S}$ of $\Omega \times \mathbb{R} \times \mathbb{R}^n$ if the coefficient matrix $a^{ij}(X,z,p)$ is positive definite for all $(X,z,p) \in \mathcal{S}$, and we use λ and Λ to denote the smallest and largest eigenvalues of the matrix (a^{ij}). Hence

$$\lambda(X,z,p)\,|\xi|^2 \le a^{ij}(X,z,p)\xi_i\xi_j \le \Lambda(X,z,p)\,|\xi|^2 \tag{8.2}$$

for all $\xi \in \mathbb{R}^n$, and P is *parabolic* in $\mathcal{S}$ if $\lambda > 0$ on $\mathcal{S}$. If the ratio Λ/λ is uniformly bounded on $\mathcal{S}$, we say that P is *uniformly parabolic* on $\mathcal{S}$. Two particular forms for $\mathcal{S}$ are of special significance here. If $\mathcal{S} = \Omega \times \mathbb{R} \times \mathbb{R}^n$, we say that P is *parabolic* (*uniformly parabolic*) in Ω, and if $\mathcal{S} = \{(X,z,p) : z = u(X), p = Du(X)\}$ for some C^1 function u, we say that P is *parabolic* (*uniformly parabolic*) at u.

An important scalar function is the Bernstein $\mathcal{E}$ function, defined by

$$\mathcal{E}(X,z,p) = a^{ij}(X,z,p)p_ip_j. \tag{8.3}$$

Clearly $\lambda\,|p|^2 \le \mathcal{E} \le \Lambda\,|p|^2$, and a key element in the existence theory is the placement of $\mathcal{E}$ in this range. For example, if

$$a^{ij}(X,z,p) = \delta^{ij} - \frac{p_ip_j}{1+|p|^2}, \tag{8.4}$$

then $\lambda = 1/(1+|p|^2)$, $\Lambda = 1$, and $\mathcal{E} = |p|^2/(1+|p|^2) = \lambda\,|p|^2$. On the other hand, if

$$a^{ij}(X,z,p) = [\delta^{ij} + p_ip_j]/(1+|p|^2), \tag{8.5}$$

then $\lambda = 1/(1+|p|^2)$, $\Lambda = 1$, and $\mathcal{E} = |p|^2 = \Lambda\,|p|^2$. As we shall see, the operators corresponding to these two different choices of a^{ij} give rise to very different existence theories even though they have the same maximum and minimum eigenvalues.

Corresponding to a parabolic operator P, there is an elliptic operator E defined by

$$Eu = a^{ij}(X,u,Du)D_{ij}u + a(X,u,Du).$$

(Strictly speaking, E is a family of operators indexed by t, but we shall not worry about this technical distinction.) We call this operator the *elliptic part* of P. Given two parabolic operators P_1 and P_2, we say that their elliptic parts are equivalent if there is a (positive) function g defined on $\Omega \times \mathbb{R} \times \mathbb{R}^n$ such that $a_1^{ij} = g a_2^{ij}$ and $a_1 = g a_2$. Although equivalent elliptic operators give rise to the same elliptic theory, we shall also see that the corresponding parabolic operators may have very different properties.

Finally, we say that P is of *divergence form* if there are a differentiable vector field A and a scalar function B defined on $\Omega \times \mathbb{R} \times \mathbb{R}^n$ such that

$$Pu = -u_t + \operatorname{div} A(X,u,Du) + B(X,u,Du).$$

Such operators occupy a special place in the quasilinear theory because it is possible to define weak solutions and use the theory of Chapter VI in suitably modified form.

1. The Schauder fixed point theorem

Our extension of the Brouwer fixed point theorem replaces the convex subset of $\mathbb{R}^n$ by a suitable analog in a Banach space.

THEOREM 8.1. *Let $\mathscr{S}$ be a compact, convex subset of a Banach space $\mathscr{B}$ and let J be a continuous map of $\mathscr{S}$ into itself. Then J has a fixed point.*

PROOF. For a positive integer k, let $(B_i)_{i=1}^N$ be a finite collection of balls of radius $1/k$ covering $\mathscr{S}$. We write x_i for the center of B_i and $\mathscr{S}_k$ for the convex hull of the points $x_1,\dots,x_N$. (Note that, in fact, the family of balls and the points involved will be different for different values of k.) Then we define a mapping $I_k : \mathscr{S} \to \mathscr{S}_k$ by

$$I_k x = \frac{\sum_{i=1}^N \operatorname{dist}(x, \mathscr{S} \setminus B_i) x_i}{\sum_{i=1}^N \operatorname{dist}(x, \mathscr{S} \setminus B_i)}.$$

It is easily seen that I_k is continuous and that

$$\|I_k x - x\| \le \frac{\sum_{i=1}^N \operatorname{dist}(x, \mathscr{S} \setminus B_i) \|x_i - x\|}{\sum_{i=1}^N \operatorname{dist}(x, \mathscr{S} \setminus B_i)} < 1/k$$

for any $x \in \mathscr{S}$ because the i-th summand in the numerator or denominator is zero whenever $\|x_i - x\| \ge 1/k$. Since $\mathscr{S}_k$ is homeomorphic to a closed ball in $\mathbb{R}^N$, it follows that J_k, the restriction of $I_k \circ J$ to $\mathscr{S}_k$, has a fixed point y_k. The compactness of $\mathscr{S}$ implies that there is a convergent subsequence $(y_{k(m)})$ with

$$\left\|y_{k(m)} - Jy_{k(m)}\right\| = \left\|J_{k(m)} y_{k(m)} - Jy_{k(m)}\right\| < 1/k(m)$$

by virtue of (1). It follows that $x = \lim y_{k(m)}$ is a fixed point of J. □

2. Applications of the Schauder theorem

Theorem 8.1 leads rather easily to a local existence theorem for quasilinear equations under very weak hypotheses on the coefficients. In addition, suitable estimates lead to global existence. To state the local theorem, we need a few hypotheses. For P defined by (8.1), we suppose that the coefficients are smooth and parabolic in a certain sense, namely, for any bounded subset $\mathscr{K}$ of $\Omega \times \mathbb{R} \times \mathbb{R}^n$, we suppose that there is a positive constant $\lambda_{\mathscr{K}}$ such that

$$\lambda_{\mathscr{K}} |\xi|^2 \leq a^{ij}(X,z,p)\xi_i\xi_j \tag{8.6a}$$

for any $(X,z,p) \in \mathscr{K}$ and any $\xi \in \mathbb{R}^n$. The appropriate smoothness of the coefficients is determined by the linear theory. The maps $X \to a^{ij}(X,u,Du)$ and $X \to a(X,u,Du)$ should be Hölder continuous if u is smooth enough. This requirement is met if there is $\alpha \in (0,1)$ such that

$$a^{ij} \text{ and } a \text{ are in } H_\alpha(\mathscr{K}) \tag{8.6b}$$

for any bounded subset $\mathscr{K}$ of $\Omega \times \mathbb{R} \times \mathbb{R}^n$. When the coefficients are independent of u and Du, the linear theory provides a unique $H_{2+\alpha}^{(-\delta)}$ solution of the Cauchy-Dirichlet problem for any $\delta \in (1,2)$ when the boundary data are smooth enough. If also the boundary data are $H_{2+\alpha}$ and the compatibility condition is satisfied, the solution is $H_{2+\alpha}$. The same results are true for quasilinear problems if Ω is small enough in the time direction. We shall prove an equivalent statement in a fixed domain Ω and study the problem in a smaller domain Ω_ε defined by

$$\Omega_\varepsilon = \{X \in \Omega : t < t_0 + \varepsilon\},$$

where we assume that $B\Omega \subset \{t = t_0\}$.

THEOREM 8.2. *Suppose $\mathscr{P}\Omega \in H_\delta$ and $\varphi \in H_\delta(\mathscr{P}\Omega)$ for some $\delta \in (1,2)$. Then there is a positive constant ε such that the problem*

$$Pu = 0 \text{ in } \Omega_\varepsilon,\ u = \varphi \text{ on } \mathscr{P}\Omega_\varepsilon, \tag{8.7}$$

has a solution $u \in H_{2+\alpha}^{(-\delta)}$. If $\mathscr{P}\Omega \in H_{2+\alpha}$ and $\varphi \in H_{2+\alpha}$ and if $P\varphi = 0$ on $C\Omega$, then $u \in H_{2+\alpha}$.

PROOF. Let $\theta \in (1,\delta)$, set $M_0 = 1 + |\varphi|_\theta$, and for $\varepsilon > 0$ to be chosen, set

$$\mathscr{S} = \{v \in H_\theta(\Omega_\varepsilon) : |v|_\theta \leq M_0\}.$$

We then define the map $J : \mathscr{S} \to H_\theta$ by $u = Jv$ if

$$-u_t + a^{ij}(X,v,Dv)D_{ij}u + a(X,v,Dv) = 0 \text{ in } \Omega_\varepsilon,\ u = \varphi \text{ on } \mathscr{P}\Omega_\varepsilon,$$

noting that, for each v, this problem has a unique solution in $H_{2+\alpha(\theta-1)}^{(-\delta)}$ by Theorem 5.15 and

$$|u|_1 \leq |u|_\delta \leq C|u|_{2+\alpha(\theta-1)}^{(-\delta)} \leq C(M_0).$$

It follows that $|u-\varphi| \le C\varepsilon$ in Ω_ε and then $|u-\varphi|_\theta \le C\varepsilon^{(\delta-\theta)/\delta}$ by interpolation. Therefore $|u|_\theta \le M_0$ if ε is sufficiently small, and hence J maps $\mathscr{S}$ into itself for such an ε. Since $\mathscr{S}$ is a convex, compact subset of H_1, it follows that J has a fixed point u, which is clearly in $H^{(-\delta)}_{2+\alpha(\theta-1)}$ and hence solves (8.7). Theorem 5.15 now shows that $u \in H^{(-\delta)}_{2+\alpha}$.

Under the additional hypotheses made at the end of the statement of this theorem, we use Theorem 5.14 to conclude first that $u \in H_{2+\alpha(\delta-1)}$ and then that $u \in H_{2+\alpha}$. □

Note that Theorem 8.2 makes no assertion about the uniqueness of the solution. We shall prove a uniqueness theorem in Chapter IX. More significant is the question of global existence, which is closely tied to the establishment of *a priori* estimates in the following five chapters. Our next theorem demonstrates this connection.

THEOREM 8.3. *Suppose that Ω, φ, and P are as in Theorem 8.2. If there is a constant M_δ (independent of ε) such that any solution u of* (8.7) *obeys the estimate*

$$|u|_\delta \le M_\delta, \tag{8.8}$$

then there is a solution of $Pu = 0$ in Ω, $u = \varphi$ on $\mathscr{P}\Omega$.

PROOF. Write (t_0, t_1) for $I(\Omega)$, and let ε_0 be the largest number in $[0, t_1 - t_0]$ such that (8.7) has solution for $\varepsilon < \varepsilon_0$. Theorem 8.2 shows that $\varepsilon_0 > 0$. We show by contradiction that $\varepsilon_0 = t_1 - t_0$. If not, then let (ε_i) be an increasing sequence of positive numbers with $\varepsilon_i \to \varepsilon_0$, and denote by u_i the solution of (8.7) with $\varepsilon = \varepsilon_i$. Condition (8.8) along with the Arzela-Ascoli theorem provides a uniformly convergent subsequence, which we also denote by (u_i), with limit u. By interpolation, we see that $u_i \to u$ in $H^{(-\tau\delta)}_{(2+\alpha)\tau}$ for $\tau \in (0,1)$. In particular, if $\tau > \max\{\frac{1}{\delta}, \frac{2}{2+\alpha}\}$, we see that $u_i \to u$ and $Du_i \to Du$ uniformly in Ω and $u_{i,t} \to u_t$ and $D^2u_i \to D^2u$ uniformly in any cylinder Q with $\overline{Q} \subset \Omega$. Therefore u solves (8.7) with $\varepsilon = \varepsilon_0$ and then Theorem 8.2 can be used to solve

$$Pu^* = 0 \text{ in } \Omega^*, u^* = \varphi \text{ on } S\Omega^*, u^* = u \text{ on } B\Omega^*,$$

where $\Omega^* = \{X \in \Omega : t_0 + \varepsilon_0 < t < t_0 + \varepsilon_0 + \eta\}$ for some $\eta > 0$. It is easy to check that w defined by

$$w(X) = \begin{cases} u(X) & \text{if } X \in \Omega_{\varepsilon_0} \\ u^*(X) & \text{if } X \in \overline{\Omega^*} \end{cases}$$

is a solution of (8.7) with $\varepsilon > \varepsilon_0$, contradicting the definition of ε_0. □

The proof of estimate (8.8) proceeds in four steps. Chapter IX deals with an estimate of the maximum of the solution u, the maximum of the gradient on the parabolic boundary appears in Chapter X, the gradient is estimated over the entire domain in Chapter XI, and Chapter XII presents a Hölder gradient estimate. As we shall see, different structure conditions are needed to infer the various estimates.

To illustrate the basic ideas, we consider the simple problem

$$-u_t + \operatorname{div} A(Du) = 0 \text{ in } \Omega, u = \varphi \text{ on } \mathscr{P}\Omega,$$

when $\Omega = B(0,R) \times (0,T)$ and $A \in H_{2,\mathrm{loc}}(\mathbb{R}^n)$, and $\varphi \in H_{2+\alpha}$. We also assume that $\varphi_t = 0$ on $S\Omega$. By writing the differential equation in the form $-u_t + a^{ij}(X)D_{ij}u = 0$ with

$$a^{ij}(X) = \frac{\partial A^i}{\partial p_j}(Du(X)),$$

we infer from the maximum principle that $\sup|u| \le \sup|\varphi|$. Next, for k a positive constant to be chosen, we define

$$h^{\pm}(X) = k[R^2 - |x|^2] \pm \varphi(X).$$

A simple calculation gives

$$-h_t^{\pm} + a^{ij}D_{ij}h^{\pm} = 2\mathscr{T}k \pm a^{ij}D_{ij}\varphi \le 0$$

provided $k \ge \sup\left|D^2\varphi\right|$. With this choice of k, the maximum principle gives $h^+ \ge u \ge -h^-$ because this inequality is easily checked on $\mathscr{P}\Omega$. In this way, we obtain the boundary gradient estimate for u.

To prove the global gradient estimate and the Hölder gradient estimate, we note that $v = v_k = D_k u$ satisfies the differential equation $-v_t + D_i(a^{ij}D_j v) = 0$ for $k = 1, \dots, n$. The boundary gradient estimate implies that $|v_k| \le C$ on $\mathscr{P}\Omega$ for each k so the maximum principle Corollary 6.26 gives the same bound for $|v_k|$ in Ω. For the Hölder estimate, we first use Lemma 7.47 (after a flattening of the boundary) to $u - \varphi$ to infer that $D_n u$ is Hölder continuous on $\mathscr{P}\Omega$. (See Chapter XII on the details of this argument near $C\Omega$.) Then the Hölder estimate Theorem 6.28 gives a Hölder estimate for Du in Ω. Therefore, under these hypotheses, there is a solution of our Cauchy-Dirichlet problem $u \in H_{2+\alpha}^{(-\delta)}$ for any $\delta \in (1,2)$ and $\alpha \in (0,1)$. If $\operatorname{div} A(D\varphi) = 0$ on $\partial B(0,R)$, then $u \in H_{2+\alpha}$.

3. A theorem of Caristi and its applications

For the oblique derivative problem with fully nonlinear boundary condition, the method of the previous section is no longer applicable. There is no simple analog of the map J from Lemma 8.2, which introduced a linear problem whose solvability is known to us. Instead, we solve a different although related linear problem for the oblique derivative problem. We also use a fixed point theorem, but its connection to the original initial-boundary value problem is not so straightforward, and the map considered in the fixed point theorem is not necessarily compact. We start with the following theorem of J. Caristi.

LEMMA 8.4. *Let (V,ρ) be a complete metric space and let g be a map from V to itself. If there is a lower semicontinuous function $\varphi : V \to [0,\infty)$ such that*

$$\rho(v,g(v)) \leq \varphi(v) - \varphi(g(v)) \tag{8.9}$$

for any $v \in V$, then g has a fixed point.

PROOF. Choose $u_0 \in V$ such that

$$\varphi(u_0) \leq \inf_V \varphi + \frac{1}{2},$$

and define u_n and S_n inductively as follows. Given u_n, we set

$$S_n = \{w \in V : \varphi(w) \leq \varphi(u_n) - \frac{1}{2}\rho(u_n,w)\},$$

and then choose $u_{n+1} \in S_n$ so that

$$\varphi(u_{n+1}) - \inf_{S_n} \varphi \leq \frac{1}{2}(\varphi(u_n) - \inf_{S_n} \varphi).$$

Since we always have $u_n \in S_n$, this procedure generates a sequence of points (u_n). If there is an n such that $S_n = \{u_n\}$, then we have $u_m = u_n$ for $m \geq n$ and hence (u_n) is Cauchy. On the other hand, if S_n contains points other than u_n for every n, then (u_n) is an infinite sequence. In this case, from the definition of S_n, we have

$$\frac{1}{2}\rho(u_n,u_{n+1}) \leq \varphi(u_n) - \varphi(u_{n+1}),$$

and then the triangle inequality yields

$$\frac{1}{2}\rho(u_n,u_m) \leq \varphi(u_n) - \varphi(u_m) \tag{8.10}$$

as long as $n \leq m$. Since the sequence $(\varphi(u_n))$ is decreasing and bounded from below, it follows that $\varphi(u_n)$ tends to a limit L and hence (u_n) is Cauchy. The completeness of V implies that (u_n) has a limit, which we denote by v. The lower semicontinuity of φ implies that

$$\varphi(v) \leq L \leq \varphi(u_n) \leq \inf_V \varphi + \frac{1}{2}.$$

Next, we observe that $S_{n+1} \subset S_n$ since $w \in S_{n+1}$ implies that

$$\begin{aligned}
\frac{1}{2}\rho(u_n,w) &\leq \frac{1}{2}\rho(u_n,u_{n+1}) + \frac{1}{2}\rho(u_{n+1},w) \\
&\leq [\varphi(u_n) - \varphi(u_{n+1})] + [\varphi(u_{n+1}) - \varphi(w)] \\
&= \varphi(u_n) - \varphi(w)
\end{aligned}$$

and therefore $w \in S_n$.

By taking the limit as $m \to \infty$ in (8.10), we see that $v \in \cap S_n$. Conversely if $w \in \cap S_n$, then

$$\frac{1}{2}\rho(u_n, w) \leq \varphi(u_n) - \varphi(w) \leq \varphi(u_n) - \inf_{S_{n-1}} \varphi.$$

By construction,

$$\varphi(u_n) - \inf_{S_{n-1}} \varphi \leq \frac{1}{2}[\varphi(u_{n-1}) - \inf_{S_{n-1}} \varphi],$$

and, because $S_{n-1} \subset S_{n-2}$, we have

$$\varphi(u_n) - \inf_{S_{n-1}} \varphi \leq \frac{1}{2}[\varphi(u_{n-1}) - \inf_{S_{n-2}} \varphi].$$

An easy induction argument leads to the inequality $\rho(u_n, w) \leq 2^{1-n}$. Sending $n \to \infty$ shows that $w = v$.

On the other hand, if $w \neq v$, then there is a positive integer N such that

$$\varphi(w) > \varphi(u_n) - \frac{1}{2}\rho(u_n, w)$$

for all $n \geq N$. Sending $n \to \infty$ implies that $\varphi(w) \geq L - \frac{1}{2}\rho(v, w)$. Since $L \geq \varphi(v)$, it follows that

$$\varphi(w) \geq \varphi(v) - \frac{1}{2}\rho(v, w).$$

(If $w = v$, this inequality is immediate.) In particular, if $w = g(v)$, we infer that $\varphi(g(v)) \geq \varphi(v) - \frac{1}{2}\rho(v, g(v))$ and hence $\frac{1}{2}\rho(v, g(v)) \geq \varphi(v) - \varphi(g(v))$. From (8.9), it follows that $\frac{1}{2}\rho(v, g(v)) \geq \rho(v, g(v))$ and therefore $\rho(v, g(v)) = 0$, so v is the desired fixed point. □

To apply this theorem, we first note the following useful calculation.

LEMMA 8.5. *Let U be a (nonempty) metric space, let $\mathscr{B}$ be a Banach space and let $J : U \to \mathscr{B}$. If JU is a closed subset of $\mathscr{B}$ and if for every $u \in U$, there are a point $u_1 \in U$ and a number $\varepsilon \in (0, 1)$ such that*

$$\|Ju_1 - (1-\varepsilon)Ju\| \leq \frac{\varepsilon}{2}\|Ju\|, \tag{8.11}$$

then $0 \in JU$.

PROOF. For $v \in JU$, choose $u \in J^{-1}(v)$ and define g by $g(v) = Ju_1$. To show that g has a fixed point, we use the triangle inequality twice. First, we note that

$$\|Ju_1 - Ju\| - \varepsilon\|Ju\| \leq \|Ju_1 - (1-\varepsilon)Ju\|$$

to infer from (8.11) that

$$\|Ju_1 - Ju\| \leq \frac{3\varepsilon}{2}\|Ju\|.$$

Then we note that

$$\|Ju_1\| - (1-\varepsilon)\|Ju\| \leq \|Ju_1 - (1-\varepsilon)Ju\|$$

and conclude that

$$\|Ju_1\| \leq (1-\frac{\varepsilon}{2})\|Ju\|.$$

Combining these inequalities yields

$$\|Ju_1 - Ju\| \leq 3(\|Ju\| - \|Ju_1\|),$$

and setting $\varphi(u) = 3\|v\|$, we see that $\rho(v, g(v)) \leq \varphi(v) - \varphi(g(v))$. Then Lemma 8.4 shows that g has a fixed point v_0. From (8.11), we have that

$$\varepsilon\|v_0\| = \|g(v_0) - (1-\varepsilon)v_0\| \leq \frac{\varepsilon}{2}\|v_0\|,$$

so $v_0 = 0$. Since $v_0 \in JU$, it follows that $0 \in JU$. □

In using Lemma 8.5, we need a simple condition which implies (8.11). For our purposes, such a condition is most conveniently stated in terms of the Gateaux variation of a function. Let J be a map from a subset U of one Banach space $\mathscr{B}_1$ to another Banach space $\mathscr{B}$. For a fixed $u \in U$ and $\psi \in \mathscr{B}_1$ such that $u + \varepsilon\varphi \in U$ for all sufficiently small ε, we define the *Gateaux variation* $J_u(\psi)$ of J at u in the direction of ψ by

$$J_u(\psi) = \lim_{\varepsilon\to 0} \frac{J(u+\varepsilon\psi) - Ju}{\varepsilon}$$

provided this limit exists. In general, this limit may exist only for certain choices of u and ψ. Suppose now that for $u \in U$, there is a $\psi \in U$ such that $J_u(\psi)$ exists and

$$J_u(\psi) + Ju = 0. \tag{8.12}$$

then $0 \in JU$. By the definition of the Gateaux variation, there is an $\varepsilon > 0$ such that

$$\|[J(u+\varepsilon\psi) - Ju] - \varepsilon J_u(\psi)\| \leq \frac{\varepsilon}{2}\|Ju\|.$$

Then (8.12) gives

$$\|[J(u+\varepsilon\psi) - Ju] + \varepsilon Ju\| \leq \frac{\varepsilon}{2}\|Ju\|,$$

which is just (8.11) with $u_1 = u + \varepsilon\psi$.

Our next step is to calculate the Gateaux variation for the operators appearing in the oblique derivative problem.

LEMMA 8.6. *Define P by* (8.1) *and N by*

$$Nu = b(X, u, Du) \tag{8.13}$$

for some function b. Suppose that there is an $\alpha \in (0,1)$ such that $\mathscr{P}\Omega \in H_{2+\alpha}$ and

$$a^{ij}, a_z^{ij}, a_p^{ij}, a, a_z, \text{ and } a_p \text{ are in } H_\alpha(\mathscr{K}), \tag{8.14a}$$

$$b, b_z, \text{ and } b_p \text{ are in } H_{1+\alpha}(\Sigma) \tag{8.14b}$$

for any compact subsets $\mathcal{K}$ *and* Σ *of* $\Omega\times\mathbb{R}\times\mathbb{R}^n$ *and* $S\Omega\times\mathbb{R}\times\mathbb{R}^n$*, respectively. Let* $\theta\in(0,\alpha)$ *and define* $J\colon H_{2+\theta}\to H_\theta(\Omega)\times H_{1+\theta}(S\Omega)$ *by* $Ju=(Pu,Nu)$*. Then the Gateaux variation of* J *exists and we have* $J_u(\psi)=(P_u(\psi),N_u(\psi))$ *with* $P_u(\psi)$ *and* $N_u(\psi)$ *defined by*

$$P_u(\psi)=-\psi_t+a^{ij}D_{ij}\psi+(a_p^{ij}D_{ij}u+a_p)\cdot D\psi+(a_z^{ij}D_{ij}u+a_z)\psi, \tag{8.15a}$$

$$N_u(\psi)=b_p\cdot D\psi+b_z\psi, \tag{8.15b}$$

where a^{ij}*, a, b and their derivatives are all evaluated at* (X,u,Du)*.*

PROOF. By virtue of the continuity of the derivatives, it is easy to see that the difference quotients in the definition of the Gateaux variation converge pointwise to the appropriate limits. It is also elementary to verify that the difference quotients are uniformly bounded in the $H_{2+\alpha}(\Omega)\times H_{1+\alpha}(S\Omega)$ norm, so the interpolation inequality Proposition 4.2 shows the convergence in the Hölder spaces with exponent θ. □

In this case, equation (8.12) is a linear parabolic differential equation with a linear boundary condition, and the boundary condition is *oblique* if, for every compact subset Σ of $S\Omega\times\mathbb{R}\times\mathbb{R}^n$, there is a positive constant χ_Σ such that

$$b_p\cdot\gamma\geq\chi_\Sigma. \tag{8.16}$$

In studying the problem

$$Pu=0 \text{ in } \Omega,\ Nu=0 \text{ on } S\Omega,\ u=\varphi \text{ on } B\Omega, \tag{8.17}$$

we need to have $N\varphi=0$ on $C\Omega$ if we want to consider solutions with continuous spatial gradient. Theorem 5.18 then gives a unique solution $\psi\in H_{2+\alpha}$ of the boundary value problem

$$P_u(\psi)=-Pu \text{ in } \Omega,\ N_u(\psi)=-Nu \text{ on } S\Omega,\ \psi=0 \text{ on } B\Omega. \tag{8.18}$$

More generally, if $u\in H_{2+\theta}$ for $\theta\in(0,\alpha)$, then ψ is in $H_{2+\theta}$.

It is often convenient to deal only with functions that satisfy the boundary condition. For the Cauchy-Dirichlet problem, it was simple to restrict our attention in this way. For the oblique derivative problem, we use the preceding observation to do so.

LEMMA 8.7. *Let P, N,* Ω *and* φ *be as in Lemma 8.6 with* $N\varphi=0$ *on* $C\Omega$*, and set*

$$U=\{u\in H_{2+\theta}(\Omega): Nu=0 \text{ on } S\Omega, u=\varphi \text{ on } B\Omega\}. \tag{8.19}$$

For any $u\in U$*, there are a positive* $\varepsilon\in(0,1)$ *and a function* $u_1\in U$ *such that*

$$|Pu_1-(1-\varepsilon)Pu|_\theta\leq\frac{\varepsilon}{2}|Pu|_\theta. \tag{8.20}$$

PROOF. For $u \in U$, let ψ be the solution of (8.18), define J as in Lemma 8.6, and for $\eta > 0$ to be chosen, take $\varepsilon \in (0,1)$ such that

$$\|(J(u+\varepsilon\psi) - Ju) - \varepsilon J_u(\psi)\| \leq \eta\varepsilon\|Ju\|,$$

so $\|J(u+\varepsilon\psi) - (1-\varepsilon)Ju\| \leq \eta\varepsilon\|Ju\|$. For simplicity, we write q for $\|Ju\| = |Pu|_\theta$. Then we have

$$|N(u+\varepsilon\psi)|_{1+\theta} \leq \eta\varepsilon q. \tag{8.21}$$

Now set

$$\Sigma = \{(X,z,p) : X \in S\Omega, |z-u(X)| \leq |\psi|_0, |p - Du(X)| \leq 2 + |D\psi|_0\}$$

and suppose $\eta < \chi_\Sigma/q$. Then

$$\pm b(X, u+\varepsilon\psi, D(u+\varepsilon\psi) \pm \frac{2\eta\varepsilon q}{\chi_\Sigma}\gamma) \geq \pm b(X, u+\varepsilon\psi, D(u+\varepsilon\psi)) + 2\eta\varepsilon q > 0$$

by (8.21), so the implicit function theorem gives a function $f \in H_{1+\theta}(S\Omega)$ such that $|f| \leq 2\eta\varepsilon q/\chi_\Sigma$ and

$$b(X, u+\varepsilon\psi, D(u+\varepsilon\psi) + f\gamma) = 0$$

on $S\Omega$. Since $\psi = 0$ on $B\Omega$ and $Nu = 0$, we can use Lemma 4.24 to construct a function ψ_1 with $\psi_1 = 0$ on $\mathscr{P}\Omega$ and $D\psi_1 = f\gamma$ on $S\Omega$. In addition, we have $|\psi_1|_{2+\theta} \leq c_1(\Omega)|f|_{1+\theta}$. We then take $u_1 = u + \varepsilon\psi + \psi_1$ for suitably chosen η and the corresponding ε. Since $u_1 \in U$, we need only verify (8.20).

For this verification, we estimate $|f|_{1+\theta}$. First we set $g = N(u+\varepsilon\psi)$ and we work locally, so that we may assume that we are working on the hyperplane $\{x^n = 0\}$. By differentiating the equations for Nu and $N(u+\varepsilon\psi)$, we obtain

$$BD_k f = \beta^i D_{ik}(u+\varepsilon\psi) + \beta^0 D_k(u+\varepsilon\psi) - D_k g,$$

where

$$\begin{aligned} B &= b^n(X, u+\varepsilon\psi, D(u+\varepsilon\psi) + f\gamma), \\ \beta^i &= b^i(X, u+\varepsilon\psi, D(u+\varepsilon\psi)) - b^i(X, u+\varepsilon\psi, D(u+\varepsilon\psi) + f\gamma), \\ \beta^0 &= b_z(X, u+\varepsilon\psi, D(u+\varepsilon\psi)) - b_z(X, u+\varepsilon\psi, D(u+\varepsilon\psi) + f\gamma), \end{aligned}$$

and $b^1, \ldots, b^n$ denote the components of the vector b_p. It follows that $|Df|_0 \leq c\eta\varepsilon q$ and then that $|Df|_\theta \leq c\eta\varepsilon q$. A similar difference quotient argument gives $\langle f\rangle_{1+\theta} \leq c\eta\varepsilon q$ and then $|f|_{1+\theta} \leq c_2\eta\varepsilon q$. Therefore, $|\psi_1|_{2+\theta} \leq c_1c_2\eta\varepsilon q$, and then a simple calculation based on our choice of ε and ψ gives (8.20). □

With this observation, it is a simple matter to prove an existence theorem for the oblique derivative problem.

THEOREM 8.8. *Suppose P, N, and Ω are as in Lemma 8.6 with conditions* (8.6a) *and* (8.16) *satisfied. Let $\theta \in (0,\alpha)$ and $\varphi \in H_{2+\alpha}(B\Omega)$ with $N\varphi = 0$ on $C\Omega$. If for any $u \in H_{2+\theta}(\Omega)$ with $u = \varphi$ on $B\Omega$, the estimate*

$$|u|_{2+\theta} \le C(P,N,\Omega,|\varphi|_{2+\theta},|Pu|_{\theta}) \tag{8.22}$$

holds, then there is a solution $u \in H_{2+\theta}$ of (8.17).

PROOF. We take U as in Lemma 8.7, $\mathscr{B} = H_\theta(\Omega)$, and $J = P$. From Lemmata 8.6 and 8.7 and the remarks immediately preceding this theorem, we need only show that JU is closed. Let (v_m) be a sequence in JU which converges, say to v, in $\mathscr{B}$. For each m choose $w_m \in U$ so that $Jw_m = v_m$. Since (Jw_m) is bounded, it follows from (8.22) that (w_m) is bounded in U and hence the Arzela-Ascoli theorem implies that there is a subsequence $(w_{m(k)})$ which converges in $C^{2,1}$. It is straightforward to check that the limit function w is also in U and that $Jw = v$ and hence JU is closed. □

In fact, if $\varphi \in H_{2+\alpha}$, we expect the solution of (8.17) to lie in $H_{2+\alpha}$. Our next result not only shows this to be the case but also that the $H_{2+\alpha}$ norm of the solution of (8.17) can be estimated in terms of a lower norm.

THEOREM 8.9. *Define P by* (8.1) *and N by* (8.13) *and suppose that conditions* (8.6) *and* (8.16) *hold. Suppose also that $b \in H_{1+\alpha}(\Sigma)$ for any bounded subset Σ of $S\Omega \times \mathbb{R} \times \mathbb{R}^n$, that $\mathscr{P}\Omega \in H_{2+\alpha}$ and that $\varphi \in H_{2+\alpha}(B\Omega)$ satisfies the condition $N\varphi = 0$ on $C\Omega$. Suppose $u \in C^{2,1}(\overline{\Omega})$ with $Pu \in H_{2+\alpha}$, $Nu \in H_{1+\alpha}$, and $u = \varphi$ on $B\Omega$. Then $u \in H_{2+\alpha}$ and*

$$|u|_{2+\alpha} \le C(P,N,\Omega,|\varphi|_{2+\alpha},|u|_{1+\delta},|Pu|_{\alpha},|Nu|_{1+\alpha},\delta) \tag{8.23}$$

for any $\delta \in (0,1]$.

PROOF. We prove the estimate via local considerations. If $\Omega' \subset\subset \Omega \cup B\Omega$, then the linear theory shows that we can estimate $|u|_{2+\alpha\delta,\Omega'}$ in terms of the quantities on the right side of (8.21). If Ω' is a small neighborhood of a point $X_0 \in S\Omega$ with $\overline{\Omega}' \cap B\Omega$ empty, then by a simple change of variables, we may assume that u satisfies $Pu = 0$ in $Q^+(R)$ and $Nu = 0$ on $Q^0(R)$ for some positive R. If $k = 1,\dots,n$, then $w = D_k u$ solves the linear boundary value problem $-w_t + D_i(A^{ij}D_j w + f^i) = 0$ in $Q^+(R)$, $\beta \cdot Dw = \psi$ on $Q^0(R)$ with

$$A^{ij} = a^{ij}[0], \beta = b_p[0],$$
$$f^i = \delta^{ik}[(a^{jm}[X] - a^{jm}[0])D_{jm}u(X) + a[X] - a[0])],$$
$$\psi = (b_p[0] - b_p[X]) \cdot Dw(X) - b_z[X]D_k u(X) - b_k[X],$$

where we have used $a^{ij}[0]$ and $a^{ij}[X]$ as shorthand for $a^{ij}(0,u(0),Du(0))$ and $a^{ij}(X,u,Du)$, respectively (and similarly for a, b_p, etc.). By imitating the proofs of Theorems 4.9 and 4.22, we see that $|D_k u|_{1+\alpha\delta,\Omega'} \le C$ for $k < n$. For $k = n$, we solve the boundary condition $b(X,u,Du) = 0$ for $D_n u$ to obtain $D_n u =$

$b^*(X,u,D'u)$ for some $H_{1+\alpha}$ function b^*. The estimate for $D_k u$ then implies that $D_n u \in H_{1+\theta}(Q^0(R))$ for $\theta = \alpha\delta^2$, so $D_n u \in H_{1+\theta}(Q^+(R/2))$ and $|D_n u|_{1+\theta} \le C$. It follows that $|Du|_{1+\theta,\Omega'} \le C$, and a similar argument works in a neighborhood of $C\Omega$. As in Theorems 4.9 and 4.22, we conclude that $|u|_{2+\theta} \le C$ for $\theta = \alpha\delta^2$. In particular, $|u|_2 \le C$, so we can repeat the argument with $\delta = 1$ to conclude that $|u|_{2+\alpha} \le C$. □

Combining this regularity estimate with the results of Theorem 8.8 gives the following analog of Theorem 8.3 for the oblique derivative problem.

COROLLARY 8.10. *Let P, N, φ, and Ω be as in Theorem 8.8. If U is nonempty and if there are positive constants $\theta < \alpha$ and $\delta \le 1$ such that*

$$|u|_{1+\delta} \le C(P,N,\Omega,|\varphi|_{2+\theta},|Pu|_\theta) \tag{8.22$'$}$$

for all $u \in U$, then there is a solution $u \in H_{2+\theta}$ of (8.17). *Moreover, if $\varphi \in H_{2+\alpha}$, then this solution is in $H_{2+\alpha}$.*

The direct application of this theorem to a typical quasilinear equation with nonlinear boundary condition is complicated by the details of the estimates even if the operators P and N have a simple form; on the other hand, the verification that U is nonempty will be trivial. For this reason, we shall investigate a slightly different problem which will be useful in our complete study of oblique derivative problems in Chapter XIII. We begin by recalling the following definitions from Chapter IV. For positive constants R and $\bar\mu$ with $\bar\mu < 1$ and $X_0 \in \mathbb{R}^{n+1}$ with $x_0^n = -\bar\mu R$, we define

$$\Sigma^+ = \{X \in Q(X_0,R) : x^n > 0\},\ \Sigma^0 = \{X \in Q(X_0,R) : x^n = 0\},$$

and we write σ for $\mathscr{P}\Sigma^+ \setminus \Sigma^0$.

THEOREM 8.11. *Suppose that (a^{ij}) is a positive definite matrix-valued function with $a^{ij} \in H_3(\Sigma^+)$, and let $f \in H_{1+\alpha}^{(2-\alpha)}(\Sigma^+ \cup \Sigma^0)$, $\varphi \in H_{3+\alpha}^{(-\alpha)}(\Sigma^+ \cup \Sigma^0)$, and $g \in H_{3+\theta,loc}(\mathbb{R}^{n-1})$ for some constants α and θ in $(0,1)$. Suppose also that $|Dg| \le \mu$. If $\bar\mu > \mu/(1+\mu^2)^{1/2}$, then there is a unique solution $u \in H_{3+\alpha}^{(-\alpha)}$ of*

$$-u_t + a^{ij}D_{ij}u = f \text{ in } \Sigma^+, \tag{8.24a}$$

$$D_n u - g(D'u) = 0 \text{ on } \Sigma^0, u = \varphi \text{ on } \sigma. \tag{8.24b}$$

PROOF. From Lemma 4.16, we immediately infer a bound for $\sup|u-\varphi|$. Then Corollary 7.51 gives local Hölder bounds for $D'u$ on Σ^+. The boundary condition then gives the corresponding bound for $D_n u$ on Σ^0, and this bound on Σ^+ follows from Corollary 7.45. From the explicit form of these estimates, we infer that $|Du|_\delta^{(1-\alpha)} \le C$ for some $\delta > 0$. Applying the linear theory (as in Theorem 8.9), we see that $|u|_{3+\alpha}^{(-\alpha)} \le C$ and the existence of a solution follows from the argument in Corollary 8.10. □

Note that a simple approximation argument then gives this theorem if we only assume that $g \in H_{1+\alpha,\text{loc}}$.

We close by showing that when we have an initial function which satisfies the boundary condition on $C\Omega$, then we can find a function satisfying the boundary condition on $S\Omega$ and equal to the given initial function more simply.

THEOREM 8.12. *Let $\mathscr{P}\Omega \in H_{2+\alpha}$, let N have the form* (8.13) *with $b \in H_{1+\alpha}$ and suppose that N is oblique. Suppose that $\varphi \in H_{2+\alpha}$ satisfies $N\varphi = 0$ on $C\Omega$. If there is a positive constant μ_1 such that*

$$\pm b(X,\varphi,D\varphi \pm \mu_1\gamma) > 0, \tag{8.25}$$

then there is a function $g \in H_{2+\alpha}$ such that $Ng = 0$ on $S\Omega$ and $g = \varphi$ on $\mathscr{P}\Omega$.

PROOF. Now we use (8.25) and the obliqueness of N to infer from the implicit function theorem that there is a (unique) function $g_1 \in H_{1+\alpha}$ such that $g_1 = 0$ on $B\Omega$ and $b(X,\varphi,D\varphi + g_1\gamma) = 0$ on $S\Omega$. Lemma 4.24 completes the proof. □

As an example to illustrate Theorem 8.12, we suppose that we have a vector-valued function A and a scalar function ψ such that

$$b(X,z,p) = A(X,z,p)\cdot\gamma + \psi(X,z),$$

and that there are positive constants M_2, μ_2 and μ_3 with

$$p\cdot A(X,z,p) \geq \mu_2|p|\,|A(X,z,p)|,$$
$$p\cdot A(X,z,p) \geq (1+\mu_3)|p|\,|\psi(X,z)|$$

for $|p| \geq M_2$. Then, for $M \geq M_2$ a constant to be further specified, we have

$$\begin{aligned}\pm b(X,\varphi,D\varphi \pm M\gamma) &= \pm\gamma\cdot A \pm \psi \\ &= \frac{1}{M}(D\varphi \pm M\gamma)\cdot A - \frac{D\varphi}{M}\cdot A \pm \psi \\ &\geq \frac{p\cdot A}{|p|}\left(\frac{|p|}{M} - \frac{|D\varphi|}{M\mu_2} - \frac{1}{1+\mu_3}\right),\end{aligned}$$

where we have suppressed the arguments $(X,\varphi,D\varphi \pm M\gamma)$ from A and (X,φ) from ψ, and we have written $p = D\varphi \pm M\gamma$ and noted that $|p| \geq M \geq M_2$. By choosing M sufficiently large, we can make the expression in parentheses positive, thus obtaining (8.25).

Notes

In the early days (1950-1964), existence of solutions for nonlinear initial-boundary value problems was usually proved on an ad hoc basis. See [153, 154] and the discussion in the introduction of [271]. (It should be noted that at the time of [271], the word quasilinear was generally applied to equations in which a^{ij} was independent of Du.) As indicated in the introduction to this chapter, local existence theorems are crucial to blow-up results (see [197]). Typically the equations are semilinear, so the contraction mapping principle can be used to good effect here. Another approach, similar to that described in Exercise 5.4, was laid out in [165, 166] for local solutions of the Cauchy problem. (See, especially [166, Teorema 4].)

Schauder proved his fixed point theorem (our Theorem 8.1) in [287] and used it in [288] to study nonlinear equations. For quasilinear elliptic equations, a more general result is needed to prove existence of solutions. This result is a special case of the Leray-Schauder fixed point theorem [196], and it is used in many other works, such as [183], to study the Cauchy-Dirichlet problem. Our local existence theorem, Theorem 8.2, or at least the form for $H_{2+\alpha}$ solutions, is a commonly known folk theorem, but does not seem to be explicitly formulated or proved elsewhere.

The Leray-Schauder theorem has a major drawback in its application to the Cauchy-Dirichlet problem for general quasilinear parabolic equations. To use this theorem, one must introduce a family of related Cauchy-Dirichlet problems and, if the solution is to lie in $H_{2+\alpha}$, the compatibility condition $P\varphi = 0$ on $C\Omega$ must be satisfied for the entire family of problems. This compatibility condition is easy to deal with for simple equations, but these conditions are difficult to verify for the general class of equations which we wish to study. This difficulty was first noted by Edmunds and Peletier [71] (see the remarks preceding Theorems 2 and 15 of that work). In the 1970's, Ivanov [117] (see the remark preceding Theorem 0.1) proposed a method to avoid this compatibility condition via global $W_p^{2,1}$ and interior $H_{2+\alpha}$ estimates for linear equations. Ten years later, I was able to show that the L^p Schauder theory could be avoided by the introduction of weighted Hölder spaces [210] (see Section 5 of that work). The local existence method presented here sidesteps the compatibility issue because it does not introduce any additional nonlinear problems.

Caristi's fixed point theorem (our Lemma 8.4) was proved by him in [44, Theorem 2.1′] as part of a program concerning mappings satisfying inwardness conditions. The proof given here is adapted from that given by Ekeland [74, Theorem 1bis], who noticed its connection to variational problems; this connection is clear from Ekeland's proof and from our modification. The major difference between Ekeland's proof and ours is that we give a direct proof of the inequality $\varphi(w) \geq \varphi(v) - \frac{1}{2}\rho(v,w)$. Lemma 8.5 comes from [158], The remainder of Section VIII.3 is based on [202, 206] with two exceptions: Theorem 8.9, which is a more careful reworking of [215, Theorem 2], and Theorem 8.11, which was new in the first edition of this book. An alternative approach to the nonlinear oblique derivative problem is provided by a nonlinear version of the method of continuity; see [101, Section 17.2]. This alternative method has some technical advantages in its application since one only needs estimates for a family of problems $Pu = \varepsilon P\psi$ for a fixed function ψ satisfying the initial conditions and ε ranging from 1 down to 0; however, the method itself uses the Frechét derivative of a map rather than the Gateaux variation and requires some smoothness of the derivative.

Besides the methods already mentioned, there are several other approaches to the existence of solutions of nonlinear problems. The most common one is nonlinear semigroup theory ([4,246]) which emphasizes various regularity aspects of the solution. In particular, Amann's approach [4] is mostly useful for systems of equations, and the strong estimates for solutions of nonlinear equations do not generally have analogs for systems.

Exercises

8.1 Suppose that a^{ij} and a are operators mapping $H_{1+\alpha\delta}$ to H_α for some fixed $\alpha \in (0,1)$ and $\delta \in (0,1]$. Prove existence theorems corresponding to Theorems 8.2 and 8.3 for the operator

$$Qu = a^{ij}[u]D_{ij}u + a[u].$$

8.2 Prove a local existence theorem similar to Theorem 8.2 when conditions (8.6a,b) hold for $\mathscr{K}$ consisting of all (X,z,p) in a neighborhood of $(X,\varphi,D\varphi)$.

CHAPTER IX

COMPARISON AND MAXIMUM PRINCIPLES

Introduction

The first step in our existence program is an estimate on the maximum of the solution of the Cauchy-Dirichlet problem. Our estimates will be proved by modifying the ideas in Chapters II, VI, and VII for estimating solutions of linear equations. In addition, we prove comparison principles which will be important for the estimates in later chapters.

1. Comparison principles

We recall that Corollary 2.5 says that if $Lu \geq Lv$ in Ω and if $u \leq v$ on $\mathcal{P}\Omega$, then $u \leq v$ in Ω. The main result of this section is that we have the same result for quasilinear operators under suitable hypotheses.

THEOREM 9.1. *Let P be the quasilinear operator defined by*

$$Pu = -u_t + a^{ij}(X,u,Du)D_{ij}u + a(X,u,Du). \tag{9.1}$$

Suppose that a^{ij} is independent of z and that there is an increasing positive constant k such that $a(X,z,p)+k(M)z$ is a decreasing function of z on $\Omega \times [-M,M] \times \mathbb{R}^n$ for any $M > 0$. If u and v are functions in $C^{2,1}(\overline{\Omega} \setminus \mathcal{P}\Omega) \cap C(\overline{\Omega})$ such that $Pu \geq Pv$ in $\overline{\Omega} \setminus \mathcal{P}\Omega$ and $u \leq v$ on $\mathcal{P}\Omega$ and if P is parabolic with respect to u or v, then $u \leq v$ in $\overline{\Omega}$.

PROOF. Let $M = \max\{\sup|u|, \sup|v|\}$, and set $w = (u-v)e^{\lambda t}$ for λ a constant at our disposal, so $w \leq 0$ on $\mathcal{P}\Omega$. In addition, at a positive maximum X_0 of w, we have $Du = Dv$, $D^2(u-v) \leq 0$, and $(u-v)_t \geq -\lambda[u-v-\varepsilon]$. For simplicity, we set $R = (X_0, u(X_0), Du(X_0))$ and $S = (X_0, v(X_0), Du(X_0))$ and then note that

$$\begin{aligned} 0 \leq Pu(X_0) - Pv(X_0) &= a^{ij}(R)D_{ij}(u-v) + [a(R)-a(S)] - (u-v)_t \\ &\leq (k(M)+\lambda)[u-v]. \end{aligned}$$

Hence, for $\lambda < -k(M)$, w cannot have an interior positive maximum, so $w \leq 0$, which implies that $u - v \leq 0$ in Ω. □

The argument in Lemma 2.3 allows us to reduce the size of the set on which u and v are $C^{2,1}$ further.

COROLLARY 9.2. *The result of Theorem 9.1 remains true if u and v are only in $C^{2,1}(\Omega)\cap C(\overline{\Omega})$.*

We also infer the uniqueness of solutions to the Cauchy-Dirichlet problem in this case.

COROLLARY 9.3. *Suppose that P is as in Theorem 9.1 and that u and v are in $C^{2,1}(\Omega)\cap C(\overline{\Omega})$. If $Pu = Pv$ in Ω and if $u = v$ on $\mathscr{P}\Omega$, then $u = v$ in Ω.*

For comparison purposes, it is also useful to have the following quasilinear version of Lemma 2.1.

LEMMA 9.4. *Suppose that u and v are in $C^{2,1}(\overline{\Omega}\setminus\mathscr{P}\Omega)\cap C(\overline{\Omega})$ and that P is parabolic at u or at v. If $Pu > Pv$ in $\overline{\Omega}\setminus\mathscr{P}\Omega$ and if $u < v$ on $\mathscr{P}\Omega$, then $u < v$ in $\overline{\Omega}$.*

PROOF. As in Lemma 2.1, we assume that there is a point X_0 in $\overline{\Omega}\setminus\mathscr{P}\Omega$ such that $u = v$ at X_0 and $u < v$ at X if $t < t_0$. Writing $R = (X_0, u(X_0), Du(X_0))$ and $w = u - v$, and noting that $Du(X_0) = Dv(X_0)$, we see that

$$0 < Pu(X_0) - Pv(X_0) = -w_t(X_0) + a^{ij}(R)D_{ij}w(X_0).$$

Since $w_t \geq 0$ and D^2w is nonnegative at X_0. This inequality cannot occur and hence $u < v$. □

2. Maximum estimates

We also have a quasilinear version of the estimate in Theorem 2.10.

THEOREM 9.5. *Let $u \in C^{2,1}(\Omega)\cap C(\overline{\Omega})$, let P be parabolic at u in Ω, and suppose that there are constants k and b_1 such that*

$$za(X,z,0) \leq kz^2 + b_1. \tag{9.2}$$

If $Pu \geq 0$ in Ω and if $I(\Omega) = (0,T)$, then

$$\sup_{\Omega} u \leq e^{(k+1)T}(\sup_{\mathscr{P}\Omega} u^+ + b_1^{1/2}). \tag{9.3}$$

PROOF. For $\lambda = -k-1$, let X be a point in $\overline{\Omega}$ at which $v = e^{\lambda t}u$ attains a positive maximum. If $X \in \mathscr{P}\Omega$ or if $v < 0$, we are done, so we may assume that $X \in \Omega$. At X, we have

$$\begin{aligned} 0 &\leq -u_t + a^{ij}D_{ij}u + a \leq -u_t + ku + \frac{b_1}{u} \\ &= -e^{-\lambda t}v_t + (k+\lambda)u + \frac{b_1}{u} \leq (k+\lambda)u + \frac{b_1}{u} \\ &= -u + \frac{b_1}{u}. \end{aligned}$$

At a maximum, we must have $u \le b_1^{1/2}$. From these observations, we easily infer (9.3). □

For our existence theory, Theorem 9.5 suffices; however, for studying large-time behavior of solutions, it is useful to have maximum estimates which do not depend on T. Our next theorem gives an example of such an estimate which shows how the quantity $\mathscr{E}$ can be used.

THEOREM 9.6. *Let P be parabolic and suppose that there are nonnegative constants μ_1 and μ_2 such that*

$$\frac{a(X,z,p)\operatorname{sgn} z}{\mathscr{E}(X,z,p)} \le \frac{\mu_1|p|+\mu_2}{|p|^2}. \tag{9.4}$$

If $Pu \ge 0$ in Ω, then

$$\sup_{\Omega} u \le \sup_{\mathscr{P}\Omega} u^+ + C(\mu_1, \operatorname{diam}\Omega)\mu_2. \tag{9.5}$$

PROOF. Define $\bar{P}$ by

$$\bar{P}v = -v_t + a^{ij}(X,u,Dv)D_{ij}v + a(X,u,Dv),$$

and suppose without loss of generality that $0 < x^1 < R = \operatorname{diam}\Omega$ in Ω. Similarly to the case in Theorem 2.11, we set

$$v = \mu_2(e^{\alpha R} - e^{\alpha x_1}) + \sup_{\mathscr{P}\Omega} u^+$$

for $\alpha = \mu_1 + 1$ and we suppose first that $\mu_2 > 0$. In $\Omega^+ = \{u > 0\}$, we have

$$\begin{aligned}\bar{P}v &= -\mu_2\alpha^2 a^{11}(X,u,Dv)e^{\alpha x^1} + a(X,u,Dv)\\ &\le -\frac{e^{\alpha x^1}}{\mu_2}\mathscr{E}(X,u,Dv)[1 - \frac{\mu_1}{\alpha} - \frac{e^{-\alpha x^1}}{\alpha}] < 0\end{aligned}$$

by (9.4) and the definition of $\mathscr{E}$. Since $\bar{P}u = Pu \ge 0$ in Ω, Theorem 9.1 gives $u \le v$ in Ω and then (9.5) follows immediately. For $\mu_2 = 0$ we just take the limit as $\mu_2 \to 0$. □

Other versions of estimates which do not depend on T are given in Exercises 9.1 through 9.3.

3. Comparison principles for divergence form operators

Now we suppose that the operator P is in divergence form,

$$Pu = -u_t + \operatorname{div} A(X,u,Du) + B(X,u,Du).$$

A comparison principle holds for weak subsolutions and supersolutions even if we relax the regularity hypotheses on the coefficients A and B and on the functions u

and v. As in Chapter VI, we say that $Pu \geq 0$ in Ω if $A(X,u,Du)$ and $B(X,u,Du)$ are defined and locally integrable in Ω and if

$$\int_\Omega [A(X,u,Du)\cdot D\eta - B(X,u,Du)\eta - u\eta_t]\,dX \leq 0$$

for all nonnegative $\eta \in C^1_{\mathscr{P}}$ which vanish on $B\Omega$. Similarly, we say that $Pu \geq Pv$ in Ω if $A(X,u,Du)$, $A(X,v,Dv)$, $B(X,u,Du)$ and $B(X,v,Dv)$ are defined and locally integrable in Ω and if

$$\int_\Omega (A(X,u,Du)\cdot D\eta - B(X,u,Du)\eta - u\eta_t)\,dX \\ \leq \int_\Omega (A(X,v,Dv)\cdot D\eta - B(X,v,Dv)\eta - v\eta_t)\,dX$$

for all nonnegative $\eta \in C^1_{\mathscr{P}}$ which vanish on $B\Omega$. We then infer the following comparison principle.

THEOREM 9.7. *Suppose that A and B are continuously differentiable with respect to z and p and that the matrix* $(A^{ij}) = (\partial A^i/\partial p_j)$ *is positive definite. If u and v are* $C^1(\overline{\Omega})$ *functions such that* $Pu \geq Pv$ *in* Ω *and* $u \leq v$ *on* $\mathscr{P}\Omega$, *then* $u \leq v$ *in* Ω.

PROOF. From the definition of $Pu \geq Pv$, we have

$$0 \leq \int_\Omega (A(X,u,Du) - A(X,v,Dv))\cdot D\eta\,dX \\ + \int_\Omega (B(X,u,Du) - B(X,v,Dv))\eta + (u-v)\eta_t\,dX$$

for any nonnegative $\eta \in C^1_{\mathscr{P}}$. From the mean value theorem, we can rewrite this inequality as

$$0 \leq \int_\Omega (a^{ij}(X)D_jw + b^i(X)w)D_i\eta - (c^i(X)D_iw + c^0(X)w)\eta - w\eta_t\,dX,$$

where $w = (u-v)^+$ and the functions a^{ij}, b^i, c^i and c^0 are uniformly bounded. Since the matrix (a^{ij}) is also uniformly positive definite, we infer from Corollary 6.16 that $w \leq 0$ in Ω as required. Alternatively, we can imitate the proof of (6.27) for $q = 0$ to infer that $w \leq 0$. □

4. The maximum principle for divergence form operators

An L^∞ estimate for solutions of equations in divergence form is quite simple and general. The first step, analogous to that in Theorem 6.15 for linear equations is a bound on the maximum of the solution in terms of a suitable L^p norm.

THEOREM 9.8. *Suppose that there are positive constants a_0, a_1, b_0, b_1, $m \geq 1$, and M such that A and B satisfy the conditions*

$$p \cdot A(X,z,p) \geq a_0 |p| - a_1 |z|^m, \tag{9.6a}$$

$$zB(X,z,p) \leq b_0(p \cdot A(X,z,p))^+ + b_1 |z|^m \tag{9.6b}$$

for all $z \geq M$. If $Pu \geq 0$ in Ω and if $u \leq M$ on $\mathscr{P}\Omega$, then

$$\sup_\Omega u \leq 2M + C \int_\Omega (u^+)^{(n+1)(m-1)}\, dX \tag{9.7}$$

for some constant C determined only by a_0, a_1, b_0, b_1, m, and n.

PROOF. For $q > \max\{1, 2b_0 - 2n\}$ we use the test function

$$\eta = \left[\left(1 - \frac{M}{u}\right)^+\right]^{(n+1)q-n} u^{q+nm-n-1}.$$

We then define

$$v = (q + nm - n - 1)\left(1 - \frac{M}{u}\right)^+ + [(n+1)q - n]\frac{M}{u},$$

and note that

$$D\eta = v\left[\left(1 - \frac{M}{u}\right)^+\right]^{(n+1)(q-1)} |u|^{q+nm-n-2} Du,$$

and we define

$$U = \int_M^u s^{q+nm-n-1}\left[\left(1 - \frac{M}{s}\right)^+\right]^{(n+1)q-n} ds.$$

If we also write $\sigma(t)$ and $\Sigma(t)$, for the subsets of $\omega(t)$ and $\Omega(t)$, respectively, on which $u > M$, then we can write the integral form of $Pu \geq 0$ as

$$\int_{\sigma(\tau)} U\, dx + \int_{\Sigma(\tau)} Du \cdot Au^{q+nm-n-2}\left(1 - \frac{M}{u}\right)^{(n+1)(q-1)} v\, dX$$
$$\leq \int_{\Sigma(\tau)} uBu^{q+nm-n-2}\left(1 - \frac{M}{u}\right)^{(n+1)q-n} dX.$$

If we also define $\chi(s) = s^{q+nm-n-1}(1 - M/s)^{(n+1)q-n}$, then a simple calculation shows that χ satisfies (6.25) with $k_1 = (2mn+2)q$ and hence that $U \geq \chi(u)(u -$

$M)/((2mn+3)q)$. Our choice of q implies that $b_0(1-M/u)\le\nu/2$ and therefore

$$\sup_\tau \int_{\sigma(\tau)} u^{q+nm-n}\left(1-\frac{M}{u}\right)^{(n+1)(q-1)} dx + \int_\Sigma |Du|\, u^{q+nm-n-2}\left(1-\frac{M}{u}\right)^{(n+1)(q-1)} \nu\, dX \le c_1 q^2 \int_\Sigma u^{q+nm-n-2+m}\left(1-\frac{M}{u}\right)^{(n+1)(q-1)} dX \tag{9.8}$$

with c_1 a constant determined by the same quantities as C and Σ the subset of Ω on which $u>M$. We now apply Theorem 6.8 with η in place of u and $u^{1/n}$ in place of λ. Since $N=n$ for $p=1$, it follows that

$$\left(\int_\Sigma w^{q\kappa}\,d\mu\right)^{1/\kappa} \le c_1c_2(n)q^3\int_\Sigma w^q\,d\mu,$$

where $\kappa=1+1/n$,

$$w=u\left(1-\frac{M}{u}\right)^{n+1}, \text{ and } d\mu=u^{nm-n-2+m}\left(1-\frac{M}{u}\right)^{-n-1} dX.$$

Iteration now gives (9.7). □

Since $|p|^r\ge|p|-1$ for $r\ge1$, we see that the maximum of u can be estimated if the structure condition (9.6a) is modified to

$$p\cdot A(X,z,p)\ge a_0|p|^r-a_1|z|^m. \tag{9.6a$'$}$$

If $m=1$, this theorem gives a bound on $\sup u$ directly. To estimate the integral of $u^{(n+1)(m-1)}$ when $m>1$, we need some further restrictions on our structure. At first, we suppose that $r=1$ and $m=2$ (which includes the cases $r\ge1$ and $m\le2$.)

THEOREM 9.9. *Suppose that there are positive constants a_0, a_1, b_0, b_1, and M such that A and B satisfy conditions (9.6a,b) with $m=2$ whenever $|z|\ge M$. If $Pu\ge0$ in Ω, if $u\le M$ on $\mathscr{P}\Omega$, and if $q>1+b_0$, then*

$$\int_\Omega (u^+)^q\,dX\le CM^q \tag{9.9}$$

for some constant C determined only by a_0, a_1, b_0, b_1, n, q, T, and $|\Omega|$.

PROOF. Now we use the test function

$$\eta=(u^{q-1}-M^{q-1})^+\operatorname{sgn}u.$$

In place of (9.7), we find, after eliminating the integral involving Du, that

$$\int_{\sigma(\tau)}\frac{u^q}{q}-\frac{M^{q-2}u^2}{2}\,dx\le C(q)\int_0^\tau\int_{\sigma(t)}u^2(u^{q-2}-M^{q-2})\,dX.$$

Simple rearrangement along with Young's inequality yields

$$\int_{\sigma(\tau)} u^q\,dx \le C[\int_0^\tau \int_{\sigma(t)} u^q\,dX + M^q],$$

and Gronwall's inequality (along with some simple algebra) completes the proof. □

These two theorems immediately give a two-sided estimate for solutions, which we record here for future use.

COROLLARY 9.10. *Suppose P satisfies conditions* (9.6a,b) *with $m = 2$. If $Pu = 0$ in Ω and $|u| \le M$ on $\mathscr{P}\Omega$, then $|u| \le C(M + M^{n+1})$ in Ω with C determined by the same quantities as in Theorem 9.9*

If $r > 2$ in (9.6a)$'$, then we can improve the integral estimate in Theorem 9.9.

THEOREM 9.11. *Suppose that there are positive constants a_0, a_1, b_0, b_1, $m > 2$, and M such that A and B satisfy the conditions*

$$p\cdot A(X,z,p) \ge a_0\,|p|^m - a_1\,|z|^m, \tag{9.10a}$$

$$zB(X,z,p) \le b_0(p\cdot A(X,z,p))^+ + b_1\,|z|^m \tag{9.10b}$$

for all $z \ge M$. If $Pu \ge 0$ in Ω and if $u \le M$ on $\mathscr{P}\Omega$, and if $q > 1 + b_0$, then there are constants C determined only by a_0, a_1, b_0, b_1, m, n, T, and $|\Omega|$ and ε_0 determined only by a_0, q, and diam Ω *such that $a_1 + b_1 \le \varepsilon_0$ implies* (9.9).

PROOF. In analogy with the proof of Theorem 9.9, we use the test function

$$\varphi = (u^{q-m+1} - M^{q-m+1})^+.$$

This time we eliminate the integral over $\sigma(\tau)$ to infer that

$$\int_\Sigma u^{q-m}\,|Du|^m\,dX \le c_1 \int_\Sigma u^q\,dX$$

for $c_1 = (qa_1 + b_1)/a_0$. If we now set $h = (u^q - M^q)^+$, then we have

$$\begin{aligned}\int_\Omega |Dh|\,dX &= q\int_\Sigma u^{q-1}\,|Du|\,dX \\ &\le q\left(\int_\Sigma u^q\,dX\right)^{(m-1)/m}\left(\int_\Sigma u^{q-m}\,|Du|^m\,dX\right)^{1/m} \\ &\le qc_1^{1/m}\int_\Sigma u^q\,dX.\end{aligned}$$

On the other hand, Poincaré's inequality implies that

$$\int_\Omega h\,dX \le \operatorname{diam}\Omega \int |Dh|\,dX,$$

and therefore

$$(1 - (\operatorname{diam}\Omega)qc_1^{1/m})\int_\Sigma u^q\,dX \le |\Omega|\,M^q.$$

If $(\operatorname{diam}\Omega)qc_1^{1/m} \le 1/2$, then this inequality implies (9.8). □

If we weaken the growth condition on B with respect to z, we can improve the growth somewhat with respect to p.

THEOREM 9.12. *Suppose that there are positive constants a_0, a_1, b_0, b_1, and M such that A and B satisfy the conditions*

$$p \cdot A(X,z,p) \ge a_0 |p| - a_1, \tag{9.6a)$''$}$$

$$\operatorname{sgn} z B(X,z,p) \le b_0 p \cdot A(X,z,p) + b_1 \tag{9.6b)$''$}$$

for all $z \ge M$. If $Pu \ge 0$ in Ω and $u \le M$ on $\mathscr{P}\Omega$, then there is a constant C determined only by a_0, a_1, b_0, b_1, M, n, and Ω such that $u \le C$ in Ω.

PROOF. Let $k = b_0$ and set $v = e^{ku}$. For

$$A^*(X,z,p) = kzA(X, \frac{\ln z}{k}, \frac{p}{kz}),$$

$$B^*(X,z,p) = kz[B(X, \frac{\ln z}{k}, \frac{p}{kz}) - k\frac{p}{z} \cdot A(X, \frac{\ln z}{k}, \frac{p}{kz})],$$

we see that $-v_t + \operatorname{div} A^*(X,v,Dv) + B^*(X,v,Dv) \ge 0$ in Ω and $v \le M^* = \exp(kM)$ on $\mathscr{P}\Omega$. An easy calculation shows that $p \cdot A^* \ge a_0 |p| - a_1 kz^2$ and $zB \le kb_1 z^2$ for $z \ge M^*$, provided $M \ge 0$. Then Corollary 9.10 gives an estimate on v, and hence on u. □

Note that the constants a_1 and b_1 can be replaced by nonnegative functions in suitable $L^{q,r}$ spaces. We leave the details to the reader in Exercise 9.7.

Notes

The maximum principles for quasilinear equations in nondivergence form are easy extensions of those for linear equations. (See the discussion, or lack thereof, in [101, 183].) Exercise 9.3, taken from [290], relies more on the quasilinear structure.

Estimates for quasilinear equations in divergence form take better advantage of the nonlinear structure of the equation. Ladyzhenskaya and Ural'tseva [183, Section V.2] prove *a priori* estimates under the conditions

$$p \cdot A \ge a_0 |p|^2 - a_1 |z|^m, \ zB \le b_0 |p|^2 + b_1 |z|^m.$$

If $m = 2$, they obtain a bound on the maximum of the solution in terms of the constants and a bound on φ; if $m > 2$, this bound depends also on an integral norm of u. Such results also follow from ours. In fact, they allow the coefficients a_1 and b_1 to lie in suitable $L^{p,q}$ spaces. Aronson and Serrin [10] obtained an *a priori* bound under the structure

$$p \cdot A \ge a_0 |p|^m - a_1 |z|^m, \ zB \le b_0 |p|^m + b_1 |z|^m$$

assuming that $\Omega = \omega \times (0,T)$ with ω having sufficiently small measure (if $m > 2$). Such an estimate follows by a simple variation of our proof. The estimates in Theorem 9.8 and 9.9 were first noted by me [204] in the special case $m = 2$ although they are very similar to [183, Theorem V.2.2].

Theorem 9.12 was first proved by Vespri [327] in slightly different form. He assumed that

$$p \cdot A \geq a_0 |p|^m - \varphi(X),\ |B| \leq b_0 |p|^m + \varphi(X)$$

with $\varphi \in L^q$ for some $q > (n+m)/n$. Orsina and Porzio [274, Theorem 4.1] also proved such a bound under slightly different assumptions:

$$A(X,z,p) = \bar{A}(X,z,p) + f(X),\ B(X,z,p) = b(X,z,p) + H(X,z,p)$$

where

$$\bar{A} \cdot p \geq \alpha |p|^m,\ zb \leq -\alpha_0 |z|^m,\ |H| \leq c(X) + b_0 |p|^m$$

for some positive constants α, α_0, and b_0 and nonnegative functions $f \in L^{\infty,r}$ for $r > (n+m)/(m-1)$ and $c \in L^{\infty,q}$ with q as before. Their proofs differ from ours in certain minor technical details (primarily concerning the cases $m > 2$ and $m < 2$) and in the use of deGiorgi iteration rather than Moser iteration. Any of these techniques can be used to prove a maximum bound under any of these sets of hypotheses.

It is also possible to consider non-power growth in the structure conditions. The techniques developed in [226] for elliptic equations are easily modified to cover the parabolic case; such an extension was done in [243, Theorem 3.1].

All the estimates in this chapter are valid for arbitrary bounded domains, although the size of the domain may affect the estimate quantitatively. On the other hand, some estimates are only valid if the domain is sufficiently small in the time-direction. For example, if we consider the blow-up problem

$$-u_t + \Delta u = u^m \text{ in } \Omega, u = 0 \text{ on } S\Omega, u = \varphi \text{ on } B\Omega,$$

with $I(\Omega) \subset (0,T)$, then $v = [(T-t)(m-1)]^{-1/(m-1)}$ is a supersolution of the equation which is infinite when $t = T$. If also $\sup|\varphi| \leq [(m-1)T]^{-1/(m-1)}$, then the comparison principle gives an upper bound for u, and a lower bound is proved similarly. Hence the maximum bound for u ties together the size of $I(\Omega)$ and the size of φ.

Exercises

9.1 Show that Theorem 9.6 holds if we replace (9.3) by

$$\frac{a(X,z,p)\operatorname{sgn} z}{\mathscr{D}^*(X,z,p)} \leq \mu_1 |p| + \mu_2.$$

9.2 Suppose we replace (9.3) by the condition

$$\frac{a(X,z,p)\operatorname{sgn} z}{(n+1)\mathscr{D}^*(X,z,p)} \leq \frac{h(X)}{g(p)}$$

for some nonnegative functions $h \in L^{n+1}(\Omega)$ and $g \in L^{n+1}_{\text{loc}}(\mathbb{R}^n)$. Set $R = \operatorname{diam}\Omega$ and suppose that

$$\int_\Omega h(X)^{n+1}\,dX < \int_0^\infty \int_{\{|p|\le h/2R\}} g(p)^{n+1}\,dp\,dh.$$

(Note that the integral on the right side of this inequality could be infinite.) If $u \in W^{2,1}_{n+1}(\Omega)$ and $Pu \ge 0$ in Ω, show that

$$\sup_\Omega u \le \sup_{\mathscr{P}\Omega} u^+ + C,$$

where C is determined by g, h, and $\operatorname{diam}\Omega$.

9.3 Suppose there are positive numbers L, M, and R such that $\operatorname{diam}\Omega \le R$ and

$$a(X,z,p)\operatorname{sgn} z \le \frac{|p|\,\mathscr{T}(X,z,p)}{R}$$

for $|z| \ge M$ and $|p| \ge L$. If $Pu \ge 0$ in Ω, show that

$$\sup_\Omega u \le \max\{M, \sup_{\mathscr{P}\Omega} u^+\} + 2LR.$$

9.4 Show that the estimate (9.2) in Theorem 9.5 can be sharpened to

$$\sup_\Omega u \le \inf_{s>k} e^{sT} \max\{\sup_{\mathscr{P}\Omega} u^+, \left(\frac{b_1}{s-k}\right)^{1/2}\}.$$

9.5 Show that if condition (9.1) is replaced by

$$za(X,z,0) \le \Phi(|z|)\,|z| + b_1$$

in Theorem 9.5 with some increasing positive function Φ defined on $[0,\infty)$ such that

$$\int_0^\infty \frac{1}{\Phi(\tau)}\,d\tau = \infty$$

then

$$\sup_\Omega u \le \inf_{s>1} \varphi^{-1}\left(e^{sT}\max\{1, \varphi(\sup_{\mathscr{P}\Omega} u^+), \varphi\left(\frac{b_1}{(s-1)\Phi(0)}\right)\}\right),$$

where φ is defined by

$$\int_0^s \frac{1}{\Phi(\tau)}\,d\tau = \varphi(s).$$

(Hint: As in [183], find a differential equation satisfied by the function $v = \varphi(u)$. See also [55].)

9.6 If (9.6a) is replaced by

$$p \cdot A \ge a_0|p|^2 - a_1|z|^2$$

for $|z| \ge M$ in Corollary 9.10, show that $u \le CM$.

9.7 Show that the results in Section IX.4 remain true provided the constants a_1 and b_1 are replaced by nonnegative functions in suitable $L^{q,r}$ spaces. Determine the appropriate values of q and r.

CHAPTER X

BOUNDARY GRADIENT ESTIMATES

Introduction

We now turn to the estimate of the gradient of a solution of the initial-boundary value problem

$$Pu = 0 \text{ in } \Omega,\ u = \varphi \text{ on } \mathscr{P}\Omega \tag{10.1}$$

on $S\Omega$. (Note that $Du = D\varphi$ on $B\Omega$.) As we shall see, this estimate is the crucial one for the current state of the existence theory in the following sense. This chapter presents a number of sets of conditions under which such an estimate is possible, and examples will be given to show that a solution need not exist if any of the conditions are violated. For example, we shall give two conditions for the solvability of the prescribed mean curvature problem, that is, (10.1) with

$$Pu = -u_t + \operatorname{div} \frac{Du}{(1+|Du|^2)^{1/2}}.$$

One of these is a geometric condition on $\mathscr{P}\Omega$ and the other is a restriction on the boundary data φ. If the geometric condition is violated, then there are choices of φ (satisfying the previously mentioned restriction) such that (10.1) has no solution. Similarly if the geometric condition is satisfied, then there are choices of φ which do not satisfy this restriction and for which there is no solution of (10.1).

In fact, rather than bound $|Du|$ directly, we shall estimate the quantity

$$[u]'_1 = \sup_{\substack{X\in S\Omega \\ Y\in\Omega \\ s\leq t}} \frac{|u(X)-u(Y)|}{|X-Y|}.$$

Under natural assumption on P, Ω, and φ, we shall bound $[u]'_1$. When $Du \in C(\overline{\Omega})$, it is easy to check that $|Du| \leq [u]'_1$ on $S\Omega$.

Our method of proof is based on the maximum principle. First we introduce an auxiliary operator $\bar{P}$, defined by

$$\bar{P}w = -w_t + \bar{a}^{ij}(X,Dw)D_{ij}w + \bar{a}(X,w,Dw) \tag{10.2}$$

with $\bar{a}^{ij}$ and $\bar{a}$ satisfying

$$\bar{a}^{ij}(X,Du) = a^{ij}(X,u,Du),\ \bar{a}(X,u,Du) = a(X,u,Du),$$

and $\bar{a}(X,\cdot,p)$ is a Lipschitz function. In particular, $\bar{a}^{ij}(X,p)=a^{ij}(X,u(X),p)$, but some freedom in the exact definition of $\bar{a}$ will be essential to the development of this method.

To prove our estimate on $[u]_1'$, we first look in a parabolic neighborhood N of a point $X_0\in S\Omega$, that is there is a positive R such that $Q(X_0,R)\subset N$. For convenience, we set $M=\sup|u-\varphi|$. If there are functions $w^{\pm}\in C(\overline{N\cap\Omega})\cap C^{2,1}(N\cap\Omega)$ such that

$$\pm\bar{P}w^{\pm}<0 \text{ in } N\cap\Omega, \tag{10.3a}$$

$$w^{+}\ge\varphi\ge w^{-} \text{ on } N\cap\mathscr{P}\Omega, \tag{10.3b}$$

$$w^{+}-M\ge\varphi(X_0)\ge w^{-}+M \text{ on } \mathscr{P}N\cap\Omega, \tag{10.3c}$$

$$w^{\pm}(X_0)=\varphi(X_0), \tag{10.3d}$$

then the comparison principle Corollary 9.2 implies that

$$w^{+}\ge u\ge w^{-} \text{ in } N\cap\Omega.$$

If there are constants $L^{\pm}$ such that

$$\frac{w^{+}(Y)-w^{+}(X_0)}{|Y-X_0|}\le L^{+},\quad \frac{w^{-}(X_0)-w^{-}(Y)}{|Y-X_0|}\le L^{-}$$

for all $Y\in N\cap\Omega$ with $s=t_0$, then

$$[u]_{1;X_0}'=\sup_{\substack{Y\in\Omega\\ s\le t_0}}\frac{|u(X_0)-u(Y)|}{|X_0-Y|}\le\max\{L^{+},L^{-},\frac{M}{R_0}\}.$$

We shall see that, under fairly general conditions, the numbers $L^{\pm}$ and R_0 can be taken independent of X_0 and hence this estimate yields an *a priori* bound for $[u]_1'$.

To facilitate our discussion, we discuss some of the technical details here. For simplicity, we only consider the supersolution w^{+}, which we henceforth denote by w. The comparison functions w will all have the form

$$w=\bar{\varphi}+f(d),$$

where $\bar{\varphi}$ is a suitable function such that $\bar{\varphi}(X_0)=\varphi(X_0)$, f is a C^2 function of one variable which is increasing and concave, and d is essentially the distance to a suitable comparison surface. Our choice of d will always satisfy $d\in C^{2,1}(N\cap\Omega)$, and then

$$\bar{P}w=\bar{a}^{ij}D_{ij}\bar{\varphi}-\bar{\varphi}_t+f''\bar{a}^{ij}D_idD_jd+f'[\bar{a}^{ij}D_{ij}d-d_t]+\bar{a}.$$

Most of our effort will be concentrated on showing that $\bar{P}w<0$. Since $f'>0$ and $f''<0$, this can be accomplished if $\bar{a}^{ij}D_idD_jd$ is large and positive or if $\bar{a}^{ij}D_{ij}d-d_t$ is large and negative. The exact meaning of large in this context will become clear from our later discussion.

For our estimates, it is useful to introduce the quantity

$$\mathscr{E}_d=\bar{a}^{ij}D_idD_jd$$

and to note that

$$\mathscr{E} = (f')^2\mathscr{E}_d + 2f'\bar{a}^{ij}D_idD_j\bar{\varphi} + \bar{a}^{ij}D_i\bar{\varphi}D_j\bar{\varphi} \le 2(f')^2\mathscr{E}_d + 2\Lambda|D\bar{\varphi}|^2. \tag{10.4a}$$

Therefore, if

$$4\Lambda|D\bar{\varphi}|^2 \le \mathscr{E}, \tag{10.4b}$$

then

$$\mathscr{E} \le 4(f')^2\mathscr{E}_d. \tag{10.4c}$$

1. The boundary gradient estimate in general domains

To demonstrate the application of the ideas mentioned in the introduction, we consider a simple geometric situation in which we can prove our estimate. Suppose first that for some point X_0 of $S\Omega$, there are a positive number R and a point $y \in \mathbb{R}^n$ such that $|x_0 - y| = R$ and the infinite cylinder

$$Q = \{X \in \mathbb{R}^{n+1} : |x-y| < R, t \le t_0\}$$

does not intersect Ω. For brevity, we say that Ω satisfies an *exterior infinite cylinder condition* at X_0. In this case, we take $d(x) = |x-y| - R$, and note that $Dd = [x-y]/|x-y|$, $d_t = 0$ and

$$D_{ij}d = |x-y|^{-3}[|x-y|^2\delta_{ij} - (x_i-y_i)(x_j-y_j)].$$

Suppose now that u is a solution of (10.1) with $\varphi \in H_2$. We then define $\bar{P}$ by (10.2) with $\bar{a}(X,z,p) = a(X,u(X),p)$ and we take $\bar{\varphi} = \varphi$. For α a positive constant at our disposal, we take $N = \{R < |x-y| < R+\alpha\}$, in which case (10.3b) and (10.3d) are immediate. Then, since $|Dd| = 1$ in N, we have

$$|Dw| = \left|f'Dd + D\varphi\right| \le f' + |D\varphi| \le 2f',$$
$$|Dw| \ge f' - |D\varphi| \ge \frac{1}{2}f'$$

provided $f' \ge 2\sup|D\varphi|$. For our structure conditions, we assume that there are positive constants p_0 and μ such that

$$|p|\Lambda,\ 1,\ a \le \mu\mathscr{E} \tag{10.5}$$

for $|p| \ge p_0$. From this assumption, it follows that

$$4\Lambda|D\bar{\varphi}|^2 \le 8\frac{|D\varphi|^2}{f'}|Dw|\Lambda \le 8\mu\frac{|D\varphi|^2}{f'}\mathscr{E},$$

and hence (10.4b) holds if $f' \geq 8\sup|D\varphi|^2/\mu$. It then follows that

$$\begin{aligned}\bar{P}w &\leq \Lambda|D^2\varphi| + |\varphi_t| + \bar{a} + f''\mathscr{E}_d + f'\Lambda/|x-y| \\ &\leq \mu\mathscr{E}\left[\frac{2(|D^2\varphi| + |\varphi_t|)}{f'} + 1 + \frac{1}{R}\right] + f''\mathscr{E}_d \\ &< \mathscr{E}_d[f'' + \mu_0(f')^2]\end{aligned}$$

for $\mu_0 = (2|\varphi|_2 + 1 + \frac{1}{R})\mu$.

To continue, we solve the differential equation

$$f'' + \mu_0(f')^2 = 0$$

to see that

$$f(d) = \frac{1}{\mu_0}\ln(1 + \frac{\mu_0}{k_1}d),\ f'(d) = \frac{1}{k_1 + \mu_0 d}$$

for k_1 a constant at our disposal. It follows that $f(\alpha) = M$ (and hence that (10.3c) holds) if $\frac{\mu_0}{k_1}\alpha = e^{\mu_0 M} - 1$, in which case

$$f'(d) \geq f'(\alpha) = \frac{1}{k_1 e^{\mu_0 M}}.$$

If k_1 is sufficiently small (and positive), we have

$$f' \geq 2\sup|D\varphi|,\ f' \geq 8\sup|D\varphi|^2/\mu,\ f' \geq 2p_0,$$

and hence $w^+ = w$ satisfies (10.3a–d). By following a similar argument for w^-, we infer the following pointwise boundary gradient estimate.

LEMMA 10.1. *Suppose that u is a solution of* (10.1) *with $\varphi \in H_2$ and that Ω satisfies an exterior infinite cylinder condition at $X_0 \in S\Omega$ with radius R. Suppose also that there are positive constants p_0 and μ such that*

$$|p|\Lambda + 1 + |a| \leq \mu\mathscr{E} \tag{10.6}$$

for $|p| \geq p_0$. Then there is a constant C determined only by $|\varphi|_2$, p_0, μ and R such that

$$[u]'_{1;X_0} \leq C. \tag{10.7}$$

If Ω satisfies an exterior infinite cylinder condition at every point of $S\Omega$ with radius R, and if (10.6) *holds, then*

$$[u]'_1 \leq C \tag{10.7$'$}$$

with the same constant C as in (10.7).

It is not difficult to see that the quantity $|\varphi|_2$ can be replaced by a slightly different expression, namely, $|\varphi^*|_2$, where $\varphi^* = \varphi$ on $S\Omega$, and $\varphi^* \geq \varphi$ on $B\Omega$. In particular, the estimate does not require any information on $|\varphi|_{2;B\Omega}$.

More significantly, we can obtain a boundary gradient estimate under weaker regularity hypotheses on the exterior surface, which was $\{|x-y| = R, t \leq t_0\}$ in

Lemma 10.1, and on the boundary data φ. In addition, the exterior surface need not be cylindrical.

First, to define $\tilde{\varphi}$, we use the following extension lemma.

LEMMA 10.2. *Let $\varphi \in C(Q(2R))$, let $M \geq \operatorname{osc} \varphi$, and set $Q = Q(R)$. Then there is a function $\tilde{\varphi}$ which is continuous on $\overline{Q} \times [0,1]$ and in $C^\infty(Q \times (0,1))$ such that*

$$\tilde{\varphi}(X_0, 0) = \varphi(X_0), \tag{10.8a}$$

$$\tilde{\varphi}(\cdot, 0) \geq \varphi \text{ in } Q, \tag{10.8b}$$

$$\tilde{\varphi} = \varphi(X_0) + M \text{ on } \mathscr{P}Q \times [0,1], \tag{10.8c}$$

$$\tilde{\varphi} - \varphi(X_0) \leq M \text{ on } Q \times [0,1]. \tag{10.8d}$$

Suppose also that $\varphi \in H_\delta(Q(2R))$ for some $\delta \in (1,2]$ and write $\tilde{\varphi}$ for $\tilde{\varphi}(X, x^0)$. Then

$$\frac{1}{R}|D_0\tilde{\varphi}| + \max_{1 \leq i \leq n} |D_i\tilde{\varphi}| \leq C(n)\left[\frac{M}{R} + [\varphi]_1\right], \tag{10.8e}$$

and there is a constant $\Phi_0(M, n, R, \delta, |\varphi|_\delta)$ such that

$$\sum_{i,j=0}^{n} |D_{ij}\tilde{\varphi}| + |\tilde{\varphi}_t| \leq \Phi_0 (x^0)^{\delta-2}. \tag{10.8f}$$

PROOF. Let $\psi \in C^2(\mathbb{R}^n \times (-\infty, t_0))$ be a nonnegative function which vanishes outside $Q(R/2)$ with $\psi(X_0) = \max \psi = 1$. We then define

$$\varphi^*(X) = \varphi(X)\psi(X) + [\varphi(X_0) + M][1 - \psi(X)].$$

Next, we let $\zeta \in C^\infty(\mathbb{R}^{n+1})$ be a nonnegative function with L^1 norm equal to 1 with $\zeta(Y) = 0$ if $s \leq 0$, $s \geq R^2/4$ or $|y| \geq R/2$. The appropriate choice for $\tilde{\varphi}$ is then

$$\tilde{\varphi}(X, x^0) = \int \varphi^*(x - yx^0, t - s(x^0)^2)\zeta(Y)\, dY.$$

Then (10.8a–f) are easy to check provided $|D\psi| \leq 4/R$, $|D^2\psi| + |\psi_t| \leq C(n)/R^2$ and similar inequalities for the same derivatives of ζ. As shown in Chapter IV, these inequalities can be satisfied by functions satisfying the other hypotheses. □

A useful variant occurs when we have stronger information on the t-regularity of φ.

PROPOSITION 10.3. *Let $\varphi \in C(Q(2R))$, let $M \geq \operatorname{osc} \varphi$, set $Q = Q(R)$, and suppose that there is a nonnegative constant Φ_1 such that*

$$\varphi(x,t) - \varphi(x,t_0) \leq \Phi_1[t_0 - t] \tag{10.9}$$

for all $x \in B(x_0, 2R)$. *Then there is a function* $\tilde{\varphi} \in C(\overline{Q} \times [0,1]) \cap C^\infty(Q \times (0,1))$ *satisfying* (10.8a,b) *and*

$$\tilde{\varphi} \geq \varphi(X_0) + M \text{ on } \mathscr{P}Q \times [0,1], \tag{10.8c$'$}$$

$$\tilde{\varphi} - \varphi(X_0) \leq M + \Phi_1[t_0 - t] \text{ in } Q \times [0,1], \tag{10.8d$'$}$$

$$\tilde{\varphi}_t \geq -\Phi_1. \tag{10.10}$$

Suppose also that $\varphi \in H_\delta(Q(2R))$ *for some* $\delta \in (1,2]$ *and write* $\tilde{\varphi}$ *for* $\varphi(X, x^0)$. *Then* (10.8e) *holds, and there is a constant* $\Phi_0(M, n, R, \delta, |\varphi|_\delta)$ *such that*

$$\sum_{i,j=0}^{n} \left|D_{ij}\tilde{\varphi}\right| \leq \Phi_0 (x^0)^{\delta-2}. \tag{10.8f$'$}$$

PROOF. We define

$$\varphi^*(X) = \varphi(x, t_0)\psi(x, t_0) + \Phi_1(t_0 - t) + [\varphi(X_0) + 3M][1 - \psi(x, t_0)].$$

Then (10.8a,b,e) and (10.8c,d,f)$'$ are easily verified and (10.10) follows since $\varphi_t^* = -\Phi_1$. □

Now, let $X_0 \in S\Omega$ and $\zeta \in (1,2)$. If there is a function $g \in H_\zeta(Q(2R))$ such that $g > 0$ in $Q(R) \cap \Omega$, and $|Dg| \neq 0$ in $Q(2R)$, and $g(X_0) = 0$, we say that Ω satisfies an *exterior* H_ζ *condition* at X_0 with constants R and $G = |g|_\zeta + 1/\min|Dg|$. (In particular, Ω satisfies an exterior H_2 condition at X_0 if and only if there is an exterior paraboloid frustrum at X_0, as in Theorem 2.12.) Now we take d to be the regularized distance from the set $\{g = 0\}$ in the cylinder $Q(X_0, R)$. From Chapter IV, we infer that $\left|D^2 d\right| + |d_t| \leq C d^{\zeta-2}$ in $Q(R) \cap \Omega$ for some C determined only by G, n, and R. A boundary gradient estimate for general domains follows, via Lemma 10.2, under only slightly stronger hypotheses than those of Lemma 10.1. An interesting element of this particular estimate is that the regularity of the comparison surface, specifically the coefficient ζ there, affects the structure conditions while the regularity of the boundary values does not.

THEOREM 10.4. *Let* $\delta \in (1,2]$ *and suppose that u is a solution of* (10.1) *with* $\varphi \in H_\delta$. *Let* $\zeta \in [\delta, 2]$ *and suppose that* Ω *satisfies an exterior* H_ζ *condition at* X_0 *with constants R and G. Suppose also that there are positive constants* μ *and* p_0 *such that*

$$|p|^{3-\zeta}\Lambda + |a| \leq \mu\mathscr{E} \tag{10.11}$$

for $|p| \geq p_0$ *and that either*

$$|p|^{3-\zeta} \leq \mu\mathscr{E}, \tag{10.12a}$$

or

$$1 \leq \mu\mathscr{E} \text{ and } g \text{ is increasing with respect to } t. \tag{10.12b}$$

Then there is a constant C determined only by G, n, R, ζ, δ, μ, and $|\varphi|_\delta$ such that (10.7) *holds. If the conditions hold with uniform constants at every point $X_0 \in S\Omega$, then* (10.7)$'$ *holds.*

PROOF. We assume without loss of generality that d has been normalized so that $0 \le d \le R/2$ in $Q(R) \cap \Omega$, and for $\alpha \in (0, R/2]$ a constant to be further specified, we take $N = \{X \in Q(X_0, R) : d(X) < \alpha\}$.

If (10.12a) holds, then we take $\bar{\varphi}(X) = \tilde{\varphi}(X, d)$. If f' is sufficiently large, then $f' > d$ so $(f')^{\delta-\zeta} < d^{\delta-\zeta}$ and hence

$$\begin{aligned}\bar{P}w &\le [\Lambda+1]d^{\delta-2} + \bar{a} + f''\mathscr{E}_d + f'[\Lambda+1]d^{\zeta-2} \\ &\le C\mu\mathscr{E}(f')^{\zeta-3}d^{\delta-2} + \mu\mathscr{E} + f''\mathscr{E}_d + C\mu(f')^{\zeta-2}d^{\zeta-2}\mathscr{E} \\ &< \mathscr{E}_d[f'' + \mu_0(f')^\delta d^{\delta-2}]\end{aligned}$$

with $\mu_0 = C\mu$. The differential equation

$$f'' + \mu_0(f')^\delta d^{\delta-2} = 0$$

is easily solved by

$$f(d) = \int_0^d [\mu_0 s^{\delta-1} + k^{\delta-1}]^{1/(1-\delta)}\, ds$$

with k a constant at our disposal. From the change of variables $\sigma = s/\alpha$, we see that $f(d) = H(d/\alpha; k/\alpha)$ where

$$H(w;\kappa) = \int_0^w [\mu_0\sigma^{\delta-1} + \kappa^{\delta-1}]^{1/(1-\delta)}\, d\sigma.$$

To achieve $f(\alpha) = M$, we need $H(1;\kappa) = M$. Since $H(1;\kappa)$ is a monotone function of the parameter κ with $H(1;\kappa) \to 0$ as $\kappa \to \infty$ and $H(1;\kappa) \to \infty$ as $\kappa \to 0$, we see that there is a unique positive value κ such that $H(1;\kappa) = M$. Since $f'(d) = H'(d/\alpha;\kappa)/\alpha$, it follows that we can make f' sufficiently large by taking α small enough. Hence (10.3a) holds. We then infer (10.3b) from (10.8b) since $f \ge 0$; (10.3c) follows from (10.8c) and the equation $f(\alpha) = M$. Finally, (10.3d) follows from (10.8a).

If (10.12b) holds, then we take $\bar{\varphi}(X) = \tilde{\varphi}(X, f(d)/M)$. So $d_t \ge 0$ and we estimate

$$\bar{P}w \le [\Lambda+1]f^{\delta-2} + \bar{a} + f''\mathscr{E}_d + f'\Lambda d^{\zeta-2}.$$

Since $f'' \le 0$, we have $f \ge f'd$ and hence

$$\bar{P}w < \mathscr{E}_d[f'' + \mu_0(f')^\delta d^{\delta-2}].$$

From this inequality, we proceed as before. □

By choosing f' large enough, we can guarantee that $\bar{\varphi} \ge \varphi$. Hence we can weaken (10.11) slightly to

$$|p|^{3-\zeta}\Lambda + \operatorname{sgn}(z-\varphi)a \le \mu\mathscr{E}$$

whenever $|p| \geq p_0$. In certain cases, especially when $\varphi \equiv 0$, this slightly weaker condition can be very useful. See the Notes for details.

When φ is independent of t, we can assert a boundary gradient estimate for cylindrical domains since Proposition 10.3 gives $\bar{\varphi}_t = 0$.

COROLLARY 10.5. *Suppose $\Omega = \omega \times (0,T)$ for some domain $\omega \in \mathbb{R}^n$ and that Ω satisfies an exterior H_ζ condition at some $X_0 \in S\Omega$. Suppose u satisfies* (10.1) *with φ independent of t. If* (10.11) *holds, then* (10.7) *holds at X_0. In fact,* (10.7) *holds at (x_0,t) for any $t \in [0,T]$. If the conditions hold with uniform constants at every point $X_0 \in S\Omega$, then* (10.7)$'$ *holds.*

2. Convex-increasing domains

As we have already seen, the relevant aspect of the geometry of Ω is not the regularity of $\mathscr{P}\Omega$ but rather the shape of an exterior comparison surface. A natural generalization of the exterior infinite cylinder condition is an exterior half-space condition which is the parabolic analog of convexity of a domain for elliptic equations. Specifically, we say that Ω is *convex-increasing* at $X_0 \in \mathscr{P}\Omega$ if there are a constant unit vector $\nu \in \mathbb{R}^n$ and a positive constant R such that $\nu \cdot x \geq 0$ whenever $X \in \Omega$ and $|X - X_0| < R$. Clearly, Ω is convex-increasing at any point of $B\Omega$. If Ω is convex-increasing at every point of $S\Omega$, we say that Ω is convex-increasing. A cylindrical domain $\Omega = \omega \times (0,T)$ is convex-increasing if and only if ω is convex. For convex-increasing domains, our structure conditions for a boundary gradient estimate can be relaxed.

THEOREM 10.6. *Suppose Ω is convex-increasing at X_0 and that u is a solution of* (10.1) *with $\varphi \in H_\delta$. Suppose also that* (10.4b) *holds for $|p| \geq p_0$ and that*

$$\Lambda + |a| \leq \mu \mathscr{E} \tag{10.13}$$

for $|p| \geq p_0$. If also

$$1 \leq \mu \mathscr{E} \tag{10.14a}$$

or else

$$\Omega \text{ is cylindrical and } \varphi \text{ is independent of } t, \tag{10.14b}$$

then there is a constant depending only on M, n, R, δ, and $|\varphi|_\delta$ such that (10.7) *holds. If the conditions are satisfied for all $X_0 \in S\Omega$ with uniform constants, then* (10.7)$'$ *holds.*

PROOF. In either case, we take $d(X) = \nu \cdot x$ and note that $\bar{a}^{ij} D_{ij} d - d_t = 0$. Since (10.4c) holds, it follows that

$$\bar{P} w \leq \Lambda |D^2 \bar{\varphi}| + |\bar{\varphi}_t| + \mathscr{E}_d [f'' + 4\mu (f')^2].$$

To proceed, we take

$$\bar{\varphi}(X) = \tilde{\varphi}(X, f(d)/M).$$

If (10.14a) holds, then

$$D_i\bar{\varphi} = D_i\tilde{\varphi} + D_0\tilde{\varphi}D_id,\ \bar{\varphi}_t = \tilde{\varphi}_t,$$

and

$$D_{ij}\bar{\varphi} = D_{ij}\tilde{\varphi} + D_{i0}\tilde{\varphi}D_jd + D_{0j}\tilde{\varphi}D_jd + D_0\tilde{\varphi}D_idD_jd,$$

so

$$|D^2\bar{\varphi}| + |\bar{\varphi}_t| \le Cf^{\delta-2} \le C(f')^\delta d^{\delta-2}$$

since $f \le M$ and $f(d) \ge f'(d)d$ since $f'' \le 0$. It follows that

$$\bar{P}w < \mathscr{E}_d[f'' + \mu_0(f')^\delta d^{\delta-2}]$$

for a suitable constant μ_0. As before, this differential inequality leads to a boundary gradient estimate.

When (10.14b) holds, we can take $\tilde{\varphi}$ and hence $\bar{\varphi}$ independent of t and a similar argument works. □

Condition (10.4b) is easily seen to be a consequence of either of the conditions

$$\Lambda = O(\lambda\,|p|^2) \text{ or } \Lambda = o(\mathscr{E})$$

as $|p| \to \infty$. In fact, (10.4b) says that $\Lambda = O(\mathscr{E})$ is sufficient provided the constant in the O condition is small enough.

The next step is to consider a parabolic analog of strictly convex domains. We say that Ω is *strictly convex-increasing* at $X_0 \in S\Omega$ if there are positive constants R, R_0, and η and a point $y \in R^n$ such that $\Omega[X_0, R_0]$ is a subset of the parabolic frustum

$$PF = \{X : |x-y|^2 + \eta(t_0 - t) < R^2, 0 < \eta(t_0 - t) < R^2/4\}.$$

In this case, we set

$$r(X) = [|x-y|^2 + \eta(t_0 - t)]^{1/2} \text{ and } d(X) = R - r(X).$$

Then $Dd = (y-x)/r$, so $|Dd| \le 1$. If $\alpha \le R/4$, then $d \le \alpha$ implies that $r \ge 3R/4$, and hence $r \le 2|x-y|$, so that $|Dd| \ge 1/2$. By direct calculation, we then have that

$$D_{ij}d = \frac{1}{r}[-\delta^{ij} + \frac{(x_i - y_i)(x_j - y_j)}{r^2}], d_t = \frac{\eta}{2r}.$$

It follows that

$$a^{ij}D_{ij}d - d_t = \frac{1}{r}[-\mathscr{T} + \mathscr{E}_d - \frac{\eta}{2}],$$

and therefore

$$\bar{P}w \le \mathscr{T}[|D^2\varphi| - \frac{f'}{r}] + |\bar{\varphi}_t| - \frac{\eta f'}{2r} + \bar{a} + \mathscr{E}_d[f'' + \frac{f'}{r}].$$

We now have several cases to consider for our boundary gradient estimate. In the first case, we don't need any hypotheses on the size of $\mathscr{E}$ compared to Λ, $|p|$, or 1.

THEOREM 10.7. *Suppose u is a solution of* (10.1) *with $\varphi \in H_2$ and suppose that Ω is strictly convex-increasing at some $X_0 \in S\Omega$ with constants η and R. If*

$$\eta(t_0 - t) < R^2 \text{ in } \Omega \tag{10.15a}$$

and if there are positive constants $R_1 > R$, p_0, and μ such that

$$|a| + \langle\varphi\rangle_2 \leq \frac{|p|\eta}{2R_1} + \frac{|p|\mathscr{T}}{R_1} + \mu\mathscr{E} \tag{10.15b}$$

for $|p| \geq p_0$, then (10.7) *holds with C determined only by M, p_0, R, R_1, η, and μ.*

PROOF. Since $r \leq R$, we have $f'/r \geq f'/R$. Moreover,

$$|Dw| = |f'Dd + D\bar{\varphi}| \leq f' + |D\bar{\varphi}| \leq f'[\frac{R+R_1}{2R_1}]$$

if f' is sufficiently large and hence

$$\frac{f'}{R} > \frac{|Dw|}{R_1} + |D^2\bar{\varphi}|_0 + 2|D\bar{\varphi}|_0^2$$

by making f' even larger. Since $\mathscr{E} \leq 2(f')^2\mathscr{E}_d + 2\Lambda|D\bar{\varphi}|^2$, it follows from (10.15) that

$$\bar{P}w \leq [2\mu(f')^2 + \frac{f'}{R} + f'']\mathscr{E}_d < \mathscr{E}_d[f'' + 3\mu(f')^2]$$

for $f' > R/\mu$. As before, this inequality leads to the desired estimate. □

For general H_δ boundary values, we need to include conditions of the form $1 + \Lambda = O(\mathscr{E})$, but we can relax (10.15) a little.

THEOREM 10.8. *Suppose u is a solution of* (10.1) *with $\varphi \in H_\delta$ for some $\delta \in (1,2]$, and suppose Ω is strictly convex-increasing at X_0 with constants R and η satisfying* (10.15a). *Suppose also that conditions* (10.4b) *and*

$$\Lambda \leq \mu\mathscr{E} \tag{10.12$'$}$$

hold for $|p| \geq p_0$. If

$$|a| \leq \frac{\eta}{R} + \frac{|p|\mathscr{T}}{2R} + \mu\mathscr{E}, 1 \leq \mu\mathscr{E} \tag{10.15b$'$}$$

or if $\langle\varphi\rangle_2$ is finite with

$$|a| + \langle\varphi\rangle_2 \leq \frac{|p|}{R}\left[\frac{\eta}{2} + \mathscr{T}\right] + \mu\mathscr{E}, \tag{10.15b$''$}$$

then (10.7) *holds.*

PROOF. In either case, it follows from the argument in Theorem 10.6 that

$$\bar{P}w \leq \Lambda c f^{\delta-2} + \mu\mathscr{E} + \mathscr{E}_d[f'' + \frac{f'}{r}],$$

and (10.12)′ and (10.4b) imply that

$$\bar{P}w \leq \mathscr{E}_d[f'' + \mu_0(f')^\delta d^{\delta-2}].$$

As before, this inequality gives (10.7). □

Note that Theorems 10.7 and 10.8 are still valid if $\eta = 0$. In this case, we obtain a boundary gradient estimate for cylinders with strictly convex cross-section. Moreover, if we replace the expression $|p|\mathscr{T}/R$ by $|p|\mathscr{T}/R_1$ for some $R_1 > R$ in (10.15b)′ or (10.15b)″, 'then we can eliminate the hypothesis (10.4b) from Theorem 10.8.

3. The spatial distance function

For a more careful examination of the boundary gradient estimate, we introduce another function which is equivalent to the usual parabolic distance d^* to $S\Omega$ given by

$$d^*(X) = \min_{\substack{Y\in S\Omega \\ s\leq t}} |X-Y|.$$

We define the *spatial distance* to $S\Omega$ by

$$d(X) = \min_{\substack{Y\in S\Omega \\ s=t}} |X-Y|.$$

It is easy to check that d and d^* are equivalent for $\mathscr{P}\Omega \in H_1$, which is our usual minimal smoothness assumption on Ω.

Most of our study of d centers on its spatial behavior. Since $d(X)$ is just the Euclidean distance from (x,t) to $\partial\omega(t)$, we begin by examining the distance function in $\mathbb{R}^n$ given by

$$d(x) = \min_{y\in\partial\omega} |x-y|,$$

for an open subset ω of $\mathbb{R}^n$. (Because the one-dimensional situation is much simpler, we assume in this section that $n \geq 2$.) We say that $y \in \partial\omega$ is a *nearest point* to x if $d(x) = |x-y|$. If x and x_0 are in Ω, we take y a nearest point to x to infer that

$$d(x_0) \leq |x_0 - y| \leq |x_0 - x| + |x-y| = |x_0 - x| + d(x).$$

By reversing the roles of x and x_0, we find that

$$d(x) \leq |x_0 - x| + d(x_0)$$

and therefore d is Lipschitz with Lipschitz constant no more than 1.

Next, fix $x \in \omega$ and take y to be a nearest point to x. For z on the line segment connecting x and y, we note that $d(z) = |z-y|$ and that y is the unique nearest point to z because the ball $B(x,d(x))$ is a subset of ω. In particular if $\partial\omega \in C^1$, then z is on the normal to y. Conversely, if there is a ball $B(z,r) \subset \omega$ with $y \in \partial\omega$ on the boundary of this ball, then y is the unique nearest point to any point in this ball on the line segment joining y and z.

Now suppose that there is a positive number R such that ω satisfies an interior sphere condition with radius R at each point $y \in \partial\omega$. If x is a point in ω with $d(x) < R$, let y be a nearest point to x in $\partial\omega$. As in the preceding paragraph, the nearest point to x is unique. We write Γ for the subset of Ω on which $d < R$. We now define $d_z(x) = R - |x-z|$ and infer by direct calculation (using Taylor's theorem) that

$$d_z(x+h) + d_z(x-h) - 2d_z(x) \geq -|h|^2/R$$

as long as $|x-z| < R$ and $|h| < d_z(x)$. In particular, if y is the unique nearest point to x in $\partial\omega$ and if $z \in \omega$ is chosen on the line through x and y so that $|z-y| = R$, then $d_z(x) = d(x)$ while $d_z(x \pm h) \geq d(x \pm h)$ provided $|h| < d(x)$. Therefore

$$d(x+h) + d(x-h) - 2d(x) \geq -|h|^2/R. \tag{10.16a}$$

A similar argument shows that

$$d(x+h) + d(x-h) - 2d(x) \leq |h|^2/R \tag{10.16b}$$

for $|h| < d(x)$ if ω satisfies an exterior sphere condition of radius R at each point of $\partial\omega$. From (10.16a,b), we infer that $d \in H_2$ and *a fortiori* that $d \in C^1$. The directional derivative of d in the direction of the line segment from y to x is the same as that for d_z, so this directional derivative is 1. Hence $|Dd| = 1$ on $\partial\omega$, which means that $\partial\omega \in C^1$. Therefore $Dd(x) = \gamma(y)$ for $x \in \Gamma$, where y is the unique nearest point to x in $\partial\omega$ and γ is the inner normal to ω at y.

From the preceding paragraph, it follows that ω satisfies interior and exterior sphere conditions with a fixed radius if and only if $\partial\omega \in H_2$. If $\partial\omega \in C^2$, we can say even more about d. For a fixed point $y_0 \in \partial\omega$, we suppose that ω can be written locally as the graph of the inequality $x^n > \psi(x')$ for some C^2 function ψ with $D\psi(0) = 0$. Then for any $y \in \partial\omega$ close enough to y_0, we have

$$\gamma(y) = \frac{(-D\psi(y'), 1)}{\sqrt{1+|D\psi(y')|^2}}.$$

The eigenvalues of the matrix $D^2\psi(y_0')$ are called the *principal curvatures* of $\partial\omega$ at y_0 and they are written as $\kappa_1, \ldots, \kappa_{n-1}$. The *mean curvature* of $\partial\omega$ at y_0 is given by

$$H(y_0) = \frac{1}{n-1}\sum \kappa_i = \frac{1}{n-1}\Delta\psi.$$

If we rotate axes so that the x^i-axis points along the eigenvector corresponding to κ_i, we have a *principal coordinate system* at y_0, in which case

$$D^2\psi(y_0') = D^2\gamma(y_0) = -\operatorname{diag}[\kappa_i].$$

Further, if $x \in \Gamma$ and if y is the unique nearest point to x, then $x = y + \gamma(y)d$. This equation defines a map from $\partial\omega \times (0,R)$ into $\mathbb{R}^n$. To analyze this map, we introduce a principal coordinate system at y_0 and write points in $\partial\omega$ near y_0 as $y = (y', \psi(y'))$. Then the equation

$$x = (y', \psi(y')) + \gamma(y', \psi(y'))d$$

defines a map J from $B \times (0,R)$ for a suitably small ball B centered at y_0'. The Jacobian matrix of this map is just $\operatorname{diag}[1 - \kappa_i d, 1]$, so J is a C^1 invertible map provided $R\kappa_i < 1$ for all i. It follows that (y', d) is a C^1 function of x. In addition,

$$D^2 d(x) = \operatorname{diag}\left[\frac{-\kappa_i(y)}{1 - \kappa_i(y)d}, 0\right] \tag{10.17}$$

in a principal coordinate system centered at y.

The behavior of d as a function of t is easier to describe. If $\mathscr{P}\Omega \in H_2$, then the argument leading to (10.16) shows that $d \in H_2$. In particular, d_t exists almost everywhere. If $\mathscr{P}\Omega \in C^{2,1}$, we fix $X_0 \in \Omega$ such that $d_t(X_0)$ exists and take y_0 to be the nearest point to x_0 in $\partial\omega(t_0)$. In a principal coordinate system centered at y_0, we have $X_0 = (0, x_0^n, t_0)$ and $d(X_0) = x_0^n$. For $t > t_0$, $d(x_0, t) \le x_0^n - \psi(0,t)$, and therefore

$$\frac{d(x_0,t) - d(X_0)}{t - t_0} \le \frac{-\psi(0,t)}{t - t_0}.$$

Sending $t \to t_0$, we see that $d_t(X_0) \le -\psi_t(0,t_0)$. A similar argument with $t < t_0$ such that $d_t(X_0) \ge -\psi_t(0,t_0)$ and therefore $d_t(X_0) = -\psi_t(X_0)$ in a principal coordinate system centered at Y_0. In general,

$$d_t = \frac{-\psi_t}{\sqrt{1 + |D\psi|^2}},$$

wherever d_t exists. Since the right side of this equation is everywhere continuous, it follows that d_t is also continuous. In other words $\mathscr{P}\Omega \in C^{2,1}$ implies that $d \in C^{2,1}$ near $S\Omega$. We write κ_0 for $\frac{-\psi_t}{\sqrt{1+|D\psi|^2}}$ and we call the n-vector $(\kappa_0, \ldots, \kappa_{n-1})$ the *space-time curvature* of $S\Omega$.

4. Curvature conditions

In the previous sections, barriers were constructed from the distance function d in terms of one-sided estimates on all curvatures of the comparison surface. In this section, we construct barriers when only some of the curvatures are negative.

The basis for these constructions is work of Jenkins and Serrin [141] who showed that a barrier exists for the elliptic minimal surface equation when the boundary of the domain has nonpositive mean curvature. Serrin [290] later developed general analogs of these results, which we expand upon in this section.

To describe the ideas, we first suppose that $\Omega = \omega \times (0,T)$ is cylindrical and we write H' for the mean curvature of $\partial\omega$. For the parabolic mean curvature operator, we have

$$a^{ij} = \frac{\delta_{ij} - \nu_i\nu_j}{\sqrt{1+|p|^2}},$$

where $\nu = p/\sqrt{1+|p|^2}$. Taking $w = f(d)$, we see that

$$\begin{aligned}\bar{a}^{ij}D_{ij}d &= (1+(f')^2)^{-1/2}\left[\Delta d - \frac{(f')^2 D_i d D_j d D_{ij} d}{1+(f')^2}\right] \\ &= (1+(f')^2)^{-1/2}\Delta d \\ &\le -(n-1)(1+(f')^2)^{-1/2}H',\end{aligned}$$

and $\mathscr{E}_d = (1+(f')^2)^{-3/2}$ by virtue of (10.17). Hence, if $\varphi \equiv 0$ and $a = 0$, then

$$\bar{P}w \le (1+(f')^2)^{-1/2}[\frac{f''}{(f')^2} - (n-1)f'H'].$$

It follows that $\bar{P}w < 0$ if $H' \ge 0$, in other words, $\partial\omega$ has nonnegative mean curvature.

Rather than analyze this particular situation in greater detail, we now present a more general setting. We suppose that there are functions a_∞^{ij}, a_0^{ij}, a_∞, and a_0 such that

$$a^{ij}(X,u,p) = a_\infty^{ij}(X,\frac{p}{|p|})\Lambda(X,u,p) + a_0^{ij}(X,u,p), \tag{10.18a}$$

$$a(X,z,p) = a_\infty(X,z,\frac{p}{|p|})\,|p|\,\Lambda(X,z,p) + a_0(X,z,p) \tag{10.18b}$$

whenever $|p| \ge p_0$. We use Λ_0 to denote an upper bound for the absolute value of the eigenvalues of a_0^{ij} and we set

$$\sigma(X,p) = |p|\,\Lambda_0(X,u,p) + |a_0(X,u,p)|\,.$$

Next, for a fixed X_0, we suppose that there is a $C^{2,1}$ surface $\mathscr{S}$ such that $X_0 \in \mathscr{S}$, $Q(X_0,R) \cap \Omega \cap \mathscr{S}$ is empty for some $R > 0$. By decreasing R as needed, we may also assume that every point in $Q(X_0,R) \cap \Omega$ has a unique nearest point in $\mathscr{S}$. For simplicity, we write Q for $Q(X_0,R)$, d for spatial distance to $\mathscr{S}$, and $\gamma = Dd$. Finally, for $Y \in Q \cap S\Omega$, we define

$$\kappa^{\pm} = -a_\infty^{ij}(Y, \pm\gamma(Y))D_{ij}d(Y).$$

Under suitable hypotheses, we now obtain a number of boundary gradient estimates by choosing

$$\bar{a}(X,z,p) = a_\infty(X,z,\frac{p}{|p|})\,|p|\,\Lambda(X,z,p) + a_0(X,u,p).$$

The first estimate requires very little of the curvatures of $\mathscr{S}$, but it uses the strongest hypotheses on the coefficients.

THEOREM 10.9. *Suppose that u is a solution of* (10.1) *with* $\varphi \in H_\delta$ *for some* $\delta \in (1,2]$. *Fix* $X_0 \in S\Omega$ *and suppose that the coefficients* a^{ij} *and a can be decomposed according to* (10.18), *that* a_∞ *is decreasing with respect to z, and that there is a constant* K_1 *such that*

$$\left|a_\infty^{ij}(X,\zeta_1) - a_\infty^{ij}((y,t),\zeta_2)\right| \leq K_1[|x-y| + |\zeta_1 - \zeta_2|], \tag{10.19a}$$

$$\left|a_\infty(X,z,\zeta_1) - a_\infty^{ij}((y,t),z,\zeta_2)\right| \leq K_1[|x-y| + |\zeta_1 - \zeta_2|], \tag{10.19b}$$

for $X \in Q \cap \Omega$, $(y,t) \in Q \cap S\Omega$, *and* ζ_1 *and* ζ_2 *in* S^{n-1}. *Suppose also that conditions* (10.4b) *and* (10.12)′ *are satisfied along with either* (10.14a) *or* (10.14b). *If also*

$$\kappa^\pm(Y) \geq \pm a_\infty(Y,\varphi(Y),\pm\gamma(Y)) \tag{10.20a}$$

for $Y \in Q \cap S\Omega$ *and*

$$\kappa_0 \leq 0, \ \sigma \leq \mu\mathscr{E}, \tag{10.20b}$$

then (10.7) *holds with C determined only by* K_1, p_0, R, μ, $|\varphi|_\delta$, *and the* $C^{2,1}$ *norm of* $\mathscr{S}$.

PROOF. For $X \in Q \cap \Omega$, write X^* for the nearest point to X on $\mathscr{S}$ and Y for the the point of intersection of the line segment joining X to X^* with $S\Omega$. (If these sets intersect in more that one point, take Y to be the closest such point to $\mathscr{S}$.) Then $Dd(Y) = Dd(X) = \gamma(Y)$, and

$$|Dw - |Dw|\,\gamma| \leq \left|Dw - f'\gamma\right| + \left||Dw| - f'\right| = |D\bar{\varphi}| + \left||Dw| - f'\right|.$$

Since $||Dw| - f'| \leq |D\bar{\varphi}|$, it follows that

$$\left|\frac{Dw}{|Dw|} - \gamma\right| \leq \frac{4|D\bar{\varphi}|}{f'}. \tag{10.21}$$

Next, we define

$$S_0 = a_\infty^{ij}(X,\frac{Dw}{|Dw|})D_{ij}d(X) + a_\infty(X,w,\frac{Dw}{|Dw|}),$$

and note that

$$\begin{aligned} f'[-d_t + \bar{a}^{ij}D_{ij}d] + \bar{a} = &-f'd_t + |Dw|\,\Lambda S_0 + (f' - |Dw|)\bar{a}^{ij}D_{ij}d \\ &+ |Dw|\,a_0^{ij}D_{ij}d + a_0. \end{aligned}$$

From (10.19), (10.20), and (10.21), we can estimate

$$\begin{aligned}
S_0 &\le a_\infty^{ij}(X, \frac{Dw}{|Dw|})D_{ij}d(Y) + a_\infty(X, w, \frac{Dw}{|Dw|}) \\
&= -\kappa^+(Y) + a_\infty(Y, \varphi(Y), \gamma(Y)) \\
&+ [a_\infty^{ij}(X, \frac{Dw}{|Dw|}) - a^{ij}(Y, \gamma(Y))]D_{ij}d(Y) \\
&+ [a_\infty(X, w(X), \frac{Dw}{|Dw|}) - a_\infty(Y, \varphi(Y), \gamma(Y))] \\
&\le K_1[|x-y| + \left|\frac{Dw}{|Dw|} - \gamma\right| [\sup D^2 d + 1] \\
&+ [a_\infty(Y, w(X), \gamma(Y)) - a_\infty(Y, \varphi(Y), \gamma(Y))].
\end{aligned}$$

Now we note that $|x-y| = d(X) - d(Y) \le d(X)$, so

$$w(X) = \bar{\varphi}(X) + f(d) \ge \bar{\varphi}(Y) - C|x-y| + f(d) \ge \varphi(Y)$$

if f' is sufficiently large. Taking (10.21) into account again, we infer that

$$S_0 \le C/f'.$$

It follows that

$$f'[-d_t + \bar{a}^{ij}D_{ij}d] + \bar{a} \le C\Lambda + C\sigma \le C\mathscr{E}.$$

With this estimate, we infer as before that

$$\bar{P}w \le [f'' + C(f')^\delta d^{\delta-2}]\mathscr{E}_d,$$

and then (10.7) follows easily. □

When the inequality in (10.20a) is strict, we can relax the regularity hypotheses on the coefficients to continuity.

THEOREM 10.10. *Suppose that u is a solution of* (10.1) *with $\varphi \in H_\delta$ for some $\delta \in (1,2]$. Fix $X_0 \in S\Omega$ and suppose that the coefficients a^{ij} and a can be decomposed according to* (10.18), *that a_∞ is decreasing with respect to z, and that a_∞^{ij} and a_∞ are uniformly continuous on $(Q \cap \Omega) \times S^{n-1}$. Suppose also that conditions* (10.4b) *and* (10.12)$'$ *are satisfied along with* (10.14a) *or* (10.14b)). *If also there are positive constants $R_0 > R_1$ such that*

$$\kappa^\pm(Y) \ge \pm a_\infty(Y, \varphi(Y), \pm\gamma(Y)) + \frac{1}{R_0} \tag{10.20a$'$}$$

for $Y \in Q \cap S\Omega$ and

$$\kappa_0 \le 0,\ |p|a_0^{ij}D_{ij}d + a_0 \le \frac{|p|\Lambda}{R_1} + \mu\mathscr{E}, \tag{10.20b$'$}$$

then (10.7) *holds with C determined only by R, R_0, R_1, μ, $|\varphi|_\delta$, a modulus of continuity estimate for a^{ij} and a, and the $C^{2,1}$ norm of $\mathscr{S}$.*

PROOF. Now the uniform modulus of continuity gives a continuous, increasing function ω with $\omega(0) = 0$ such that

$$\left| a_\infty^{ij}(X, \frac{Dw}{|Dw|}) - a_\infty^{ij}(Y, \gamma(Y)) \right| \leq \omega(\frac{C}{f'}),$$
$$a_\infty(X, w(X), \frac{Dw}{|Dw|}) - a_\infty(Y, \varphi(Y), \gamma(Y)) \leq \omega(\frac{C}{f'}).$$

It follows that

$$S_0 \leq -\frac{1}{R_0} + C\omega(\frac{C}{f'}) \leq -\frac{1}{R_1}$$

provided f' is large enough. Hence

$$f'[-d_t + \bar{a}^{ij} D_{ij} d] + \bar{a} \leq \mu \mathscr{E},$$

and the proof is completed in the usual way. □

The incorporation of κ_0 into conditions (10.20a) and (10.20a)$'$ would be very useful, but no such result is known to the author. Also, a comparison surface $\mathscr{S}$ satisfying the conditions of Theorem 10.9 or 10.10 at every point of $S\Omega$ exists if and only if $S\Omega$ satisfies these conditions. As in the previous sections, we can replace condition (10.14) by a suitable restriction on $\langle\varphi\rangle_2$ in either theorem.

COROLLARY 10.11. *Let u be a solution of* (10.1) *with $\varphi \in H_\delta$ for some $\delta \in (1,2]$. Fix $X_0 \in S\Omega$ and suppose that the coefficients a^{ij} and a can be decomposed according to* (10.18), *and that a_∞ is decreasing with respect to z. Suppose also either*
(a) that there is a constant K_1 such that (10.19a,b) *hold for $X \in Q \cap \Omega$, $(y,t) \in Q \cap S\Omega$, and ζ_1 and ζ_2 in S^{n-1}, that* (10.20a) *holds for $Y \in Q \cap S\Omega$, that* (10.20b) *holds, and that*

$$\langle\varphi\rangle_2 \leq |p|\Lambda[\kappa^\pm \mp a_\infty(X, \varphi, \pm\gamma)] + \mu\mathscr{E} \tag{10.22}$$

or
(b) that there are positive constants $R_0 > R_1$ such that (10.20a,b)$'$ *hold and that*

$$\langle\varphi\rangle_2 \leq |p|\Lambda\varepsilon_0 + \mu\mathscr{E} \tag{10.22$'$}$$

for some positive constant $\varepsilon_0 < 1/R_1 - 1/R_0$.
If conditions (10.4b) *and* (10.12)$'$ *hold, then* (10.7) *holds with C determined only by K_1, p_0, R, μ, $|\varphi|_\delta$, and the $C^{2,1}$ norm of $\mathscr{S}$.*

Moreover, we can remove the conditions (10.4b) and (10.12)$'$ in Theorem 10.10 and Corollary 10.11(b) if $\delta = 2$. We refer the reader to Exercise 10.1 for details.

5. Nonexistence results

We now show that the conditions outlined in the preceding sections are essential to the existence theory. Specifically, whenever the geometric conditions on $S\Omega$ or the structure conditions on the operator fail to hold, there are choices of the function φ for which (10.1) has no classical solution. (Note that this statement does not violate the local existence results of Chapter VIII because we are specifying the set on which the solution is expected to exist.)

Throughout this section, we adopt the following simplifying conventions. First, $\bar{P}$ is as before with $\bar{a}(X,z,p) = a(X,u,p)$. Also, we use K_0, M, β, and θ to denote positive constants with $\beta = \frac{\theta}{1+\theta}$. Finally, we assume without loss of generality that $t = 0$ on $B\Omega$.

We start with a simple comparison principle.

LEMMA 10.12. *Let Γ be a nonempty, relative open H_δ portion of $S\Omega$ with $\delta > 1$, and let $u \in C(\overline{\Omega}) \cap C^{2,1}(\Omega \cup \Gamma)$ and $v \in C(\overline{\Omega}) \cap C^{2,1}(\Omega)$. If*

$$\bar{P}v < \bar{P}u \text{ in } \Omega, v \geq u \text{ on } \mathscr{P}\Omega \setminus \Gamma, \tag{10.23a}$$

and if

$$\frac{\partial v}{\partial \gamma} = -\infty \text{ on } \Gamma, \tag{10.23b}$$

then $v \geq u$ in Ω.

PROOF. By Corollary 9.2, if $u - v$ has a nonnegative maximum, it must occur on Γ, but (10.23b) precludes this possibility. □

For our applications of Lemma 10.12, we write $\delta = \operatorname{diam}\Omega$ and we fix positive constants α and α_1 with $\alpha < \delta/2$. Now fix $X_0 \in S\Omega$ and suppose that $\alpha_1 \leq t_0$. We also define

$$\Omega_1 = \{X \in \Omega : t < t_0 - \alpha_1\},\ \Omega_2 = \{X \in \Omega : |x - x_0| > \alpha, t_0 - \alpha_1 < t < t_0\},$$

and $m_1 = \sup_{\mathscr{P}\Omega_1} \varphi$. If $a < 0$ for $z > M$ and $|p| > p_0$, we take $v = m_1 + (\varepsilon + p_0)x^1$ in the maximum principle to infer that

$$u \leq \max\{M, m_1 + (\varepsilon + p_0)\delta\} \text{ in } \Omega_1$$

for any $\varepsilon > 0$. Sending $\varepsilon \to 0$, we conclude that

$$u \leq \max\{M, p_0\delta\} \text{ in } \Omega_1. \tag{10.24}$$

Now, for $\psi \in C^2(\alpha, \delta)$ and $m \in \mathbb{R}$ to be further specified, we take

$$v(X) = m + \psi(|x - x_0|).$$

A simple calculation shows that

$$\begin{aligned}\bar{P}v &= \frac{\psi''}{(\psi')^2}\mathscr{E} + \frac{\psi'}{|x-x_0|}\left[\mathscr{T} - \frac{\mathscr{E}}{(\psi')^2}\right] + \bar{a} \\ &\le \frac{\psi''}{(\psi')^2}\mathscr{E} + \bar{a}\end{aligned}$$

on $\Omega^* = \Omega_1 \cap \{u > M\}$ provided $\psi' \le 0$. If we suppose that

$$a \le -|p|^\theta \mathscr{E} \tag{10.25}$$

for $z > M$ and $|p| > p_0$, then we have

$$\bar{P}v \le (\frac{\psi''}{(\psi')^2} - |\psi'|^\theta)\mathscr{E}$$

on Ω^*, so the choice

$$\psi(r) = K[(\delta-\alpha)^\beta - (r-\alpha)^\beta] \tag{10.26}$$

gives

$$\bar{P}v \le \mathscr{E}(r-\alpha)^{-\beta}[\frac{1-\beta}{K\beta} - (K\beta)^\theta] < 0$$

on Ω^* if K is taken sufficiently large. We now set

$$m_2 = \sup_{\mathscr{P}\Omega_1 \cap \{|x-x_0|>\alpha\}} u,$$

and take $m = \max\{M, m_2, m_1 + p_0\delta\}$. From (10.24) and Lemma 10.12, with Ω replaced by Ω^* and

$$\Gamma = \{X \in \Omega : |x-x_0| = \alpha, t < t_0\},$$

it follows that

$$\sup_{\Omega_2} u \le \max\{M, m_1 + p_0\delta, m_2\} + K\delta^\beta. \tag{10.27}$$

In other words, we can estimate the maximum of u on a subset of Ω in terms of a proper subset of $\mathscr{P}\Omega$ when (10.25) holds. From this estimate, we can prove our first non-existence result.

THEOREM 10.13. *Suppose $X_0 \in S\Omega$ and suppose that there are positive constants R and α_1 and a point $y \in \mathbb{R}^n$ with $|x-y| = R$ such that the cylinder $Q = \{|x-y| < R, t_0 - \alpha_1 < t < t_0\}$ is a subset of Ω. If there are positive constants M, p_0, R_0, and θ with $R_0 < R$ such that*

$$a \le -|p|^\theta \mathscr{E} - \frac{|p|\mathscr{T}}{R_0} \tag{10.28}$$

for $z > M$ and $|p| > p_0$, then there is a function $\varphi \in C^\infty$ such that (10.1) *has no solution.*

PROOF. Set

$$Q^* = \{|x-x_0| < R-R_0, t_0-\alpha_1 < t < t_0\} \cap \Omega.$$

From (10.27) with $\alpha = R - R_0$, we can estimate

$$m_0 = \sup_{\mathscr{P}Q^* \cap \Omega} u$$

in terms of $\sup_{\mathscr{P}\Omega \setminus \mathscr{Q}^*} \varphi$. Our goal now is to estimate $u(X_0)$ in terms of m_0.

To this end, we fix $\varepsilon \in (0, \alpha)$, we define

$$\psi(r) = K[(R-\varepsilon)^\beta - (R-\varepsilon-r)^\beta], \tag{10.26$'$}$$

and we consider the function $v = m_0 + \psi(|x-y|)$. An easy calculation shows that $\bar{P}v < 0$ in

$$Q'_\varepsilon = \{|x-y| < R-\varepsilon, |x-x_0| < R-R_0, t_0-\alpha_1 < t < t_0, u(X) > M\},$$

and that $\partial v/\partial \gamma = -\infty$ on

$$\Gamma = \{|x-y| = R\} \cap \mathscr{P}Q'_\varepsilon,$$

so $u \le v$ in Q'_ε by Lemma 10.12. In particular,

$$u(x, t_0) \le m_0 + KR^\beta$$

if $|x-y| = R-\varepsilon$ and $(x, t_0) \in \mathscr{P}Q'_\varepsilon$. Sending $\varepsilon \to 0$, we see that

$$u(X_0) \le m_0 + KR^\beta.$$

It follows that $\varphi(X_0) = u(X_0)$ cannot be prescribed arbitrarily. For example, we can take $\varphi \equiv 0$ on $\mathscr{P}\Omega \setminus \mathscr{P}Q^*$ with $\varphi(X_0)$ sufficiently large. □

Note that, when $\alpha_1 = t_0$, we can take the function φ to be independent of t. Hence Corollary 10.5 with $\zeta = 2$ is sharp in the sense that $\mathscr{E}$ cannot be replaced by $|p|^\theta \mathscr{E}$ with $\theta > 0$. It is now a simple matter to extend this result to show that Theorem 10.4 is sharp.

We suppose now that there is a positive constant η_0 such that

$$-\eta_0 + a \le -|p|^\theta \mathscr{E}$$

when $z > M$ and $|p| > p_0$. With ψ given by (10.26) and $v = m + \psi(|x-y|) + \eta_0 t$, we infer that

$$\sup_{\Omega_2} u \le \max\{M, m_1 + p_0\delta, m_2\} + K\delta^\beta + \eta_0 t_0. \tag{10.27$'$}$$

with m_1 and m_2 as in (10.27). From this estimate, we can show the non-existence of solutions to (10.1) under a different set of conditions.

THEOREM 10.14. *Suppose $X_0 \in S\Omega$ and suppose that there are positive constants R and α_1 and a point $y \in \mathbb{R}^n$ with $|x-y| = R$ such that the cylinder $Q^* = \{|x-y| < R, t_0 - \alpha_1 < t < t_0\}$ is a subset of Ω. If there are positive constants M, p_0, R_0, θ, and Φ with $R_0 < R$ such that*

$$-\Phi + a \leq -|p|^\theta - \frac{|p|\mathscr{T}}{R_0}, \; a \leq 0 \tag{10.29}$$

for $z > M$ and $|p| > p_0$, then there is a function $\varphi \in C^\infty$ such that (10.1) *has no solution. Moreover, if $\alpha_1 = t_0$, then for any $\eta_0 > 0$, there is such a function φ with $|\varphi_t| \leq \Phi + \eta_0$.*

PROOF. Let $\eta_0 \in (0, \Phi)$. Since $a \leq 0$, we have

$$-\eta_0 + a \leq \frac{\eta_0}{\Phi}[-\Phi + a] \leq -|p|^{\theta/2}\mathscr{E}$$

if $|p| > p_0 + (\Phi/\eta_0)^{2/\theta}$. From the proof of (10.27)′, we infer that

$$\sup_{\Omega_2} u \leq \max\{M, m_1 + p_0\delta, m_2\} + K\delta^{\theta/(2+\theta)} + \eta_0 t_0.$$

Writing m_0 for the right side of this inequality and defining ψ by (10.26)′, we take $v = m_0 + \psi(|x-y|) + \Phi t$ to infer that

$$u(X_0) \leq m_0 + KR^\beta + \Phi t.$$

As before, this inequality gives the desired result. □

Hence Theorem 10.4 with (10.12b) is sharp in the sense that $\mathscr{E}$ cannot be replaced by $|p|^\theta \mathscr{E}$ in (10.11) or in (10.12b); the sharpness of this theorem with (10.12a) is currently unknown. Since Ω can be taken convex-increasing, we also conclude the sharpness of Theorem 10.6 from this result. We leave it to the reader (Exercise 10.2) to show that Theorems 10.7 and 10.8 are sharp.

Finally, a slight modification of the preceding argument shows that the curvature conditions of Section X.5 are also sharp.

THEOREM 10.15. *Suppose that the coefficients of P can be decomposed according to* (10.18) *with a_∞ independent of z and $\sigma = o(|p|\Lambda)$ as $|p| \to \infty$. If*

$$a \leq 0, \; \mathscr{E} \leq |p|^{1-\theta}\Lambda \tag{10.30}$$

for $z > M$ and $|p| > p_0$ and if

$$\kappa^-(X_0, \gamma(X_0)) + a_\infty(X_0, \gamma(X_0)) > 0 \tag{10.31}$$

for some comparison surface $\mathscr{S} \subset \Omega$, then there is a function $\varphi \in C^\infty$ such that (10.1) *has no solution. If Ω is cylindrical, then φ can be taken independent of t.*

PROOF. From the first inequality of (10.30), we infer (10.24). With $v = m + \psi(|x - x_0|)$, we now have

$$\begin{aligned}\bar{P}v &\le \left(\frac{\psi''}{(\psi')^2} + \frac{\psi'}{|x-x_0|}[|\psi'|^{\theta-1} - (\psi')^{-2}]\right)\mathscr{E} \\ &< \left(\frac{\psi''}{(\psi')^2} - \frac{|\psi'|^{\theta}}{2|x-x_0|}\right)\mathscr{E}\end{aligned}$$

in Ω^* provided $\psi' < -2^{1/(1+\theta)}$. Hence, if we take

$$\psi(r) = \int_r^{\delta}\left(\frac{1+\theta}{2}\ln\frac{\tau}{\alpha}\right)^{-1/(1+\theta)} d\tau,$$

then $\bar{P}v < 0$ in Ω^* and therefore

$$\sup_{\Omega_2} u \le \max\{M, m_1 + p_0\delta, m_2\} + \psi(\alpha).$$

Using the change of variables $s = \tau/\alpha$, we have that

$$\psi(\alpha) = \alpha\int_1^{\delta/\alpha}\left(\frac{1+\theta}{2}\ln s\right)^{-1/(1+\theta)} d\tau,$$

and a simple application of l'Hôpital's rule shows that $\psi(\alpha) \to 0$ as $\alpha \to 0$. Therefore, for any $\rho > 0$, there is an $\alpha_0 > 0$ such that $\alpha \le \alpha_0$ implies that

$$\sup_{\Omega_2} u \le \max\{M, m_1 + p_0\delta, m_2\} + \rho.$$

To proceed, we fix $R > 0$ and take

$$v = m + \psi(R - d(X))$$

for some function ψ to be further specified and α so small that $d(X) < R$ and X has a unique nearest point on $\mathscr{S}$ for $X \in \Omega'$, which is the connected subset of

$$\{|x - x_0| < \alpha, t_0 - \alpha_1 < t < t_0\} \setminus \mathscr{S}$$

which lies in Ω. For $\varepsilon > 0$ sufficiently small, we write Ω'_ε for the subset of Ω' on which $d > \varepsilon$ and $u > M$. It follows from (10.31) that there is a positive constant ε_0 such that

$$\begin{aligned}\bar{P}v &= \frac{\psi''}{(\psi')^2}\mathscr{E} + \psi'\Lambda[-a_\infty^{ij}(X,\gamma(X))d_{ij}d(X) + a_\infty(X,\gamma(X))] + \psi' a_0^{ij} D_{ij}d + a_0 \\ &\le \frac{\psi''}{(\psi')^2}\mathscr{E} + \psi'\Lambda\varepsilon_0\end{aligned}$$

in Ω'_ε if R is small enough and $-\psi'$ is large enough. Using the second inequality of (10.30) and ψ given by (10.26)$'$, we conclude that $\bar{P}v < 0$ and hence we can estimate

$$\varphi(X_0) \le \max\{M, m_1 + p_0\delta, m_2\} + \rho + KR^{\beta}$$

for ρ and R arbitrarily small. □

An easy refinement shows that our restrictions on $\langle\varphi\rangle_2$ are sharp.

THEOREM 10.16. *Suppose that the coefficients of P can be decomposed according to* (10.18) *with a_∞ independent of z and $\sigma = o(|p|\Lambda)$ as $|p| \to \infty$. If* (10.30) *holds for $z > M$ and $|p| > p_0$ and if*

$$|p|\Lambda[\kappa^-(X_0,\gamma(X_0)) + a_\infty(X_0,\gamma(X_0))] + \Phi > 0 \tag{10.31$'$}$$

for some comparison surface $\mathscr{S} \subset \Omega$, then there is a function $\varphi \in C^\infty$ such that (10.1) *has no solution. If Ω is cylindrical, then for any $\varepsilon > 0$, φ can be taken so that $|\varphi_t| \le \Phi + \varepsilon$.*

PROOF. In estimating $\varphi(X_0)$ at the end of the proof of Theorem 10.15, we take

$$v = m + \psi(R - d(X)) + \Phi t.$$

□

6. The case of one space dimension

When $n = 1$, the boundary gradient estimate is much simpler because equations are now automatically uniformly parabolic. We start by collecting the necessary estimates.

THEOREM 10.17. *Let $1 < \delta \le \zeta \le 2$ and suppose $\Omega \subset \mathbb{R}^2$ satisfies an exterior H_ζ condition at $X_0 \in S\Omega$. Let u be a solution of* (10.1) *with $\varphi \in H_\delta$. If*

$$|a| \le \mu a^{11}|p|^2 \tag{10.32}$$

for $|p| \ge p_0$ and if one of the sets of conditions

$$1 \le \mu |p|^{\zeta-1} a^{11}, \tag{10.33a}$$

$$1 \le \mu a^{11}|p|^2 \text{ and } g \text{ is increasing with respect to } t \tag{10.33b}$$

$$g \text{ is increasing with respect to } t \text{ and } \varphi \text{ is independent of } t \tag{10.33c}$$

is satisfied, then (10.7) *holds with C determined only by M, MR, δ, and $|\varphi|_\delta$.*

PROOF. When (10.33a) or (10.33b) holds, we use Theorem 10.4 and when (10.33c) holds, we use Corollary 10.5. □

Non-existence proofs are also easier. If there are positive constants η_0 and θ such that

$$-\eta_0 + a \le -|p|^{2+\theta} a^{11}, \tag{10.34}$$

then sending $\alpha \to 0$ in (10.26)$'$ implies that there are boundary values φ such that (10.1) has no solution. If $\alpha_1 = t_0$, then we can choose this φ so that $|\varphi| < \eta_0 + \varepsilon$ for any $\varepsilon > 0$. If also $a \le 0$, then (10.34) holds with η replaced by any smaller positive number (and θ decreased), so we obtain a non-existence result in this case with φ_t arbitrarily small. If (10.34) holds with $\eta_0 = 0$ and $\alpha_1 = t_0$, then we can take φ to be time-independent.

7. Continuity estimates

If the boundary function φ is merely continuous, then it is possible to prove a boundary modulus of continuity estimate. To this end, we suppose that φ is continuous at $X_0 \in S\Omega$. Fix $\varepsilon > 0$ and choose $\delta > 0$ such that $|X - X_0| < \delta$ implies that $|\varphi(X) - \varphi(X_0)| < \varepsilon$. We then define functions $\varphi^{\pm}$ by

$$\varphi^{\pm}(X) = \varphi(X_0) \pm \left(\varepsilon + \frac{2\sup|\varphi|}{\delta^2}|X - X_0|^2\right).$$

Then $\varphi^{\pm} \in C^{2,1}$ and $\varphi^{-} \leq \varphi \leq \varphi^{+}$ on $\mathscr{P}\Omega$. If P and $\varphi^{\pm}$ satisfy the hypotheses of the theorems in our boundary gradient estimates, our barrier constructions provide a function w, continuous at X_0 such that

$$\varphi^{-} - w \leq u \leq \varphi^{+} + w$$

in some parabolic neighborhood of X_0, and hence u is continuous at X_0. We thus infer the following continuity estimates.

THEOREM 10.18. *Let u be a solution of* (10.1) *with φ continuous having modulus of continuity ω. If P and φ satisfy the hypotheses of Lemma 10.1, Theorem 10.4, Corollary 10.5, Theorem 10.6, Theorem 10.7, Theorem 10.8, Theorem 10.9, Theorem 10.10, or Corollary 10.11, then there is a modulus of continuity ω^* determined only by ω,* $\sup|u|$, $\sup|\varphi|$, *P, and Ω such that*

$$|u(X) - \varphi(Y)| \leq \omega^*(|X - Y|) \tag{10.35}$$

for any $X \in \Omega$ and $Y \in \mathscr{P}\Omega$.

Note that the restrictions on $\langle\varphi\rangle_2$ in Theorems 10.7 and 10.8 and Corollaries 10.5 and 10.11 must be satisfied in this theorem.

Notes

The basic ideas of the boundary gradient estimate go back to Bernstein's investigation of two-dimensional elliptic problems [17] (see, in particular, Theorem 2 of the second part of [17]). Bernstein's ideas were further developed for the minimal surface equation by Finn (see the introduction to [83]) and by Gilbarg [99, Lemma 2] for equations in convex domains. Serrin [290, Sections 7–11] expanded on the ideas of these and other authors to create a general existence theory for elliptic equations in an arbitrary number of space dimensions. We refer to [290] for a more detailed history of the boundary gradient estimate for elliptic equations.

Parabolic analogs of these results were discovered by many different people, but in the special case of cylindrical domains. Ladyzhenskaya and Ural′tseva studied uniformly parabolic equations in great detail in the early 1960's but they relegate the boundary gradient estimate to a simple remark that the Bernstein method for elliptic equations is also applicable to parabolic equations. (See page 271 of the English translation of [184].) Following some of Serrin's work, Edmunds and Peletier [71] (see Section 4 for the boundary gradient estimate) examined a slight but significant generalization of uniformly parabolic equations known as regularly parabolic equations; they form a subclass of the class of equations we studied in Section X.1. At about the same time, Ivanov [117, Section 2] looked at cylindrical domains with strictly convex cross-sections in analogy with [99]. Then Trudinger [307] (Theorems 5 through 8 and Corollary 4 are devoted to parabolic equations) gave a complete analog of the Serrin conditions for a cylindrical domain, with H_2 boundary values; however, unlike the previously mentioned authors, he did not pursue the sharpness of these conditions. In [210, Section 2], I provided a complete analog of Serrin's results for cylindrical domains even with H_δ boundary values, including the sharpness of the conditions (in [210, Section 3]). In fact, the boundary gradient estimates and the non-existence results in that paper (when specialized to the case $\delta = 2$) are proved by adding the simplest possible time-dependence to the elliptic arguments in [290, Section 18] and [101, Section 14.4]; non-existence in the one-dimensional case was first studied by Fillipov [82]. The extension to $\delta < 2$ in [210] follows ideas of Giusti [103, Section 3] and modified by me [200, Chapter II], both for elliptic problems. The investigation of non-cylindrical domains in this work is new, but all other considerations in this chapter, in particular the restrictions on $\langle\varphi\rangle_2$ come from [210]. Gradient bounds for linear equations in noncylindrical domains were also proved by Kamynin and Khimchenko [151, Theorem 1]. In fact, they also relaxed the Hölder modulus of continuity could be relaxed to a Dini modulus, and they showed that a boundary gradient estimate essentially implies a Dini modulus of continuity for the gradients of the boundary data (see the corollary to [151, Theorem 2]). The corresponding result for quasilinear equations depends on a more detailed analysis of the ordinary differential equation $f'' + h(f') = 0$ for general functions h, with zero initial condition. This analysis was carried in the elliptic case by Serrin [290] (see the proof of Theorem 8.1) for H_2 boundary data and by me [209, Section 2] for C^1 boundary data. The modifications needed to apply these results to parabolic equations are quite straightforward.

The arguments in Section X.3 are a combination of the Appendix of [227] (for the interior and exterior sphere condition) and [101, Section 14.6] (for the discussion of curvatures; note that Serrin [290, Section 3] assumed that $\partial\omega \in C^3$). Further information on the C^1 nature of the distance function can be found in [245] and the references therein.

When the modulus of continuity ω in Theorem 10.18 is a power function (so that φ is Hölder continuous), the same is true for ω^* under suitable additional conditions. For linear equations, this implication was proved by Cannon [43, Theorem 3] for cylindrical domains and by Kamynin [146, Theorem 2] for noncylindrical domains. For quasilinear elliptic equations, it was proved by Arkhipova [9] and Lieberman [201], and the method used in these papers easily carries over to parabolic equations.

Although the structure conditions described in this chapter are quite general, there are several detailed examinations of situations not covered by the results in this chapter. For example, Tersenov and Tersenov [301] show that the conditions $a = (\mathscr{E})$ can be replaced, for uniformly parabolic equations, by an inequality that allows a to grow arbitrarily in certain p directions provided an appropriate monotonicity condition is satisfied. Schulz and Williams [289] showed that, in the elliptic case, when the curvature conditions are violated, additional hypotheses still lead to boundary continuity estimates. Bell, Friedman, and Lacey [15] considered the equation $-u_t + \exp(u_x)u_{xx} = 0$ in $(0,L) \times (0,T)$ with zero initial data and time-dependent boundary data; for such an problem, pointwise bounds are needed on the boundary data. Fila and Lieberman [81] studied the equation

$$-u_t + u_{xx} + f(u_x) = 0 \text{ in } (0,L) \times (0,T)$$

with zero boundary data and nonnegative initial data, assuming that f is nonnegative and $f(s)s^{-2} \to \infty$ fast enough as $s \to \infty$; the model case being $f(s) = e^s$. In this case, solutions do not exist for all time despite a uniform bound on the solution. The gradient can only become infinite at $x = 0$, which it does at some finite time T. After this time, the solution continues to exist in a generalized sense, satisfying the equation and boundary condition only at $x = L$. This sort of generalized solution has also been studied (for other equations) by Iannelli and Vergara Caffarelli [114], Lichnewsky and Temam [199], Marcellini and Miller [248], Kawohl and Kutev [156], and Hayasida and Ikeda [110]. Bertsch, Dal Passo, and Franchi [18] looked at operators of the form $-u_t + (f(u_x))_x$, where f is increasing but bounded from above on $\mathbb{R}$; their main concern is with the behavior near a "top" point of the domain, such as the north pole of a space-time ball.

In [81], the nonlinearity in the equation fights against the sign of the solution in a sense which was exploited in the other direction by Crandall, Lions, and Souganidis [58]. They studied nonnegative solutions of the equation

$$-u_t + \Delta u + g(x, Du) = 0 \text{ in } \omega \times (0,\infty)$$

when g is nonpositive for large p and $g(x,0) \geq 0$. Provided g grows fast enough, they produce a unique solution which vanishes on $\partial\omega \times (0,\infty)$ and is infinite at $t = 0$. We investigate some of the properties of solutions of this equation in Exercise 10.3.

Finally, Oliker and Ural′tseva [273] study the equation

$$-u_t + [\delta^{ij} - v^i v^j] D_{ij} u = 0$$

in a cylindrical domain without the corresponding curvature condition on $\partial\omega$. They prescribe a generalized version of the boundary condition $u = 0$ on $S\Omega$ and show that there is a unique solution. Of particular relevance for this chapter is the boundary gradient estimate from [273, Theorem 2.3], which shows that, after a sufficiently long time, the solution attains the boundary values continuously and satisfies a boundary gradient estimate.

Exercises

10.1 Show that we can remove conditions (10.4b) and (10.14) from Theorem 10.10 when $\delta = 2$.

10.2 Show that Theorems 10.7 and 10.8 are sharp in the sense that $\mathscr{E}$ cannot be replaced by $|p|^\theta \mathscr{E}$ in conditions (10.15b), (10.15b)′, and (10.15b)″. In addition R cannot be replaced by a smaller constant in these conditions, and R_1 cannot be smaller than R.

10.3 Consider the initial-boundary value problem

$$-u_t + \Delta u + H(x, Du) = 0 \text{ in } \Omega, u = 0 \text{ on } S\Omega, u = \varphi \text{ on } B\Omega,$$

for $\mathscr{P}\Omega \in H_\delta$, $\varphi \in H_\delta$, $\varphi \geq 0$ in $B\Omega$, and $\varphi = 0$ on $C\Omega$. If $H(x,0) \geq 0$ for all x near $S\Omega$ and $H(x,p) \leq 0$ for all x near $S\Omega$ and all p with $|p| \geq p_0$, show that (10.7) holds with C depending only on n, φ and Ω. (Hint: note that $w \equiv 0$ is a subsolution.) What happens if the Laplace operator is replaced by a more general operator (of the form $a^{ij}(Du)D_{ij}u$)?

CHAPTER XI

GLOBAL AND LOCAL GRADIENT BOUNDS

Introduction

In this chapter, we discuss *a priori* estimates for the magnitude of the gradient of a solution of a quasilinear equation in terms of our previous estimates. These estimates correspond to the third step of our existence program. The method used here starts by differentiating the equation $Pu = 0$, and hence the structure conditions imposed in order to obtain a gradient bound usually involves the derivatives of the coefficients a^{ij} and a. In the divergence-structure case, some of the derivative conditions can be relaxed and other structures are used because of the different type of argument invoked. Again the Bernstein $\mathscr{E}$ function will play an important role in our estimates.

We also prove interior gradient estimates for solutions which do not necessarily have globally bounded gradient. Such estimates lead to existence theorems for the Cauchy-Dirichlet problem when the boundary values are only continuous. They will also reappear in our study of the oblique derivative problem.

1. Global gradient bounds for general equations

In Chapters III and IV, we obtained gradient bounds for simple equations by first differentiating the differential equation with respect to x^k for $k = 1, \dots, n$, then multiplying by $D_k u$, and finally summing over k. In this section, we generalize this idea. Suppose that u is a $C^{2,1}$ solution of

$$-u_t + a^{ij}(X,u,Du)D_{ij}u + a(X,u,Du) = 0 \tag{11.1}$$

in some domain Ω. To simplify the notation, we set

$$m = \inf_{\Omega} u, \ M = \sup_{\Omega} u.$$

Next we consider a strictly increasing function ψ_0 defined on $[m,M]$ and we write ψ for the inverse of ψ_0. Assuming that ψ and ψ_0 are C^3 on their domains, we define $\bar{u} = \psi_0(u)$ and $\omega = \psi''/(\psi')^2$, where ψ and its derivatives are evaluated at $\bar{u}$. Then we can rewrite (11.1) as

$$-\bar{u}_t + \psi' a^{ij} D_{ij}\bar{u} + \frac{1}{\psi'}(a + \omega\mathscr{E}) = 0. \tag{11.2}$$

Supposing that $Du \in C^{2,1}$, we differentiate (11.2) with respect to x^k, multiply by $D_k u$, and sum on k. If we set $v = |Du|^2$ and $\bar{v} = |D\bar{u}|^2$, we thus obtain

$$\begin{aligned}0 = &-\frac{1}{2}\bar{v}_t + \psi' a^{ij} D_{ijk}\bar{u} D_k \bar{u} \\ &+ a^{ij,r} D_{ij}\bar{u} D_{rk} u D_k \bar{u} + \psi' a_x^{ij} D_{ij}\bar{u}\bar{v} + a_k^{ij} D_{ij}\bar{u} D_k \bar{u} \\ &+ \frac{1}{\psi'}[a^r D_{rk} u D_k \bar{u} + \psi' a_z \bar{v} + a_k D_k \bar{u} + \omega(\mathscr{E}^r D_{rk} u D_k \bar{u} + \mathscr{E}_z \bar{v} + \mathscr{E}_k D_k \bar{u})] \\ &+ \left[\frac{\omega'}{\psi'}\mathscr{E} - \omega a - \omega^2 \mathscr{E}\right]\bar{v}\end{aligned}$$

where the superscript r denotes differentiation with respect to p_r, and the subscripts z and k denote differentiation with respect to z and x^k, respectively. To simplify this equation, we define the operators

$$\delta = D_z + |p|^{-2} p \cdot D_x, \; \bar{\delta} = p \cdot D_p,$$

and the vector b_0 by

$$b_0^r = \psi' a^{ij,r} D_{ij}\bar{u} + a^r + \omega \mathscr{E}^r.$$

It follows that

$$\begin{aligned}0 = &-\frac{1}{2}\bar{v}_t + \frac{1}{2}a^{ij} D_{ij}\bar{v} - a^{ij} D_{ik}\bar{u} D_{jk}\bar{u} \\ &+ \psi' \bar{v}(\omega \bar{\delta} a^{ij} + \delta a^{ij}) D_{ij}\bar{u} + \frac{1}{2} b_0^r D_r \bar{v} \\ &+ \left[\frac{\omega'}{\psi'}\mathscr{E} + \omega^2(\bar{\delta} - 1)\mathscr{E} + \omega(\delta \mathscr{E} + (\bar{\delta} - 1)a) + \delta a\right]\bar{v}.\end{aligned} \tag{11.3}$$

To simplify this expression still further, we assume the existence of a positive definite matrix-valued function (a_*^{ij}) and a vector-valued function (f_i) such that (a^{ij}) can be decomposed as

$$a^{ij}(X,z,p) = a_*^{ij}(X,z,p) + \frac{1}{2}[p_i f_j(X,z,p) + p_j f_i(X,z,p)]. \tag{11.4}$$

We always asume that a_*^{ij} and f_j are differentiable with respect to (x,z,p). Such a decomposition is always possible if we take $a_*^{ij} = a^{ij}$ and $f_j = 0$ (as we shall do for uniformly parabolic equations) but, for some equations, a suitable choice of these functions is crucial to our gradient estimate. With this decomposition, we can rewrite (11.3) as

$$\begin{aligned}0 = &-\frac{1}{2}\bar{v}_t + \frac{1}{2}a^{ij} D_{ij}\bar{v} + \frac{1}{2}b^r D_r \bar{v} \\ &+ \psi' \bar{v}(\omega \bar{\delta} a_*^{ij} + \delta a_*^{ij}) D_{ij}\bar{u} - a_*^{ij} D_{ik}\bar{u} D_{jk}\bar{u} \\ &+ \left[\frac{\omega'}{\psi'}\mathscr{E} + \omega^2(\bar{\delta} - 1)\mathscr{E} + \omega(\delta \mathscr{E} + (\bar{\delta} - 1)a) + \delta a\right]\bar{v},\end{aligned} \tag{11.5}$$

where

$$b^r = b_0^r + \psi'[-f_i D_{ir}\bar{u} + \omega\bar{\delta} f_r + (\delta+1) f_r].$$

Now we use Cauchy's inequality to estimate the terms involving second derivatives of $\bar{u}$:

$$\psi'\bar{v}\omega\bar{\delta}a_*^{ij}D_{ij}\bar{u} \leq \frac{1}{2}\lambda_*\sum|D_{ij}\bar{u}|^2 + \frac{v\bar{v}}{2\lambda_*}\sum(\bar{\delta}a_*^{ij})^2\omega^2,$$

and

$$\psi'\bar{v}\delta a_*^{ij}D_{ij}\bar{u} \leq \frac{1}{2}\lambda_*\sum|D_{ij}\bar{u}|^2 + \frac{v\bar{v}}{2\lambda_*}\sum(\delta a_*^{ij})^2.$$

Since

$$\lambda_*\sum|D_{ij}\bar{u}|^2 \leq a_*^{ij}D_{ik}\bar{u}D_{jk}\bar{u},$$

it follows from (11.5) that

$$-\bar{v}_t + a^{ij}D_{ij}\bar{v} + b^r D_r\bar{v} + c^0\mathscr{E}\bar{v} \geq 0, \tag{11.6}$$

where $c^0 = \omega'/\psi' + A\omega^2 + B\omega + C$, and the coefficients A, B, and C are given by

$$A = \frac{1}{\mathscr{E}}\left(\frac{v}{2\lambda_*}\sum(\bar{\delta}a_*^{ij})^2 + (\bar{\delta}-1)\mathscr{E}\right), \tag{11.7a}$$

$$B = \frac{1}{\mathscr{E}}\left(\delta\mathscr{E} + (\bar{\delta}-1)a\right), \tag{11.7b}$$

$$C = \frac{1}{\mathscr{E}}\left(\frac{v}{2\lambda_*}\sum(\delta a_*^{ij})^2 + \delta a\right). \tag{11.7c}$$

We can now describe conditions under which (11.6) provides a global gradient bound for solutions of (11.1). Suppose there is a function ψ such that $c^0\mathscr{E}$ is bounded above by some nonnegative constant k, at least for $X \in \Omega$, $z = u(X)$ and $|p| \geq L$ for some positive constant L. If we now define $\Omega_L = \{X \in \Omega : |Du| > L\}$, then the weak maximum principle implies that

$$\sup_{\Omega_L}\bar{v} \leq e^{kT}\sup_{\mathscr{P}\Omega_L}\bar{v},$$

and hence

$$\sup_{\Omega}|Du| \leq e^{kT/2}\left[\frac{\max\psi'}{\min\psi'}\right]\max\{L, \sup_{\mathscr{P}\Omega}|Du|\}. \tag{11.8}$$

In fact, (11.8) continues to hold even if we only assume that $u \in C^{2,1}$. To see why this is so, we rewrite (11.3) in the weak form

$$\begin{aligned}
0 = & \frac{1}{2}\int_{\omega(\tau)} \bar{v}\eta\,dx - \frac{1}{2}\int_{\Omega(\tau)} \bar{v}\eta_t\,dX \\
& + \int_{\Omega(\tau)} (\frac{1}{2}a^{ij}D_j\bar{v}D_i\eta + \eta a^{ij}D_{ik}\bar{u}D_{jk}\bar{u})\,dX \\
& + \frac{1}{2}\int_{\Omega(\tau)} (D_j(a^{ij})D_i\bar{v} - (\psi')^{-2}(\psi' a^{jk,r}D_{jk}\bar{u} + \omega\mathscr{E}^r)D_r v)\eta\,dX \\
& - \int_{\Omega(\tau)} [\psi'(\omega\bar{\delta}a^{ij} + \delta a^{ij})D_{ij}\bar{u} - \omega a + \omega^2\mathscr{E} + \frac{\omega'}{\psi'}\mathscr{E}]\bar{v}\eta\,dX \\
& - \int_{\Omega(\tau)} aD_r(D_r u\bar{v}\eta)\,dX.
\end{aligned}$$

If we approximate u in $C^{2,1}$ by a sequence of smooth functions (u_k), then this integral inequality is true with u_k replacing u in the arguments of a^{ij}, $\mathscr{E}$ and all their derivatives and a replaced by some function $a_k(X, u_k, Du_k)$. By the assumed $C^{2,1}$ convergence, we have that $a_k(X, u_k, Du_k) \to a$ uniformly in Ω and hence this integral inequality holds for $u \in C^{2,1}$. It follows that (11.6) holds in the weak form

$$-\bar{v}_t + D_i(a^{ij}D_j\bar{v}) + (b^i - D_j(a^{ij}))D_i\bar{v} + c^0\mathscr{E}\bar{v} \geq 0,$$

and then (11.8) follows from the weak maximum principle for weak solutions, Corollary 6.26.

Next, we examine conditions implying the upper bound for $c^0\mathscr{E}$. We first discuss conditions which guarantee that this quantity is nonpositive, which is equivalent to the condition $c^0 \leq 0$. To achieve this inequality, we first note that A, B, and C will be bounded above if and only if the numbers

$$A_\infty, B_\infty, C_\infty = \lim_{|p|\to\infty} \sup_{\Omega\times[m,M]} A, B, C \tag{11.7$'$}$$

are all finite. Assuming these quantities to be finite, we define $\chi = \omega \circ \psi^{-1}$ and note that $c^0 \leq 0$ for $|p|$ sufficiently large if there is a nonnegative solution of the Ricatti inequality

$$\chi' + A_\infty\chi^2 + B_\infty\chi + C_\infty + \varepsilon \leq 0 \tag{11.9}$$

on the interval $[m, M]$ for some $\varepsilon > 0$. Given a solution of (11.9), we see that $\chi = (\ln\phi)'$ for $\phi = \psi' \circ \psi^{-1}$ and then

$$\frac{\max\psi'}{\min\psi'} = \frac{\max\phi}{\min\phi}.$$

If $A_\infty < 0$, $C_\infty < 0$ or $B_\infty < -2(|A_\infty C_\infty|)^{1/2}$, then (at least for ε small enough) there is a positive constant κ which solves (11.9), in which case we have

$$\phi(z) = \exp(\kappa(z - m)).$$

If $A_\infty = 0$, $B_\infty > 0$, and $C_\infty > 0$, then (11.9) is solved for $\varepsilon = C_\infty$ by taking χ to be a solution of the linear ordinary differential equation $\chi' + B_\infty \chi + 2C_\infty = 0$. We find that

$$\chi = \frac{2C_\infty}{B_\infty}(\exp(B_\infty(M-z)-1)),$$

and hence

$$\phi = \exp(\frac{2C_\infty}{B_\infty}(M-z) + \frac{2C_\infty}{B_\infty^2}e^{B_\infty(M-z)} - \frac{2C_\infty}{B_\infty^2}e^{B_\infty(M-m)}).$$

If $A_\infty > 0, B_\infty > 0$ and $C_\infty = 0$, we set

$$\chi(z) = \exp((A_\infty + B_\infty + 1)(m-z)),$$

in which case (11.9) holds for $\varepsilon = \exp(-(A_\infty + B_\infty + 1)\operatorname{osc} u)$. Hence

$$\phi = \exp(-\frac{e^{(A_\infty+B_\infty+1)(m-z)}}{(A_\infty + B_\infty + 1)}).$$

The case $A_\infty > 0$, $B_\infty = -2(A_\infty C_\infty)^{1/2}$, $C_\infty > 0$ is easily reduced to the previous case. We leave the details to the reader.

A final useful case is when the oscillation of u is small. In this case, we take $\chi(z) = 4|C_\infty|(M-z)$ so

$$\phi(z) = \exp(-2|C_\infty|(M-z)^2)$$

and suppose that

$$4\operatorname{osc} u \leq (|A_\infty C_\infty| + B_\infty^2)^{-1/2}. \tag{11.10}$$

Then, for $\varepsilon = |C_\infty|$, a simple calculation yields (11.9). In this case, the gradient bound will depend also on the oscillation of u. A more careful analysis (see Exercise 11.1) can be used to increase the upper bound for $\operatorname{osc} u$ in (11.10), but this condition is adequate for our purposes.

We now collect the results of this calculations in the following theorem.

THEOREM 11.1. *Let $u \in C^{2,1}(\Omega)$ with $Du \in C(\overline{\Omega})$ and suppose that $Pu = 0$ in Ω. Suppose that P is parabolic at u and that the quantities A_∞, B_∞, and C_∞ defined by* (11.7) *and* ((11.7)′) *are finite. If either* (11.10) *holds or*

$$\min\{A_\infty, C_\infty, B_\infty + 2|A_\infty C_\infty|^{1/2}\} \leq 0, \tag{11.11}$$

then

$$\sup_\Omega |Du| \leq c_1, \tag{11.12}$$

where c_1 is a constant determined only by $\sup_{\mathcal{P}\Omega}|Du|$, A_∞, B_∞, C_∞, $\operatorname{osc} u$ and the limit behavior in (11.7)′.

The conditions implying that $c^0\mathscr{E}$ is bounded from above are much simpler to describe.

COROLLARY 11.2. *Let $u \in C^{2,1}(\Omega)$ with $Du \in C(\overline{\Omega})$ and suppose that $Pu = 0$ in Ω. If P is parabolic at u and if*

$$\lim_{|p|\to\infty} \sup_{\Omega\times[m,M]} C\mathscr{E} < \infty, \tag{11.7$''$}$$

then (11.12) *holds with c_1 depending only on* $\sup_{\mathscr{P}\Omega}|u|$, $\operatorname{osc} u$, T *and the limit behavior in* (11.7)$''$.

PROOF. Now we can take $\psi(z) = z$ and apply Corollary 2.5. □

A special case of Corollary 11.2 is when the limit in (11.7)$''$ is negative (or more generally when $C \leq 0$ for large p). In this situation, (11.8) holds with $k = 0$. In other words, $|Du|$ attains its maximum on $\mathscr{P}\Omega$.

2. Examples

In this section, we consider some examples of parabolic equations which illustrate the structure conditions described in the previous section.

2.1. Uniformly parabolic equations. Here we assume that the ratio Λ/λ of maximum to minimum eigenvalues of the matrix (a^{ij}) is bounded. We also suppose that

$$|a| + |p|^2\,|\bar{\delta}a^{ij}| + |\bar{\delta}a| = O(\lambda\,|p|^2), \tag{11.13a}$$

$$|\delta a^{ij}| = o(\lambda), \delta a \leq o(\lambda\,|p|^2) \tag{11.13b}$$

as $|p| \to \infty$. If we replace (11.13b) by the weaker conditions

$$|\delta a^{ij}| = O(\lambda), \delta a \leq O(\lambda\,|p|^2), \tag{11.13b$'$}$$

we have what are sometimes called the natural conditions for the operator P (See [183, Section VI.3]). If (11.13a,b) hold, then a gradient bound follows from Theorem 11.1 by taking $a_*^{ij} = a^{ij}$ and observing that the conditions give $C_\infty \leq 0$. The bound depends on $\sup_{\mathscr{P}\Omega}|Du|$ and on the limit behavior in (11.13). Alternatively, if

$$\delta a \leq O(1), |p|\,|\delta a^{ij}| = O(\lambda^{1/2}) \tag{11.14}$$

as $|p| \to \infty$, then we obtain a gradient bound by virtue of Corollary 11.2.

In either case, our gradient bound is actually independent of the maximum of the ratio Λ/λ; however, for interior gradient bounds and for gradient bounds under the natural conditions, this ratio will play a key role.

2.2. Prescribed mean curvature equations. Now we consider equations of the form

$$u_t = (1+|Du|^2)^{\tau/2}(\Delta u - \nu_i\nu_j D_{ij}u + nH(X,u)(1+|Du|^2)^{1/2}),$$

where $\nu = Du/(1+|Du|^2)^{1/2}$ and $\tau \in \mathbb{R}$. If $\tau = 0$, then we take

$$a_*^{ij} = \delta^{ij}, f_i = \frac{p_i}{1+|p|^2},$$

and then compute

$$A = -1 + \frac{2}{1+|p|^2}, \qquad B = -\frac{(1+|p|^2)^{1/2}}{|p|^2}nH,$$
$$C = \frac{n(1+|p|^2)^{3/2}}{|p|^2}H_z + \frac{n(1+|p|^2)^{3/2}}{|p|^4}p\cdot H_x.$$

If $H_z \le 0$, then we have $A_\infty = -1$, $B_\infty = 0$, and $C_\infty = n\sup|H_x|$, and we obtain a gradient bound from the case $A_\infty < 0$ in Theorem 11.1.

Since we always have

$$C\mathscr{E} = n(1+|p|^2)^{(\tau+1)/2}[H_z + \frac{p\cdot H_x}{|p|^2}],$$

we can obtain a gradient bound from Corollary 11.2 provided one of the following sets of conditions holds:

(1) $\tau \le 0$ and $H_z \le 0$,
(2) $\tau \le -1$,
(3) $H_z < 0$.

In cases 1 and 2, the gradient bound depends on T. When $\tau = 0$, this equation arises in the study of flow by mean curvature; see [113]. Case 2 includes the equation

$$u_t = \operatorname{div}\frac{Du}{(1+|Du|^2)^{1/2}} + nH(X,u)$$

when H is Lipschitz with respect to x and z. We shall return to this equation in Section XI.5.

2.3. A false mean curvature equation. Our final example is the equation

$$u_t = g((1+|Du|^2)^{1/2})(\Delta u + D_iuD_juD_{ij}u + nH(X,u)(1+|Du|^2)^{1/2}).$$

As noted in Chapter VIII, the matrix (a^{ij}) for this operator has the same ratio of maximum to minimum eigenvalues as the previous one. If we take $a_*^{ij} = g((1+$

$|p|^2)^{1/2})\delta^{ij}$, this time we find that

$$A = \frac{n^2(g')^2|p|^4}{2g^2(1+|p|^2)^2} + 3 - \frac{2}{1+|p|^2} + \frac{g'|p|^2}{g(1+|p|^2)^{1/2}},$$

$$B = nH[\frac{|p|^2}{(1+|p|^2)^{1/2}} - 1 + |p|^2\frac{g'}{g}]\frac{1}{|p|^2+|p|^4},$$

$$C = \frac{n}{|p|^2(1+|p|^2)^{1/2}}(H_z + \frac{p\cdot H_x}{|p|^2}).$$

Hence A_∞ is finite as long as there are positive constants g_0 and L such that $s|g'(s)| \le g_0 g(s)$ for $s \ge L$. Under this assumption, which is true if $g(s) = s^\alpha$ for some $\alpha \in \mathbb{R}$, $B_\infty = 0$ and $C_\infty = 0$ for any choice of g. Hence Theorem 11.1 gives a gradient bound here, too. In Section XI.5, we shall consider this example when $g(s) = \exp(\frac{1}{2}s^2)$, in which case the equation can be written in divergence form. Gradient bounds will be proved in this case also.

3. Local gradient bounds

To derive gradient bounds which do not depend on the maximum of $|Du|$ over all of $\mathscr{P}\Omega$, we multiply $\bar{v}$ by a cut-off function η and analyze the resultant equation for $w = \eta\bar{v}$. We consider several different configurations since each uses slightly different additional structure conditions.

First, we examine estimates which are local in time and global in space. Specifically, we assume a known gradient estimate on $S\Omega$ but not on $B\Omega$. Then we can take η to depend only on t, so

$$Dw = \eta D\bar{v}, D^2w = \eta D^2\bar{v}, w_t = \eta_t\bar{v} + \eta\bar{v}_t.$$

From (11.6), we infer that

$$-w_t + a^{ij}D_{ij}w + b^rD_rw + (c^0 + C_1)\mathscr{E}w \ge 0$$

with b^r and c^0 as in Section XI.1 and $C_1 = \eta_t/(\eta\mathscr{E})$. If we set $\zeta_1 = t^+$ and take $\eta = \zeta_1^k$ for some $k \ge 1$ to be chosen, then

$$C_1 = \frac{k}{\zeta_1\mathscr{E}} = \frac{k\bar{v}^{1/k}}{w^{1/k}\mathscr{E}}.$$

If we suppose that there are positive constants L_1, θ, and μ with $\theta \le 1$ such that

$$|p|^{2\theta} \le \mu\mathscr{E} \tag{11.15}$$

for $|p| \ge L_1$, then

$$C_1 \le \mu k w^{-1/k}(\max\psi')^{2/k},$$

and hence $C_1 \le \varepsilon$ for any given positive ε on the set where $w \ge L'$ provided L' is sufficiently large (determined from ψ). From the considerations of Section XI.1, we conclude that

$$|Du(X)| \leq c_2[1 + t^{-1/(2\theta)}] \tag{11.16}$$

under the hypotheses of Theorem 11.1 (along with (11.15)) and c_2 is determined by L' and the same quantities as was c_1. In fact, the form of estimate (11.16) can be further refined. For this refinement, we set

$$K = (32|A_\infty| + 16)^{1/(2\theta)}, \sigma_0 = K^\theta \max\{1, |C_\infty|^{1/2}, |B_\infty|\}.$$

Now we set $\sigma = \operatorname{osc} u$ and suppose that the maximum of w occurs at a point for which $|Du| \geq K[\mu/\theta]^{1/(2\theta)}[\sigma/t^{1/2}]^{1/\theta}$. If $\sigma \geq 1/\sigma_0$, then we immediately infer from (11.16) that

$$|Du(X)| \leq c_2[1 + \left(\frac{\operatorname{osc} u}{t^{1/2}}\right)^{1/\theta}]. \tag{11.16$'$}$$

On the other hand if $\sigma < 1/\sigma_0$, then we have $B_\infty \leq 1/(K^\theta \sigma)$ and $C_\infty + C_1 \leq 2/(K^\theta \sigma)^2$, so by imitating the proof of the global gradient bound when (11.10) holds, we see that $c^0 + C_1 \leq 0$ provided

$$4\sigma \leq \left(\frac{2|A_\infty|^2}{K^\theta \sigma} + \frac{1}{(K^\theta \sigma)^2}\right)^{-1/2}.$$

Since the right side of this inequality is exactly 4σ, it follows that w cannot attain its maximum at a point where $|Du| \geq K[\mu/\theta]^{1/(2\theta)}[\sigma/t^{1/2}]^{1/\theta}$. In addition, the corresponding function ϕ is given by

$$\phi(z) = \exp(-\frac{(z-m)^2}{K^{2\theta}\sigma^2}),$$

so the ratio $\max\phi/\min\phi$ is bounded by a constant independent of σ, and hence (11.16)$'$ holds in this case also.

For a gradient bound which is local in space, but not in time, we assume that a gradient bound is already known on $B\Omega$ but not on $S\Omega$. In fact, our method assumes that Ω contains a cylinder of the form $B(R) \times (0,T)$. Now we take η to depend on x but not on t and we compute

$$w_t = \eta \bar{v}_t, D_i w = \eta D_i \bar{v} + \bar{v} D_i \eta,$$
$$D_{ij} w = \eta D_{ij} v + D_i \eta D_j \bar{v} + D_j \eta D_i \bar{v} + \bar{v} D_{ij} \eta.$$

Using these identities in (11.5), we obtain

$$\begin{aligned}
0 = &-\frac{1}{2}w_t + \frac{1}{2}a^{ij}D_{ij}w + \frac{1}{2}(b^i - \frac{2}{\eta}a^{ij}D_j\eta)D_jw \\
&- \eta a_*^{ij}D_{ik}\bar{u}D_{jk}\bar{u} + \psi' w[\omega\bar{\delta}a_*^{ij} + \delta a_*^{ij}]D_{ij}\bar{u} \\
&+ \left[\frac{\omega'}{\psi'} + \omega^2(\bar{\delta}-1)\mathscr{E} + \omega(\delta\mathscr{E} + (\delta-1)a) + \delta a\right]w \\
&+ \left[\frac{1}{\eta^2}a^{ij}D_i\eta D_j\eta - \frac{1}{2\eta}a^{ij}D_{ij}\eta - \frac{1}{2\eta}b^iD_i\eta\right]w.
\end{aligned}$$

Next, we set $f_j^i = \partial f_j/\partial p_i$, and we write b^i in the form

$$\begin{aligned}
b^i = \psi' a_*^{kj,i}D_{kj}\bar{u} + \frac{1}{\eta}f_k^iD_kw - \frac{\bar{v}}{\eta}f_k^iD_k\eta \\
+ \omega(\delta f_i + \mathscr{E}^i) + (\delta+1)f_i + a^i.
\end{aligned}$$

Substituting this expression into the previous equation and defining

$$\delta_1 = -\frac{D\eta}{2\eta}\cdot D_p$$

and

$$b_2^i = b^i - \frac{2}{\eta}a^{ij}D_j\eta - v\delta_1 f_i$$

yields

$$-w_t + a^{ij}D_{ij}w + b_2^iD_iw + 2(c^0 + B_0\omega + C_0)\mathscr{E}w \geq 0$$

for c^0 as before, and

$$\begin{aligned}
B_0 &= -\frac{1}{\mathscr{E}}\frac{|p|^2}{2\eta}D_i\eta(\bar{\delta}f_i + \mathscr{E}^i), \\
C_0 &= \frac{1}{\mathscr{E}}[\frac{|p|^2}{4\lambda_*}\sum(\delta_1a_*^{ij})(\delta_1a_*^{ij} + 2\delta a_*^{ij}) \\
&\quad + \frac{v}{2\eta}D_i\eta(\delta f_i + D\eta\cdot f_{i,p}) + \frac{1}{\eta^2}a^{ij}D_i\eta D_j\eta - \frac{1}{2\eta}a^{ij}D_{ij}\eta].
\end{aligned}$$

Now we suppose that there is a constant $\theta \in (0,1]$ such that

$$|p|^{\theta+1}\left|a_{*,p}^{ij}\right| = O((\lambda_*\mathscr{E})^{1/2}), \tag{11.17a}$$

$$|p|^{2\theta}\Lambda + |p|^\theta\left(\left|\mathscr{E}_p\right| + \left|a_p\right|\right) = O(\mathscr{E}), \tag{11.17b}$$

$$|p|^{2\theta}\left|f_p\right| + |p|^\theta\left(\left|(\delta+1)f\right| + \left|\bar{\delta}f\right|\right) = O(\mathscr{E}). \tag{11.17c}$$

Then we take $\zeta_2 = 1 - |x-x_0|^2/R^2$ and $\eta = \zeta_2^{2/\theta}$ to obtain

$$B_0 \leq \frac{c}{Rw^{\theta/2}}, C_0 \leq c(\frac{1}{R^2w^\theta} + \frac{1}{Rw^{\theta/2}}).$$

As before, these inequalities lead to the estimate

$$|Du(x_0,t)| \leq c_3(1+\left(\frac{\operatorname{osc} u}{R}\right)^{1/\theta}) \tag{11.18}$$

with c_3 determined by the same quantities as c_1 and also by the limit behavior in (11.17) assuming all the hypotheses in Theorem 11.1 are satisfied.

Finally if conditions (11.15) and (11.17) hold in some cylinder $Q(2R)$, then we can use $\eta = (\zeta_1\zeta_2)^{2/\theta}$ to infer that

$$|Du| \leq c_4(1+\left(\frac{\operatorname{osc} u}{R}\right)^{1/\theta}) \tag{11.19}$$

in $Q(R)$.

The next theorem collects our various interior gradient bounds.

THEOREM 11.3. *Let $u \in C^{2,1}(\Omega)$ with $Pu = 0$ in Ω and P parabolic at u. Suppose that the quantities A_∞, B_∞, and C_∞ defined by* (11.7) *and* (11.7)$'$ *are finite and that* (11.10) *or* (11.11) *holds.*
(a) If also (11.15) *holds, then*

$$\sup_{\Omega(t_2)} |Du(\cdot,t_2)| \leq c_2[1+\frac{\sigma_1}{(t_2-t_1)^{1/2}}]^{1/\theta} \tag{11.20}$$

for some c_2 determined by $\sup_{\mathscr{P}\Omega\cap\{t_1<t<t_2\}}|Du|$, *$A_\infty$, B_∞, C_∞ and the limit behavior in* (11.7), (11.7)$'$.
(b) If also (11.17a,b,c) *hold in $Q = B(x_0,R)\times(0,T)$ for some positive R and T, holds, then* (11.18) *holds with c_3 determined by* $\sup_{BQ}|Du|$, *A_∞, B_∞, C_∞ and the limit behavior in* (11.7), (11.7)$'$.
(c) If also (11.15) *and* (11.17a,b,c) *hold in $Q(2R)$, then* (11.18) *holds with c_3 determined by A_∞, B_∞, C_∞ and the limit behavior in* (11.7), (11.7)$'$.

We now see how these conditions pertain to the examples in Section XI.2. For example (i), we infer a local gradient bound in space if conditions (11.13) and

$$|p|\,|a^{ij}_p| + |p|^{-1}\,|a_p| = O(\lambda) \tag{11.21}$$

hold or if (11.14) and (11.21) hold. For a gradient bound which is local in time, we also require $1 = O(\lambda)$. Then (11.15) holds with $\theta = 1$, and hence so does (11.19).

If we assume that conditions (11.13a), (11.13b)$'$, and (11.21) hold and if there are constants λ_0 and λ_1 such that $\lambda_0 \leq \lambda \leq \lambda_1$, then a gradient bound is still valid. The proof uses the following Hölder estimate.

LEMMA 11.4. *Let $u \in W^{2,1}_{n+1}(Q(R))$ satisfy the inequality*

$$|Lu| \leq \lambda\mu_0|Du|^2 + f \text{ in } Q(R) \tag{11.22}$$

for some function $f \in L^{n+1}$ and some nonnegative constant μ_0, where $Lu = -u_t + a^{ij}(X)D_{ij}u$ and the matrix (a^{ij}) satisfies the inequalities

$$\lambda |\xi|^2 \le a^{ij}\xi_i\xi_j \le \Lambda |\xi|^2 \tag{11.23}$$

for some positive constants λ and Λ. Then there are positive constants α and c_5 determined only by n, Λ, λ, and $\mu_0 \sup|u|$ such that

$$\operatorname*{osc}_{Q(r)} u \le c_5 \left(\frac{r}{R}\right)^\alpha \left(\operatorname*{osc}_{Q(R)} u + R^{n/(n+1)} \|f\|_{n+1,Q(R)}\right) \tag{11.24}$$

for any $r \in (0,R)$.

PROOF. Suppose first that v is nonnegative and satisfies $Lv \le \lambda\mu_0 |Dv|^2 + f$ in $Q(r)$. If we set $w = (1 - \exp(-\mu_0 v))/\mu_0$, it follows that $Lw \le f$ in $Q(r)$ and then the weak Harnack inequality Theorem 7.38 implies that

$$\left(\frac{1}{|\Theta(r)|}\int_{\Theta(r)} w^p\,dX\right)^{1/p} \le c(\inf_{Q(r)} w + r^{n/(n+1)} \|f\|_{n+1}).$$

Setting $M = \sup v$, we have that $w \le v \le \exp(\mu_0 M)w$ in $Q(r)$ and hence

$$\left(\frac{1}{|\Theta(r)|}\int_{\Theta(r)} v^p\,dX\right)^{1/p} \le c(\inf_{Q(r)} v + r^{n/(n+1)} \|f\|_{n+1})$$

with c now depending also on $\mu_0 M$. From this estimate, the Hölder estimate follows by the argument of Theorem 6.28. □

Combining this Hölder estimate with our previous gradient bounds, we conclude the following global gradient bound for uniformly elliptic equations.

THEOREM 11.5. *Suppose $u \in C^{2,1}(\Omega)$ satisfies (11.1) in Ω. If there are positive constants L, λ_0, and μ such that*

$$|\delta a^{ij}| + |p|\,|a^{ij}_p| \le \mu\lambda, \tag{11.25a}$$

$$\delta a \le \mu\lambda |p|^2, \tag{11.25b}$$

$$|p|\,|a_p| + |a| \le \mu\lambda |p|^2 \tag{11.25c}$$

$$\lambda_0 \le \lambda \le \Lambda \le 1/\lambda_0 \tag{11.25d}$$

for $|p| \ge L$, then there is a constant c_6 determined only by $\operatorname{osc} u$, λ_0, and μ such that for any $X \in \Omega$, we have

$$|Du(X)| \le c_6 \left(L + \frac{\operatorname{osc}_{Q(d(X))} u}{d(X)}\right). \tag{11.26}$$

If also $Du \in C(\overline{\Omega})$, then

$$\sup_\Omega |Du| \le c_7 (L + \sup_{\mathcal{P}\Omega} |Du|) \tag{11.27}$$

with c_7 depending on $\operatorname{osc} u$, λ_0 and μ.

In fact, a global gradient bound holds without the assumption of global continuity for Du; instead, we can use the boundary Lipschitz estimate of Chapter X along with estimate (11.26) in an imitation of the proof of Theorem 6.33. We defer the details to Exercise 11.2.

For example (ii), condition (11.15) fails if $\tau \le 0$, so we cannot prove a local in time estimate unless $\delta H \le 0$ and $\tau > 0$. In addition, (11.17b) always fails, so we cannot prove a local in space gradient estimate using this method. In Section XI.6, we prove such a local gradient bound for the case $\tau = -1$ by writing the equation in divergence form.

In example (iii), let us suppose that $g(s) = s^\alpha$ for some real number α. Then (11.15) holds if and only if $\alpha > -4$ so this inequality implies local in time estimates, while (11.17a–c) always hold with $\theta = 1$, so we always have local in space estimates.

4. The Sobolev theorem of Michael and Simon

In this section, we prove a Sobolev-type inequality for functions defined on manifolds in $\mathbb{R}^m$ which is particularly useful in deriving gradient bounds for nonuniformly parabolic equations.

Let m and n be integers with $2 \le n \le m$, let U be an open subset of $\mathbb{R}^m$ and let M be a subset of U. Let μ be a (nonnegative) measure defined on sets of the form $M \cap B$ for B a Borel subset of $\mathbb{R}^m$, and suppose $\mu(M \cap K)$ is finite whenever K is compact. Finally, let (g^{ij}) and $\mathscr{H}$ be, respectively, a symmetric-matrix-valued and a vector-valued $L^1_{\text{loc}}(M; d\mu)$ function, and define the operator δ by $\delta_i h = g^{ij} D_j h$ for any $h \in W^{1,2}(U)$. Suppose also that

$$\sum_{i=1}^{m} g^{ii}(x) = n, \tag{11.28a}$$

$$0 \le g^{ij}(x)\xi_i\xi_j \le |\xi|^2 \tag{11.28b}$$

for almost all $x \in M$ (with respect to the measure μ) and all $\xi \in \mathbb{R}^m$,

$$\int_M [\delta h + h\mathscr{H}]\, d\mu = 0 \tag{11.29}$$

for any $h \in C^1(U)$ with compact support, and

$$\limsup_{\rho \to 0^+} \frac{\mu(B(\xi,\rho))}{\omega \rho^n} \ge 1 \tag{11.30}$$

for all $\xi \in M$.

A simple example of U, M, μ, (g^{ij}), and $\mathscr{H}$ satisfying all these hypotheses (which is the only one we actually need) is created by considering a function $u \in C^2(\Omega)$ for some open set $\Omega \in \mathbb{R}^n$. We then set $m = n+1$ and $U = \Omega \times \mathbb{R}$ and take M to be the graph of u. If we define $v = (1 + |Du|^2)^{1/2}$ and $\nu = (Du, -1)/v$ (so that ν is the downward pointing unit normal to M), then we define $g^{ij} = \delta^{ij} - \nu_i \nu_j$

and $\mathscr{H} = g^{ij}D_{ij}u\nu/(nv)$, and μ is defined by $d\mu = v\,dx$. Note that δ is just differentiation in the tangent plane to M and $\mathscr{H}$ is the mean curvature vector of the graph.

To prove our Sobolev-type inequalities, we begin by examining the growth rate of the integrals of certain functions.

LEMMA 11.6. *Let $\lambda \in C^1(\mathbb{R})$ be an increasing function which vanishes on the negative real axis. For h a nonnegative function in $C^1(U)$ with compact support and $\xi \in M$, define $r = r(x) = |x - \xi|$,*

$$\varphi(\rho) = \int_M h(x)\lambda(\rho - r)\,d\mu(x), \tag{11.31a}$$

$$\psi(\rho) = \int_M [|\delta h(x)| + h(x)\,|\mathscr{H}(x)|]\lambda(\rho - r)\,d\mu(x). \tag{11.31b}$$

Then

$$-\frac{d}{d\rho}\left(\frac{\varphi(\rho)}{\rho^n}\right) \leq \frac{\psi(\rho)}{\rho^n}. \tag{11.32}$$

PROOF. Set

$$I = \int_M \delta_i((x_i - \xi_i)\lambda(\rho - r)h)\,d\mu,$$

and write λ for $\lambda(\rho - r)$. Then by direct calculation,

$$\begin{aligned} I &= n\varphi(\rho) + \int_M (-r\lambda' r^{-2} g^{ij}(x_i - \xi_i)(x_j - \xi_j) + \lambda(x_i - \xi_i)\delta_i h)\,d\mu \\ &\geq n\varphi - \rho\varphi' - \rho\int_M |\delta h|\,\lambda\,d\mu \end{aligned}$$

by (11.28) and (11.31) since $r \leq \rho$ wherever λ is nonzero. On the other hand, (11.29) gives

$$I = -\int_M h\lambda[(x_i - \xi_i)\mathscr{H}_i]\,d\mu \leq \rho\int_M h\lambda\,|\mathscr{H}|\,d\mu.$$

Combining these estimates and using the definition of ψ gives $n\varphi - \rho\varphi' \leq \rho\psi$, and (11.32) is a simple algebraic consequence of this inequality. □

Next we show that the integral of h near a point where it is large can be estimated in terms of the integrals of $|\delta h|$ and $h|\mathscr{H}|$.

LEMMA 11.7. *With h as in Lemma 11.6, suppose that $h(\xi) \geq 1$, and define*

$$\bar{\varphi}(\rho) = \int_{M\cap B(\xi,\rho)} h\,d\mu, \tag{11.33a}$$

$$\bar{\psi}(\rho) = \int_{M\cap B(\xi,\rho)} [|\delta h| + h|\mathscr{H}|]\,d\mu, \tag{11.33b}$$

$$R = 2\left(\frac{1}{\omega_n}\int_M h\,d\mu\right)^{1/n}. \tag{11.34}$$

Then there is $\rho \in (0,R)$ *such that*

$$\bar{\varphi}(4\rho) \le 4^n R \bar{\psi}(\rho). \tag{11.35}$$

PROOF. With λ, φ, and ψ as in Lemma 11.6 and $\sigma \in (0,R)$, we integrate (11.32) from σ to R to obtain

$$\sigma^{-n}\varphi(\sigma) \le R^{-n}\varphi(R) + \int_0^R \rho^{-n}\psi(\rho)\,d\rho.$$

Now for $\varepsilon \in (0,\sigma)$, choose λ so that $\lambda(t) = 1$ for $t \ge \varepsilon$. It follows that

$$\sigma^{-n}\bar{\varphi}(\sigma-\varepsilon) \le R^{-n}\bar{\varphi}(R) + \int_0^R \rho^{-n}\bar{\psi}(\rho)\,d\rho.$$

Since σ and ε are arbitrary, we can send ε to zero in this inequality and then take the supremum over σ to infer that

$$\sup_{0<\sigma<R} \sigma^{-n}\bar{\varphi}(\sigma) \le R^{-n}\bar{\varphi}(R) + \int_0^R \rho^{-n}\bar{\psi}(\rho)\,d\rho.$$

Let us denote the supremum here by S. If (11.35) is not true for all $\rho \in (0,R)$, then

$$\int_0^R \rho^{-n}\bar{\psi}(\rho)\,d\rho \le \frac{2}{4^n R}\int_0^R \bar{\varphi}(4\rho)\,d\rho \le \frac{S}{2} + \frac{1}{2R}\int_R^\infty s^{-n}\bar{\varphi}(s)\,ds.$$

Now we note that $\bar{\varphi}(s) \le 2^{-n}\omega_n R^n$, so

$$\int_R^\infty s^{-n}\bar{\varphi}(s)\,ds \le 2^{-n}\omega_n R/(n-1),$$

and hence (after some rearrangement),

$$S \le 2\omega_n\Big(2^{-n} + \frac{1}{2^n(n-1)}\Big) < \omega_n.$$

This inequality contradicts (11.30) because $h(\xi) \ge 1$ and hence (11.35) holds. □

We are now ready to prove our Sobolev inequality.

THEOREM 11.8. *Under the hypotheses* (11.28), (11.29), *and* (11.30) *we have*

$$\Big(\int_M h^{n/(n-1)}\,d\mu\Big)^{(n-1)/n} \le \frac{4^{n+1}}{\omega_n^{1/n}}\int_M (|\delta h| + h|\mathscr{H}|)\,d\mu \tag{11.36}$$

for any nonnegative $h \in C^1(U)$ *with compact support.*

PROOF. Write A for the subset of M on which $h \ge 1$, define R by (11.34), and for i a positive integer, set $\rho_i = R2^{1-i}$. With $\bar{\varphi}$ and $\bar{\psi}$ as in Lemma 11.7, define

$$A_i = \{\xi \in A : \bar{\varphi}(4\rho) \le 2^{2n-1}R\bar{\psi}(\rho) \text{ for some } \rho \in (\tfrac{1}{2}\rho_i, \rho_i]\},$$

and note that Lemma 11.7 implies that $A = \bigcup_{i\ge 1} A_i$.

Next, define sets B_k inductively as follows. Let B_1 be a finite subset of A_1 such that

$$A_1 \subset \bigcup_{\xi \in B_1} B(\xi, 2\rho_1)$$

and $B(\xi_1, \rho_1)$ and $B(\xi_2, \rho_1)$ are disjoint for distinct points ξ_1 and ξ_2 in B_1. If $B_1, \dots, B_{k-1}$ are given, we take B_k to be a finite subset of

$$A_k \setminus \cup_{\xi \in B_{k-1}} B(\xi, 2\rho_{k-1})$$

such that

$$A_k \setminus \cup_{\xi \in B_{k-1}} B(\xi, 2\rho_{k-1}) \subset \bigcup_{\xi \in B_k} B(\xi, 2\rho_k),$$

and $B(\xi_1, \rho_k)$ and $B(\xi_2, \rho_k)$ are disjoint for distinct points ξ_1 and ξ_2 in B_k. It then follows that $B_k \subset A_k$, that

$$A \subset \cup_{k=1}^{\infty} \cup_{\xi \in B_k} B(\xi, 2\rho_k),$$

and that $B(\xi_1, \rho_j)$ and $B(\xi_2, \rho_k)$ are disjoint if $\xi_1 \in B_j$ and $\xi_2 \in B_k$ are distinct points. (These points are automatically distinct if $j \neq k$.) Finally, define

$$S(k) = \cup_{\xi \in B_k} B(\xi, 2\rho_k), s(k) = \cup_{\xi \in B_k} B(\xi, \rho_k).$$

Hence, for any $\xi \in B_k$, there is $\rho \in (\frac{1}{2}\rho_k, \rho_k]$ such that $\bar{\varphi}(4\rho) \leq 2^{2n-1} R \bar{\psi}(\rho)$. Since $\bar{\varphi}$ and $\bar{\psi}$ are both increasing, it follows that

$$\bar{\varphi}(2\rho_k) \leq 2^{2n-1} R \bar{\psi}(\rho_k).$$

If we now sum this inequality over all $\xi \in B_k$ and then sum the resulting inequality over k, we see that

$$\mu(A) \leq \int_A h \, d\mu \leq 2^{2n-1} R \int_M (|\delta h| + h|\mathscr{H}|) \, d\mu \tag{11.37}$$

because $s(k)$ and $s(j)$ are disjoint if $k \neq j$, and $h \geq 1$ on A.

Now, let α and ε be positive constants and let λ be an increasing function on $\mathbb{R}$ such that $\lambda(s) = 0$ if $s \leq -\varepsilon$ and $\lambda(s) = 1$ if $s \geq 0$. Replacing h by $\lambda(h - \alpha)$ in(11.37) and writing A_α for the subset of M on which $h \geq \alpha$, we see that

$$\mu(A_\alpha) \leq 4^n \omega_n^{-1/n} \left(\int_M \lambda \, d\mu\right)^{1/n} \int_M (\lambda' |\delta h| + \lambda |\mathscr{H}|) \, d\mu$$

because $\lambda(h - \alpha) = 1$ whenever $h \geq \alpha$. Now we multiply this inequality by $\alpha^{1/(n-1)}$ and note that $\lambda \leq 1$ on M and $\lambda = 0$ whenever $h \leq \alpha - \varepsilon$, so

$$\alpha^{1/(n-1)} \left(\int_M \lambda \, d\mu\right)^{1/n} \leq \left(\int_M (h + \varepsilon)^{n/(n-1)} \, d\mu\right)^{1/n}.$$

It follows that

$$\alpha^{1/(n-1)} \mu(A_\alpha) \leq 4^n \omega_n^{-1/n} \left(\int_M (h + \varepsilon)^{n/(n-1)} \, d\mu\right)^{1/n} \int_M (\lambda' |\delta h| + \lambda |\mathscr{H}|) \, d\mu.$$

Now we integrate this inequality with respect to α and note that

$$\int_0^\infty \lambda'(h-\alpha)\,d\alpha = 1,$$

$$\int_0^\infty \lambda(h-\alpha)\,d\alpha \leq \int_0^{h+\varepsilon} 1\,d\alpha = h+\varepsilon.$$

Combining all these estimates and using (7.9) gives

$$\frac{n-1}{n}\int_M h^{n/(n-1)}\,d\mu \leq 4^n\omega_n^{-1/n}\left(\int_M (h+\varepsilon)^{n/(n-1)}\,d\mu\right)^{1/n} \times \int_M [|\delta h| + (h+\varepsilon)\,|\mathcal{H}|]\,d\mu.$$

Now we send ε to zero and simplify the resultant inequality, taking into account that $n \geq 2$. □

The proof of Corollary 6.10 then gives the following weighted analog of Theorem 11.8.

COROLLARY 11.9. *Suppose that g is a nonnegative $L^\infty(M)$ function and let $N > 2$ if $n = 2$ and $N = n$ if $n > 2$. Then*

$$\int_M h^{2(N+2)/N} g\,d\mu \leq c(N)(\int_M h^2 g^{N/2}\,d\mu)^{2/N} \times (\int_M (|\delta h|^2 + h^2\,|\mathcal{H}|^2)\,d\mu)^{n/N}(\int_M h^2\,d\mu)^{1-n/N} \tag{11.38}$$

for any $h \in C^1(U)$ with compact support.

5. Estimates for equations in divergence form

When the parabolic equation (11.1) can be written in divergence form, it is possible to infer gradient bounds by suitably modifying the hypotheses from Sections XI.1 and XI.3. To illustrate this point in a particularly simple case, we suppose that u is a $C^{2,1}$ solution of

$$-u_t + \operatorname{div} A(X,u,Du) + B(X,u,Du) = 0 \text{ in } \Omega \tag{11.39}$$

with $Du \in C(\overline{\Omega})$. Now define

$$C_k^i = A_z^i D_k u + A_k^i + B\delta_k^i$$

and suppose that there are nonnegative constant β and L such that

$$\sum_{i,k} |C_k^i|^2 \leq \beta\lambda\,|Du| \tag{11.40}$$

for $|Du| \geq L$. By differentiating (11.39) with respect to x^k, multiplying the resultant equation by $D_k u$ and summing on k, we have that $v = |Du|^2$ is a weak solution of the equation

$$-v_t + D_i(a^{ij}D_j v + 2C^i_k D_k u) + 2a^{ij}D_{jk}uD_{ik}u + 2C^i_k D_{ik}u = 0$$

in $\{v > L^2\}$, and then Cauchy's inequality gives

$$-v_t + D_i(a^{ij}D_j v + b^i v) + c^0 v \geq 0,$$

where $b^i = 2C^i_k D_k u/v$ and $c^0 = \lambda^{-1}\sum |C^i_k|^2$. From (11.40), we infer that $|b^i|$ and c^0 are bounded by $C(\beta)\lambda$ and hence (from Corollary 6.16)

$$\sup_{\Omega}|Du| \leq C(\beta,\lambda,T)\max\{L, \sup_{\mathscr{P}\Omega}|Du|\}.$$

Condition (11.40) allows less smooth coefficients than our previous structure conditions, but slower growth of them. For example, if $A(X,z,p) = p$, then Theorem 11.1 requires that $|B| = O(|p|^2)$ and some additional hypotheses on the derivatives of B while (11.40) requires $|B| = O(|p|)$ but no additional hypotheses. It is possible to modify the brief argument just given to include this faster growth (see [101, Section 15.4; 183, Section V.4]), but we shall expand our considerations to cover non-uniformly parabolic equations as well.

In order to state our results, we define quantities related to the Sobolev inequality of the previous section. We suppose that u is a solution of(11.39) and we define

$$v = (1+|Du|^2)^{1/2}, \nu = Du/v, g^{ij} = \delta^{ij} - \nu_i \nu_j,$$
$$H = \frac{1}{nv}g^{ij}D_{ij}u, d\mu = v\,dx.$$

(Note that our definition of ν is actually the projection onto $\mathbb{R}^n$ of the vector ν from Section XI.4 and similarly for g^{ij}. In addition, H denotes the scalar mean curvature rather than the mean curvature vector, so $\mathscr{H} = H\nu$ and $|\mathscr{H}| \leq |H|$.) For our first structure condition, we suppose that there are two matrices (C^i_k) and (D^i_k) such that (D^i_k) is differentiable with respect to (x,z,p) (but not necessarily with respect to t) and

$$C^i_k + D^i_k = A^i_z p_k + A^i_k + B\delta^i_k.$$

We define

$$A^{ij} = v\frac{\partial A^i}{\partial p_j}, D^{ij}_k = \frac{\partial D^i_k}{\partial p_j}, \mathscr{F}_k = p_i\frac{\partial D^i_k}{\partial z} + \frac{\partial D^i_k}{\partial x^i},$$

and suppose that there are nonnegative constants β_1 and τ_0 along with bounded functions $\Lambda_0(X,z,p)$ and $\bar{\mu}(X,z,p)$ such that

$$C_k^i g^{jk}\zeta_{ij} \le \beta_1\Lambda_0^{1/2}(A^{ij}\zeta_{ik}\zeta_{jk})^{1/2}, \tag{11.41a}$$

$$C_k^i v^k\xi_i \le \beta_1\Lambda_0^{1/2}(A^{ij}\xi_i\xi_j)^{1/2}, \tag{11.41b}$$

$$vD_k^{ij}v^k\zeta_{ij} \le \beta_1\Lambda_0^{1/2}(A^{ij}\zeta_{ik}\zeta_{jk})^{1/2}, \tag{11.41c}$$

$$\mathscr{F}_k v^k \le \beta_1^2\Lambda_0 \tag{11.41d}$$

$$A^{ij}\xi_i\eta_j \le (\bar{\mu}\,|\xi|^2)^{1/2}(a^{ij}\eta_i\eta_j)^{1/2} \tag{11.41e}$$

for all $n\times n$ matrices ζ, all n-vectors ξ and η, and all $(X,z,p)\in\Omega\times\mathbb{R}\times\mathbb{R}^n$ such that $z=u(X)$ and $v>\tau_0$. We also assume that $n\ge 2$ in the remainder of this section.

For this method, it is convenient to introduce two quantities involving second derivatives of u. They are

$$\mathscr{C}^2 = v^{-2}A^{ij}g^{km}D_{ik}uD_{jm}u, \mathscr{E}_1 = v^{-2}A^{ij}D_ivD_jv.$$

Note that $\mathscr{C}^2$ is always positive. In the special case of the prescribed mean curvature equation , we have $A=\nu$ and a simple calculation shows that $\mathscr{C}^2$ is just the sum of the principal curvatures of the graph of u. The quantity $\mathscr{E}_1$ is a simple variant of the Bernstein function $\mathscr{E}$. We also write Ω_τ and $\omega_\tau(t)$ for the subsets of Ω and $\omega(t)$, respectively on which $v>\tau$. The first step in proving our gradient bounds is a simple energy inequality.

LEMMA 11.10. *Let χ be a nonnegative Lipschitz function defined on $[\tau_0,\infty)$ and suppose that there are constants $\tau\ge\tau_0$ and $c(\chi)\ge 0$ such that*

$$0\le\frac{(\xi-\tau)\chi'(\xi)}{\chi(\xi)}\le c(\chi) \tag{11.42}$$

for almost all $\xi\in[\tau,\infty)$. Suppose also that conditions (11.41) *hold. Let $\zeta\in C^1(\overline{\Omega})$. If $v\le\tau$ on $\mathscr{P}\Omega\cap\operatorname{supp}\zeta$, then*

$$\frac{1}{1+c(\chi)}\sup_{t\in I(\Omega)}\int_{\omega_\tau(t)}(v-\tau)^2\chi\zeta^2\,dx+\int_{\Omega_\tau}[(1-\frac{\tau}{v})\mathscr{C}^2+\mathscr{E}_1]\chi\zeta^2\,d\mu\,dt$$
$$\le 36(1+c(\chi))\int_{\Omega_\tau}[\beta_1^2\Lambda_0\zeta^2+\bar{\mu}\,|D\zeta|^2+v|\zeta\zeta_t|]\chi\,d\mu\,dt. \tag{11.43}$$

PROOF. Fix $s\in I(\Omega)$, let η be a Lipschitz vector-valued function which vanishes on $\mathscr{P}\Omega$, and use the test function $\operatorname{div}\eta$ in the weak form of (11.39). After integrating by parts, we find that

$$\int_{\Omega(s)}[u_tD_k\eta^k+(v^{-1}A^{ij}D_{jk}u+C_k^i)D_i\eta^k]\,dX=\int_{\Omega(s)}(\mathscr{F}_k\eta^k+D_k^{ij}D_{ij}u\eta^k)\,dX.$$

Next, we take $\eta = \theta v$ for a nonnegative Lipschitz scalar-valued function θ vanishing on $\mathscr{P}\Omega$. Substituting this choice for η yields

$$\begin{aligned}\int_{\Omega(s)} [u_t D_k(\theta v^k) + (v^{-1}A^{ij}D_j v + C_k^i v^k)D_i\theta]\,dX \\ + \int_{\Omega(s)} \theta(v^{-1}A^{ij}D_{jk}u + C_k^i)D_i v^k\,dX \\ = \int_{\Omega(s)} (\mathscr{F}_k v^k + D_k^{ij}D_{ij}uv^k)\theta\,dX.\end{aligned}$$

We now note that $D_i v^k = g^{mk}D_{im}u/v$ and $A^{ij}D_{ik}uD_{jk}u = v^2(\mathscr{C}^2 + \mathscr{E}_1)$. It then follows from (11.41a,c,d) that

$$\begin{aligned}\int_{\Omega(s)} u_t D_k(\theta v^k)\,dX + \int_{\Omega(s)} [(v^{-1}A^{ij}D_j v + C_k^i v^k)D_i\theta + \frac{1}{2}\mathscr{C}^2\theta]\,dX \\ \le \int_{\Omega(s)} [\frac{1}{2}\mathscr{E}_1 + 2\beta_1^2\Lambda_0]\theta\,dX.\end{aligned}$$

Finally, we choose $\theta = (v-\tau)^+\chi(v)\zeta^2$. If $u \in C^{2,1}$ and if Du_t exists and is continuous, we have that

$$-\int_{\Omega(s)} u_t D_k(\theta v^k)\,dX = \int_{\Omega(s)} D_k u_t v^k\theta\,dX = \int_{\Omega(s)} v_t\theta\,dX.$$

Setting

$$\Xi(\sigma) = \int_\tau^\sigma (\xi-\tau)^+\chi(\xi)\,d\xi,$$

we infer that

$$-\int_{\Omega(s)} u_t D_k(\theta v^k)\,dX = \int_{\omega(s)} \Xi(v)\zeta^2\,dx + 2\int_{\Omega(s)} \Xi(v)\zeta\zeta_t\,dX,$$

and an easy approximation argument shows that this identity holds for arbitrary $u \in C^{2,1}$. As in Theorem 6.15, it follows that

$$\frac{1}{2+2c(\chi)}(v-\tau)^2\chi(v) \le \Xi(v) \le \frac{1}{2}(v-\tau)^2\chi(v),$$

and therefore (assuming without loss of generality that $\chi(\sigma)=0$ for $\sigma\leq\tau$)

$$\begin{aligned}\frac{1}{2+2c(\chi)}\int_{\omega(s)}(v-\tau)^2\chi\,dx+\int_{\Omega(s)}\zeta^2 v[\chi+(v-\tau)^+\chi']\mathscr{E}_1\,dX\\ +\int_{\Omega(s)}\frac{1}{2}\zeta^2(v-\tau)\chi\mathscr{C}^2\,dX\\ \leq\int_{\Omega(s)}\zeta^2[\frac{1}{2}\mathscr{E}_1+2\beta_1^2\Lambda_0](v-\tau)\chi\,dX\\ -\int_{\Omega(s)}\zeta^2 C_k^i v^k D_i v(\chi+(v-\tau)\chi')\,dX\\ -2\int_{\Omega(s)}[v^{-1}A^{ij}D_j vD_i\zeta\zeta+C_k^i v^k D_i\zeta\zeta]\chi\,dX\\ +\int_{\Omega(s)}v^2|\zeta\zeta_t|\chi\,dX.\end{aligned}$$

We then use (11.41b) and (11.41e) to estimate the last two integrals in this inequality, noting that $\chi'\geq 0$. Some slight rearrangement along with Cauchy's inequality gives

$$\begin{aligned}\frac{1}{4+4c(\chi)}\int_{\omega(s)}(v-\tau)^2\chi\,dx+\frac{1}{4}\int_{\Omega(s)}[(v-\tau)\mathscr{C}^2+v\mathscr{E}_1]\chi\,dX\\ \leq\int_{\Omega(s)}\beta_1^2\Lambda_0[3v\chi+4v(\chi+(v-\tau)\chi')]\,dX\\ +\int_{\Omega(s)}[9\bar{\mu}\,|D\zeta|^2 v+|\zeta\zeta_t|v^2]\chi\,dX.\end{aligned}$$

If we multiply this inequality by 4 and use the upper bound for χ' from (11.42), the desired result follows by taking the supremum over s. □

From our energy inequality and the Sobolev inequality, we can now reduce our pointwise estimate of $|Du|$ to an integral estimate of a suitable quantity. For this reduction, we introduce three positive $C^1[\tau_0,\infty)$ functions w, λ and Λ. The functions λ and Λ are not to be confused with the maximum and minimum eigenvalues of the matrix (a^{ij}); however, they are connected to these eigenvalues and this connection will become obvious from our examples. In addition to their smoothness, the functions w, λ and Λ obey the following monotonicity properties:

$$w \text{ is increasing,} \tag{11.44a}$$

$$\xi^{-\beta}w(\xi) \text{ is a decreasing function of } \xi, \tag{11.44b}$$

$$w(\xi)^{\beta}(\Lambda(\xi)/\lambda(\xi))^{N/2}/\xi \text{ is an increasing function of } \xi, \tag{11.44c}$$

$$\xi^{-\beta}(\Lambda(\xi)/\lambda(\xi))^{N/2} \text{ is a decreasing function of } \xi \tag{11.44d}$$

for some nonnegative constant β. For simplicity in notation, we also fix $X_0 \in \Omega$ and $\rho > 0$, and define

$$\Sigma(\tau,\rho) = \{X \in \Omega : |x - x_0| < \rho, t < t_0, v(X) > \tau\},$$
$$\Sigma'(\rho) = \{X \in \mathscr{P}\Omega : |x - x_0| < \rho, t < t_0\},$$
$$\Omega_\tau(\rho) = \{X \in \Omega[\rho] : v(X) > \tau\}.$$

LEMMA 11.11. *Suppose conditions* (11.41) *and* (11.44) *hold, and that*

$$v\lambda(v)\left(1 + \left(\frac{v\lambda'(v)}{\lambda(v)}\right)^2\right) g^{ij}\xi_i\xi_j \le A^{ij}\xi_i\xi_j, \tag{11.45a}$$
$$\Lambda_0 \le v\Lambda,\ \lambda \le \Lambda \tag{11.45b}$$
$$\bar{\mu} \le v\Lambda \tag{11.45c}$$

on Ω_{τ_0}. *If also* $v \le \tau$ *on* $\Sigma'(\rho)$ *for some* $\tau \ge \tau_0$, *then*

$$\sup_{\Sigma(\tau,\rho)} \left(1 - \frac{\tau}{v}\right)^{N+2} w \le C(n,\beta,\beta_1\rho)\rho^{-n-2} \int_{\Sigma(\tau,2\rho)} w(\Lambda/\lambda)^{N/2}\Lambda\, d\mu\, dt. \tag{11.46}$$

If also $v \le \tau$ *on* $\mathscr{P}\Omega[2\rho]$ *and* $\Lambda \ge 1$, *then*

$$\sup_{\Omega_\tau(\rho)} \left(1 - \frac{\tau}{v}\right)^{N+2} w \le C(n,\beta,\beta_1\rho)\rho^{-n-2} \int_{\Omega_\tau(2\rho)} w(\Lambda/\lambda)^{N/2}\Lambda\, d\mu\, dt. \tag{11.47}$$

PROOF. For $q \ge 1 + \beta$, use the function

$$\chi(v) = \left(\frac{\Lambda}{\lambda}\right)^{N/2} w^q \left[\left(1 - \frac{\tau}{v}\right)^+\right]^{(N+2)(q-1)} / v$$

in Lemma 11.10 and note that (11.42) holds with $c(\chi) = C(\beta, N)q$. Let $\zeta = (1 - |x - x_0|^2/(2\rho)^2)^+$, and replace ζ^2 by $\zeta^{(N+2)q-N}$. From (11.43) and (11.45b,c) we infer that

$$\begin{aligned}
&\sup_s \int_{\omega_\tau(s)} \left(1 - \frac{\tau}{v}\right)^{(N+2)q-N} \left(\frac{\Lambda}{\lambda}\right)^{N/2} w^q \zeta^{(N+2)q-N}\, d\mu \\
&\quad + \int_{\Omega_\tau} \left[\left(1 - \frac{\tau}{v}\right)\mathscr{C}^2 + \mathscr{E}_1\right] \chi(v)\zeta^{(N+2)q-N}\, d\mu\, dt \\
&\quad \le Cq\rho^{-2} \int_{\Omega_\tau} \Lambda \left(1 - \frac{\tau}{v}\right)^{(N+2)(q-1)} \left(\frac{\Lambda}{\lambda}\right)^{N/2} \zeta^{(N+2)(q-1)} w^q\, d\mu\, dt.
\end{aligned}$$

Now we define the function h by

$$h^2 = \chi\lambda\left(1 - \frac{\tau}{v}\right)^2 v\zeta^{(N+2)q-N}.$$

Since (11.45a) implies that

$$v\lambda\left(1 + \left(\frac{v\lambda'(v)}{\lambda(v)}\right)^2\right)|\delta v|^2 \le v^2\mathscr{E}_1,$$

it follows that

$$|\delta h|^2 \le C(n,\beta)q\chi[\mathscr{E}_1\zeta^{(N+2)q-N} + \rho^{-2}\lambda v\zeta^{(N+2)(q-1)}].$$

Moreover, we have $v\lambda H^2 \le C(n)\mathscr{C}^2$, so

$$\sup_s \int_{\omega(s)} h^2\lambda^{-1}\,d\mu + \int_\Omega (|\delta h|^2 + h^2H^2)\,d\mu\,dt$$
$$\le Cq\rho^{-2}\int_{\Omega_\tau} \Lambda\left(1-\frac{\tau}{v}\right)^{(N+2)(q-1)}\left(\frac{\Lambda}{\lambda}\right)^{N/2} w^q\,d\mu\,dt.$$

Now apply Corollary 11.9 to estimate the left side of this inequality, and set $\kappa = (N+2)/N$,

$$\bar{w} = w\left(1-\frac{\tau}{v}\right)^{N+2}\zeta^{N+2}, d\bar{\mu} = \rho^{-n-2}\left(\frac{\Lambda}{\lambda}\right)^{N/2}\left(1-\frac{\tau}{v}\right)^{-N-2}\zeta^{-N-2}v\,dX.$$

In this way, we find that

$$\left(\int_{\Sigma(\tau,2\rho)} \bar{w}^{q\kappa}\,d\bar{\mu}\right)^{1/\kappa} \le Cq^2\int_{\Sigma(\tau,2\rho)} \bar{w}^q\,d\bar{\mu},$$

and our usual iteration leads to

$$\sup_{\Sigma(\tau,2\rho)} \bar{w} \le C\int_{\Sigma(\tau,2\rho)} \bar{w}\,d\bar{\mu}.$$

Rewriting this inequality in terms of w, λ, and Λ gives (11.46).

If $v \le \tau$ on $\mathscr{P}\Omega[2\rho]$ and $\Lambda \ge 1$, we repeat the above argument with $\zeta_1 = \zeta(1+[t-t_0]/\rho^2)^+$ replacing ζ. □

Note that Lemma 11.11 gives a global gradient estimate in terms of a boundary gradient estimate and an integral estimate. In this case, the estimate can be written in a somewhat different form, which we present in Exercise 11.3.

When the structure functions w, λ and Λ are all powers of v, Lemma 11.10 gives an estimate of the maximum of v in terms of the integral of some, possibly large, power of v. Our next estimate reduces this power to a manageable size. To deal with nonpower functions, we shall assume more generally that there are nonnegative constants β_2 and β_3 such that

$$\left(\frac{\Lambda}{\lambda}\right)^{N/2}\Lambda v \le \beta_2 w^{\beta_3} Du\cdot A \tag{11.48}$$

on Ω_{τ_0}. Hence we wish to estimate $\int w^q Du\cdot A\,dX$ in terms of $\int Du\cdot A\,dX$ for $q = 2+\beta_3$. We shall prove this type of estimate for arbitrary $q \ge 2$. Some additional structure conditions are needed for this estimate and one of them uses an additional function ε which is assumed to be positive and decreasing on $[\tau_0,\infty)$.

LEMMA 11.12. *Suppose conditions* (11.41a–d) *and* (11.44a,b) *hold and that* $v < \tau$ *on* $\mathscr{P}\Omega$. *Suppose also that there are nonnegative constants* β_4 *and* β_5 *and a positive decreasing function* ε *such that*

$$w'(v)A\cdot\xi \le \beta_4(A^{ij}\xi_i\xi_j)^{1/2}(v\cdot A)^{1/2}, \tag{11.49a}$$

$$v|A_z| + |A_x| \le \beta_5 Du\cdot A, \tag{11.49b}$$

$$\bar{\mu}v \le \beta_4^2 Du\cdot A, \tag{11.49c}$$

$$\Lambda_0 v \le \varepsilon(v)w^2 Du\cdot A, \tag{11.49d}$$

$$w|A| \le Du\cdot A, \tag{11.49e}$$

on Ω_{τ_0}. *Set* $\sigma = \sup_t \operatorname{osc}_{\omega(t)} u$, *let* $q \ge 2$, *and set* $E = \exp(\beta_5\sigma)$. *If there is a constant* $\tau_1 \ge \tau$ *such that*

$$400(\beta_1\sigma)^2E^2q^2\beta_4^2\varepsilon(\tau_1)(1+\beta q)^3 \le 1, \tag{11.50}$$

then there is a constant C *determined only by* n, q, β, $\beta_1\rho$, *and* β_4 *such that*

$$\int_{\Sigma(\tau,\rho)} w^q Du\cdot A\,dX \le C\left(w(\tau_1) + E\frac{\sigma}{\rho}\right)^q \int_{\Sigma(\tau,2\rho)} Du\cdot A\,dX. \tag{11.51}$$

If also

$$v^2 \le \beta_4 w Du\cdot A, \tag{11.52}$$

then

$$\int_{\Omega_\tau(\rho)} w^q Du\cdot A\,dX \le C\left(w(\tau_1) + E\frac{\sigma}{\rho}\right)^q \int_{\Omega_\tau(2\rho)} Du\cdot A\,dX. \tag{11.53}$$

PROOF. Set $\bar{u}(X) = u(X) - \inf_{\omega(t)} u$ and let ζ be as in Lemma 11.11. Suppressing the set of integration $\Sigma(\tau,2\rho)$ on our integrals, we define

$$I = \int \zeta^q w^q Du\cdot A\,dX,$$

$$I' = \int \zeta^q \exp(\beta_5\bar{u})(1+\beta_5\bar{u})(w^q - w(\tau)^q)^+ Du\cdot A\,dX.$$

To estimate I, we integrate I' by parts to obtain

$$\begin{aligned} I' &= -\int \bar{u}\exp(\beta_5\bar{u})\operatorname{div}\left[\zeta^q(w^q - w(\tau)^q)^+A\right]dX \\ &= -q\int \bar{u}\exp(\beta_5\bar{u})\zeta^{q-1}(w^q - w(\tau)^q)^+ D\zeta\cdot A\,dX \\ &\quad - q\int \bar{u}\exp(\beta_5\bar{u})\zeta^q w^{q-1}w' Dv\cdot A\,dX \\ &\quad - \int \bar{u}\exp(\beta_5\bar{u})\zeta^q(w^q - w(\tau)^q)^+ \operatorname{div}A\,dX, \end{aligned}$$

and we write I_1, I_2, and I_3 for the three terms on the right side of this equation.

First, we have

$$I_1 \le qE\frac{\sigma}{\rho}\int \zeta^{q-1}w^{q-1}w|A|\,dX$$
$$\le qE\frac{\sigma}{\rho}\int (w\zeta)^{q-1}Du\cdot A\,dX$$

by (11.49e) since $|D\zeta| \le 1/\rho$.

Using (11.49a) and Cauchy's inequality, we have

$$I_2 \le \int (\beta_4^2 q^2 E^2\sigma^2\zeta^q w^{q-2}\mathscr{E}_1 v)^{1/2}(\zeta^q w^q Du\cdot A)^{1/2}\,dX$$
$$\le \frac{1}{4}I + \beta_4^2 q^2 E^2\sigma^2\int w^{q-2}\zeta^q\mathscr{E}_1\,d\mu\,dt.$$

For I_3, we first note that

$$\operatorname{div} A = v^{-1}A^{ij}D_{ij}u + D_iuA^i_z + A^i_i,$$

so (11.49b) gives

$$I_2 \le \sigma E\int (w^q - w(\tau)^q)^+\zeta^q v^{-1}A^{ij}D_{ij}u\,dX$$
$$+\int \beta_5\bar{u}\exp(\beta_5\bar{u})\zeta^q(w^q - w(\tau)^q)Du\cdot A\,dX.$$

For the estimate of the first integral here, we note that $v \ge 2\tau$ implies that $w^q - w(\tau)^q \le 2w^q(1-\tau/v)$ while $v < 2\tau$ implies that

$$w^q - w(\tau)^q = q\int_\tau^v w'(\xi)w(\xi)^{q-1}\,d\xi \le q\beta\int_\tau^v \frac{w(\xi)^q}{\xi}\,d\xi$$
$$\le q\beta\frac{w^q}{\tau}\int_\tau^v 1\,d\xi \le 2q\beta w^q(1-\tau/v).$$

Hence $w^q - w(\tau)^q \le 2(1+\beta q)w^q(1-\tau/v)$, so

$$\sigma E\int \zeta^q(w^q - w(\tau)^q)^+ v^{-1}A^{ij}D_{ij}u\,dX$$
$$\le 2(1+\beta q)\beta_4\sigma E\int \zeta^q w^q\left(1-\frac{\tau}{v}\right)(\mathscr{C}^2+\mathscr{E}_1)^{1/2}(v\cdot A)^{1/2}\,dX$$
$$\le \frac{1}{4}I + 4(1+\beta q)^2\beta_4^2\sigma^2E^2\int \zeta^q w^{q-2}\left[\left(1-\frac{\tau}{v}\right)\mathscr{C}^2+\mathscr{E}_1\right]\,d\mu\,dt$$

by (11.41e), (11.49c), and (11.49e) because (11.49e) implies that $w \le v$. It follows that

$$I_3 \le \frac{1}{4}I + \int \beta_5\bar{u}\exp(\beta_5\bar{u})\zeta^q(w^q - w(\tau)^q)Du\cdot A\,dX$$
$$+4(1+\beta q)^2E^2\beta_4^2\sigma^2\int \zeta^q w^{q-2}\left[\left(1-\frac{\tau}{v}\right)\mathscr{C}^2+\mathscr{E}_1\right]\,dX.$$

Using these estimates for I_1, I_2, and I_3, we find that

$$\begin{aligned} I' &\le \frac{1}{2}I + \int \beta_5 \bar{u} \exp(\beta_5 \bar{u})(w^q - w(\tau)^q) Du\cdot A\, dX \\ &+ 2qE\frac{\beta_4\sigma}{\rho}\int (w\zeta)^{q-1} Du\cdot A\, dX \\ &+ 5(1+\beta q)^2 E^2 \beta_4^2 \sigma^2 \int w^{q-2}\zeta^q \left[\left(1-\frac{\tau}{v}\right)\mathscr{C}^2 + \mathscr{E}_1\right] dX. \end{aligned}$$

Simple rearrangement, along with the observations that $\exp(\beta_5\bar{u}) \ge 1$ and $0 \le \zeta \le 1$, gives

$$\begin{aligned} \frac{1}{2}I &\le w(\tau)^q \int Du\cdot A\, dX + 2qE\frac{\beta_4\sigma}{\rho}\int (w\zeta)^{q-1} Du\cdot A\, dX \\ &+ 5(1+\beta q)^2 E^2 \beta_4^2 \sigma^2 \int w^{q-2}\zeta^q \left[\left(1-\frac{\tau}{v}\right)\mathscr{C}^2 + \mathscr{E}_1\right] dX. \end{aligned}$$

This last integral is now estimated via Lemma 11.10. If we take $\chi(v) = w^{q-2}$, then (11.42) holds with $c(\chi) = \beta q$ and then (11.43) implies that

$$\begin{aligned} &\int w^{q-2}\zeta^q \left[\left(1-\frac{\tau}{v}\right)\mathscr{C}^2 + \mathscr{E}_1\right] d\mu\, dt \\ &\qquad \le 36(1+\beta q)\int [\beta_1^2 \Lambda_0 v w^{q-2}\zeta^q + \bar{\mu} v (w\zeta)^{q-2}]\, dX. \end{aligned}$$

From this inequality and (11.49c,d), we infer that

$$\begin{aligned} \frac{1}{2}I &\le w(\tau)^q \int Du\cdot A\, dX + 180E^2\beta_4^2 q^2 \beta_1^2 \sigma^2 (1+\beta q)^3 \varepsilon(\tau) I \\ &+ C\frac{E\sigma}{\rho}\int (w\zeta)^{q-1} Du\cdot A\, dX + C\left(\frac{E\sigma}{\rho}\right)^2 \int (w\zeta)^{q-2} Du\cdot A\, dX. \end{aligned}$$

We now replace τ by τ_1 and use Young's inequality to infer from (11.50) that

$$\int_{\Sigma(\tau_1,2\rho)} (w\zeta)^q Du\cdot A\, dX \le c\left(w(\tau_1) + \frac{E\sigma}{\rho}\right)^q \int_{\Sigma(\tau_1,2\rho)} Du\cdot A\, dX.$$

Finally we note that

$$\int_{\Sigma(\tau,2\rho)\setminus\Sigma(\tau_1,2\rho)} (w\zeta)^q Du\cdot A\, dX \le w(\tau_1)^q \int_{\Sigma(\tau,2\rho)\setminus\Sigma(\tau_1,2\rho)} Du\cdot A\, dX,$$

and the proof is completed by adding these last two inequalities and recalling that $\zeta \ge 1/2$ on $\Sigma(\tau,\rho)$.

If (11.52) holds, we proceed with ζ_1 replacing ζ. □

Note that (11.50) holds if ε goes to zero as $v \to \infty$ or if a modulus of continuity is known for u. In our examples, we shall have reason to consider both possibilities.

Finally, we estimate $\int Du\cdot A\,dX$. For this estimate, we define

$$Q_\tau(\rho)=\{X\in\Omega:|x-x_0|\le\rho,\ t_0-4\rho^2<t<t_0,\ v(X)>\tau\}.$$

LEMMA 11.13. *Suppose that there are constants β_6 and β_7 such that*

$$|Du|\,|A|\le\beta_6 Du\cdot A,|B|\le\beta_7 Du\cdot A \tag{11.54}$$

on Ω_{τ_0}. Fix $X_0\in\Omega$ and $\rho>0$ and suppose that $\operatorname{osc}_{\Omega[2\rho]}u\le M$ for some nonnegative constant M. Set $\tau_2=\max\{\tau_0,4\beta_6M/\rho\}$ and

$$\Delta=\sup_{v<\tau_2}\left\{(B-\beta_7Du\cdot A)^+ +(Du\cdot A)^+ +\frac{M}{\rho}|A|\right\}. \tag{11.55}$$

If $S\Omega[2\rho]$ is empty, then

$$\int_{Q_{\tau_2}(\rho)}Du\cdot A\,dX\le C(n)\exp(\beta_7M)\rho^n[M^2+\Delta\rho^2]. \tag{11.56}$$

If $S\Omega[2\rho]$ is not empty, suppose that $\mathscr{P}\Omega\in H_1$, that $u\in C(\overline{\Omega[2\rho]})$, and that there is a function $\varphi\in H_1$ such that $u=\varphi$ on $S\Omega[2\rho]$ with $|D\varphi|\le\Phi$ and $|u-\varphi|\le\Phi d$ in $\Omega[2\rho]$, where d denotes distance to $S\Omega$. Then

$$\int_{Q_{\tau_2}(\rho)}Du\cdot A\,dX\le C(n)\exp(\beta_7M)\rho^n[M^2+(\Delta+C(\Omega)\Phi^2)\rho^2]. \tag{11.56$'$}$$

PROOF. Suppose first that $S\Omega[2\rho]$ is empty. Let ζ be a cut-off function of x only, and set

$$m=\inf_{Q(2\rho)}u,\omega=(u-m-M[1-\zeta])^+,E=\exp(\beta_7(u-m))$$

$$\Delta_1=\sup_{v<\tau_2}(B-\beta_7Du\cdot A)^+ +\frac{M}{\rho}|A|.$$

If we use the test function $\eta=E\omega$, it follows that

$$\int_\Omega u_t\eta\,dX+\int(Du\cdot A\beta_7-B)E\omega+D\omega\cdot AE\,dX=0.$$

Now we define

$$U(X)=\int_{m+M[1-\zeta(x)]}^{u(X)}(z-m-M[1-\zeta])^+\exp(\beta_7(z-m))\,dz$$

and $E_0=\exp(\beta_7M)$, so that $U_t=u_t\eta$ and $0\le U(X)\le E_0M^2$. Defining also $A^*=E\omega(\beta_7Du\cdot A-B)-ED\omega\cdot A$, we have

$$\begin{aligned}\int_{\Omega_{\tau_2}}A^*\,dX&\le\int_{Q(2\rho)\setminus\Omega_{\tau_2}}A^*\,dX\\&\quad+\int_{B(x_0,2\rho)}[U(x,t_0-4\rho^2)-U(x,t_0)]\,dX\\&\le\omega_nE_0\rho^n[\Delta_1\rho^2+M^2].\end{aligned}$$

Now choose

$$\zeta = \min\{1, 4 - \frac{|x-x_0|^2}{\rho^2}\}$$

so that $\zeta = 1$ in $Q' = Q_{\tau_2}(\rho)$. Since

$$E\omega(\beta_7 Du\cdot A - B) \geq 0,\ D\omega\cdot A = Du\cdot A + MD\zeta\cdot A \geq \frac{1}{2}Du\cdot A$$

on Q', we have $A^* \geq \frac{1}{2}Du\cdot A$ on Q' and hence

$$\frac{1}{2}\int_{Q'} Du\cdot A\,dX \leq \omega_n E_0\rho^n[\Delta\rho^2 + M^2].$$

On the other hand,

$$\int_{Q_\tau(\rho)\setminus Q'} Du\cdot A\,dX \leq \omega_n\rho^{n+2}\sup_{v<\tau_2}(Du\cdot A)^+,$$

and the proof is completed by adding these two inequalities.

If $S\Omega[2\rho]$ is not empty, we replace m by $\varphi(X)$, noting that we can assume that $|\varphi_t| \leq C(\Omega)\Phi d^{-1}$ by Lemma 4.24. Then

$$U_t = u_t\eta - \varphi_t[u - \varphi - M(1-\zeta)]^+\exp[\beta_7 M(1-\zeta)],$$

so that $U_t \geq u_t\eta - C(\Omega)\Phi^2\exp(\beta_7 M)$. We then infer a bound on

$$\int_{Q_{\tau_2}\cap\{u\geq\varphi\}} Du\cdot A\,dX.$$

Similar arguments with $-u$ replacing u give the corresponding estimate for the set on which $u \leq \varphi$. □

Note that the bound on $|u-\varphi|$ is essentially the one from Chapter X.

Although it is possible to write a theorem which incorporates our structure conditions into a single estimate for $\sup|Du|$, we shall see that is more convenient to use these three lemmata in various combinations to study the examples from Section XI.3.

First, the mean curvature equation in divergence form can be written as

$$-u_t = \operatorname{div} v + nH(X,u),$$

so we have $A = v$ and $A^{ij} = g^{ij}$. If we assume that $H_z \leq 0$ and if we take $C^i_k = 0$, then $D^{ij}_k = 0$ and (11.41), (11.44), and (11.45) hold with

$$w = v, \lambda = \frac{1}{2v}, \Lambda_0 = 1, \Lambda = \frac{1}{v}, \bar{\mu} = 1,$$
$$\beta = 1, \beta_1 = (n\sup|H_x|)^{1/2}.$$

Hence, if we take $\tau_0 \geq 2\sup_{\mathscr{P}\Omega} v$ and $\rho = \operatorname{diam}\Omega$, then Lemma 11.11 gives

$$\sup v \leq C(n, \rho^2\sup|H_x|)[\tau_0 + \rho^{-n-2}\int_{\Omega_{\tau_0}} v\,dX].$$

Since $2v \geq Du \cdot A$ on Ω_{τ_0}, we can use Lemma 11.13 to bound this integral. For

$$\tau_0 = n\sup|H| + 2\sup_{\mathscr{P}\Omega} v + M,$$

condition (11.54) holds with $\beta_6 = 1$ and $\beta_7 = 1/M$. Then

$$\Delta \leq n\sup|H| + \tau_0 + M/\rho,$$

so we infer that

$$\sup v \leq C(n, \rho^2 \sup|H_x|)(\sup|H| + \sup_{\mathscr{P}\Omega} v + M + \frac{M^2}{\rho^2}).$$

If we only assume that v is bounded on $B\Omega$, then it follows that

$$\sup_{Q'(\rho)} v \leq C(n, \rho^2 \sup|H_x|)(\sup|H| + \sup_{B\Omega} v + M + \frac{M^2}{\rho^2}),$$

for $Q'(\rho) = \{X \in \Omega : |x - x_0| < \rho\}$, as long as $\rho \leq \frac{1}{2}d(X_0)$.

In the context of equations in divergence form, uniform parabolicity means that there are positive constants $\theta_1, \cdots, \theta_5$, a constant $\alpha > -1$, and a decreasing function ε_1 such that

$$\theta_1 v^\alpha |\xi|^2 \leq a^{ij}\xi_i\xi_j \leq \theta_2 v^\alpha |\xi|^2,$$
$$Du \cdot A \geq \theta_3 v^{\alpha+2} - \theta_4, |A| \leq \theta_5 v^{\alpha+1},$$
$$v|A_z| + |A_x| + |B| \leq \varepsilon_1(v)^{1/2} v^{\alpha+2}.$$

Now conditions (11.41), (11.44), (11.45), (11.48), (11.49) and (11.52) are satisfied with $D^i_k = 0$,

$$w = \theta_5 v/(2\theta_3), \lambda = c(\alpha, \theta_1)v^\alpha, \Lambda = c(\theta_2)v^{\alpha+2}, \Lambda_0 = c(n)\varepsilon_1 v^{\alpha+3},$$
$$\bar{\mu} = \theta_2 v^{\alpha+1}, \varepsilon = c(n, \theta_3, \theta_5)\varepsilon_1,$$
$$\beta = N, \beta_1 = 1, \beta_2 = c(n, \theta_1, \theta_2, \theta_3), \beta_3 = N + 3, \beta_4 = c(\theta_1, \theta_2, \theta_3, \theta_5),$$
$$\beta_5 = \varepsilon_1(1), \beta_6 = \theta_5/2\theta_3, \beta_7 = 1$$

provided $\tau_0 \geq \max\{1, (2\theta_4/\theta_3)^{1/(\alpha+2)}\}$. In addition, $\Lambda \geq 1$, so we infer a completely local gradient bound in this case provided either $\varepsilon_1(v) \to 0$ as $v \to \infty$ or else a modulus of continuity is known for u. When $\alpha = 0$, a Hölder estimate follows by modifying Theorem 6.27 along the lines of Lemma 11.4. (In fact, a Hölder estimate is known for arbitrary $\alpha > -1$ although the proof is much more complicated. We refer to the Notes for a detailed discussion.)

We write the false mean curvature equation in the form

$$-u_t + \operatorname{div}(\exp(\frac{1}{2}v^2)Du) + B(X, u, Du) = 0$$

subject to the condition

$$|B| \leq \theta_1 v^{1/2} \exp(\frac{1}{2}v^2)$$

for some positive constant θ_1. Note that the elliptic part of this equation is not necessarily equivalent to the elliptic part of the equation in example (iii) of Section XI.2; however, the present method only gives results in this situation. Now conditions (11.41), (11.44), and (11.45) are satisfied for

$$w = v,$$
$$\lambda = v^{-4}\exp(\frac{1}{2}v^2), \Lambda_0 = \exp(\frac{1}{2}v^2), \bar{\mu} = v^3\exp(\frac{1}{2}v^2), \Lambda = v^2\exp(\frac{1}{2}v^2),$$
$$\beta = 2N, \beta_1 = \theta_1.$$

Moreover, (11.48) holds with $\beta_2 = 2$ and $\beta_3 = 3N+2$, so

$$\sup v \le C(1+\rho^{-n-2}\int v^{3N+2}Du\cdot A\,dX).$$

This integral cannot be estimated via Lemma 11.12 because, for example, (11.49c) fails. The following global estimate is then useful.

LEMMA 11.14. *Suppose conditions* (11.41a–d) *and* (11.49a,b) *hold on* Ω_{τ_0} *and that* (11.44a,b) *hold. Suppose also that*

$$A^{ij}\zeta_{ij} \le \beta_4(v^2 v\cdot A)^{1/2}(A^{ij}\zeta_{ik}\zeta_{jk})^{1/2}, \tag{11.57a}$$
$$\Lambda_0 v \le \varepsilon(v)Du\cdot A \tag{11.57b}$$

on Ω_{τ_0}. *If* q, σ, E, *and* τ *are as in Lemma 11.12 and if* $\tau_1 \ge \tau$ *is so large that* (11.50) *holds, then*

$$\int_{\Omega_\tau} w^q Du\cdot A\,dX \le 50(w(\tau_1)+1)^q \int_{\Omega_\tau} Du\cdot A\,dX. \tag{11.51$'$}$$

PROOF. In the proof of Lemma 11.12, take $\zeta \equiv 1$ so that $I_1 = 0$. In addition, we find that

$$I_2 \le \frac{1}{4}I + \frac{1}{10}\int w^{q-2}Du\cdot A\,dX.$$

To estimate I_3, we note from (11.57a) that

$$\sigma E\int (w^q - w(\tau)^q)^+ v^{-1}A^{ij}D_{ij}u\,dX \le \frac{1}{4}I$$
$$+ 144(1+\beta q)^3E^2\beta_4^2\sigma^2\beta_1^2\int w^q\Lambda_0 v\,dX$$

by using $\chi = w^q$ in Lemma 11.10. Then we estimate this last integral by using (11.57b) and finish the proof as before. □

We apply Lemma 11.14 with

$$\varepsilon = v^{-1}, \beta_5 = 0, \beta_4 = 1, \tau_1 = \max\{\sup_{\mathcal{P}\Omega} v, C(n,\theta_1)\},$$

so

$$\int v^{3N+2}Du\cdot A\,dX \le C(n,\theta_1)\sup_{\mathcal{P}\Omega} v^{3N+2}\int Du\cdot A\,dX.$$

Now Lemma 11.13 with $\beta_6 = 1$ and $\beta_7 = 1$ gives

$$\int Du \cdot A\,dX \le C(n, \Phi, \operatorname{osc} u).$$

The combination of these three estimates implies the gradient bound.

6. The case of one space dimension

When the number of space dimensions is one, the gradient bound can be proved by reducing it to a boundary gradient estimate for a problem with two space dimensions. Such a reduction was exploited by Kruzhkov in [163]. Although the boundary gradient estimate appears in Chapter X, we reprove it here in a slightly different fashion in order to bring out the exact form of the dependence of the estimate on various parameters. In what follows, G will be a domain in the three-dimensional space $\{(x,y,t) \in \mathbb{R}^3\}$ such that $x > y$ in G.

LEMMA 11.15. *Suppose that $\bar{P}$ is defined on $C^{2,1}(G)$ by*

$$\bar{P}w = -w_t + \alpha(x,y,t,Dw)w_{xx} + \beta(x,y,t,Dw)w_{yy} + \bar{a}(x,y,t,Dw) \tag{11.58}$$

for functions α, β and $\bar{a}$ defined on $G \times \mathbb{R}^2$ with α and β positive and suppose that there are positive constants b_0 and b_1 such that

$$\bar{a}(x,y,t,p) \le b_0[\alpha(x,y,t,p)p_1^2 + \beta(x,y,t,p)p_2^2] \tag{11.59}$$

for $(x,y,t) \in G$ and $|p_1|, |p_2| \ge b_1$. If $v \in C^{2,1}(G) \cap C(\bar{G})$ satisfies $\bar{P}v \ge 0$ in G and if there are positive constants M_1 and L_1 such that

$$v \le M_1 \text{ in } G,\ v \le L_1[x-y] \text{ on } \mathscr{P}G, \tag{11.60}$$

then

$$v \le \exp(b_0 M_1)(L_1 + b_1)[x-y] \tag{11.61}$$

in G.

PROOF. Set $E = \exp(b_0 M_1)$, $a = Eb_0[L_1 + b_1]$, $\sigma = (E-1)/a$, and

$$w(x,y,t) = \frac{1}{b_0}\ln(a[x-y]+1),$$

and use G' to denote the subset of G on which $x - y < \sigma$. A simple calculation shows that $w_x = -w_y$ and that $|w_x|, |w_y| \ge L_1 + b_1$ in G'. Therefore $\bar{P}w \le 0$ in G'. Moreover, $w \ge (L_1 + b_1)(x-y) \ge v$ on $\mathscr{P}G' \cap \mathscr{P}G$ and $w = M \ge v$ on $\mathscr{P}G' \setminus \mathscr{P}G$, so the comparison principle implies that $w \ge v$ in G'. Since $w \le (L_1 + b_1)E[x-y]$ in G', it follows that (11.61) holds in G'.

In addition, on $G \setminus G'$, we have

$$v \le M_1 \le \frac{M_1}{\sigma}[x-y] = (L_1 + b_1)E\frac{b_0 M_1}{E-1}[x-y].$$

Since $s \le e^s - 1$ for $s \ge 0$, we see that (11.61) also holds in $G \setminus G'$ and hence in all of G. □

From Lemma 11.15, we immediately derive the following global gradient bound.

THEOREM 11.16. *Suppose P is defined on $C^{2,1}(\Omega)$ by*

$$Pw = -w_t + a^{11}(X,w,w_x)w_{xx} + a(X,w,w_x) \tag{11.62}$$

and that there are positive constants β_0 and β_1 such that

$$|a(X,z,p)| \le \beta_0 a^{11}(X,z,p)p^2 \tag{11.63}$$

for $|p| \ge \beta_1$. If $u \in C^{2,1}(\Omega) \cap C(\overline{\Omega})$ is a solution of $Pu = 0$ in Ω and if there are constants M and L_0 such that

$$|u(x,t) - u(y,t)| \le M \tag{11.64a}$$

for all (x,y,t) such that (x,t) and (y,t) are in Ω and

$$|u(x,t) - u(y,t)| \le L_0 |x-y| \tag{11.64b}$$

for all (x,y,t) such that $(x,t) \in \mathscr{P}\Omega$ and $(y,t) \in \Omega$, then

$$\sup_\Omega |u_x| \le \exp(\beta_0 M)(L_0 + \beta_1). \tag{11.65}$$

PROOF. For (x,t) and (y,t) in Ω with $x > y$, define

$$\begin{aligned} v^+(x,y,t) &= u(x,t) - u(y,t), \\ G &= \{(x,y,t) : x > y, (x,t) \text{ and } (y,t) \text{ are in } \Omega\}, \\ \alpha(x,y,t,p) &= a^{11}(x,t,u(x,t),p_1), \\ \beta(x,y,t,p) &= a^{11}(y,t,u(y,t),p_1), \\ \bar{a}(x,y,t,p) &= a(x,t,u(x,t),p_1) - a(y,t,u(y,t),p_2). \end{aligned}$$

Then $\bar{P}v^+ \ge 0$ in G, and conditions (11.59) and (11.60) hold with $b_0 = \beta_0$, $b_1 = \beta_1$, $L_1 = L_0$, and $M_1 = M$. Hence Lemma 11.15 gives

$$v^+ \le \exp(\beta_0 M)[L_0 + \beta_1](x-y).$$

Sending $x - y$ to zero gives

$$u_x \le 2\exp(\beta_0 M)[L_0 + \beta_1],$$

and the same argument with v^+ replaced by $-v^+$ yields the same estimate for $-u_x$. □

Since, in the one dimensional case, the ratio Λ/λ is always 1 and hence bounded, condition (11.63) is exactly the definition of uniform parabolicity. Hence we have a global gradient bound for what we have previously called uniformly parabolic equations. In addition, Theorem 11.16 gives a global gradient bound for the false mean curvature equation (or any of its variants with equivalent elliptic part).

On the other hand, for the mean curvature equation , we have $a^{11}(X,z,p) = v^{-3}$ and $a = nH(X,z)$. Hence (11.63) fails to hold unless $H = 0$. To cover this case, we refer the reader to Exercise 11.8.

Local estimates are only somewhat more complicated.

THEOREM 11.17. *With P as in Theorem 11.16, and $\Omega = (-2r,2r)\times(0,T)$, suppose that u satisfies $Pu = 0$ in Ω,* (11.64a) *and*

$$|u(x,0)-u(y,0)| \le L_0|x-y| \tag{11.64b)$'$}$$

for x and y in $(-2r,2r)$. Then

$$|u_x| \le \exp(4\beta_0 M)(L_0+\beta_1+4M/r) \tag{11.66}$$

in $(-r,r)\times(0,T)$.

PROOF. This time, we set $\rho = (x^2+y^2)^{1/2}$ and

$$v^{\pm}(x,y,t) = \pm[u(x,t)-u(y,t)] - M\left[\left(\frac{\rho}{r}-1\right)^+\right]^2.$$

With the obvious choices for α, β, and $\bar{a}$, we see that (11.59) and (11.60) hold for $b_1 = \beta_1 + 4M/r$, and b_0, L_1, and M_1 as in Theorem 11.16. Applying Lemma 11.15 to $v^{\pm}$ and proceeding as before completes the proof. □

Hence a gradient bound which is local in space follows under exactly the same hypotheses as for a global bound. The fully local estimate needs one additional assumption.

THEOREM 11.18. *Suppose P is as in Theorem 11.16 and that there is a positive constant β_2 such that*

$$1 \le \beta_2 a^{11}(X,z,p)p^2 \tag{11.67}$$

for $|p| \ge \beta_1$. If u satisfies $Pu = 0$ in $Q(2r)$ and (11.64a), *then*

$$|u_x| \le 2\exp(4\beta_0 M)\exp(\beta_2 M^2/r^2)\left(\beta_1+\frac{4M}{r}\right) \tag{11.68}$$

in $Q(r)$.

PROOF. Now we set

$$v(x,y,t) = \pm[u(x,t)-u(y,t)] - M\left[\left(\frac{\rho}{r}-1\right)^+\right]^2 - M\left(\frac{t}{3r^2}-1\right)^+,$$

and observe that

$$\bar{a} \le 4\beta_0(\alpha p_1^2+\beta p_2^2) + \frac{2}{3}\frac{M}{r^2} \le (4\beta_0 + \frac{\beta_2 M}{r^2})(\alpha p_1^2+\beta p_2^2)$$

for $|p_1|, |p_2| \ge \beta_1 + 8M/r$. □

Note that condition (11.67) is exactly the condition from Chapter X which allows a boundary gradient estimate for time-dependent boundary data. As already noted, uniformly parabolic equations satisfy this condition when $\alpha \geq -2$ as does the false mean curvature equation, but the mean curvature equation does not satisfy (11.67).

7. A gradient bound for an intermediate situation

It is also possible to prove a gradient bound for problems with equations in nondivergence form under weaker regularity assumptions than used so far. The argument is basically a reworking of ideas in Section XI.5, although it can be described without reference to the details of that section. Unfortunately, a key structure condition seems to eliminate nonuniformly parabolic equations.

To take advantage of our previous work, we define v, ν, g^{ij}, H, and $d\mu$ as in Section XI.5. For a matrix-valued function (a^{ij}) defined on $\Omega \times \mathbb{R} \times \mathbb{R}^n$, we also define the following functions:

$$\bar{a}^{ijm} = \frac{\partial a^{ij}}{\partial p_m} - \frac{\partial a^{im}}{\partial p_j}, \ \bar{b}^i = -\frac{\partial a^{ij}}{\partial z} p_j - \frac{\partial a^{ji}}{\partial x^j},$$
$$c_k^{ij} = \frac{\partial a^{ij}}{\partial z} p_k + \frac{\partial a^{ij}}{\partial x^k}, \ A^{ij} = v a^{ij},$$

and then we define $\mathscr{C}^2$ and $\mathscr{E}_1$ as in Section XI.5.

We then have the following energy estimate.

LEMMA 11.19. *Let $\zeta \in C^1_{\mathscr{P}}(\Omega)$ and let $u \in C^{2,1}(\Omega)$ solve $Pu = 0$ in Ω. Suppose that conditions* (11.41a,b,c) *hold on Ω_{τ_0} with $D_k^{ij} = v^{-1} c_k^{ij}$ and $C_k^i = a\delta_k^i$ and that* (11.41e) *holds. Suppose also that*

$$v\bar{a}^{ijm}\zeta_{ij}\xi_m \leq \beta_0 (a^{ij}\zeta_{ik}\zeta_{jk})^{1/2}(a^{ij}\xi_i\xi_j)^{1/2}, \tag{11.69a}$$
$$v\bar{b}^i\xi_i \leq \beta_1 \Lambda_0^{1/2}(va^{ij}\xi_i\xi_j)^{1/2} \tag{11.69b}$$

on Ω_{τ_0}. If χ is a Lipschitz function satisfying

$$(\beta_0 - 1)^+ \leq \frac{\chi'(v)(v-\tau)}{\chi(v)} \leq c(\chi) \tag{11.70}$$

for $v \geq \tau$ and if $\tau \geq \tau_0$, then

$$\frac{1}{1+c(\chi)} \sup_{t\in I(\Omega)} \int_{\Omega_\tau(t)} (v-\tau)^2 \chi \zeta^2 \, dx + \int_{\Omega_\tau} [(1-\frac{\tau}{v})\mathscr{C}^2 + \mathscr{E}_1]\chi\zeta^2 \, d\mu \, dt$$
$$\leq 32(1+c(\chi)) \int_{\Omega_\tau} (\beta_1^2\Lambda_0\zeta^2 + \zeta|\zeta_t| + \bar{\mu}|D\zeta|^2)\chi \, d\mu \, dt. \tag{11.71}$$

PROOF. Suppose first that u and Du are in $C^{2,1}$. If we differentiate the equation $Pu = 0$ with respect to x^k, we have

$$\begin{aligned} 0 &= -D_k u_t + a^{ij} D_{ijk} u + D_k(a^{ij}) D_{ij} u + D_k a \\ &= -(D_k u)_t + D_i(a^{ij} D_{jk} u + \delta^i_k a) + \bar{a}^{ijm} D_{ij} u D_{mk} u + \bar{b}^i D_{ik} u + c^{ij}_k D_{ij} u \end{aligned}$$

Now multiply this equation by θv^k and integrate by parts to obtain

$$\begin{aligned} \int -v_t \theta \, dX + \int [a^{ij} D_{jk} u + \delta^i_k a] D_i v^k \theta + [a^{ij} D_j v + a v^i] D_i \theta \, dX \\ + \int [\bar{a}^{ijm} D_{ij} u D_m v + \bar{b}^i D_i v + c^{ij}_k D_{ij} u v^k] \theta \, dX = 0. \end{aligned}$$

Finally, we take $\theta = (v - \tau)^+ \chi(v) \zeta^2$, and estimate the terms as in Lemma 11.10. $\square$

Note that (11.69a) does not hold for equations with elliptic part equivalent to that of the mean curvature equation except when the equation can be written in divergence form. The same is true for the false mean curvature equation. However, all these conditions are satisfied for uniformly parabolic equations.

From the energy inequality, we infer a local pointwise bound for the gradient in terms of a suitable integral as in Lemma 11.11.

LEMMA 11.20. *Suppose P satisfies the hypotheses of Lemma 11.19 and also conditions* (11.44c,d), *and* (11.45) *with $w = v$. If u solves $Pu = 0$ in $Q(2r)$, then*

$$\sup_{Q(r)} \left[\left(1 - \frac{v}{\tau}\right)^+ \right]^{N+2} v^2 \le C r^{-n-2} \int_{Q(2r) \cap \Omega_\tau} v^2 \left(\frac{\Lambda}{\lambda}\right)^{N/2} \Lambda v \, dX. \qquad (11.72)$$

PROOF. From the proof of Lemma 11.11, we see that it suffices to check that the lower bound in (11.70) holds for

$$\chi = \left(\frac{\Lambda}{\lambda}\right)^{N/2} w^{2q-1} (1 - \frac{\tau}{v})^{(N+2)(q-1)}$$

with q large enough. But it is simple to see that it holds for $q > 1 + \beta + \beta_0$. $\square$

If $\left(\frac{\Lambda}{\lambda}\right)^{N/2} \le \beta_2 v^{\beta_3}$, we need only estimate the integral of v^q for q large. Here we take explicit advantage of our assumed uniformly parabolic equation. Specifically, we suppose that there are constants α and β_4, and a decreasing function ε such that

$$v^\alpha |\xi|^2 \le a^{ij} \xi_i \xi_j, \ \Lambda_0 \le \varepsilon(v) v^{\alpha+3}, \ \bar{\mu} \le \beta_4 v^\alpha. \qquad (11.73)$$

We now wish to show that if ε goes to zero or if a modulus of continuity is known for u, then we can estimate the integral of v^q. This estimate follows from a simple modification of the proof of Lemma 11.12.

LEMMA 11.21. *Suppose that conditions* (11.69), (11.41e), *and* (11.73) *hold. Suppose also that* $q \geq 2+\beta+\beta_0+\alpha^+$. *If there is a constant* $\tau_1 \geq \max\{2,\tau_0\}$ *such that* $\varepsilon(\tau_1)\beta_1^2\sigma^2q^2 \leq \frac{1}{4}$, *then*

$$\int_{Q(r)\cap\Omega_\tau} v^q\,dX \leq C(\tau_1+\frac{\sigma}{r})^q r^{n+2}. \tag{11.74}$$

PROOF. Now we set

$$I=\int_{\Omega_\tau} v^q\zeta^q\,dX, I'=\int (v^{q-2}-\tau^{q-2})^+\zeta^q\,|Du|^2\,dX.$$

We can integrate I' by parts, if we write $|Du|^2 = Du\cdot Du$, to see that

$$\begin{aligned} I' = &-(q-2)\int v^{q-3}\zeta^q\bar{u}Du\cdot Dv\,dX \\ &-q\int (v^{q-2}-\tau^{q-2})^+\bar{u}\zeta^{q-1}D\zeta\cdot Du\,dX \\ &-\int (v^{q-2}-\tau^{q-2})^+\zeta^q\bar{u}D_{ii}u\,dX. \end{aligned}$$

The integrals on the right side of this equation are estimate as in Lemma 11.12 to infer that

$$I' \leq \frac{3}{8}I + C(\beta_4)q^2\left(\frac{\sigma}{r}\right)^2\int (v\zeta)^{q-2}\,dX + C(\beta)\frac{\sigma}{r}\int (v\zeta)^{q-1}\,dX$$

for $\tau=\tau_1$. Since $|Du|^2 \geq 3v^2/4$ for $v \geq 2$, it follows that

$$\int_{\Omega_{\tau_1}} (v\zeta)^q\,dX \leq C(\tau_1+\frac{\sigma}{r})^q r^{n+2},$$

and the desired result follows from this one as in Lemma 11.12. □

For uniformly parabolic equations, the conditions are easily verified under the hypotheses

$$\begin{gathered} v^\alpha|\xi|^2 \leq a^{ij}\xi_i\xi_j \leq \theta_1 v^\alpha|\xi|^2, \\ \left|\frac{\partial a^{ij}}{\partial p_m}-\frac{\partial a^{im}}{\partial p_j}\right| \leq \theta_2 v^{\alpha-1}, \\ |a_z^{ij}| = o(v^\alpha),\ |a_x^{ij}| = o(v^{\alpha+1}),\ |a| = o(v^{\alpha+2}). \end{gathered}$$

Hence the assumptions about the derivatives of a, which were used in Sections XI.2 and XI.3, do not affect the gradient bound. As always, when a modulus of continuity is known, the little o conditions can be relaxed to big O.

Notes

The maximum principle method for proving gradient bounds is due to Bernstein [16, 17] and was expanded considerably by Ladyzhenskaya [182] and Ladyzhenskaya and Ural'tseva [186, Section 6] to yield global and local gradient bounds for uniformly parabolic equations; the same authors also were able to deal with some classes of nonuniformly parabolic equations as well in [188, Theorem 3]. Serrin [291] (see Section 6 for the discussion of parabolic equations) exploited the decomposition (11.4) to analyze large classes of equations. All these authors investigated elliptic equations as well, in which case the hypotheses can be weakened by considering equations equivalent to the original one [187, 188] or by introducing appropriate multiplier functions [291]. As in [101, Sections 15.2 and 15.3], we have followed the argument in [291] and we have weakened the assumed regularity of the solutions from $u \in C^{2,1}$ and $Du \in C^{2,1}$ to just $u \in C^{2,1}$. It is clear that this regularity can be further relaxed to $u \in W_2^{2,1}$ and Du continuous.

When the gradient bound is local in time and space, then the global regularity of the solution can be relaxed from global continuity of Du to global continuity of u along with a boundary Lipschitz modulus of continuity estimate. (See Exercise 11.2). Such a result has been known for quasilinear elliptic equations with simple structure for some time [305, Theorem 5], but the general result seems to be quite recent [209, Theorem 4.1] (again, for elliptic equations only).

Other related gradient bounds can be found in [116; 118, Section 2.3; 139, Section 3] and in the references from page 9 of the English translation of [118]. In addition, Krylov [173, 174] has developed an alternative method for proving gradient bounds, which also applies to a large class of degenerate, fully nonlinear equations. The Hölder estimate Lemma 11.4, proved by Ladyzhenskaya and Ural'tseva (see [190]) in 1980, resolved a long standing question of the sufficiency of the natural condition for a gradient bound. As already mentioned in the notes to Chapter VII, Dong [68, Theorem X.7] proved a Hölder estimate for a class of uniformly parabolic equations with λ approximately equal to $1+|p|^\alpha$ with $\alpha \in (0,1)$.

For divergence structure equations, Ladyzhenskaya and Ural'tseva [185, Section 4] proved gradient bounds when the operator is uniformly parabolic as considered here for the entire range $\alpha > -1$; they also proved the Hölder estimate for $\alpha = 0$. DiBenedetto [61, Theorem 1] extended their result to the case $\alpha > 0$, and, along with Chen [51, Theorem], he also proved the Hölder estimate for $\alpha \in (-1,0)$. (In [50, Theorem 1], they had proved the result for A independent of time. See also [52, Theorem 2] for an alternative approach to this estimate.) Our estimate (11.40) is based on work of Trudinger [303, Lemma 5] by way of [101, Theorem 15.6]. For nonuniformly parabolic equations, the first major step was taken by Bombieri, DeGiorgi and Miranda [20] in their proof of interior gradient bounds for the elliptic minimal surface equation in any number of dimensions (although Finn [83, Theorem III] had proved such bounds in the two-dimensional case). Their method depends on an isoperimetric inequality of Federer and Fleming [80, Remark 6.6] and was adapted to similar parabolic equations by Gerhardt [96, Section 3] and Ecker [70, Section 3]. The isoperimetric inequality was refined and modified by Michael and Simon [256, Theorem 2.1] to produce the Sobolev inequality Theorem 11.8. This inequality was used by Simon [292] to derive interior gradient bounds for the prescribed mean curvature equation and a large class of nonuniformly elliptic equations. Simon's bounds were extended to parabolic equations in [204] under slightly different hypotheses from those given here. The structure conditions in this book come from [232, Section 5] in which the weighted inequality Corollary 11.9 first appears.

The case of one space dimension is special. Such problems were investigated by, for example, Oleinik and Venstell′ [272] when a^{11} is independent of Du. Their results were extended to include this dependence by Ladyzhenskaya and Ural′tseva (see [183, Section VI.6]. In [163], Kruzhkov further extended these results and we have followed his proofs in Section XI.6 to prove global and local gradient bounds in very general situations. Since Kruzhkov was concerned with fully nonlinear equations as well, his hypotheses for quasilinear equations are a little stronger than the ones given here.

Although the basic idea in Lemma 11.19 of showing that $|Du|^2$ is a subsolution of on equation in divergence form with terms not involving derivatives of a is well-known from analyses of the Hölder gradient estimate as far back as 1963 [186, Sections 3 and 5], it was not applied to the global gradient bound until quite recently. The application of the idea was first published by Ivanov, Ivochkina, and Ladyzhenskaya [121] in 1980. We have modified their argument only to make it conform to the discussion for divergence structure equations. Further relaxation of the hypotheses, in particular replacement of L^∞ bounds on coefficients by L^p bounds, can be found in [189].

Unlike the situation for a boundary gradient estimate, the conditions implying global and interior gradient bounds do not seem to be necessary in any simple sense for the bounds to hold. In fact, only the recent works [5, 98] address the question of when a solution can fail to exist because a structure condition for the global gradient bound fails. For the elliptic prescribed mean curvature equation

$$\operatorname{div} v = nH(x),$$

Bombieri and Giusti [21, Section 3.3] showed that the condition $DH \in L^\infty$ can be relaxed only slightly. In particular, they allow $v \cdot DH$ to lie in a slightly larger space that L^∞ which includes all L^p spaces with finite p and they show that the equation need not have a Hölder continuous solution even if $v \cdot DH \in L^p$ for all finite p.

Recently, there has been some interest in gradient estimates for equations of divergence form when the inhomogeneous grows quickly. Chen, Nakao, and

Ohara [45,46] have shown that, for equations of the form

$$-u_t + \operatorname{div}(\sigma(|Du|)Du) + g(|Du|) = 0$$

(for suitable functions σ and g, which are essentially power functions), one can estimate the gradient of the solution even if g grows more rapidly than allowed by the results in this chapter. In addition, they obtain some estimates which are almost local in time, only requiring that the initial data have small L^q gradient for sufficiently large q. Their hypotheses include two conditions that seem unnecessary in general: they assume that the domain Ω is a cylinder with cross-section possessing nonnegative mean curvature, and the boundary values of u are always assumed to be zero.

Exercises

11.1 By solving the Ricatti equation

$$\chi' + H\chi^2 + J\chi + K = 0$$

explicitly for constants $H > 0$, $J > -2(HK)^{1/2}$, and $K > 0$, show that this equation has a nonnegative solution on the interval $[a,b]$ (in this case) if and only if

$$b - a < \frac{2}{(HK+4J^2)^{1/2}} F\Big(\frac{J}{2(HK)^{1/2}}\Big),$$

where

$$F(t) = \begin{cases} \left(\frac{1+t^2}{1-t^2}\right)^{1/2} \operatorname{arccot} \frac{t}{(1-t^2)^{1/2}} & \text{if } -1 < t < 1 \\ 2^{1/2} & \text{if } t = 1 \\ \left(\frac{t^2+1}{t^2-1}\right)^{1/2} \operatorname{arccot} \frac{t}{(t^2-1)^{1/2}} & \text{if } t > 1. \end{cases}$$

11.2 Suppose $u \in C(\overline{\Omega})$ with $Du \in C(\Omega)$ and that there are positive constants L_0 and r_0 and an increasing function F such that

$$|Du(X_0)| \le F\left(\frac{\operatorname{osc}_{Q(r)} u}{r}\right)$$

whenever $X_0 \in \Omega$ and $r \le \min\{\frac{1}{2}d(X_0), r_0\}$, and

$$|u(X) - u(Y)| \le L_0 |X - Y|$$

whenever $X \in \Omega$ and $Y \in \Omega$. Show that $|Du|$ is uniformly bounded in Ω and derive an estimate for this bound.

11.3 Suppose that the hypotheses of Lemma 11.11 are satisfied except for (11.41e) and (11.45c). If $v \le \tau$ on $\mathscr{P}\Omega$, show that

$$\sup_{\Omega_\tau} \left(1 - \frac{\tau}{v}\right)^{-n-2} w \le C(n,\beta,\beta_1) \int_{\Omega_\tau} w(\Lambda/\lambda)^{N/2}\, d\mu\, dt.$$

11.4 Show that there is a global gradient bound for equations of the form (11.39) if there are constants $\alpha > -1$ and $\sigma < 1 + \frac{1}{N+2}$ such that

$$\frac{\partial A^i}{\partial p_j}\xi_i\xi_j \geq |p|^\alpha, \quad |p|\,|A_z| + |A_x| + |B| \leq C|p|^{\alpha+\sigma}$$

if $u = 0$ on $\mathscr{P}\Omega$.

11.5 Show that the conclusion of Exercise 11.4 continues to hold if we change the second inequality to

$$|p|\,|A_z| + |A_x| + |B| = o(|p|^{\alpha+1+1/(N+2)}).$$

11.6 Show that, if we replace (11.41d,e) by

$$v D_k^{ij} v_k \zeta^{ij} \leq (\beta_1^2 \Lambda_0 + \beta_8 v)^{1/2} (A^{ij}\zeta_{ik}\zeta_{jk})^{1/2},$$
$$\mathscr{F}_k v^k \leq \beta_1^2 \Lambda_0 + \beta_8 v$$

in Lemma 11.15, then we have

$$\frac{1}{1+c(\chi)} \sup_{t\in I(\Omega)} \int_{\Omega_\tau(t)} (v-\tau)^2 \chi \zeta^2 e^{-\theta t}\, dx$$
$$+ \int_{\Omega_\tau} [(1-\frac{\tau}{v})\mathscr{C}^2 + \mathscr{E}_1]\chi\zeta^2 e^{-\theta t}\, d\mu\, dt$$
$$\leq 32(1+c(\chi)) \int_{\Omega_\tau} (\beta_1^2 \Lambda_0 \zeta^2 + \beta_8 \tau + \bar{\mu}\,|D\zeta|^2) e^{-\theta t} \chi\, d\mu\, dt$$

for ζ depending only on x, where $\theta = 4\beta_8(1+c(\chi))$.

11.7 Use Exercise 11.6 to prove an analog of Lemma 11.15 and then infer a gradient bound for the mean curvature equation if H_z is bounded above. Prove both a global bound and a bound which is local in space.

11.8 Suppose that, in the hypotheses of Theorem 11.16, the assumption on a is relaxed to $a(X,z,p) = a_1(X,z,p) + a_2(X,z)$ with a_1 satisfying (11.65) and

$$|a_{2,x}| \leq \beta_2, a_{2,z} \leq \beta_3$$

for some nonnegative constants β_2 and β_3. Show that the following gradient bound is true:

$$\sup_\Omega |u_x| \leq C(\beta_3 T)\exp((1+\varepsilon)\beta_0 M)(L_0 + \beta_1 + C(\varepsilon)\beta_2 T),$$

where $\varepsilon > 0$, and give expressions for the two constants $C(\beta_3 T)$ and $C(\varepsilon)$.

11.9 Suppose that $A(p) = f(v)p$ for some function f satisfying the inequalities

$$-1 \leq \frac{v f'(v)}{f(v)} \leq k$$

for some positive constant k. Prove a gradient bound for solutions of(11.39) if

$$|B| \leq \theta_1 v^2 f(v),\ B_z + v \cdot B_x \leq \theta_2 v f(v),\ |B_p| + |B_p \cdot p| \leq \theta_3 f(v).$$

(This generalization of the mean curvature problem is discussed in [203, 219, 292].)

11.10 Suppose that we take the following generalization of the false mean curvature equation:

$$-u_t + g((1+|Du|^2)^{1/2}(\Delta u + D_i u D_j u D_{ij} u) + a(X,u,Du) = 0.$$

What hypotheses on g and a guarantee a global gradient bound or a local gradient bound? In particular, if $g(s) = \exp(\frac{1}{2}s^2)$, show that $|a| = o(|Du|\, g)$ implies a global bound.

CHAPTER XII

HÖLDER GRADIENT ESTIMATES AND EXISTENCE THEOREMS

Introduction

We are now ready to complete our existence program by establishing interior and global Hölder estimates for the gradient of a solution of the quasilinear parabolic equation $Pu = 0$ in bounded domains assuming that the solution in question and its gradient are bounded. For divergence structure equations, that is, when P is given by

$$Pu = -u_t + \operatorname{div} A(X,u,Du) + B(X,u,Du), \tag{12.1}$$

we assume that A is differentiable with respect to the variables (x,z,p) and that A and B are continuous with respect to all variables. For nondivergence structure equations, that is, P is given by

$$Pu = -u_t + a^{ij}(X,u,Du)D_{ij}u + a(X,u,Du), \tag{12.2}$$

we assume that a^{ij} is differentiable with respect to (x,z,p) and that a^{ij} and a are continuous with respect to all variables. In the case of one space dimension, we can even remove the differentiability hypothesis. The basic tools for the interior estimates are the Hölder and weak Harnack estimates of Chapters VI and VII. The oscillation estimate Lemma 7.47 for u/x^n will be the key to the boundary estimates.

To illustrate the various structures allowed by the assumptions of the preceding three chapters and this one, we present a select list of existence theorems in the last section of this chapter. Included as special cases are the three examples from Chapter XI: uniformly parabolic equations, the mean curvature equation, and the false mean curvature equation.

1. Interior estimates for equations in divergence form

Suppose that P has the form (12.1) and that A is differentiable with respect to (x,z,p). If $u \in C^{2,1}$ satisfies $Pu = 0$ in Ω, then we showed in Chapter XI that $w = D_k u$ is a weak solution of the equation

$$-w_t + D_i(a^{ij}D_j w + f^i_k) = 0$$

in Ω for $k = 1,\dots,n$, where

$$a^{ij} = \frac{\partial A^i}{\partial p_j},\ f^i_k = \frac{\partial A^i}{\partial z} D_k u + \frac{\partial A^i}{\partial x^k} + \delta^i_k B,$$

and B and the derivatives of A are evaluated at $(X, u(X), Du(X))$. If an upper bound is known for $|u|$ and $|Du|$, then the coefficients a^{ij} and f^i_k are also bounded. More exactly, if

$$|u| + |Du| \le K \text{ in } \Omega \tag{12.3}$$

for some positive constant K, then there are positive constants λ_K, Λ_K and μ_K such that

$$\lambda_K |\xi|^2 \le a^{ij}\xi_i\xi_j,\ |a^{ij}| \le \Lambda_K,\ |f^i_k| \le \mu_K.$$

It then follows from Theorem 6.28 that w is Hölder continuous in Ω. We write this estimate as follows.

THEOREM 12.1. *Let $u \in C^{2,1}$ satisfy $Pu = 0$ in Ω and* (12.3)*, and suppose P has the form* (12.1) *with A differentiable with respect to (x, z, p) and A and B continuous. Then there is a positive constant $\alpha(n, \lambda_K, \Lambda_K)$ such that for any $\Omega' \subset\subset \Omega$, we have*

$$[Du]_{\alpha;\Omega'} \le C(n, K, \lambda_K, \Lambda_K, \mu_K, \operatorname{diam}\Omega) d^{-\alpha}, \tag{12.4}$$

where $d = \operatorname{dist}(\Omega', \mathcal{P}\Omega)$. In addition,

$$\operatorname*{osc}_{Q(X_0,r)} Du \le C(n, K, \lambda_K, \Lambda_K) \left(\frac{r}{R}\right)^{\alpha} \left(\operatorname*{osc}_{Q(X_0,R)} Du + \mu_K R\right) \tag{12.5}$$

as long as $0 < r \le R \le d(X_0)$.

2. Equations in one space dimension

When the number of space dimensions is one, the Hölder gradient estimate is also quite simple. We note now that $w = u_x$ is a weak solution of the equation

$$-w_t + (a^{11}(X, u, u_x) w_x + a(X, u, u_x))_x = 0,$$

and hence as before, w is Hölder continuous.

THEOREM 12.2. *Suppose that $n = 1$ and that a^{ij} and a are continuous with respect to all variables, and let $u \in C^{2,1}$ be a solution of $Pu = 0$ in $\Omega \subset \mathbb{R}^2$ satisfying* (12.3)*. If λ_K, Λ_K and μ_K are constants such that*

$$\lambda_K \le a^{11}(X, z, p) \le \Lambda_K,\ |a(X, z, p)| \le \mu_K \tag{12.6}$$

for $|z| + |p| \le K$, then there is a positive constant $\alpha(n, \lambda_K, \Lambda_K)$ such that for any $\Omega' \subset\subset \Omega$, we have

$$[u_x]_{\alpha;\Omega'} \le C(n, K, \lambda_K, \Lambda_K, \mu_K, \operatorname{diam}\Omega) d^{-\alpha}, \tag{12.7}$$

where $d = \operatorname{dist}(\Omega', \mathcal{P}\Omega)$. In addition, (12.5) *holds as long as $0 < r \le R \le d(X_0)$.*

This theorem could also be proved by noting that $v(x,y,t) = u(x,t) - u(y,t)$ satisfies a linear parabolic equation and then applying Lemma 7.47 (with $x-y$ in place of x^n).

3. Interior estimates for equations in general form

Hölder gradient estimates for parabolic equations of the general form (12.2) are proved by showing that certain combinations of the derivatives of the solutions are subsolutions of linear equations in divergence form. Then we apply the weak Harnack inequality Theorem 6.18 to these combinations to infer the estimates.

To begin, we recall the definitions

$$a^{ijm} = \frac{\partial a^{ij}}{\partial p_m} - \frac{\partial a^{im}}{\partial p_j},$$
$$b^i = -\frac{\partial a^{ij}}{\partial x^j} - p_j\frac{\partial a^{ij}}{\partial z},$$
$$c^{ij}_k = \frac{\partial a^{ij}}{\partial x^k} + p_k\frac{\partial a^{ij}}{\partial z},$$

from Chapter XI and conclude that $D_k u$ is a weak solution of the equation

$$-(D_k u)_t + D_i(a^{ij}D_j(D_k u) + a\delta^i_k) + a^{ijm}D_{ij}uD_m(D_k u)$$
$$+ b^i D_i(D_k u) + c^{ij}_k D_{ij}u = 0.$$

In addition, for $c^{ij} = 2c^{ij}_k p_k - 2a\delta^{ij}$, we see that $v = |Du|^2$ is a weak solution of the equation

$$-v_t + D_i(a^{ij}D_j v + 2aD_i u) + a^{ijm}D_{ij}uD_m v$$
$$+ b^i D_i v + c^{ij}D_{ij}u - 2a^{ij}D_{ik}uD_{jk}u = 0.$$

Adding these equations, we see that, for any $\varepsilon > 0$ and $k = 1,\dots,n$, the functions

$$w^{\pm} = w^{\pm}_k = \pm D_k u + \varepsilon v$$

satisfy

$$-w^{\pm}_t + D_i(a^{ij}D_j w^{\pm} + f^i_k) + a^{ijm}D_{ij}uD_m w^{\pm}$$
$$+ b^i D_i w^{\pm} + (\pm c^{ij}_k + \varepsilon c^{ij})D_{ij}u - \varepsilon a^{ij}D_{ik}uD_{jk}u = 0,$$

where $f^i_k = a[\delta^i_k + 2\varepsilon D_i u]$. Then Scharwz's inequality implies that

$$-w^{\pm}_t + D_i(a^{ij}D_j w^{\pm} + f^i_k) \geq -c_0|Dw|^2 - g \tag{12.8}$$

for

$$g = |b|^2/(\varepsilon\lambda_K) + \sum\left|c^{ij}_k\right|^2/(\varepsilon\lambda_K) + \varepsilon\sum|c^{ij}|^2/\lambda_K,$$
$$c_0 = \sup|a^{ijm}|^2/(\varepsilon\lambda_K) + \varepsilon\lambda_K.$$

Next, we suppose that $Pu = 0$ in $Q(4R)$ and define

$$W_k^{\pm} = \sup_{Q(4R)} w_k^{\pm}.$$

Note that $W^{\pm} - w^{\pm}$ is a supersolution of the equation corresponding to (12.8). In order to apply Theorem 6.18, we need to eliminate the quadratic gradient term, which we do by defining

$$\bar{w} = \frac{1}{\mu}[1 - \exp(\mu(w^{\pm} - W^{\pm}))],$$

with $\mu = 2c_0/\lambda_K$. For

$$\bar{f}_k^i = (\mu\bar{w}+1)f_k^i,\ \bar{g} = (\mu\bar{w}+1)(2c_0\sum_i (f_k^i)^2/\lambda_K^2 + g),$$

we then have

$$-\bar{w}_t + D_i(a^{ij}D_j\bar{w} + \bar{f}_k^i) - \bar{g} \le 0,$$

so the weak Harnack inequality Theorem 6.18 implies that

$$\frac{1}{|\Theta(R)|}\int_{\Theta(R)} \bar{w}\, dX \le c_1(n,\lambda_K,\Lambda_K)(\inf_{Q(R)} \bar{w} + F_K R + G_K R^2),$$

where $F_K \ge \sup\left|\bar{f}_k^i\right|$, $G_K \ge \sup \bar{g}$. Assuming that $\varepsilon = \eta/K$ with $0 < \eta \le 1$ (which turns out to be the correct form for our later considerations), we see that

$$W^{\pm} - w^{\pm} \le \bar{w} \le \exp(c_0 K)(W^{\pm} - w^{\pm}),$$

and hence

$$\frac{1}{|\Theta(R)|}\int_{\Theta(R)} (W^{\pm} - w^{\pm})\, dX \le c_2(\inf_{Q(R)} (W^{\pm} - w^{\pm}) + F_K R + G_K R^2)$$

for $c_2 = c_1 \exp(c_0 K)$.

To proceed, we note that $w_k^+ + w_k^- = 2\varepsilon v$ and hence

$$\begin{aligned} W_k^+ - w_k^+ + W_k^- - w_k^- &\ge \operatorname*{osc}_{Q(4R)} D_k u + 2\varepsilon[\inf_{Q(4R)} v - v] \\ &\ge \operatorname*{osc}_{Q(4R)} D_k u - 2\varepsilon \operatorname*{osc}_{Q(4R)} v \end{aligned}$$

in $Q(4R)$. In addition,

$$\inf_{Q(R)} (W^{\pm} - w^{\pm}) = \sup_{Q(4R)} w^{\pm} - \sup_{Q(R)} w^{\pm} \le \operatorname*{osc}_{Q(4R)} w^{\pm} - \operatorname*{osc}_{Q(R)} w^{\pm}.$$

Setting

$$\omega_k(\rho) = \operatorname*{osc}_{Q(\rho)} w_k^+ + \operatorname*{osc}_{Q(\rho)} w_k^-,$$

we infer that

$$\operatorname*{osc}_{Q(4R)} D_k u - 2\varepsilon \operatorname*{osc}_{Q(4R)} v \le c_2(\omega_k(4R) - \omega_k(R) + F_K R + G_K R^2). \tag{12.9}$$

Finally, we set $\omega = \sum_k \omega_k$ and note that

$$\operatorname*{osc}_{Q(\rho)} v \leq K \sum_k \operatorname*{osc}_{Q(\rho)} D_k u$$

for any ρ so that

$$\sum_k \operatorname*{osc}_{Q(\rho)} D_k u \leq \omega(\rho) \leq 3 \sum_k \operatorname*{osc}_{Q(\rho)} D_k u \tag{12.10}$$

provided $\varepsilon \leq 1/(2nK)$. In particular, for $\varepsilon = 1/(10nK)$, we obtain after summing on k in (12.9) that

$$\omega(4R) \leq 3nc_2(\omega(4R) - \omega(R) + F_K R + G_K R^2).$$

Now we estimate F_K and G_K as follows: Choose μ_K so that

$$\mu_K \geq K[|a_x^{ij}| + |a_z^{ij}|\,|p|] + |a|$$

for $|z| + |p| \leq K$, and note that

$$F_K \leq C(c_0 K)\mu_K, G_K \leq C(c_0 K)\mu_K^2/K.$$

Hence

$$\omega(4R) \leq c_3(\omega(4R) - \omega(R) + \mu_K R),$$

where c_3 is determined by n, λ_K, Λ_K, $\sup\left|a_p^{ij}\right| K$, and $\mu_K R/K$. As usual, it follows that

$$\omega(r) \leq C\left(\frac{r}{R}\right)^{\alpha} (\omega(R) + \mu_K R)$$

for C and α determined by c_3. Invoking (12.10) provides the estimate for Du.

THEOREM 12.3. *Let $u \in C^{2,1}$ satisfy $Pu = 0$ in Ω and* (12.3), *and suppose P has the form* (12.2) *with a^{ij} differentiable with respect to (x,z,p) and a^{ij} and a continuous. Then there is a positive constant α determined only by n, λ_K, Λ_K, $\sup\left|a_p^{ij}\right| K$, and $\mu_K R/K$ such that for any $\Omega' \subset\subset \Omega$, we have* (12.4). *In addition,* (12.5) *holds, with C depending also on $\mu_K R/K$ as long as $0 < r \leq R \leq d(X_0)$.*

4. Boundary estimates

To estimate the Hölder norm of the gradient of a solution of $Pu = 0$ near $\mathscr{P}\Omega$, we use the boundary estimate Lemma 7.47 and a related result valid near $B\Omega$. The connection between the pointwise behavior of u from Lemma 7.47 and the behavior of Du from the theorems in this chapter is seen through the following simple lemma.

LEMMA 12.4. *Let $u \in C(\bar{Q})$ and suppose $Du \in C(Q)$ for some cylinder $Q = Q(X_1,R)$. Suppose moreover that there are nonnegative constants c_1, c_2 and α with $\alpha < 1$ such that*

$$\operatorname*{osc}_{Q(X_0,r)} Du \leq c_1 \left(\frac{r}{\rho}\right)^{\alpha} (\operatorname*{osc}_{Q(X_0,\rho)} Du + c_2\rho^{\alpha}) \tag{12.5$'$}$$

whenever $r \le \rho$ and $Q(X_0,\rho) \subset Q$. Then for any $L \in \mathbb{R}^n$ and $U \in \mathbb{R}$, we have

$$\sup_{Q(X_1,R/2)} |Du - L| \le C(c_1,n,\alpha)[R^{-1}\sup_{Q}|u(X) - L\cdot x - U| + c_2 R^{\alpha}]. \tag{12.11}$$

PROOF. Define

$$v(X) = u(X) - L\cdot x - U$$

and note that (12.5)′ is the same as

$$\operatorname*{osc}_{Q(X_0,r)} Dv \le c_1 \left(\frac{r}{\rho}\right)^{\alpha} (\operatorname*{osc}_{Q(X_0,\rho)} Dv + c_2\rho^{\alpha}).$$

It follows that

$$[Dv]^{(1)}_{\alpha} \le c_1([Dv]^{(1)}_0 + c_2 R^{1+\alpha}),$$

and then the interpolation inequality (4.2c) gives

$$[Dv]^{(1)}_{\alpha} \le C(c_1,\alpha)(|v|_0 + c_2 R^{1+\alpha})$$

and

$$|Dv|^{(1)}_0 \le C(c_1,\alpha)(|v|_0 + c_2 R^{1+\alpha}).$$

In particular, if $|X - X_0| \le R/2$, then

$$|Dv(X)| \le C(c_1,\alpha)(R^{-1}|v|_0 + c_2 R^{\alpha}),$$

and writing this inequality in terms of u, L and U gives (12.11). □

From Lemma 12.4 and our interior Hölder gradient bounds, we then infer an estimate near $S\Omega$.

THEOREM 12.5. *Let $u \in C^{2,1} \cap H_1(\Omega)$ satisfy $Pu = 0$ in Ω, $u = \varphi$ on $S\Omega$, and suppose that $\mathscr{P}\Omega \in H_{1+\beta}$, $\varphi \in H_{1+\beta}$ for some $\beta \in (0,1]$ with $|\varphi|_{1+\beta} = \Phi$. Suppose also that P satisfies the hypotheses of Theorem 12.1, Theorem 12.2 or Theorem 12.3 uniformly in Ω. Then there are constants C_1 and α depending on n, λ_K, Λ_K and also $\mu_K R/K$ under the assumptions of Theorem 12.3, such that for any $X_0 \in S\Omega$, we have*

$$\operatorname*{osc}_{\Omega(X_0,r)} Du \le C[\left(\frac{r}{R}\right)^{\alpha} K + \Phi r^{\beta} + \mu_K r] \tag{12.12}$$

as long as $0 < r < R \le t_0^{1/2}$.

PROOF. Without loss of generality, we may assume $X_0 = 0$, $\mathscr{P}\Omega[X_0,R] = Q^0(R)$, and $\Omega[X_0,R] = Q^+(R)$. Then Lemma 7.47 implies that $D_n u$ exists on $Q^0(R/2)$. Let us extend φ (as in that lemma) so that $D_n\varphi = 0$ on Q^0.

Now we fix $r \in (0,R/2)$ and $X_1 \in Q^+(r)$ and we set $L = Du(X_1')$ and $U = \varphi(X_1') - L'\cdot x_1$. We then have

$$u(X) - L\cdot x - U = [u(X) - \varphi(X) - L_n x^n] \\ + [\varphi(X) - \varphi(X')] + [\varphi(X') - L'\cdot(x - x_1) - \varphi(X_1')].$$

Since $\varphi \in H_{1+\beta}$ and $D_n\varphi = 0$ on Q^0, it follows that

$$|\varphi(X)-\varphi(X')| \le C\Phi R_0^{1+\beta},\ |\varphi(X') - L'\cdot(x-x_0)-\varphi(X_0)]| \le \Phi R_0^{1+\beta},$$

where $R_0 = x_1^n/2$. Then Lemma 7.47 implies that

$$\left|L_n - \frac{u(X)-\varphi(X)}{x^n}\right| \le C[K\left(\frac{r}{R}\right)^\alpha + \Phi r^\beta + \mu_K r]$$

so that

$$|u(X)-L\cdot x-U| \le CR_0[K\left(\frac{r}{R}\right)^\alpha + \Phi r^\beta + \mu_K r].$$

Applying Lemma 12.4 in $Q(X_1,R_0)$ then yields

$$|Du(X_1)-Du(X_1')| \le C[K\left(\frac{r}{R}\right)^\alpha + \Phi r^\beta + \mu_K r]$$

since we may assume that α from Lemma 7.47 is the same as α from any one of Theorems 12.1, 12.2 and 12.3. Since Lemma 7.47 also implies that

$$|Du(X_1')-Du(0)| \le C[K\left(\frac{r}{R}\right)^\alpha + \Phi r^\beta + \mu_K r],$$

we infer (12.12). □

For regularity near $B\Omega$, we use the following variation on Lemma 7.47, for which we define

$$Q'(R) = \{|x| < R, 0 < t < R^2\}.$$

LEMMA 12.6. *Let F_0, R, Λ, and β be positive constants with $\beta \le 1$. Let (a^{ij}) be a positive definite matrix defined on $Q'(R)$ with $\mathcal{T} \le \Lambda$, define P by $Pw = -w_t + a^{ij}D_{ij}w$, and let $\varphi \in H_{1+\beta}(B(R))$. Suppose that $u \in C^{2,1}(Q')$ satisfies the conditions $|Pu| \le F_0 t^{(\beta-1)/2}$ on Q' and $u(\cdot,0) = \varphi$ on $B(R)$. Then v defined by*

$$v(X) = u(X) - D\varphi(0)\cdot x - \varphi(0) \tag{12.13}$$

can be estimated by

$$\sup_{Q'(r)} |v| \le C(\Lambda)[(F_0 + [D\varphi]_\beta)r^{1+\beta} + \sup_{Q'(R)} |v| \left(\frac{r}{R}\right)^{1+\beta}]. \tag{12.14}$$

PROOF. Set

$$M = \sup_{Q'(R)} |v|,$$

and define $w_1 = t^{(1+\beta)/2}$, $w_2 = (|x|^2 + 2\Lambda t)^{(1+\beta)/2}$, and

$$w = 2F_0 w_1 + \left(\frac{M}{R^{1+\beta}} + [D\varphi]_\beta\right) w_2.$$

A simple calculation shows that $Pw \le P(\pm v)$ in $Q'(R)$ and $w \ge \pm v$ on $\mathscr{P}Q'(R)$, so the maximum principle implies that $w \ge \pm v$ in $Q'(R)$. Therefore

$$\sup_{Q'(r)} |v| \le \sup_{Q'(r)} w = 2F_0 r^{1+\beta} + (1+2\Lambda)^{(1+\beta)/2}\left([D\varphi]_\beta + \frac{M}{R^{1+\beta}}\right) 2r^{1+\beta}.$$

Inequality (12.14) now follows easily from this one because $(1+2\Lambda)^{(1+\beta)/2} \le 1+2\Lambda$. □

With (12.14) in place of (7.66), we follow the proof of Theorem 12.5 to infer regularity near the initial surface.

THEOREM 12.7. *Let P satisfy the hypotheses of any of Theorems 12.1, 12.2 or 12.3 uniformly in $\Omega \cap B\Omega$. If also $u = \varphi$ on $B\Omega$, then* (12.12) *holds for $X_0 \in B\Omega$ as long as $0 < r < R < \operatorname{dist}(X_0, S\Omega)$.*

Finally, for regularity near a point of $C\Omega$, we use the following simple variant of Lemma 12.6. To simplify notation, we define

$$Q_+(R) = \{|x| < R, 0 < t < R^2, x^n > 0\},$$
$$Q_0(r) = \{|x| < R, 0 < t < R^2, x^n = 0\}.$$

LEMMA 12.8. *Let F_0, R, β, λ, and Λ be positive constants with $\beta \le 1$. Let (a^{ij}) be a positive definite matrix defined on $Q'(R)$ with $\mathscr{T} \le \Lambda$ and*

$$a^{ij}\xi_i\xi_j \ge \lambda|\xi|^2, \tag{12.15}$$

define P by $Pw = -w_t + a^{ij}D_{ij}w$, and let $\varphi \in H_{1+\beta}(Q_+(R))$. Suppose that $u \in C^{2,1}(Q')$ satisfies the conditions $|Pu| \le F_0 \max\{t^{(\beta-1)/2}, \lambda[x^n]^{\beta-1}\}$ on Q_+ and $u = \varphi$ on Q_0 and on $B(R) \times \{0\}$. Then v defined by (12.13) *can be estimated by*

$$\sup_{Q_+(r)} |v| \le C(\beta,\Lambda)[(F_0+F_1)r^{1+\beta} + \sup_{Q_+(R)} |v| \left(\frac{r}{R}\right)^{1+\beta}], \tag{12.14$'$}$$

where $F_1 = [D\varphi]_\beta + \langle\varphi\rangle_{1+\beta}$.

PROOF. We now choose $r < R$ and set

$$w_3 = 2t^{(1+\beta)/2} + \frac{1}{\beta}r^{1+\beta} - \frac{1}{\beta}(x^n)^{1+\beta} + \frac{r^{\beta-1}}{\beta}\left((x^n)^2 + 2\Lambda t\right).$$

Then

$$w = 2F_0 w_3 + \left(\frac{M}{R^{1+\beta}} + F_1\right) w_2$$

satisfies $Pw \le P(\pm v)$ in $Q_+(R)$ and $w \ge \pm v$ on $\mathscr{P}Q_+(R)$, so $w \ge \pm v$ in $Q_+(R)$. It follows that

$$\sup_{Q_+(r)} |v| \le \sup_{Q_+(r)} w \le \frac{4+2\Lambda}{\beta} F_0 r^{1+\beta} + \left(\frac{M}{R^{1+\beta}} + F_1\right)(1+2\Lambda)^{(1+\beta)/2} r^{1+\beta},$$

which implies (12.14)$'$ as before. □

Combining Lemma 12.8 with Theorem 12.5 gives the Hölder gradient estimate near $C\Omega$.

THEOREM 12.9. *Suppose that the hypotheses of one of Theorems 12.1, 12.2 or 12.3 hold uniformly in the intersection of Ω with a neighborhood N of a point $X_0 \in C\Omega$. If $\mathcal{P}\Omega \cap N \in H_{1+\beta}$ and $u = \varphi$ on $\mathcal{P}\Omega \cap N$ for some $\varphi \in H_{1+\beta}(\mathcal{P}\Omega \cap N)$, then* (12.12) *holds as long as $R \leq \operatorname{dist}(X_0, \Omega \setminus N)$.*

Combining Theorems 12.5, 12.7, and 12.9 gives a global Hölder gradient bound.

THEOREM 12.10. *Suppose P has one of the forms* (12.1) *or* (12.2) *in Ω with $\mathcal{P}\Omega \in H_{1+\beta}$. If it has the form* (12.1), *suppose also that A is continuously differentiable with respect to (x,z,p). If it has the form* (12.2), *suppose either that $n = 1$ or that a^{ij} is continuously differentiable with respect to (x,z,p). If $u \in C^{2,1} \cap C(\overline{\Omega})$ with $Du \in L^\infty$ satisfies $Pu = 0$ in Ω, $u = \varphi$ on $\mathcal{P}\Omega$ for some $\varphi \in H_{1+\beta}$, then there are positive constants α and C determined only by n, β, λ_K, Λ_K, and* $\operatorname{diam}\Omega$ *(and also μ_K/K if P has the form* (12.2) *and $n > 1$) such that*

$$[Du]_\alpha \leq C[K + |\varphi|_{1+\beta} + \mu_K]. \tag{12.16}$$

5. Improved results for nondivergence equations

In fact, the Hölder gradient estimate for nondivergence structure equations is true under weaker conditions on the coefficients than we have already studied. Here we show how to relax the regularity with respect to x and z. Although this result does not have any direct relation to the existence questions studied in this chapter, the argument will prove useful in our study of oblique derivative problems in Chapter XIII. As was the case for our estimates for linear equations, we first study a problem with "frozen" coefficients. In this situation, the coefficient a^{ij} will still depend on p.

LEMMA 12.11. *Suppose that a^{ij} is a Lipschitz, matrix-valued function defined on $\mathbb{R}^n$ and that there are positive constants K, λ, λ_0, and Λ such that*

$$a^{ij}(p)\xi_i\xi_j \geq \lambda|\xi|^2, \tag{12.17a}$$

$$\left|a^{ij}(p)\right| \leq \Lambda, \tag{12.17b}$$

$$\left|a^{ij}_p(p)\right| \leq \lambda_0 \text{ if } |p| \leq K, \tag{12.17c}$$

$$\left|a^{ij}_p(p)\right| = 0 \text{ if } |p| \geq K. \tag{12.17d}$$

Fix $R > 0$ and set $Q = Q(R)$. Then for any $u \in C(\bar{Q})$, there is a unique $C^{2,1}(Q) \cap C(\bar{Q})$ of

$$-v_t + a^{ij}(Dv)D_{ij}v = 0 \text{ in } Q, v = u \text{ on } \mathcal{P}Q. \tag{12.18}$$

PROOF. If $u \in H_{1+\alpha}$ for some $\alpha > 0$, then we can use Corollary 9.3 and Theorems 9.5, 10.4, 11.1, and 12.10 to infer that (12.18) is uniquely solvable. For arbitrary continuous u, let (u_m) be a sequence of $H_{1+\alpha}$ functions converging uniformly to u and use v_m to denote the solution of (12.18) with u_m replacing u. An easy application of Theorem 9.1 shows that (v_m) is uniformly Cauchy in Q. In addition, (11.22) and Theorem 12.2 imply that (Dv_m) is uniformly bounded in H_β on any $Q(r)$ with $r < R$, with β possibly depending also on r. The linear theory then implies that (v_m) is Cauchy in $C^{2,1}(Q(r))$ for any $r < R$. Hence the limit $v = \lim v_m$ exists pointwise as do Dv, D^2v and v_t. It is simple to check that v solves (12.18) and uniqueness follows from Corollary 9.3. □

Our next step is an alternative characterization of functions with Hölder continuous gradient.

LEMMA 12.12. *Let $u \in C(Q(R))$ for some $R > 0$ and suppose that there are positive constants c_1 and α with $\alpha \le 1$ such that for any $X_0 \in Q(R/2)$ and any $r \le R/2$, there is a vector $V(X_0, r)$ satisfying*

$$|u(X) - u(X_0) - V(X_0, r) \cdot (x - x_0)| \le c_1 r^{1+\alpha} \tag{12.19}$$

whenever $|X - X_0| \le r$. then $Du \in H_\alpha(Q(R/2))$ and

$$[Du]_{\alpha, Q(R/2)} \le C(n, \alpha) c_1. \tag{12.20}$$

PROOF. It's easy to check that the mollification $u(X, \tau)$ satisfies the inequalities

$$|u_{xx}| + |u_{x\tau}| + |u_{\tau\tau}| \le Cc_1 \tau^{\alpha-1}, \ |u_{xt}| + |u_{t\tau}| \le Cc_1 \tau^{\alpha-2}.$$

As in Chapter IV (see, for example, Lemma 4.3), these inequalities give (12.20). □

From these two preliminary results, we now prove a Hölder gradient bound in small cylinders.

LEMMA 12.13. *Let R and K be positive constants, and let (a^{ij}) be a matrix valued function defined on $Q(R) \times \{|p| \le K\}$ such that there are positive constants λ, Λ, and λ_0 such that* (12.15) *and*

$$\mathcal{T} \le \Lambda, \ |a^{ij}_p| \le \lambda_0 \tag{12.21}$$

hold. Suppose that $u \in C^{2,1}(Q(R))$ solves $-u_t + a^{ij}(X, Du) D_{ij} u + a(X) = 0$ in $Q(R)$ with $|Du| \le K$ in $Q(R)$. If a is bounded with $|a| \le \mu$ in $Q(R)$ for some nonnegative constant μ, then there are positive constants C, α, and σ determined only by K, n, λ, λ_0, and Λ such that $|a^{ij}(X,p) - a^{ij}(Y,p)| \le \sigma$ implies

$$\operatorname*{osc}_{Q(r)} Du \le C\left(\frac{r}{R}\right)^\alpha [\operatorname*{osc}_{Q(R)} Du + \mu R]. \tag{12.22}$$

PROOF. The idea here is to prove (12.19) with a suitable constant c_1. Hence, we fix $X_0 \in Q(R)$, and we set $R_0 = d(X_0)/2$ and $M = [Du]^*_\alpha$ for α to be determined. We also choose $r < R_0$ and suppose that $V(X_0,r)$ has been determined.

Next we extend a^{ij} to all of $Q(R) \times \mathbb{R}^n$ so that the hypotheses of Lemma 12.11 hold. First, for each X, we define a Lipschitz extension $(A^{ij}(X,\cdot))$ satisfying $|A^{ij}| \le \Lambda$ and $\left|A^{ij}_p\right| \le \lambda_0$, and note that $A^{ij}(X,p)\xi_i\xi_j \ge (\lambda/2)|\xi|^2$ for $|p| \le K + (\lambda/2)\lambda_0$. Now we define

$$\zeta(p) = \min\{1, \frac{2\lambda_0}{\lambda}(|p|-K)^+\}$$

and $a^{ij} = \zeta\lambda\delta^{ij} + (1-\zeta)A^{ij}$, so that all the hypotheses of Lemma 12.11 hold with constants determined only by K, λ, λ_0, and Λ. Hence there is a solution v of

$$-v_t + a^{ij}(X_0,Dv)D_{ij}v = 0 \text{ in } Q(X_0,r),\ v = u \text{ on } \mathcal{P}Q(X_0,r).$$

Now we define $\bar{u}$ by

$$\bar{u}(X) = u(X) - u(X_0) - V(X_0,r)\cdot(x - x_0),$$

and we set $M = \sup_{Q(X_0,r)} \bar{u}$, $Q = Q(X_0,r/2)$, and $H = M + \mu r^2$. Then Corollary 7.45 gives

$$[\bar{u}]_{\theta,Q} \le Cr^{-\theta}H$$

for positive constants C and θ determined only by n, λ, and Λ. Writing $d = \operatorname{dist}(X,\mathcal{P}Q)$, we see that there is a function $U \in H^{(-\theta)}_{2+\alpha}(Q)$ with $U = \bar{u}$ on $\mathcal{P}Q$ and

$$|U_t| + \left|D^2U\right| \le Cr^{-\theta}Hd^{\theta-2},\ |DU| \le Cr^{-\theta}Hd^{\theta-1}.$$

The proof of Lemma 4.16 implies that $|\bar{u} - U| \le CH(\frac{d}{r})^\theta$ and a similar estimate holds for $\bar{v} = \bar{u} + v - u$. For $\delta_0 \in (0,1)$ to be further specified we set $Q' = Q(X_0,(1-\delta_0)r/2)$. Since

$$\sup_S |u - v| \le \operatorname*{osc}_S(u - v) = \operatorname*{osc}_S(\bar{u} - \bar{v}) \le \operatorname*{osc}_S \bar{u} + \operatorname*{osc}_S \bar{v}$$

for any set S meeting $\mathcal{P}Q$, it follows that

$$\sup_{Q\setminus Q'} |u - v| \le C\delta_0^\theta[M + \mu r^2].$$

For $\delta \in (0,1)$ to be further specified, we take δ_0 so small that $C\delta_0^\theta \le \delta/2$. Hence we have

$$\sup_{Q\setminus Q'} |u - v| \le \frac{\delta}{2}M + C\mu r^2$$

with C independent of δ. Now on Q', we can use (11.24) and Theorem 12.2 to infer that there are positive constants $C(\delta)$ and $\varepsilon(\delta)$ (depending on δ) so that

$$[Dv]^*_{\varepsilon(\delta),Q'} \le C(\delta).$$

Now we use the linear theory to infer that

$$\sup_{Q'} |D^2 v| \le C(\delta) r^{-2} \sup_{Q'} \bar{v} \le C(\delta) r^{-2} \sup_{Q} \bar{u}.$$

With this pointwise estimate for $D^2 v$, we now prove that u and v are close on Q'. To this end, we set

$$w = A(t - t_0) + \frac{\delta}{2} M + C\mu r^2$$

with $A = \sigma C(\delta) r^{-2} M + \mu$. We then have

$$\begin{aligned} P(v+w) &= -(v+w)_t + a^{ij}(X, D(v+w)) D_{ij}(v+w) + a \\ &= -v_t + a^{ij}(X, Dv) D_{ij} v + a - A \\ &\le [a^{ij}(X, Dv) - a^{ij}(X_0, Dv)] D_{ij} v + \mu - A \\ &\le \sigma C(\delta) r^{-2} M + \mu - A = 0 \end{aligned}$$

in Q'. Since $A \ge 0$ it follows that $u \le v + w$ on $\mathscr{P}Q'$ and then the comparison principle implies that $u \le v + w$ in Q'. Similarly, $u \ge v - w$ in Q' and therefore

$$\sup_{Q} |u - v| \le [\sigma C(\delta) + \frac{\delta}{2}] M + C\mu r^2. \tag{12.23}$$

Now we analyze the behavior of Dv in Q. Since (11.24) implies that $|Dv| \le C$ in $Q(X_0, r/4)$, it follows that there are constants C and β (independent of δ) such that

$$\operatorname*{osc}_{Q(X_0,\rho)} Dv \le C \left(\frac{\rho}{r}\right)^{\beta} \operatorname*{osc}_{Q(X_0,r/4)} Dv$$

provided $\rho \le r/4$. Arguing as in Lemma 12.4, we find that

$$\sup_{Q(X_0,\rho)} |v - Dv(X_0) \cdot (x - x_0) - v(X_0)| \le C \left(\frac{\rho}{r}\right)^{1+\beta} \sup \bar{v}.$$

Combining this inequality with (12.23) then gives a constant C_1 such that

$$\begin{aligned} &\sup_{Q(X_0,\rho)} |u - Dv(X_0) \cdot (x - x_0) - u(X_0)| \\ &\qquad \le [C_1 \left(\frac{\rho}{r}\right)^{1+\beta} + \delta + \sigma C(\delta)] M + c\mu r^2. \end{aligned}$$

Now we choose $\alpha \in (0, \beta)$ and then take $\tau < \frac{1}{2}$ so that $C_1 \tau^{1+\beta} \le \frac{1}{2} \tau^{1+(\alpha+\beta)/2}$. If $\delta = \tau^{1+(\alpha+\beta)/2}/4$ and $\sigma = \delta / C(\delta)$, then, by setting $\rho = \tau r$ and $V(X_0, \tau r) =$

$Dv(X_0)$, we find that

$$\begin{aligned}\sup_{Q(X_0,\tau r)} &|u-u(X_0)-V(X_0,\tau r)\cdot(x-x_0)| \\ &\le \tau^{1+(\alpha+\beta)/2}\sup_{Q(X_0,r)}|u-u(X_0)-V(X_0,r)\cdot(x-x_0)| \\ &\quad + C\mu r^2.\end{aligned}$$

In general, we define $V(X_0,R_0)=Du(X_0)$ and then define $L(X_0,\tau^k R_0)$ inductively by the preceding description. Finally if $\tau^{k+1}R_0<r\le\tau^k R_0$ for some nonnegative integer k, we define $V(X_0,r)=V(X_0,\tau^k R_0)$. The usual iteration scheme gives

$$\begin{aligned}\sup_{Q(X_0,r)} &|u-u(X_0)-V(X_0,r)\cdot(x-x_0)| \\ &\le C\left(\frac{r}{R_0}\right)^{1+\alpha}\left[\sup_{Q(X_0,R_0)}|u-u(X_0)-V(X_0,R_0)\cdot(x-x_0)|+\mu R_0^2\right] \\ &\le CR_0^{-\alpha}r^{1+\alpha}[K_1+\mu R_0]\end{aligned}$$

for $K_1=\operatorname{osc}_{Q(R_0)}Du$. This inequality leads to (12.22) by simple algebra and Lemma 12.12. □

6. Selected existence results

In this section, we prove the existence of solutions to the Cauchy-Dirichlet problem for various quasilinear parabolic equations, recalling that Theorems 8.3 and 4.29 show that the existence is an immediate consequence of a Hölder estimate on the gradient and a global bound for u and Du. Rather than attempt to construct examples showing all possible combinations of our structure conditions, we focus mainly on the specific examples already discussed in previous chapters.

6.1. Uniformly parabolic equations. Suppose first that P has the divergence form (12.1) with $A\in H_{1+\alpha}(K)$ and $B\in H_\alpha(K)$ for any bounded subset K of $\Omega\times\mathbb{R}\times\mathbb{R}^n$. From Corollary 9.10 and Lemmata 11.11, 11.12, and 11.13, along with Theorems 10.4 and 12.10, we have the following existence result.

THEOREM 12.14. *Let $\mathscr{P}\Omega\in H_{1+\beta}$ and $\varphi\in H_{1+\beta}$ for some $\beta\in(0,1)$. Suppose that there are nonnegative constants a_0, a_1, b_0, b_1, and M_0 with $a_0>0$ such that*

$$p\cdot A(X,z,p)\ge a_0|p|^2-a_1|z|^2 \tag{12.24a}$$

$$zB(X,z,p)\le b_0(p\cdot A(X,z,p))^+ + b_1|z|^2 \tag{12.24b}$$

for $|z| \geq M_0$. *Suppose also that there is a positive function* λ *such that* $a^{ij} = \partial A^i/\partial p_j$ *satisfies*

$$a^{ij}\xi_i\xi_j \geq \lambda(|z|)\,|\xi|^2, \tag{12.25a}$$

and that

$$|p|^2\,|A_p| + |p|\,|A_z| + |A_x| + |B| = O(|p|^2) \tag{12.25b}$$

as $|p| \to \infty$. *If* A_p, A_z, A_x, *and* B *are Hölder continuous with exponent* α *as functions of* (X,z,p), *then there is a solution* $u \in H_{2+\alpha}^{(-1-\beta)}$ *of* $Pu = 0$ *in* Ω, $u = \varphi$ *on* $\mathcal{P}\Omega$.

If B is Lipschitz with respect to z and p, then uniqueness of the solution in $C^0 \cap C^{2,1}$ follows from Theorem 9.7. In addition, if $\mathcal{P}\Omega \in H_{2+\alpha}$ and $\varphi \in H_{2+\alpha}$ and if φ satisfies the compatibility condition $P\varphi = 0$ on $C\Omega$, then the linear theory implies that $u \in H_{2+\alpha}$.

Our existence result is easily generalized to other divergence structure equations which retain the uniform parabolicity.

THEOREM 12.15. *Let* $\mathcal{P}\Omega \in H_{1+\beta}$ *and* $\varphi \in H_{1+\beta}$ *for some* $\beta \in (0,1)$. *Suppose that there are nonnegative constants* a_0, a_1, b_0, b_1, m, m', *and* M_0 *with* $a_0 > 0$ *and* $m > \max\{1, m'\}$ *such that*

$$p \cdot A(X,z,p) \geq a_0\,|p|^m - a_1\,|z|^{\max\{2,m'\}} \tag{12.24a$'$}$$

$$zB(X,z,p) \leq b_0(p \cdot A(X,z,p))^+ + b_1\,|z|^{\max\{2,m'\}} \tag{12.24b$'$}$$

for $|z| \geq M_0$. *Suppose also that there is a positive function* λ *such that* $a^{ij} = \partial A^i/\partial p_j$ *satisfies*

$$a^{ij}\xi_i\xi_j \geq \lambda(|z|)(1+|p|)^{m-2}\,|\xi|^2, \tag{12.25a$'$}$$

and that

$$|A_p| = O(|p|^{m-2}),\ |p|\,|A_z| + |A_x| + |B| = o(|p|^m) \tag{12.25b$'$}$$

as $|p| \to \infty$. *If* A_p, A_z, A_x, *and* B *are Hölder continuous with exponent* α *as functions of* (X,z,p), *then there is a solution* $u \in H_{2+\alpha}^{(-1-\beta)}$ *of* $Pu = 0$ *in* Ω, $u = \varphi$ *on* $\mathcal{P}\Omega$.

Note that conditions (12.24a,b)$'$ can be weakened in case $m > 2$. According to Theorem 9.11, we can assume that $m' = m$ as long as a_1 and b_1 are sufficiently small. In addition, the term $o(|p|^m)$ in hypothesis (12.25b)$'$ can be relaxed to $O(|p|^m)$ by the Hölder estimates of DiBenedetto [61] for $m > 2$ and Chen and DiBenedetto [50,51] for $m \in (1,2)$.

When P has the form (12.2) with (a^{ij}) satisfying (12.25a), we use Theorems 9.5, 10.4, 12.10 along with Lemmata 11.20 and 11.21 to infer existence of solutions to the Cauchy-Dirichlet problem.

THEOREM 12.16. *Let $\mathscr{P}\Omega \in H_{1+\beta}$ and $\varphi \in H_{1+\beta}$ for some $\beta \in (0,1)$. Suppose that there are nonnegative constants b_1 and k such that*

$$za(X,z,0) \leq b_1 |z|^2 + k \tag{12.26}$$

Suppose that (a^{ij}) satisfies (12.25a) *and that $a^{ij} \in H_1(K)$ and $a \in H_\alpha(K)$ for any bounded subset K of $\Omega \times \mathbb{R} \times \mathbb{R}^n$. Suppose finally that*

$$|p|^3 |a^{ij}_p| + |p| |a^{ij}_z| + |a^{ij}_x| + |a| = O(|p|^2). \tag{12.27}$$

Then there is a solution in $H^{(-1-\beta)}_{2+\alpha}$ to $Pu = 0$ in Ω, $u = \varphi$ on $\mathscr{P}\Omega$.

If a is Lipschitz with respect to z and p and if a^{ij} is independent of z, then this solution is unique.

When (12.25a) is replaced by ((12.25a)$'$), a similar existence result is true.

THEOREM 12.17. *Let $\mathscr{P}\Omega \in H_{1+\beta}$ and $\varphi \in H_{1+\beta}$ for some $\beta \in (0,1)$. Suppose that there are nonnegative constants b_1 and k such that* (12.26) *holds. Suppose that $a^{ij} \in H_1(K)$ and $a \in H_\alpha(K)$ for any bounded subset K of $\Omega \times \mathbb{R} \times \mathbb{R}^n$, and that (a^{ij}) satisfies* (12.25a)$'$. *Suppose finally that*

$$|a^{ij}_p| = O(|p|^{m-2}),\ |p| |a^{ij}_z| + |a^{ij}_x| + |a| = o(|p|^m). \tag{12.27$'$}$$

Then the problem $Pu = 0$ in Ω, $u = \varphi$ on $\mathscr{P}\Omega$ has a solution in $H^{(-1-\beta)}_{2+\alpha}$.

When $2 < m < 3$, the Hölder estimate of Dong [68] allows us to replace the $o(|p|^m)$ term in 12.27' by $O(|p|^m)$.

6.2. Mean curvature equations. When the condition $\Lambda \leq C\lambda$ is dropped from our hypotheses, the geometry of the domain Ω plays a major role in the nature of the existence results. To illustrate this role, we first define

$$\mathscr{M}u = \operatorname{div}(1 + |Du|)^{-1/2} Du,$$

and then consider the operator P given by

$$Pu = -u_t + (1 + |Du|^2)^{\tau/2} \mathscr{M}u \tag{12.28}$$

for a real parameter τ. We then have the following theorem for cylindrical domains.

THEOREM 12.18. *Let $\Omega = \omega \times (0,T)$ with $\partial\omega \in C^2$ and denote by H' the mean curvature of $\partial\omega$. Then the Cauchy-Dirichlet problem $Pu = 0$ in Ω, $u = \varphi$ on $\mathscr{P}\Omega$ with P given by* (12.28) *is solvable for arbitrary $\varphi \in H_{1+\beta}$ if and only if $\tau \geq 1$ and $H' \geq 0$. If $0 < \tau < 1$, the Cauchy-Dirichlet problem is solvable for arbitrary $\varphi \in H_{1+\beta}$ with $\varphi_t \in L^\infty$ if and only if $H' > 0$. If $\tau = 0$, then the Cauchy-Dirichlet problem is solvable for arbitrary $\varphi \in H_{1+\beta}$ with $\varphi_t \in L^\infty$ if and only if $\sup|\varphi_t| \leq (n-1)\inf H'$. If $\tau < 0$, then the Cauchy-Dirichlet problem is solvable for arbitrary time-independent $\varphi \in H_{1+\beta}$ if and only if $H' \geq 0$.*

PROOF. First, we note from Corollary 9.2 (with $v = \pm\max\varphi$ and u replaced by $\pm u$) that $\sup|u| = \sup|\varphi|$. When $\tau \geq 1$ or else $\tau < 0$ and φ is time-independent, the boundary gradient estimate follows from Theorem 10.9 by setting

$$a_\infty^{ij}(X,p) = \delta^{ij} - \frac{p_i p_j}{|p|^2}, a_\infty = a_0 = 0,$$

so $\kappa^\pm = -H'$. For $0 < \tau \leq 1$, we use Corollary 10.11. A global gradient bound and Hölder gradient bound follow from Corollary 11.2 and Theorem 12.10.

The sharpness of the conditions was already noted in Section X.5. □

We can go one step further by considering the operator

$$Pu = -u_t + (1+|Du|^2)^{\tau/2}[\mathscr{M}u + H(X,u)]$$

with H Lipschitz with respect to x and z, assuming that there are nonnegative constants k and b_1 such that $zH(X,z) \leq kz^2 + b_1$. From Theorem 10.9 and Corollary 10.11, we see that instead of $H' \geq 0$, we must assume

$$(n-1)H'(X) \geq |H(X,\varphi)| \text{ for } X \in S\Omega.$$

Additional hypotheses are needed which depend on τ. If $\tau \geq 1$, we assume also that $H_z \leq 0$. If $0 < \tau < 1$, we assume also that $H_z \leq 0$, that $\langle\varphi\rangle_2 < \infty$, and that

$$(n-1)H'(X) > |H(X,\varphi)| \text{ for } X \in S\Omega.$$

If $\tau = 0$, we assume also that that $\langle\varphi\rangle_2$ is finite with

$$(n-1)\inf H' \geq \sup|H(\cdot,\varphi)| + \langle\varphi\rangle_2.$$

Finally, if $\tau < 0$, we assume also that $\varphi_t = 0$.

6.3. False mean curvature equations. For operators of the form

$$Pu = -u_t + (1+|Du|^2)^{\tau/2}[\Delta u + D_iuD_juD_{ij}u + H(X,u)(1+|Du|^2)^{1/2}] \quad (12.29)$$

with τ a real parameter, we have the following existence result.

THEOREM 12.19. *Let $\mathscr{P}\Omega \in H_{1+\beta}$ and $\varphi \in H_{1+\beta}$, let $H \in H_1(K)$ for any compact subset K of $\Omega \times \mathbb{R}$, and suppose that there are nonnegative constants k and b_1 such that*

$$zH(X,z) \leq kz^2 + b_1. \quad (12.30)$$

Let P have the form (12.29). If $\tau < -3$, we suppose also that Ω is cylindrical. If $\tau < -4$, we suppose further that φ is time-independent. Then there is a solution $u \in H_{2+\alpha}^{(-1-\beta)}$ of $Pu = 0$ in Ω, $u = \varphi$ on $\mathscr{P}\Omega$.

PROOF. Theorem 9.5 gives the bound on $|u|_0$, and the boundary gradient estimate follows from Theorem 10.4 if $\tau \geq -4$ and from Corollary 10.5 if $\tau < -4$. A global gradient estimate is always true here because $C_\infty = 0$, and a Hölder gradient estimate follows from Theorem 12.10. □

For the false mean curvature operator in divergence form

$$Pu = -u_t + \operatorname{div}(\exp(\frac{1}{2}(1+|Du|^2)Du)) + B(X,u,Du), \tag{12.29$'$}$$

we have a corresponding result.

THEOREM 12.20. *Let P have the form (12.29)$'$ with $B \in H_\alpha(K)$ for any bounded subset K of $\Omega \times \mathbb{R} \times \partial R^n$ and let $\mathscr{P}\Omega \in H_{1+\beta}$ and $\varphi \in H_{1+\beta}$. If*

$$zB(X,z,p) \le b_0 \exp(\frac{1}{2}(1+|p|^2))\,|p|^2 + b_1\,|z|^m \tag{12.31}$$

for $|z| \ge M$ and if $|B| = O(\exp(\frac{1}{2}(1+|Du|^2))\,|Du|)$, then there is a solution $u \in H_{2+\alpha}^{(-1-\beta)}$ of $Pu = 0$ in Ω, $u = \varphi$ on $\mathscr{P}\Omega$.

6.4. Equations in convex-increasing domains. When Ω is strictly convex-increasing, then Theorems 9.5, 10.7, 12.10 and Corollary 11.2 give an existence result under suitable structure conditions.

THEOREM 12.21. *Suppose Ω is a strictly convex-increasing domain with constants R and η and $\mathscr{P}\Omega \in H_{1+\beta}$ for some $\beta \in (0,1)$. Suppose condition (12.7) holds. Let $\varphi \in H_2$, and suppose there are constants k, p_0, and μ such that the structure conditions*

$$|a| + \langle\varphi\rangle_2 \le \frac{\eta}{2R_1} + \frac{|p|\,\mathscr{T}}{R_1} + \mu\mathscr{E}, \tag{12.32}$$

$$a_z + \frac{p}{|p|^2}\cdot a_x \le k \tag{12.33}$$

are satisfied for $|p| \ge p_0$ and $R_1 > R$. If also a^{ij} is independent of x and z, then there is a solution of $Pu = 0$ in Ω, $u = \varphi$ on $\mathscr{P}\Omega$ in $H_{2+\alpha}^{(-1-\beta)}$.

6.5. Continuous boundary values. By means of the interior gradient estimates, some of the existence results can be extended to include boundary values which are merely continuous. In this case, we approximate φ by a sequence of $H_{1+\beta}$ functions (φ_m) and use the appropriate estimates to infer that the corresponding solutions u_m of $Pu_m = 0$ in Ω, $u_m = \varphi_m$ on $\mathscr{P}\Omega$ converge to a solution of the limit problem. In particular, once we know that the solutions are uniformly bounded, Theorem 10.18 guarantees a uniform modulus of continuity up to $\mathscr{P}\Omega$.

For uniformly parabolic equations, we use Lemmata 11.19 and 11.20 (if the equation is in general form) or Lemmata 11.11, 11.12, and 11.13 (if the equation is in divergence form) to infer the following existence theorem.

THEOREM 12.22. *Let P satisfy the hypotheses of Theorem 12.14, 12.15, 12.16, or 12.17. Let Ω satisfy an exterior H_ζ condition for some $\zeta \in (1,2]$. Then for any continuous function φ defined on $\mathscr{P}\Omega$, there is a solution $u \in C(\overline{\Omega}) \cap C^{2,1}(\Omega)$ of $Pu = 0$ in Ω, $u = \varphi$ on $\mathscr{P}\Omega$.*

For the mean curvature equations, the interior gradient bound is given in terms of a gradient bound for the initial function, so our existence theorem takes this condition into account.

THEOREM 12.23. *Suppose that the hypotheses of Theorem 12.18 on P and Ω are satisfied. Then the Cauchy-Dirichlet problem $Pu = 0$ in Ω, $u = \varphi$ on $\mathcal{P}\Omega$ is solvable for any $\varphi \in C(\mathcal{P}\Omega)$ such that $D\varphi$ is bounded on compact subsets of $B\Omega$ and φ_t satisfies the restrictions of Theorem 12.18.*

In the case of the false mean curvature equation, we only have proved interior gradient estimates when P has the form (12.29), so our existence result only includes this case.

THEOREM 12.24. *Suppose that the hypotheses of Theorem 12.19 on P and Ω are satisfied. If $\tau \geq -3$, then the Cauchy-Dirichlet problem has a solution for any continuous boundary function φ. If $\tau < -3$, the Cauchy-Dirichlet problem has a solution if φ is continuous on $\mathcal{P}\Omega$ and $D\varphi$ is bounded on compact subsets of $B\Omega$.*

6.6. One-dimensional problems. If the number of space dimensions is one, then the existence result is much more straightforward.

THEOREM 12.25. *Let $0 < \beta < 1$ and $1 + \beta \leq \zeta \leq 2$, let $\Omega \subset \mathbb{R}^2$ and suppose that $\mathcal{P}\Omega \in H_\zeta$. Let $\varphi \in H_{1+\beta}$. If (12.26) holds, if*

$$|a| = O(a^{11}|p|^2), \tag{12.34}$$

and if one of the sets of conditions

$$1 = O(|p|^{\zeta-1}a^{11}), \tag{12.35a}$$

$$1 = O(a^{11}|p|^2) \text{ and } \Omega \text{ is a cylinder} \tag{12.35b}$$

$$\Omega \text{ is a cylinder and } \varphi \text{ is independent of } t \tag{12.35c}$$

is satisfied, then there is a solution $u \in H_{2+\alpha}^{(-1-\beta)}$ of $Pu = 0$ in Ω, $u = \varphi$ on $\mathcal{P}\Omega$. Moreover if (12.35a) or (12.35b) holds, then there is a solution $u \in C^{2,1} \cap C(\overline{\Omega})$ of this problem for arbitrary $\varphi \in C(\mathcal{P}\Omega)$. If (12.35c) holds, then there is a solution if also φ_x is bounded on compact subsets of $B\Omega$.

Notes

The basis for our Hölder gradient estimates is the Hölder estimate of Nash [264]. The application to parabolic divergence structure equations was first explicitly noted by Ladyzhenskaya and Ural′tseva [184, Section 5], who also introduced the functions $w_k^{\pm}$ for studying equations in general form [186, Section 2]. Our proof differs from theirs in two respects: We use the weak Harnack inequality as in [304, Theorem 2.3] (as was also done in [101, Section 13.3]), and we sum the estimates for $w_k^{\pm}$ rather than introducing a new iteration scheme. This summation was a key step in the second derivative Hölder estimates of Evans [76, Section 4] and Krylov [169, Theorem 2.1] for fully nonlinear equations, and its application to Hölder gradient estimates comes from [238, Theorem 4.1].

Boundary Hölder gradient estimates were first proved by Ladyzhenskaya and Ural'tseva [185, Section 2; 186, Section 5] under the additional hypothesis that all coefficients in the equation (whether or not the equation is in divergence form) were differentiable with respect to (t,z,p), thus replacing the parabolic equation by an elliptic equation since this assumption leads to an estimate on u_t. The crucial estimate of Krylov [170, Theorem 4.2] (which we proved in a slightly different form as Lemma 7.47) shows that this additional hypothesis is not needed. In [170, Theorem 4.2], Krylov also proves a corresponding result near $C\Omega$, analogous to our Lemma 12.8, by a different argument. Our proof of Lemmata 12.6 and 12.8 comes from work of Cannon [43] on initial regularity for linear equations, although the iteration scheme is based on more recent work [224, Lemma 2.1]. The connection between the interior and boundary estimates is based on Lemma 4.11; a different account of this connection is given in [210, Theorem 4.6]. Note that the pointwise estimates from Lemmata 12.6 and 12.8 can also be used to infer that the full gradient is Hölder on an appropriate portion of $\mathscr{P}\Omega$ and then the proofs of the interior estimates can be repeated at the boundary with suitable modifications.

Our proof of the Hölder gradient estimate for equations in general form with the coefficients a^{ij} being differentiable only with respect to p is based on ideas of Caffarelli [30, Theorem 2] for fully nonlinear elliptic equations. The underlying principles are the same as for the linear case of Chapter IV, but the implementation is quite different. Wang [333, Theorem 1.3] gave a proof for viscosity solutions of fully nonlinear parabolic equations which, when specialized to classical solutions of quasilinear equations, assume that the coefficients a^{ij} are just continuous with respect to all variables, but this proof seems to have some serious gaps. Trudinger [315] proves a Hölder gradient estimate for viscosity solutions of fully nonlinear elliptic equations, which, for quasilinear equations, assumes the coefficients a^{ij} are differentiable with respect to p. Although Lemma 12.13 is not used for the existence theory for quasilinear equations with Dirichlet boundary conditions, it will become important for our study of oblique derivative problems in the next chapter. Applications to fully nonlinear problems will appear in Chapter XIV.

More recently, several authors have investigated the Hölder gradient estimate under weaker assumptions on the regularity of a^{ij} with respect to p. Using techniques developed by Chen [49] for viscosity solutions of elliptic equations, Zhu [347] showed that if a^{ij} is Hölder continuous with exponent $\beta > 1/2$ (actually a slightly weaker assumption is made), then a Hölder gradient estimate can be proved in terms of $[u]_1$ and $\langle u\rangle_\theta$ for some $\theta > 1$. Bourgoing [22] showed how to get the $\langle u\rangle_\theta$ estimate, but only for solutions of the Cauchy problem and the estimate depends on the regularity of the initial data. However, Han [108, Theorem 1] proved the same estimate as Zhu except without the dependence on $\langle u\rangle_\theta$.

Our list of existence theorems is based on the existence theorems in [101, Sections 15.5 and 15.6] for quasilinear elliptic equations, but it can hardly be considered exhaustive. Other existence theorems can be found, for example, in [71, Section 7; 210, Section 5; 307], etc. For uniformly parabolic equations, Ladyzhenskaya and Ural′tseva [183, Section V.6; 189, Theorem 1.3] showed that existence and regularity continue to hold when the boundedness of certain structure functions is relaxed to their membership in suitable $L^{p,q}$ classes. This work was further expanded by Apushkinskaya and Nazarov [7] (in which the appropriate estimates are proved without an explicit existence theorem being given).

Exercises

12.1 Show that Theorem 12.1 holds if, instead of assuming that μ_K is a constant, we assume that it is a function in the Morrey space $M^{2,n+\beta}$ provided α is also allowed to depend on β. Prove the corresponding version of Theorem 12.2.

12.2 Show that Theorem 12.3 continues to hold if, instead of assuming that μ_K is bounded, we assume that it is in the Lebesgue space $L^{2q,2r}$ with $1/r+n/2q<1$.

12.3 Show that the hypotheses of Theorem 12.1 can be relaxed along the lines of Section XII.4. Specifically, if $\lambda\,|\xi|^2 \le a^{ij}\zeta_i\zeta_j$, $|A_p| \le \Lambda$,

$$|A(x,t,z,p)-A(y,t,w,p)| \le \Lambda_1(1+|p|)[|x-y|^\alpha+|z-w|^\alpha],$$

$$|B| \le \Lambda_2[1+|p|^2]$$

for positive constants λ, Λ, Λ_1, and Λ_2, show that there are constants $\theta(n,\lambda,\Lambda)\in(0,1)$ and C determined also by Λ_1, Λ_2, $\operatorname{diam}\Omega$, and a modulus of continuity estimate for u such that

$$[Du]_{\theta,\Omega'} \le C\operatorname{dist}(\Omega',\mathcal{P}\Omega)^{-\theta}.$$

(See [97] and [217].)

12.4 Show that Lemma 12.13 holds if the L^∞ bound on a in (12.24) is relaxed to a bound in the Morrey space $M^{n+1,\beta}$ with α depending also on β.

CHAPTER XIII

THE OBLIQUE DERIVATIVE PROBLEM FOR QUASILINEAR PARABOLIC EQUATIONS

Introduction

When the Dirichlet boundary condition $u = \varphi$ on $S\Omega$ is replaced by a nonlinear boundary condition $b(X,u,Du) = 0$ there, many of the previous results remain true for the resultant initial-boundary value problem, and the proofs are similar to those for the Dirichlet case. The main differences are in the technical details, which tend to be significant. On the other hand, the boundary estimates are proved by modifying the methods used for interior estimates and, consequently, the geometry of the domain Ω is not as important for such problems. For example, we have seen that the Cauchy-Dirichlet problem for the parabolic minimal surface equation,

$$-u_t + \operatorname{div}((1+|Du|^2)^{-1/2}Du) = 0, \tag{13.1}$$

is solvable in a smooth cylindrical domain $\Omega = \omega \times (0,T)$ for arbitrary Dirichlet boundary data if and only if the mean curvature of the boundary of ω is nonnegative, and, even then, the boundary data must be independent of t. As we shall see, with the conormal boundary condition

$$(1+|Du|^2)^{-1/2}Du\cdot\gamma + \psi(x) = 0,$$

equation (13.1) is solvable in arbitrary smooth cylindrical domains assuming only that ψ is sufficiently smooth and $|\psi| < 1$. The necessity of this last condition is obvious since $\left|(1+|Du|^2)^{-1/2}Du\right| < 1$.

We prove some basic maximum bounds and comparison principles for solutions of the general problem

$$-u_t + a^{ij}(X,u,Du)D_{ij}u + a(X,u,Du) = 0 \text{ in } \Omega, \tag{13.2a}$$

$$b(X,u,Du) = 0 \text{ on } S\Omega,\ u = \varphi \text{ on } B\Omega \tag{13.2b}$$

in Section XIII.1, and we always assume the boundary condition to be *oblique* by which we mean that the function b is differentiable with respect to p and $b_p\cdot\gamma > 0$ on $S\Omega$. These estimates are the analogs of those in Chapter IX. Then gradient bounds (analogous to those of Chapter XI) for solutions are given in Sections XIII.2 and XIII.3. In Section XIII.2, we prove them for conormal problems in cylindrical domains, that is, for problems in which there is a vector function $A(X,z,p)$ such that $a^{ij}(X,z,p) = \partial A/\partial p$ (so that the equation is in divergence form) and also $b(X,z,p) = A(X,z,p)\cdot\gamma + \psi(X,z)$ for some scalar function ψ, where γ is the inner normal to $\partial\omega$. This structure is analogous to the linear divergence structure already encountered in Section VI.10. In Section XIII.3, we prove a gradient bound for problems in the general form (13.2). Next, Section XIII.4 provides a Hölder gradient estimate for conormal problems. For general oblique problems, we first prove a Hölder gradient estimate and some existence theorems when the equation is linear in Section XIII.5, and then a Hölder gradient estimate for oblique derivative problems with quasilinear equations in Section XIII.6. Some simple existence theorems are given in Section XIII.7.

1. Maximum estimates

Our first step is an estimate on the maximum of the solution of problem (13.2) without assuming that it is in conormal form.

THEOREM 13.1. *Suppose that $b_p\cdot\gamma > 0$ and that there are nonnegative constants M_0 and M_1 such that*

$$\operatorname{sgn} z\, b(X,z,-\operatorname{sgn} z M_1\gamma) < 0 \tag{13.3a}$$

for $|z| \ge M_0$. Suppose also that $(a^{ij}) \ge 0$ and that there are increasing functions a_0, a_1 and Λ_0 such that

$$\operatorname{sgn} z\, a(X,z,p) \le a_0(|p|)\,|z| + a_1(|p|), \tag{13.3b}$$

$$\Lambda(X,z,p) \le \Lambda_0(|p|). \tag{13.3c}$$

If u is a solution of (13.2) *with $\mathscr{P}\Omega \in H_2$, then*

$$\sup_\Omega |u| \le C(a_0,a_1,M_0,M_1,\sup|\varphi|,\Omega,\Lambda_0). \tag{13.4}$$

PROOF. We shall only prove an upper bound for u since the lower bound is proved similarly. In addition, without loss of generality, we may assume that $|\varphi| \le M_0$. Since $\mathscr{P}\Omega \in H_2$, there is a regularized distance function $\rho \in H_2 \cap C^{2,1}$ with $D\rho = |D\rho|\gamma$ and $\frac{1}{2} \le |D\rho| \le 2$ on $S\Omega$. In addition, $|D\rho| \le 2$ in Ω. With A a constant to be determined, we now define

$$v^+(X) = \exp(-2M_1\rho) + \exp(At) + M_0.$$

Our goal is to show that there cannot be a point X_0 such that $u - v^+ = 0$ at X_0 and $u - v^+ \le 0$ at X with $t \le t_0$. By calculation,

$$Pv^+ = -2M_1E[-\rho_t + a^{ij}D_{ij}\rho] + 4M_1Ea^{ij}D_i\rho D_j\rho + a - A\exp(At),$$

where $E = \exp(-M_1\rho)$ and a^{ij} and a are evaluated at (X,v^+,Dv^+). Since $|Dv^+| = |-2M_1ED\rho| \le 4M_1$, we have

$$Pv^+ \le C_1(a_0(4M_1),a_1(4M_1),M_1,M_0,\Lambda_0(4M_1),\Omega) + [a_0(4M_1) - A]\exp(At),$$

so $Pv^+ < 0$ in Ω if $A > C_1 + a_0(4M_1)$, and X_0 is not in Ω. In addition $u - v^+ < 0$ on the closure of $B\Omega$ by construction, so X_0 is not in the closure of $B\Omega$, either.

Finally, on $S\Omega$, we have

$$b(X, v^+, Dv^+) = b(X, v^+, -2M_1 |D\rho| \gamma) < 0$$

since b is oblique. On the other hand, if $X_0 \in S\Omega$, we have $u(X_0) = v^+(X_0)$ and $Du(X_0) = Dv^+(X_0) + \tau\gamma$ for some nonnegative number τ, so

$$b(X_0, v^+, Dv^+) \leq b(X_0, u, Du) = 0,$$

and so X_0 is not in $S\Omega$.

It follows that there is no such point X_0 and hence $u \leq v^+$. This inequality gives (13.4). □

Maximum estimates for solutions of (13.2) under different hypotheses are given in Exercises 13.1 and 13.2.

For our estimates for conormal problems, we need the following variant of Lemma 6.36.

LEMMA 13.2. *Let $\omega \subset \mathbb{R}^n$ and suppose that $\partial\omega \in C^2$. Then there is a $C^1(\overline{\Omega})$ vector field γ with $|\gamma| \leq 1$ in ω and γ is the unit inner normal to $\partial\Omega$. Moreover, if $-\operatorname{div}\gamma \leq K_0$ for some constant K_0, then*

$$\int_{\partial\omega} h\,ds \leq K_0 \int_\omega h\,dx + \int_\omega |Dh|\,dx \tag{13.5}$$

for all nonnegative Lipschitz functions h.

PROOF. Let ρ be a C^2 regularized distance for ω, so $D\rho \in C^1(\overline{\omega})$ and there are positive constants δ and ε such that $|D\rho| \geq \delta$ for $d \leq \varepsilon$. Then

$$\gamma = \frac{D\rho}{(|D\rho|^2 + \rho^2)^{1/2}}$$

is the desired function. To prove (13.5), we write

$$\begin{aligned} \int_{\partial\omega} h\,ds &= \int_{\partial\omega} h\gamma\cdot\gamma\,ds = -\int_\omega \operatorname{div}(h\gamma)\,dx \\ &= -\int_\omega h\operatorname{div}\gamma\,dx - \int_\omega \gamma\cdot Dh\,dx \end{aligned}$$

by the divergence theorem. The proof is completed by noting that $-\operatorname{div}\gamma \leq K_0$ and $|\gamma| \leq 1$. □

From this inequality, we infer a simple Sobolev inequality.

LEMMA 13.3. *Under the hypotheses of Lemma 13.2, if $\lambda \in L^\infty(\omega)$, then*

$$\int_\omega h^{(n+1)/n}\lambda\,dx \leq C(n)\left(\int_\omega h\lambda^n\,dx\right)^{1/n}\left(\int_\omega [|Dh| + K_0 h]\,dx\right). \tag{13.6}$$

PROOF. For $\varepsilon > 0$, we define $f_\varepsilon(x) = \min\{1, d(x)/\varepsilon\}$ and note from Theorem 6.11 that

$$\int_\omega (f_\varepsilon h)^{(n+1)/n}\lambda\,dx \le C(n)\left(\int_\omega f_\varepsilon h\lambda^n\,dx\right)^{1/n}\left(\int_\omega |D(f_\varepsilon h)|\,dx\right).$$

We now estimate

$$\int_\omega |D(f_\varepsilon h)|\,dx \le \int h|Df_\varepsilon|\,dx + \int f_\varepsilon|Dh|\,dx.$$

Since $Df_\varepsilon = Dd/\varepsilon$ on $\{d<\varepsilon\}$ and $Df_\varepsilon = 0$ elsewhere, it follows that

$$\int h|Df_\varepsilon|\,dx = \int_{\{d<\varepsilon\}} \frac{h}{\varepsilon}\,dx \to \int_{\partial\omega} h\,d\sigma$$

as $\varepsilon \to 0$. Since $f_\varepsilon \to 1$ as $\varepsilon \to 0$, it follows that

$$\lim_{\varepsilon\to 0}\int |D(f_\varepsilon h)|\,dx \le \int |Dh|\,dx + \int_{\partial\omega} h\,d\sigma,$$

and hence

$$\int_\omega h^{(n+1)/n}\lambda\,dx \le C(n)\left(\int_\omega h\lambda^n\,dx\right)^{1/n}\left(\int_\omega |Dh|\,dx + \int_{\partial\omega} h\,d\sigma\right).$$

The desired result follow from this estimate by virtue of Lemma 13.2. □

With these preliminary results, we can state the maximum estimate for the conormal problem in the following way.

LEMMA 13.4. *Suppose* $\Omega = \omega\times(0,T)$ *for some domain* ω *with* $\partial\omega\in C^2$. *Suppose also that there are positive constants* a_0, a_1, b_0, b_1, c_0 *and* M_0 *with* $c_0 < 1$ *such that*

$$p\cdot A(X,z,p) \ge a_0|p| - a_1|z|^2, \tag{13.7a}$$

$$zB(X,z,p) \le b_0(p\cdot A(X,z,p))^+ + b_1|z|^2, \tag{13.7b}$$

$$z\psi(X,z) \le a_0c_0|z| \tag{13.7c}$$

for $|z|\ge M_0$. *Suppose also that* $|\varphi|\le M_0$. *If* $u\in H_2(\Omega)$ *is a weak solution of*

$$-u_t + \operatorname{div}A(X,u,Du) + B(X,u,Du) = 0 \text{ in } \Omega, \tag{13.8a}$$

$$A(X,u,Du)\cdot\gamma + \psi(X,u) = 0 \text{ on } S\Omega, \tag{13.8b}$$

$$u = \varphi \text{ on } B\Omega, \tag{13.8c}$$

then

$$\sup_\Omega |u| \le C(a_0,a_1,b_0,b_1,c_0,n)\int_\Omega |u|^{n+1}\,dX + 2M, \tag{13.9a}$$

$$\int_\Omega |u|^q\,dX \le C(a_0,a_1,b_0,b_1,c_0,n,q,T)M^q|\omega| \tag{13.9b}$$

for $M = M_0 + c_0K_0$ *with* K_0 *as in Lemma 13.2, and* $q\ge 1$.

PROOF. For $q > \max\{2n, (b_0+1)/a_0(1-c_0)\}$, we use the test function

$$\eta = \left[\left(1-\frac{M}{|u|}\right)^+\right]^{(n+1)q-n} |u|^{q+n-2} u,$$

and observe that

$$D\eta = \left[\left(1-\frac{M}{|u|}\right)^+\right]^{(n+1)(q-1)} |u|^{q+n-2} q_1 Du,$$

for

$$q_1 = [(n+1)q-n]\frac{M}{|u|} + (q+n-1)\left(1-\frac{M}{|u|}\right).$$

Setting

$$U = \int_M^u \left[\left(1-\frac{M}{\tau}\right)^+\right]^{(n+1)(q-1)} |\tau|^{q+n-1}\tau\, d\tau,$$

$$\Sigma = \{X \in \Omega : |u| \geq M\}, \sigma = \{X \in S\Omega : |u| \geq M\},$$

we have that

$$\begin{aligned}\sup_{0<t<T} \int_\omega U(x,t)\,dx + \int_\Sigma \left(1-\frac{M}{|u|}\right)^{(n+1)(q-1)} |u|^{q+n-2} q_1 Du \cdot A\,dX \\ \leq \int_\Sigma \left(1-\frac{M}{|u|}\right)^{(n+1)q-n} |u|^{q+n-2} uB\,dX \\ + \int_\sigma \left(1-\frac{M}{|u|}\right)^{(n+1)q-n} |u|^{q+n-2} u\psi\,ds\,dt.\end{aligned}$$

Next, we use Lemma 13.2 and (13.7c) to estimate the boundary integral:

$$\begin{aligned}\int_\sigma \left(1-\frac{M}{|u|}\right)^{(n+1)q-n} |u|^{q+n-2} u\psi\,ds\,dt \\ \leq a_0 \int_\Sigma \left(1-\frac{M}{|u|}\right)^{(n+1)(q-1)} |u|^{q+n}\,dX \\ + a_0c_0 \int_\Sigma \left(1-\frac{M}{|u|}\right)^{(n+1)(q-1)} |u|^{q+n-2} q_1 |Du|\,dX,\end{aligned}$$

noting that $c_0K_0 \leq M$. Therefore, if we set $h = |\eta|$ and note that $q/2 \leq q_1 \leq (n+1)q$, we see that

$$\begin{aligned}\sup_{0<t<T} \int_\omega h|u|(x,t)\,dx + \int_\Omega |Dh|\,dX \\ \leq (a_1+b_a+a_0)q^2 \int_\Sigma \left(1-\frac{M}{|u|}\right)^{(n+1)(q-1)} |u|^{q+n-1}\,dX,\end{aligned}$$

and then Lemma 13.3 with $\lambda = |u|^{1/n}$ gives us

$$\left(\int_\Sigma w^{q\kappa}\,d\mu\right)^{1/\kappa} \le Cq^2 \int_\Sigma w^q\,d\mu,$$

where

$$w = \left(1 - \frac{M}{|u|}\right)^{n+1} |u|,\, d\mu = \left(1 - \frac{M}{|u|}\right)^{-n-1} |u|^n\,dX.$$

Iterating this inequality for different values of q yields (13.9a).

To prove (13.9b)), we use the test function $\eta = (|u|^{q-2} - M^{q-2})^+ u$. With

$$U = \int_M^u (|\tau|^{q-2} - M^{q-2})^+ \tau\,d\tau,$$

we have that

$$\int_\omega U(x,s)\,dx \le C(a_1,b_1,q) \int_0^s \int_\omega (|u|^{q-2} - M^{q-2})^+ |u|^2\,dX$$

and hence

$$\int_\omega |u(x,s)|^q\,dx \le C \int_0^s \int_\omega |u(x,t)|^q\,dX + CM^q|\omega|s.$$

The desired result follows from this inequality via Gronwall's inequality. □

Condition (13.7c) is not as restrictive as it might first appear. For example, if

$$p\cdot A \ge f(|p|) - a_1|z|^2,\, z\psi \le c_2|z|,$$

for a function f with $f(\tau)/\tau \to \infty$ as $\tau \to \infty$ and an arbitrary constant c_2, then conditions (13.7a,b,c) hold with $c_0 = \frac{1}{2}$ and suitable M_0 and a_0. In Exercise 13.3, we shall indicate how to obtain a maximum estimate if $z\psi$ is superlinear provided the assumptions on A and B are suitably modified.

2. Gradient estimates for the conormal problem

As it happens, gradient estimates are relatively straightforward for solutions of the conormal problem once we know the technique for proving interior gradient bounds for equations in divergence form. There are two new ingredients. The first is a pair of integral inequalities: a connection among certain boundary and interior integrals, analogous to Lemma 13.2, and an extension of the Sobolev inequality Corollary 11.9 to functions which do not necessarily vanish on $\partial\omega$. The second ingredient is the construction of a function, here called v_1, which behaves similarly to the function $v = (1 + |Du|^2)^{1/2}$ as far as the differential equation is concerned but which also satisfies a useful boundary condition.

We start with a version of the integration by parts formula, Lemma 13.2, which uses integration on surfaces. We use all the definitions from Section XI.5 without further comment.

LEMMA 13.5. *Let ω be a subset of $\mathbb{R}^n$ with $\partial\omega \in C^2$ and let $h \in H_1(\omega)$. Suppose that there is a constant $c_0 < 1$ such that $|\nu \cdot \gamma| \le c_0$ on $\partial\omega \cap \{h > 0\}$. Extend γ to all of ω as in Lemma 13.2 and suppose $-\operatorname{div}\gamma$ is less than or equal to some constant K_0. Then*

$$\int_{\partial\omega} h\nu\, ds \le \frac{1}{1-c_0^2}\int_\omega (|\delta h| + |hH| + K_0|h|)\nu\, dx. \tag{13.10}$$

PROOF. We integrate by parts to see that

$$\begin{aligned}\int_\omega \gamma^i \delta_i h\nu\, dx &= \int_\omega (\gamma^i D_i h - \nu\cdot\gamma\nu^i D_i h)\nu\, dx \\ &= -\int_\omega [D_i(\nu\gamma^i - \nu\cdot\gamma\nu^i\nu)]h\, dx - \int_{\partial\omega}[1-(\nu\cdot\gamma)^2]h\nu\, ds.\end{aligned}$$

Now we estimate the right side of this equation. First,

$$D_i(\nu\gamma^i - \nu\cdot\gamma\nu^i\nu) = D_i(\nu g^{ij}\gamma_j) = \nu g^{ij}D_i\gamma_j + \gamma_j D_i(\nu g^{ij})$$

and

$$D_i(\nu g^{ij}) = D_i(\nu\delta^{ij} - \nu_i D_j u) = D_j\nu - HD_j u - \nu_i D_{ij}u = -HD_j u.$$

It follows that

$$\int_{\partial\omega} h\nu[1-(\nu\cdot\gamma)^2]\, ds = \int_\omega \gamma^i\delta_i h\, d\mu - \int_\omega h[g^{ij}D_i\gamma_j + H\nu\cdot\gamma]\, d\mu.$$

In fact, this equation holds for any C^1 extension of the inner normal field γ. If we choose the one described in the hypotheses of this lemma, then we have $-g^{ij}D_i\gamma_j \le K_0$ and $|\nu\cdot\gamma| \le 1$ in ω. Since we also have $1-(\nu\cdot\gamma)^2 \le 1-c_0^2$ on $\partial\omega$, (13.10) follows immediately. □

Next, we have a weighted Sobolev inequality for arbitrary Lipschitz functions in ω.

LEMMA 13.6. *Under the hypotheses of Lemma 13.5, there is a constant C determined only by n and c_0 such that*

$$\begin{aligned}\int_\omega h^{2(N+2)/N} g\, d\mu \le C\Big(\int_\omega h^2 g^{N/2}\, d\mu\Big)^{2/N}\Big(\int_\omega h^2\, d\mu\Big)^{1-n/N} \\ \times\Big(\int_\omega |\delta h|^2 + h^2H^2 + K_0^2h^2\, d\mu\Big)^{n/N}\end{aligned} \tag{13.11}$$

for any nonnegative Lipschitz function h.

PROOF. Define $f_\varepsilon(x)$ as in Lemma 13.3 and apply Corollary 11.9 to $f_\varepsilon h$. The proof follows that of Lemma 13.3 (with Lemma 13.5 in place of Lemma 13.2) once we note that $\delta_i f_\varepsilon = (g^{ij}D_j d)/\varepsilon$, so

$$\varepsilon|\delta f_\varepsilon| = (g^{ij}g^{ik}D_i dD_j d)^{1/2} \le (g^{ij}D_i dD_j d)^{1/2} = (1-(\nu\cdot Dd)^2)^{1/2}$$

and therefore

$$\int_\omega h|\delta f_\varepsilon|\,d\mu \le \frac{1}{\varepsilon}\int_{\{d<\varepsilon\}} (1-(\nu\cdot Dd)^2)^{1/2} h\nu\,dx \to \int_{\partial\omega} (1-(\nu\cdot\gamma)^2)^{1/2} h\nu\,ds$$

as $\varepsilon \to 0$. □

It is no longer possible to estimate v directly because it does not satisfy a useful boundary condition. Instead, we introduce an auxiliary function v_1 which behaves similarly to v as far as the differential equation is concerned but which also satisfies a simple boundary condition. The key to constructing v_1 is a variant of the function b which defines the boundary condition. In fact, the construction does not assume a conormal boundary condition. For simplicity of notation, we define $c^{ij} = \delta^{ij} - \gamma^i\gamma^j$, where γ is the usual extension of the normal field. It is also useful to introduce the following hypothesis for a function f. We say that f is *k-decreasing* for a positive constant k if f is decreasing and $f(s)s^k$ is an increasing function of s.

LEMMA 13.7. *Let b be a scalar function in $C^1(S\Omega\times\mathbb{R}^n\times\mathbb{R})$ and suppose that $b_p\cdot\gamma > 0$ whenever $b(X,z,p) = 0$. Suppose also that there are constants $c_0 \in (0,1)$ and $p_0 \ge 1$ such that $b(X,z,p) = 0$ and $|p| \ge p_0$ imply*

$$|p\cdot\gamma| \le c_0(1+|p|^2)^{1/2}. \tag{13.12}$$

Suppose further that there are k-decreasing functions ε_k for $k = 1,2$, an increasing function Λ_0 and a nonnegative constant b_0 such that $b(X,z,p) = 0$ and $|p| \ge p_0$ imply

$$|b_p(X,z,p)| \le b_0 b_p(X,z,p)\cdot\gamma, \tag{13.13a}$$
$$|b_z(X,z,p)| \le \varepsilon_1(|p|)\,|p|\,b_p(X,z,p)\cdot\gamma, \tag{13.13b}$$
$$|b_x(X,z,p)| \le \varepsilon_2(|p|)\,|p|^2\, b_p(X,z,p)\cdot\gamma, \tag{13.13c}$$
$$|b_t(X,z,p)| \le \Lambda_0(|p|) b_p(X,z,p)\cdot\gamma. \tag{13.13d}$$

Then there is a function $\bar b \in C^1(\overline{\Omega}\times\mathbb{R}^n\times\mathbb{R})$ with $\bar b_p \in C^1((\overline{\Omega}\times\mathbb{R}^n\times\mathbb{R})\setminus\{b=0\})$ such that $\bar b(X,z,p) = 0$ for $X \in S\Omega$ if $b(X,z,p) = 0$ and $|p| \ge C(c_0)p_0$. Moreover,

$$\frac{2}{3} \le \bar b_p\cdot\gamma \le \frac{4}{3} \tag{13.14}$$

on $S\Omega\times\mathbb{R}\times\mathbb{R}^n$, and

$$|\bar b_p| \le 2(b_0+1), \tag{13.15a}$$
$$|\bar b_z| \le C(b_0,c_0)\varepsilon_1(|p|)\,|p| \tag{13.15b}$$
$$|\bar b_x| \le C(b_0,c_0)\varepsilon_2(|p|)\,|p|^2 \tag{13.15c}$$
$$|\bar b_t| \le C(b_0,c_0)\Lambda_0(C(c_0)\,|p|) \tag{13.15d}$$

for $|p| \geq C(c_0)p_0$. *Also*

$$\left|\bar{b}\bar{b}_{pp}\right| \leq C(b_0, c_0, n), \tag{13.16a}$$

$$\left|\bar{b}\bar{b}_{pz}\right| \leq C(b_0, c_0, n)\varepsilon_1(|p|)\,|p|, \tag{13.16b}$$

$$\left|\bar{b}\bar{b}_{px}\right| \leq C(b_0, c_0, n)\varepsilon(|p|)\,|p|^2. \tag{13.16c}$$

Moreover, if we define

$$v_1 = (1 + c^{ij}D_iuD_ju + \eta\bar{b}^2)^{1/2} \tag{13.17a}$$

$$\nu_1 = \frac{1}{v_1}(p - (p\cdot\gamma)\gamma + \eta\bar{b}\bar{b}_p), \tag{13.17b}$$

$$b^{km} = c^{km} + \eta(\bar{b}^k\bar{b}^m + \bar{b}\bar{b}^{km}), \tag{13.17c}$$

for a positive constant η, *then there is a positive constant* η_0 *determined only by* b_0 *and* c_0 *such that* $\eta \leq \eta_0$ *implies*

$$b^{km}\xi_k\xi_m \geq \frac{\eta}{8}|\xi|^2, \tag{13.18a}$$

$$\frac{\eta v}{2} \leq v_1 \leq 2v, \tag{13.18b}$$

$$\nu_1\cdot p \geq \frac{v_1}{4} \text{ if } v_1 \geq 4, \tag{13.18c}$$

$$|\nu_1| \leq 2, \tag{13.18d}$$

where $v = (1+|p|^2)^{1/2}$.

PROOF. By the implicit function theorem, for each fixed (X,z) we can write the equation $b(X,z,p) = 0$ as

$$p\cdot\gamma = g(X,z,p')$$

with p' the vector having components $c^{ij}p_j$. Moreover, as long as $|p'| \geq p_0$, we have

$$\left|\frac{\partial g}{\partial p'}\right| \leq b_0, \tag{13.19a}$$

$$|g_z| \leq C(c_0)\varepsilon_1(|p'|)\,|p'|, \tag{13.19b}$$

$$|g_x| \leq C(c_0)\varepsilon_2(|p'|)\,|p'|^2, \tag{13.19c}$$

$$|g_t| \leq \Lambda_0(C(c_0)\,|p|), \tag{13.19d}$$

$$|g| \leq C(c_0)\,|p'|. \tag{13.19e}$$

Now we note that the normal field γ can be chosen so that $|\gamma| = 1$ in a neighborhood of $S\Omega$; with this extension, we extend g so that conditions (13.19a–e) hold in this neighborhood (and for $|p'| < p_0$ as well with ε_1 and ε_2 extended to be constant on $(0, p_0]$) and $g \equiv 0$ outside this neighborhood. For φ as in Lemma 4.3, we define

$$\bar{g}(X,z,p,\tau) = \int_{\mathbb{R}^n} g(X,z,c^{ij}(p_i - \frac{\tau q_i}{2b_0+4+2C(c_0)}))\varphi(q)\,dq.$$

Since $|\bar{g}_\tau| \le 1/2$, there is a unique function $\Phi(X,z,p)$ such that

$$\Phi = p\cdot\gamma - \bar{g}(X,z,p,\Phi).$$

Following the construction of regularized distance in Section IV.5, we see that

$$\frac{1}{2}\left|p\cdot\gamma - g(X,z,p')\right| \le |\Phi(X,z,p)| \le 2\left|p\cdot\gamma - g(X,z,p')\right| \le C(c_0)(|p|+1),$$

so $\Phi(X,z,p) = 0$ if and only if $b(X,z,p) = 0$. It is then easy to check that

$$\left|p' - \frac{\theta\Phi(X,z,p)q'}{2b_0+4+C(c_0)}\right| \le 2(|p|+1)$$

for any $\theta \in (0,1]$. Since ε_k is k-decreasing, we then see that

$$\begin{aligned}
|\bar{g}_z| &\le C(c_0)\varepsilon_1(|p|)\,|p|,\\
|\bar{g}_x| &\le C(c_0)\varepsilon_2(|p|)\,|p|^2,\\
|\bar{g}_t| &\le \Lambda_0(C(c_0)\,|p|),\\
|\bar{g}| &\le C(c_0)\left|p'\right|,
\end{aligned}$$

where $\bar{g}$ and its derivatives are evaluated at $(X,z,p',\theta\Phi(X,z,p))$ provided $|p| \ge p_0$. We now continue our analysis in the neighborhood of $S\Omega$ on which $|\gamma| = 1$, as everything we have asserted is obvious outside this neighborhood.

We then take

$$\bar{b}(X,z,p) = p\cdot\gamma - \bar{g}(X,z,p',\theta\Phi(X,z,p))$$

with $\theta \in (0,1/2)$ to be further specified. From now on, Φ and all its derivatives are to be evaulated at (X,z,p) and $\bar{g}$ and its derivatives are to be evaluated at $(X,z,p',\theta\Phi)$ unless otherwise indicated. Since $\bar{g}_p\cdot\gamma = 0$, we have

$$\bar{b}_p\cdot\gamma = 1 - \theta g_\tau \Phi_p\cdot\gamma. \tag{13.20}$$

By differentiation, we have

$$\Phi_p = \gamma - \bar{g}_p(X,z,p',\Phi) - \theta\bar{g}_\tau(X,z,p',\Phi)\Phi_p,$$

so $\Phi_p\cdot\gamma = 1 - \theta\bar{g}_\tau(X,z,p',\Phi)\Phi_p\cdot\gamma$ and hence, since $\theta \in (0,1/2)$, $\Phi_p\cdot\gamma \in (4/5,4/3)$. Using this inequality in (13.20) gives (13.14). Similar arguments give (13.15a–d).

For the second derivative estimates, we have

$$\bar{b}^{km} = -\bar{g}^{km} - \theta(\bar{g}^k_\tau\Phi^m + \bar{g}^m_\tau\Phi^k + \bar{g}_\tau\Phi^{km} + \theta\bar{g}_{\tau\tau}\Phi^k\Phi^m).$$

We now estimate the terms in this expression. First $\left|\bar{g}_{\zeta\zeta}\right| \le C/(\theta\Phi), \left|\bar{g}_\zeta\right| \le C$ for $\zeta = (p,\tau)$ and $\left|\Phi_p\right| \le C, \left|\Phi_{pp}\right| \le C/|\Phi|$, and hence $\left|\bar{b}\bar{b}_{pp}\right| \le C/\theta$. Once θ has been chosen, this inequality implies (13.16a), and (13.16b–d) are proved similarly.

In addition, $\bar{g}^k_\zeta \gamma_k = 0$, so $\left|\bar{b}\bar{b}^{km}\gamma_m\right| \le C$ and $\left|\bar{b}\bar{b}^{km}\gamma_k\gamma_m\right| \le C\theta$. Hence,

$$\left|\bar{b}\bar{b}^{km}\xi_k\xi_m\right| \le C\left|\xi'\right|^2/\theta + C\theta\left|\xi\cdot\gamma\right|^2,$$

and so

$$b^{km}\xi_k\xi_m \ge \left|\xi'\right|^2(1 - C\frac{\eta}{\theta}) + \eta\left|\xi\cdot\gamma\right|^2(\frac{1}{4} - C\theta).$$

We infer (13.18a) from this inequality by first choosing θ and then η sufficiently small.

Now we write

$$v_1^2 = 1 + c^{ij}D_iuD_ju + \eta\bar{b}^2 \le 1 + \left|p'\right|^2(1 + C\eta) + 2\eta\left|p\cdot\gamma\right|^2 \le 4v^2$$

provided η is sufficiently small. Also, if η is sufficiently small, we have

$$\begin{aligned} v_1^2 &\ge 1 + \left|p'\right|^2 + \eta\left|p\cdot\gamma\right|^2 - C\eta\left|p\right|\left|p'\right| \\ &\ge 1 + \left|p'\right|^2(1 - C\eta) + \frac{\eta}{2}\left|p\cdot\gamma\right|^2 \ge \frac{\eta}{2}v^2, \end{aligned}$$

thus proving (13.18b).

Next, we have

$$v_1\cdot p = \frac{1}{v_1}(\left|p'\right|^2 + \eta\bar{b}\bar{b}_p\cdot p),$$

and

$$\left|\bar{b}_p\cdot p - \bar{b}\right| = \left|\bar{g} - \bar{g}_p\cdot p - \bar{g}_\tau\Phi_p\cdot p\right| \le C(\left|p'\right| + 1) + C\theta\left|p\cdot\gamma\right|,$$

so

$$v_1\cdot p \ge \frac{1}{v_1}(\left|p'\right|^2(1 - C\eta) + \eta\left|p\cdot\gamma\right|^2(1 - C\theta) - C\eta) \ge \frac{v_1}{2} - \frac{2}{v_1}$$

for θ and η sufficiently small, which implies (13.18c). Finally, (13.18d) follows by noting that

$$\begin{aligned} \left|v_1\right|^2 &= \frac{1}{v_1^2}(1 + \left|p' + \eta\bar{b}\bar{b}_p\right|^2) \le \frac{1}{v_1^2}(1 + 2\left|p'\right|^2 + 2\eta^2\bar{b}^2\left|\bar{b}_p\right|^2) \\ &\le \frac{1}{v_1^2}(2 + 2\left|p'\right|^2 + 2\eta\bar{b}^2) = 2 \end{aligned}$$

provided $\eta(2b_0 + 2) \le 1$. □

The significance of the quantities v_1 and b^{km} is seen by setting

$$a_0 = \frac{1}{v_1}\eta\bar{b}\bar{b}_z,\ a_i = \frac{1}{v_1}[D_i(c^{jk})D_juD_ku + \eta\bar{b}\bar{b}_i],$$

$$a_0^k = \frac{\eta}{v_1}[\bar{b}_z\bar{b}^k + \bar{b}\bar{b}_z^k],\ a_i^k = \frac{1}{v_1}(D_i(c^{jk})D_ju + \eta[\bar{b}_i\bar{b}^k + \bar{b}\bar{b}_i^k]).$$

Then

$$D_iv_1 = v_1^jD_{ij}u + a_0D_iu + a_i,$$

and
$$D_i v_1^k = \frac{1}{v_1} b^{km} D_{im} u - \frac{1}{v_1} D_i v_1 v_1^k + a_0^k D_i u + a_i^k.$$

To apply Lemma 13.6 to the conormal problem, we assume that there are nonnegative constants β_1, β_2, β_3 and τ' with $\beta_2 < 1$ such that
$$\left(\sum_j |a^{ij}\gamma_i|^2\right)^{1/2} \leq \beta_1 a^{ij}\gamma_i\gamma_j, \tag{13.21a}$$
$$|\psi(X,z)| \leq \beta_2 \nu\cdot A, |A| \leq \beta_3 \nu\cdot A \tag{13.21b}$$
on the subset of $S\Omega$ where $v \geq \tau'$. Here $a^{ij} = \partial A^i/\partial p_j$. It is clear that (13.21a) is just (13.13a) and the assumption of ellipticity for A guarantees that $b_p\cdot\gamma > 0$. To see that (13.12) holds, we consider the quantity
$$F_0 = (A+\psi\gamma)\cdot(\gamma-(\nu\cdot\gamma)\nu).$$
Schwarz's inequality implies that
$$|F_0| \leq (|A|+|\psi|)\,|\gamma-(\nu\cdot\gamma)\nu|,$$
and a direct calculation gives
$$|\gamma-(\nu\cdot\gamma)\nu|^2 = 1+(|\nu|^2-2)(\nu\cdot\gamma)^2 \leq 1-(\nu\cdot\gamma)^2$$
because $|\nu| \leq 1$. It then follows from (13.21b) that
$$|F_0| \leq (\beta_2+\beta_3)A\cdot\nu(1-(\nu\cdot\gamma)^2)^{1/2}.$$
On the other hand, the boundary condition implies that
$$F_0 = -(\nu\cdot\gamma)(A\cdot\nu+\psi\nu\cdot\gamma),$$
so
$$|F_0| \geq |\nu\cdot\gamma| A\cdot\nu(1-\beta_2).$$
Combining the two estimates for $|F_0|$ and rearranging yields
$$|\nu\cdot\gamma| \leq \left(1-\frac{(1-\beta_2)^2}{(1-\beta_2)^2+(\beta_2+\beta_3)^2}\right)^{1/2} < 1$$
for $v \geq \tau'$. Hence (13.12) follows from (13.21b).

We are now ready to study the initial-boundary value problem (13.8) when $\Omega = \omega\times(0,T)$ for some bounded domain $\omega \subset \mathbb{R}^n$ with $\partial\omega \in C^2$. The proof of the gradient bound follows the outline presented in Section XI.5 with suitable modification. We suppose that there are two matrices (C_k^i) and (D_k^i) such that (D_k^i) is differentiable with respect to (x,z,p) (but not necessarily with respect to t) and
$$C_k^i + D_k^i = A_z^i + A_k^i + B\delta_k^i,$$
we define
$$A^{ij} = v\frac{\partial A^i}{\partial p_j},\ D_k^{ij} = \frac{\partial D_k^i}{\partial p_j},\ \mathscr{F}_k = p_i\frac{\partial D_k^i}{\partial z} + \frac{\partial D_k^i}{\partial x^i},$$

and we introduce the quantities

$$\mathscr{C}_1^2 = A^{ij}b^{km}D_{ik}uD_{jm}u/vv_1, \mathscr{E}_2 = A^{ij}D_iv_1D_jv_2/vv_1,$$

and we write Ω_τ, $\omega_\tau(t)$, and $S\Omega_\tau$ to denote the subsets of Ω, $\omega(t)$, and $S\Omega$, respectively, on which $v_1 \geq \tau$. In addition to (13.21a,b), we assume that there are functions λ_0, Λ_0, ε_0, ε_1, and ε_2, and a nonnegative constant β_4 such that

$$|A_z| + |\psi_z| \leq \beta_4\lambda_0\varepsilon_1, \tag{13.22a}$$

$$|A_x| + |\psi_x| \leq \beta_4\lambda_0\varepsilon_2 v_1, \tag{13.22b}$$

$$|A_t| + |\psi_t| \leq \beta_4^2\lambda_0\Lambda_0(|p|/C(c_0))/v_1, \tag{13.22c}$$

where

$$c_0 = \left(1 - \frac{(1-\beta_2)^2}{1-(1-\beta_2)^2+(\beta_2+\beta_3)^2}\right)^{1/2}$$

and $C(c_0)$ is the constant from (13.15) (here all functions are evaluated at v_1 rather than at v) on $S\Omega_{\tau_0}$ and

$$\lambda_0 g^{ij}\xi_i\xi_j \leq A^{ij}\xi_i\xi_j \tag{13.23}$$

on Ω_{τ_0}; we assume also that ε_1 and ε_2 are 1-decreasing. Note that for $b(X,z,p) = A(X,z,p)\cdot\gamma + \psi(X,z)$, conditions (13.22a,b,c) are just (13.13b,c,d).

We introduce the following additional hypotheses on Ω_{τ_0}:

$$\lambda_0 \leq \Lambda_0,\ \bar\mu \leq \Lambda_0, \tag{13.24a}$$

$$A^{ij}\xi_i\eta_j \leq (\bar\mu|\xi|^2)^{1/2}(A^{ij}\eta_i\eta_j)^{1/2}, \tag{13.24b}$$

$$A^{ij}\xi_i\eta_j \leq (\bar\mu|\eta|^2)^{1/2}(A^{ij}\xi_i\xi_j)^{1/2}, \tag{13.24c}$$

$$C_k^i b^{jk}\zeta_{ij} \leq \beta_4\Lambda_0^{1/2}(A^{ij}\zeta_{ik}\zeta_{jk})^{1/2}, \tag{13.24d}$$

$$C_k^i v_i\xi^k \leq \beta_4\Lambda_0^{1/2}(A^{ij}\xi_i\xi_k)^{1/2}, \tag{13.24e}$$

$$\sum_{i,k}|C_k^i|\,\varepsilon_2|p| \leq \beta_4\Lambda_0, \tag{13.24f}$$

$$\varepsilon_1 C_k^i p_i\xi^k \leq \beta_4\Lambda_0|\xi|, \tag{13.24g}$$

$$vD_k^{ij}v_1^k\zeta_{ij} \leq \beta_4\Lambda_0^{1/2}(A^{ij}\zeta_{ik}\zeta_{jk})^{1/2}, \tag{13.24h}$$

$$\mathscr{F}_k v_1^k \leq \beta_4^2\Lambda_0, \tag{13.24i}$$

$$\varepsilon_1 A^{ij}\xi_j p_i \leq \Lambda_0^{1/2}(A^{ij}\xi_i\xi_j)^{1/2}, \tag{13.24j}$$

$$\varepsilon_1 A^{ij}\xi_i p_j \leq \Lambda_0^{1/2}(A^{ij}\xi_i\xi_j)^{1/2}, \tag{13.24k}$$

$$\varepsilon_2^2\bar\mu|p|^2 \leq \Lambda_0, \tag{13.24l}$$

$$\varepsilon_1\left|p\cdot A_z + A_i^i + B\right| \leq \beta_4^2\Lambda_0, \tag{13.24m}$$

$$|D\gamma| \leq \beta_4. \tag{13.24n}$$

Finally, we define $D_T = D - \gamma(\gamma\cdot D)$ and

$$g_k = D_k\psi + \psi_z D_k u + A\cdot D_T\gamma_k - \gamma_k D_i^k,$$

and we assume that there is a C^1 function Λ_1 such that

$$g_k v_1^k \le \Lambda_1, \tag{13.25a}$$

$$(v_1\Lambda_1')^2 + \Lambda_1^2 \le \beta_4^2\lambda_0\Lambda_0 \tag{13.25b}$$

on $S\Omega_{\tau_0}$.

LEMMA 13.8. *Let χ be a nonnegative, Lipschitz function satisfying*

$$0 \le \frac{\chi'(\xi)(\xi-\tau)}{\chi(\xi)} \le c(\chi) \tag{13.26}$$

for $\xi \ge \tau$ and $\tau \ge \tau_0$. Let $\zeta \in C^1(\overline{\Omega})$, and suppose that $|v_1(x,0)| \le \tau_0$ wherever $\zeta(x,0)$ is nonzero and that conditions (13.21) – (13.25) *are satisfied. Then there is a constant c_1 determined only by n, β, β_1, β_2, and β_3 such that*

$$\sup_s \frac{1}{1+c(\chi)}\int_{\omega_\tau(s)} (v_1-\tau)^2\chi\zeta^2\,dx + \int_{\Omega_\tau}\left[\left(1-\frac{\tau}{v_1}\right)\mathscr{C}_1^2 + \mathscr{E}_2\right]\chi\zeta^2\,d\mu\,dt$$
$$\le c_1[1+c(\chi)]^2\int_{\Omega_\tau}[\beta_4^2\Lambda_0\zeta^2 + |\zeta\zeta_t| + \bar\mu|D\zeta|^2]\chi\,d\mu\,dt. \tag{13.27}$$

PROOF. If we use the test function $D_k(\theta v_1^k)$ with $\theta\equiv 0$ wherever $v_1 \le \tau_0$, we find that

$$\int_\Omega \theta\mathscr{C}_1^2\,dX - \int_\Omega u_t D_k(\theta v_1^k)\,dX + \int_\Omega [v^{-1}A^{ij}D_{jk}uv_1^k + C_k^i v_1^k]D_i\theta\,dX$$
$$= \int_\Omega [D_k^{ij}v_1^k - v_1^{-1}C_k^i b^{kj}]D_{ij}u\theta\,dX + \int_\Omega [\mathscr{F}_k v_1^k - C_k^i a_i^k]\theta\,dX$$
$$+ \int_\Omega a_0^k C_k^i D_i u\theta\,dX + \int_\Omega \frac{1}{vv_1}A^{ij}D_{jk}uv_1^k D_i v_1\theta\,dX$$
$$- \int_\Omega [v^{-1}A^{ij}D_{jk}ua_i^k + v^{-1}A^{ij}D_{jk}D_i ua_0^k]\theta\,dX$$
$$+ \int_{S\Omega} g_k v_1^k\theta\,d\sigma\,dt,$$

where $d\sigma$ denotes surface measure on $\partial\omega$.

From (13.24d) and (13.24h), we have

$$[D_k^{ij}v_1^k - v_1^{-1}C_k^i b^{kj}]D_{ij}u \le \frac{1}{\eta}\beta_4\Lambda_0^{1/2}[(\frac{v}{v_1})^{1/2} + (\frac{v_1}{v})^{1/2}][\mathscr{C}_1^2]^{1/2},$$

and (13.24f) and (13.24i) imply that $\mathscr{F}_k v_1^k - C_k^i a_i^k \leq 2\beta_4^2\Lambda_0$. Now we write

$$A^{ij}D_{jk}uv_1^kD_iv_1 = \left(1-\frac{\eta}{32}\right)[vv_1\mathscr{E}_1 - A^{ij}a_jD_iv_1 - a_0A^{ij}D_iv_1D_ju] + \frac{\eta}{32}[A^{ij}D_{jk}uv_1^kD_{im}uv_1^m + A^{ij}D_{jk}uv_1^ka_i + a_0a^{ij}D_{jk}uv_1^kD_iu].$$

Since

$$A^{ij}D_{jk}uv_1^kD_{im}v_1^m \leq \frac{8}{\eta}vv_1\mathscr{C}_1^2,$$

and

$$|a_i| \leq C\beta_4\varepsilon_1|Du|^2,\ |a_0| \leq C\beta_4\varepsilon_2|Du|$$

by virtue of (13.22a,b) and (13.24n), it follows that

$$\int_\Omega \frac{1}{vv_1}a^{ij}D_{jk}uv_1^kD_iu\theta\,dX \leq \frac{1}{3}\int_\Omega \mathscr{C}_1^2\theta\,dX + \left(1-\frac{\eta}{64}\right)\int_\Omega \mathscr{E}_2\theta\,dx + C\int_\Omega \beta_4^2\Lambda_0\theta\,dX.$$

Then (13.18a), (13.22a), and (13.24j) imply that

$$-v^{-1}A^{ij}D_{jk}uD_iua_0^k \leq C\Lambda_0^{1/2}(\mathscr{C}_1^2)^{1/2},$$

and (13.22b), (13.24b), and (13.24l) imply that

$$-v^{-1}A^{ij}D_{jk}ua_i^k \leq C\Lambda_0^{1/2}(\mathscr{C}_1^2)^{1/2}.$$

Combining all these estimates and taking $\theta = (v_1-\tau)^+\chi\zeta^2$, we infer that

$$\begin{aligned}\int_\Omega -u_tD_k\theta\,dX &+ \frac{1}{2}\int_\Omega (v_1-\tau)\mathscr{C}_1^2\chi\zeta^2\,dX + (\tfrac{\eta}{64}-1)\int_\Omega v_1\mathscr{E}_2\chi\zeta^2\,dX \\ &+ \int_\Omega v^{-1}A^{ij}D_{jk}uv_1^kD_iv_1[\chi+(v_1-\tau)\chi']\zeta^2\,dX \\ &\leq -\int_\Omega C_k^iv_1^kD_iv_1[\chi+(v_1-\tau)\chi']\zeta^2\,dX \\ &\quad -2\int_\Omega v^{-1}A^{ij}D_{jk}uv_1^kD_i\zeta\zeta(v_1-\tau)\chi\,dX \\ &\quad -2\int_\Omega C_k^iv_1^kD_i\zeta\zeta\chi(v_1-\tau)\,dX + C\int_\Omega \beta_4^2\Lambda_0(v_1-\tau)\chi\zeta^2\,dX \\ &\quad + \int_{S\Omega}\Lambda_1\chi\zeta^2(v_1-\tau)\,d\sigma\,dt.\end{aligned}$$

To proceed, we estimate the boundary integral:

$$\begin{aligned}\int_{\partial\omega}\Lambda_1\chi\zeta^2(v_1-\tau)\,d\sigma &\le 2\int_{\partial\omega}\Lambda_1\chi\zeta^2(1-\frac{\tau}{v_1})v\,d\sigma\\ &\le C\int_\omega\left|\delta(\Lambda_1\chi\zeta^2(1-\frac{\tau}{v_1}))\right|d\mu\\ &+C\int_\omega\left|\Lambda_1\chi\zeta^2(1-\frac{\tau}{v_1})\right||H|\,d\mu\\ &+C\int_\omega\beta_4\Lambda_1\chi\zeta^2\,d\mu\end{aligned}$$

by Lemma 13.5. Each of these integrals is estimated by using (13.25b). In this way, we conclude that

$$\begin{aligned}-\int_\Omega u_tD_k(v_1^k(v_1-\tau)\chi\zeta^2)\,dX+\frac{\eta}{64}\int_\Omega[(1-\frac{\tau}{v_1})\mathscr{C}_1^2+\mathscr{E}_2]v_1\chi\zeta^2\,dX&\\ \le C\int_\Omega[\beta_4^2\Lambda_0\zeta^2+\bar\mu\,|D\zeta|^2]v_1\chi\,dX.&\end{aligned}$$

Now we estimate the first integral on the left side of this inequality. First, integration by parts with respect to x and the approximation argument from Lemma 11.10 give

$$\begin{aligned}-\int_\Omega u_tD_k(v_1^k(v_1-\tau)\chi\zeta^2)\,dX&=\int_{\omega(T)}\Xi(v_1)\zeta^2\,dX-2\int_\Omega\Xi(v_1)\zeta\zeta_t\,dX\\ &-\int_\Omega\eta\bar b\bar b_zu_t(1-\frac{\tau}{v_1})\chi\zeta^2\,dX-\int_\Omega\eta\bar b\bar b_t(1-\frac{\tau}{v_1})\chi\zeta^2\,dX\end{aligned}$$

with Ξ defined as in Lemma 11.10. The integrals involving Ξ are estimated as before.

Next, we note that

$$u_t=v^{-1}A^{ij}D_{ij}u+[A_z^iD_iu+A_i^i+B],$$

so (13.24m) implies that

$$\begin{aligned}\int_\Omega\eta\bar b\bar b_zu_t(1-\frac{\tau}{v_1})\chi\zeta^2\,dX&\le C\int_\Omega\beta_4^2\Lambda_0\chi\zeta^2\,d\mu\,dt\\ &+C\int_\Omega[(1-\frac{\tau}{v_1})\mathscr{C}_1^2]^{1/2}[\varepsilon_2^2v^2\bar\mu\beta_4^2]^{1/2}\chi\zeta^2\,d\mu\,dt.\end{aligned}$$

Finally, (13.22c) implies that $\eta\bar b\bar b_t(1-\frac{\tau}{v_1})\chi\zeta^2\le C\beta_4^2\Lambda_0\chi\zeta^2v$. Combining all these estimates and using Cauchy's inequality yields (13.27). □

From the energy inequality Lemma 13.8 and the Sobolev inequality Lemma 13.6, we infer a pointwise bound just as in Lemma 11.11. As before, we set

$$\Omega(\tau,\rho)=\Omega_\tau\cap\{|x-x_0|<\rho\}$$

for $X_0\in\Omega$ and $\rho>0$.

LEMMA 13.9. *Let $X_0 \in \Omega \cup S\Omega$, and let $\rho > 0$. Suppose, in addition to the hypotheses of Lemma 13.8, that there are nonnegative C^1 functions w, λ, and Λ such that*

$$w \text{ is increasing,} \tag{13.28a}$$

$$\xi^{-\beta} w(\xi) \text{ is a decreasing function of } \xi \tag{13.28b}$$

$$w(\xi)^\beta (\Lambda(\xi)/\lambda(\xi))^{N/2}/\xi \text{ is an increasing function of } \xi, \tag{13.28c}$$

$$\text{and } \xi^{-\beta}(\Lambda(\xi)/\lambda(\xi))^{N/2} \text{ is a decreasing function of } \xi \tag{13.28d}$$

for some nonnegative constant β. If

$$v_1 \lambda(v_1)\left(1 + \left(\frac{v_1 \lambda'(v_1)}{\lambda(v_1)}\right)^2\right) g^{ij}\xi_i\xi_j \le A^{ij}\xi_i\xi_j, \tag{13.29a}$$

$$\Lambda_0 \le v_1 \Lambda \tag{13.29b}$$

in $\Omega(\tau, 2\rho)$ for some $\tau \ge \tau_0$, then there is a constant c_2 determined only by n, β, β_1, β_2, β_3, and $\beta_4\rho$ such that

$$\sup_{\Omega(\tau,\rho)} \left(1 - \frac{\tau}{v}\right)^{-N-2} w^2 \le c_2 \rho^{-n-2} \int_{\Omega(\tau,2\rho)} w^2 (\Lambda/\lambda)^{N/2} \Lambda \, d\mu \, dt. \tag{13.30}$$

The right side of (13.30) is estimated in much the same way as the right side of (11.47).

LEMMA 13.10. *Let X_0 and ρ be as in Lemma 13.9 and suppose, in addition to the hypotheses of Lemma 13.8, that* (13.28a,b) *hold and that*

$$w' A \cdot \xi \le \beta_5 (A^{ij}\xi_i\xi_j)^{1/2} (v \cdot A)^{1/2}, \tag{13.31a}$$

$$w|A| \le Du \cdot A, \tag{13.31b}$$

$$A^{ij}\zeta_{ij} \le \beta_5 (A^{ij}\zeta_{ik}\zeta_{jk})^{1/2} (v \cdot A)^{1/2}, \tag{13.31c}$$

$$\bar{\mu} v \le \beta_5^2 Du \cdot A, \tag{13.31d}$$

$$\sup |\psi| \le \beta_5 \min\{|A|, \lambda_0\}, \tag{13.31e}$$

$$v|A_z| + |A_x| \le \beta_6 Du \cdot A, \tag{13.32}$$

$$\Lambda_0 v \le \varepsilon(v) w^2 Du \cdot A. \tag{13.33}$$

Set $\sigma = \sup_t \operatorname{osc}_{\omega(\rho)\times\{t\}} u$ and $E = \exp \beta_6 \sigma$. If $q \ge 2$, then there are constants $c_3 = c_3(n, q, \beta, \beta_1, \beta_2, \beta_3)$ and $c_4 = c_4(n, q, \beta, \beta_1, \beta_2, \beta_3, \beta_4\rho, \beta_5)$ such that

$$c_3 \beta_5^2 \beta_4^2 \sigma^2 E^2 \varepsilon(\tau_1) \le 1 \tag{13.34}$$

for some $\tau_1 \ge \tau$ implies that

$$\int_{\Omega(\tau,\rho)} w^q Du \cdot A \, dX \le c_4 \left(w(\tau_1) + E\frac{\sigma}{\rho}\right)^q \int_{\Omega(\tau,2\rho)} Du \cdot A \, dX. \tag{13.35}$$

PROOF. Following the proof of Lemma 11.12, we see that we only need to estimate the boundary integral $I_b = \int_0^T I_0(s)\,ds$, where

$$I_0(t) = -\int_{\partial\omega\times\{t\}} \bar{u}\exp(\beta_5\bar{u})(w^q - w(\tau)^q)^+\zeta^q A\cdot\gamma\,ds.$$

Now we note that $A\cdot\gamma = -\psi$ and set $\Psi = \sup|\psi|$ to conclude that

$$\begin{aligned}I_0(t) &\le \sigma E\Psi\int_{\partial\omega\times\{t\}}(w^q - w(\tau)^q)^+\zeta^q\,ds\\ &\le 2\beta q\sigma E\Psi\int_{\partial\omega\times\{t\}} w^{q-1}(1-\frac{\tau}{v_1})^+\zeta^q v\,ds.\end{aligned}$$

Now we use Lemma 13.5 to estimate this integral. For simplicity, we also use K to denote any constant determined only by q, β, β_1, β_2, and β_3 and we suppress the region of integration $\omega(t)\cap\{v_1 > \tau\}$. Then

$$\begin{aligned}I_0(t) &\le KE\Psi\sigma\int\left(w^{q-2}w' + w^{q-1}\frac{\tau}{v_1}\right)|\delta v_1|\,\zeta^q\,d\mu\\ &+KE\Psi\sigma\int(w\zeta)^{q-1}|\delta\zeta|\,v\,dx\\ &+KE\Psi\sigma\int w^{q-1}\left(1-\frac{\tau}{v_1}\right)H\zeta^q\,d\mu\\ &+KE\Psi\beta_4\sigma\int w^{q-1}\zeta^q v\,dx.\end{aligned}$$

To proceed, we set

$$J' = \int(w\zeta)^q Du\cdot A\,dx,$$

and we write $J_1,\dots,J_4$ for the four terms on the right side of this inequality. First, we observe that $\Psi \le \beta_5(\lambda_0)^{1/2}(\beta_3 v\cdot A)^{1/2}$ and $w' \le \beta w/v_1$ to see that

$$\begin{aligned}J_1 &\le KE\beta_5\sigma\int(w^{q-2}\lambda_0|\delta v_1|^2/(v_1^2))^{1/2}(w^q v\cdot A)^{1/2}\,d\mu\\ &\le KE\beta_5\sigma(\int w^{q-2}\mathscr{E}_2\zeta^2\,d\mu)^{1/2}(J')^{1/2}.\end{aligned}$$

Next, we note that $|\delta\zeta| \le |D\zeta|$ to see that

$$J_2 \le KE\beta_5\frac{\sigma}{\rho}\int(w\zeta)^{q-1}Du\cdot A\,dx.$$

To estimate J_3, we recall that $\lambda_0 H^2 \le K\mathscr{C}_1^2$ so that

$$J_3 \le KE\beta_5\sigma\left(\int w^{q-2}\left(1-\frac{\tau}{v_1}\right)\mathscr{C}_1^2\zeta^q\,d\mu\right)^{1/2}(J')^{1/2}.$$

Finally,

$$J_4 \le KE\beta_5\sigma\beta_4\int(w\zeta)^{q-1}Du\cdot A\,dx.$$

With these estimates, we proceed as in Lemma 11.12. □

The final bound, on the integral of $Du \cdot A$, follows by another simple modification of previous arguments. To state this bound more easily, we define

$$Q(r,\tau) = \{|x - x_0| < r, t_0 - 4r^2 < t < t_0, X \in \Omega, v > \tau\},$$
$$K_2 = \sup_{\rho>0} \rho^{1-n} \left|B_\rho \cap \partial\omega\right|,$$

where the supremum is also taken over all balls B_ρ in $\mathbb{R}^n$.

LEMMA 13.11. *Suppose that there are nonnegative constants β_7, β_8, and τ_0 such that*

$$|B| \le \beta_7 Du \cdot A, v|A| \le \beta_8 Du \cdot A \tag{13.36}$$

on $Q(2\rho, \tau_0)$. If $M \ge \operatorname{osc}_{\Omega(2\rho)} u$ and $\tau_2 \ge \max\{\tau_0, 8\beta_8 M/\rho\}$ and

$$\Delta_2 = \sup_{\{v \le \tau_2\}} (B - \beta_7 Du \cdot A)^+ + (Du \cdot A)^+ + \frac{M}{\rho}|A|, \tag{13.37}$$

then

$$\int_{Q(\rho,\tau_0)} Du \cdot A\, dX \le C(n) \exp(\beta_7 M) \rho^n [M^2 + \Delta_2 \rho^2 + K_2 M \rho \sup|\psi|]. \tag{13.38}$$

PROOF. Using the test function η from Lemma 11.13, we find that

$$\int_{Q(\rho,\tau_2)} Du \cdot A\, dX \le C(n) \exp(\beta_7 M) \rho^n [M^2 + \Delta_1 \rho^2] + C(n) \int_{S\Omega(\rho)} \psi \eta \, d\sigma\, dt.$$

This boundary integral is estimated in the obvious way to infer (13.38). □

As in Section XI.5, our method provides gradient bounds for the conormal problem associated with various equations. In all examples, we assume that $\Omega = \omega \times (0,T)$ for some domain $\omega \subset \mathbb{R}^n$ with $\partial\omega \in C^2$, and we set $\Gamma = \sup\left|D^2 d\right|$, where d is distance to $\partial\omega$.

2.1. The prescribed mean curvature equation. We first consider problem (13.8) when $A = v$, $\psi_t = 0$, and $\psi_z = 0$. We suppose that there are nonnegative constants $\Psi < 1$, Ψ_1, $\theta_0, \dots, \theta_3$ such that

$$|\psi| \le \Psi, \psi_z = 0, |\psi_x| \le \Psi_1,$$
$$|B| \le \theta_0 v, v B_z \le \theta_1, |B_x| \le \theta_2, \left|B_p\right| + \left|p \cdot B_p\right| \le \theta_3 / v.$$

Then conditions (13.21)–(13.25) hold with

$$\beta_1 = \frac{1}{1-\Psi}, \beta_2 = \frac{1+\Psi}{2}, \beta_3 = \frac{1}{2},$$
$$\beta_4 = C(\Psi)[\Gamma + \Psi_2 + \theta_1 + \theta_2 + \theta_3],$$
$$\varepsilon_1 \equiv 0, \varepsilon_2 = \frac{1}{v},$$
$$\lambda_0 = \Lambda_0 = \bar{\mu} = 1, \Lambda_1 = \Psi + \Gamma$$

provided τ_0 is sufficiently large (determined only by Ψ) and $\tau_0 \geq v_1(\varphi)$. The only condition that needs any elaboration is (13.21). To prove this condition, we note that

$$\left(\sum_i (a^{ij}\gamma_j)^2\right)^{1/2} = \frac{1}{v}|\gamma - (v \cdot \gamma)v| \leq \frac{1}{v},$$

while

$$a^{ij}\gamma_i\gamma_j = v^{-1}[1-(v\cdot\gamma)^2] \geq [1-\Psi^2]/v$$

wherever $A \cdot \gamma + \psi = 0$. By choosing

$$w = v_1^{1/2}, \Lambda = \frac{1}{v_1}, \lambda = \frac{1}{2v_1},$$

we see that (13.28a,b,c) are valid with $\beta = 2$, so Lemma 13.9 gives

$$\sup_{\Omega\cap\{|x-x_0|<\rho\}} v_1 \leq c_2\rho^{-n-2}\int_{\Omega\cap\{|x-x_0|<2\rho\, v_1\geq\tau\}} Du \cdot A\, dX + 2^{N+2}\tau$$

for any $\tau \geq \tau_0$ and c_2 is determined only by n, Ψ, and $\beta_4\rho$. Since (13.36) holds with

$$\beta_7 = 2\theta_0, \beta_8 = 2,$$

we infer that

$$\sup_{\Omega\cap\{|x-x_0|<\rho\}} v_1 \leq c_3 \exp(2\theta_0 M)\frac{T}{\rho^2}[\frac{M^2}{\rho^2} + 1 + |D\varphi|_0].$$

A global estimate for $|Du|$ follows from this inequality by choosing ρ appropriately and taking advantage of the interior estimates from Chapter XI. By suitably modifying our arguments, we can also allow ψ to depend on t and we need not assume $\psi_z \leq 0$; see Exercise 13.4 for details.

2.2. Uniformly parabolic equations. Now, we assume that there are nonnegative constants $\theta_0, \ldots, \theta_5$, Ψ, and Ψ_1 such that

$$\frac{\partial A^i}{\partial p_j}\xi_i\xi_j \geq \theta_0|\xi|^2$$

for all $\xi \in \mathbb{R}^n$,

$$\begin{gathered}
|A_p| \le \theta_1, |A| \le \theta_2(1+|Du|^2)^{1/2}, Du \cdot A \ge \theta_0 |Du|^2 - \theta_3, \\
\nu |A_z| + |A_x| + |B| \le \theta_4(1+|Du|^2), |A_t| \le \theta_5(1+|Du|^3), \\
|\psi| \le \Psi,\ |\psi_x| + |\psi_z| + |\psi_t| \le \Psi_1.
\end{gathered}$$

Then conditions (13.21)–(13.25) hold with

$$\begin{gathered}
\beta_1 = \theta_1/\theta_0, \beta_2 = 1/2, \beta_3 = 2\theta_2\theta_0, \\
\beta_4 = K[\Psi_1 + \Gamma + \theta_4 + \theta_5], \\
\varepsilon_1 = \varepsilon_2 \equiv 1, \\
\Lambda_0 = K\nu_1^3, \lambda_0 = \min\{1, \theta_0\}\nu, \Lambda_1 = K[\Psi_1 + \Psi_2 + \Gamma]\nu, \bar{\mu} = \theta_1 \nu,
\end{gathered}$$

where K denotes a constant determined only by θ_0, θ_1, and n provided τ_0 is sufficiently large determined only by n, θ_0, θ_3, and Ψ. Choosing

$$w = \nu_1, \lambda = \min\{1, \omega_0\}, \Lambda = K\nu_1^2$$

gives (13.28) and (13.29) with $\beta = N$. In addition, (13.31)–(13.34) hold for

$$\beta_5 = K[\frac{\theta_2}{\theta_0} + \frac{\theta_1}{\theta_2}],\ \beta_6 = K\theta_4,\ \varepsilon \equiv \frac{K}{\theta_3}$$

provided ρ is small enough because we have an *a priori* Hölder estimate for u. Since (13.36) holds with $\beta_7 = 2\theta_4/\theta_0$ and $\beta_8 = 2\theta_2/\theta_0$, we conclude that

$$\sup |Du| \le C(n, \theta_0 \ldots, \theta_5, \Psi, \Psi_1, \Psi_2, \sup |u|, \sup |D\varphi|, \Omega).$$

We leave to the reader the formulation of a gradient bound for the more general class of uniformly parabolic operators discussed in Chapter XI.

3. Gradient bounds for uniformly parabolic problems in general form

A gradient bound for solutions of problems in the general form (13.2) requires additional work because there is no connection between the boundary function b and the coefficient matrix (a^{ij}). This difficulty can be overcome in several different ways, which we discuss in the notes. Here we focus on a modification of the approach from Section XIII.2. The key idea is a suitable extension of the function b to $\Omega \times \mathbb{R} \times \mathbb{R}^n$. To avoid some of the more involved calculations, we shall work only with uniformly parabolic equations in a domain with a flat boundary portion.

The first step is a suitable version of Lemma 13.6.

LEMMA 13.12. *Let b be as in Lemma 13.6 with $Q^0(2)$ in place of $S\Omega$. Suppose also that ε_1 and ε_2 are constants and that $\Lambda_0(|p|) = b_3 |p|^3$ for some constant b_3. Then there is a function $\bar{b} \in C^1(\overline{Q^+(1)} \times \mathbb{R} \times \mathbb{R}^n) \cap C^2(\overline{Q^+(1)} \times \mathbb{R} \times \mathbb{R}^n) \setminus \{b =$*

$0\})$ such that $\bar{b}(X,z,p) = 0$ if and only if $b(X,z,p) = 0$. Moreover, conditions (13.14), (13.15a–d), *and* 13.16a–c) *hold. In addition*

$$|\bar{b}\bar{b}_{zz}| \leq C(b_0,c_0,n)\varepsilon_1^2|p|, \tag{13.39a}$$

$$|\bar{b}\bar{b}_{zx}| \leq C(b_0,c_0,n)\varepsilon_1\varepsilon_2|p|^2, \tag{13.39b}$$

$$|\bar{b}\bar{b}_{xx}| \leq C(b_0,c_0,n)\varepsilon_2^2|p|^2 \tag{13.39c}$$

for $|p| \geq C(c_0)p_0$. Finally, if v_1, v_1, and b^{km} are defined by (13.17a–c), *then there is positive constant $\eta_0(b_0,c_0,n)$ such that $\eta \leq \eta_0$ implies* (13.18a–d).

PROOF. With g chosen as in Lemma 13.6, we define

$$\bar{g}(X,z,p',\tau) = \int g(x - \frac{\tau\xi}{L_1\|p'\|^2}, t, z - \frac{\tau\zeta}{L_2\|p'\|}, p' - \frac{\tau\pi}{L_3})\varphi(\Xi)\,d\Xi,$$

where $\Xi = (\xi,\zeta,\pi) \in \mathbb{R}^{n-1}\times\mathbb{R}\times\mathbb{R}^{n-1}$,

$$\|p'\| = (1+\tau^2+|p'|^2)^{1/2}$$

and L_1, L_2, L_3 are positive constants. Since

$$\begin{aligned}\bar{g}_\tau = &\int\left[\frac{2\tau^2-\|p'\|^2}{L_1\|p'\|^4}\right]\xi^i g_i\varphi(\Xi)\,d\Xi\\ &+\int\left[\frac{\tau^2-\|p'\|^2}{L_2\|p'\|^3}\right]\xi^i g_i\varphi(\Xi)\,d\Xi\\ &-\int\frac{1}{L_3}g^k\pi_k\varphi(\Xi)\,d\Xi,\end{aligned}$$

it follows that

$$|\bar{g}_\tau| \leq C[\frac{\varepsilon_2}{L_1}+\frac{\varepsilon_1}{L_2}+\frac{b_0}{L_3}],$$

and hence we can choose L_1, L_2, and L_3 so that $|\bar{g}_\tau| \leq 1/2$. With these constants fixed, it's easy to see that

$$|\bar{g}_{xx}| \leq \frac{C\varepsilon_2^2}{\tau}\|p'\|^4, \qquad |\bar{g}_{xz}| \leq \frac{C\varepsilon_1\varepsilon_2}{\tau}\|p'\|^3,$$

$$|\bar{g}_{xp}| \leq \frac{C\varepsilon_2}{\tau}\|p'\|^2, \qquad |\bar{g}_{x\tau}| \leq \frac{C\varepsilon_2}{\tau}\|p'\|^2,$$

$$|\bar{g}_{zz}| \leq \frac{C\varepsilon_1^2}{\tau}\|p'\|^2, \qquad |\bar{g}_{zp}| \leq \frac{C\varepsilon_1}{\tau}\|p'\|,$$

$$|\bar{g}_{z\tau}| \leq \frac{C\varepsilon_1}{\tau}\|p'\|, \qquad |\bar{g}_{pp}| \leq \frac{C}{\tau},$$

$$|\bar{g}_{p\tau}| \leq \frac{C}{\tau}, \qquad |\bar{g}_{\tau\tau}| \leq \frac{C}{\tau}.$$

We now take Φ to be the fixed point of the equation

$$\Phi = p_n - g(X,z,p',\Phi),$$

and set

$$\bar{b}(X,z,p) = p_n - \bar{g}(X,z,p',\eta_1\Phi)$$

with η_1 a sufficiently smalll, positive constant. Noting that $|\Phi| \le c[1+|p|]$ leads to (13.14), (13.15), (13.16), and (13.39) just as before.

Next, we calculate

$$\bar{b}^{km} - \bar{g}^{km} - \eta_1[\Phi^m \bar{g}^k_\tau + \Phi\bar{g}^m_\tau + \Phi^{km}\bar{g}_\tau + \eta_1\Phi^k\Phi^m\bar{g}_{\tau\tau}],$$

and hence $\left|\bar{b}^{km}\right| \le C/(\eta_1\tau)$. Since $\bar{g}^{km} = 0$ if $k = n$ or $m = n$, we also have

$$\left|\bar{b}^{kn}\right| \le C/\Phi, \left|\bar{b}^{nn}\right| \le C\eta_1/\Phi,$$

and (13.17) follows from these inequalities. □

With this construction, we can now prove a boundary gradient bound.

THEOREM 13.13. *Suppose that there are positive constants b_0, b_1, M, λ, Λ, Λ_1, Λ_2, and Λ_3 such that a^{ij} and a satisfy*

$$a^{ij}\xi_i\xi_j \ge \lambda|\xi|^2, \tag{13.40a}$$

$$\left|a^{ij}\right| \le \Lambda, \tag{13.40b}$$

$$|p|^2\left|a^{ij}_p\right| + |p|\left|a^{ij}_z\right| + \left|a^{ij}_x\right| \le \Lambda_1|p|, \tag{13.40c}$$

$$|a| \le \Lambda_2|p|^2, \tag{13.40d}$$

$$|p|^2|a_p| + |a_x| \le \Lambda_3|p|^3, \tag{13.40e}$$

$$|p|a_z \le \Lambda_3|p|^2 \tag{13.40f}$$

$$\left|b_p\right| \le b_0 b^n, \tag{13.40g}$$

$$|p|^2|b_z| + |p||b_x| + |b_t| \le b_1|p|^3 b^n, \tag{13.40h}$$

for $|p| \ge M$. Let $u \in C^{2,1}(Q^+(R) \cup Q^0(R))$ satisfy

$$-u_t + a^{ij}(X,u,Du)D_{ij}u + a(X,u,Du) = 0 \ \text{in}\ Q^+(R), \tag{13.41a}$$

$$b(X,u,Du) = 0 \ \text{on}\ Q^0(R) \tag{13.41b}$$

with $R < 1$. Then there are positive constants θ and C determined only by b_0, b_1, n, λ, Λ, Λ_1, Λ_2, and Λ_3 such that $\operatorname{osc}_{Q^+(R)} u \le \theta$ implies that

$$\sup_{Q^+(R/4)} |Du| \le C\left[\frac{\operatorname{osc}_{Q^+(R)} u}{R} + M\right]. \tag{13.42}$$

PROOF. Set $\alpha = \operatorname{osc}_{Q^+(R)} u$, $\rho = \min\{R/4, \alpha\}$,

$$\Sigma = \{X \in Q^+(R/2) : x^n < \rho\},$$

$$\zeta(X) = \left(1 - \frac{4|x|^2}{R^2}\right)^2 \left(1 + \frac{4t}{R}^2\right)^2,$$

$$M_0 = \sup u, M_1 = \sup \zeta v_1^2.$$

For α_1 and α_2 positive constants to be further specified, we define

$$u^* = \exp(\frac{u - M_0}{\alpha}), w = \zeta v_1^2 + \alpha_1 M_1 u^* + \alpha_2 M_1 v x^n.$$

By using Theorem 11.5, we see that

$$|Du(X)| \le C\left[\frac{\alpha}{x^n} + M\right]$$

in $Q^+(R/2)$, and hence $vx^n \le C_1 M\alpha$. If we suppose that $\alpha_1 \le 1/4$ and $C_1 M\alpha\alpha_2 \le 1/4$, then $\zeta v_1^2 \ge \frac{1}{2}M_1$ and $\zeta v^2 \ge \frac{1}{8}M_1$ at a maximum point X_0 of w.

Suppose first that $x_0^n = 0$. Noting that $\bar{b} = 0$ and $x^n = 0$ on Q^0, we see that

$$\begin{aligned}
b^i D_i w &= v_1^2 b^i D_i\zeta - 2\zeta(b_z|D'u|^2 + b_x \cdot D'u) + \frac{\alpha_1}{\alpha} M_1 u^* p \cdot b_p + \alpha_2 M_1 v b^n \\
&\ge b^n[-b_0|D\zeta|v_1^2 - 2\zeta(b_1 v_1^2 v - c\frac{\alpha_1}{\alpha} M_1 u^* b_0 + \alpha_2 M_1 v] \\
&\ge b^n M_1 v(-b_0\frac{2}{\zeta^{1/2}Rv} - b_0\frac{\alpha - 1}{\alpha} - 4b_1 + \alpha_2).
\end{aligned}$$

If we suppose that $M_1 \ge (8\alpha/\alpha_1 R)^2$, then $\zeta^{1/2}Rv \ge 2RM_1^{1/2} \ge \alpha_1/4\alpha$ at X_0. It follows that w cannot have a maximum on $Q^0 \cap \{|Du| \ge M\}$ if

$$\alpha_2 = 4b_1 + 3b_0\frac{\alpha_1}{\alpha}.$$

Note that $C_1 M\alpha\alpha_2 \le CM[\theta + \alpha_1] \le 1/4$ if θ and α_1 are small enough.

With this choice for α_2, we now determine α_1 so that the maximum cannot occur in $Q^+ \cap \{|Du| \ge M\}$. First, we set

$$B^k = a^{ij,k}D_{ij}u + a^k - \frac{2}{\zeta}a^{kj}D_j\zeta,$$

and we define the operator $L = -D_t + a^{ij}D_{ij} + B^k D_k$. We then have

$$\begin{aligned}
Lw &= v_1^2 L\zeta + \zeta L(v_1^2) + 2a^{ij}D_i(v_1^2)D_j\zeta \\
&+ \frac{\alpha_1}{\alpha}M_1 u^* Lu + \frac{\alpha_1}{\alpha^2}M_1 u^* \mathscr{E} \\
&+ \alpha_2 M_1 x^n Lv + \alpha_2 M_1 v B^n + 2\zeta_2 M_1 a^{nj} D_j v.
\end{aligned}$$

From the definition of B^k along with conditions (13.40c) and (13.40e), we have

$$L\zeta \geq -\frac{C}{R^2} - \frac{C|Du|\zeta^{1/2}}{R} - \frac{C|D^2u|\zeta^{1/2}}{R|Du|}.$$

Next, we define the operators δ_T and δ_0 by

$$\delta_T f = f_z + |p'|^{-2} p' \cdot f_x, \; \delta_0 f = f_z + p \cdot f_x / p \cdot b_p,$$

and we note that

$$\begin{aligned}
L(v_1^2) = {} & 2|D'u|^2 [\delta_T a^{ij} D_{ij}u + \delta_T a] \\
& + 2\sum_{k<n} a^{ij} D_{ik}u D_{jk}u + 2\eta \bar{b}\bar{b}_z[-u_t + a^{ij}D_{ij}u] \\
& - 2\eta \bar{b}\bar{b}_p \cdot Du[\delta_0 a^{ij} D_{ij}u + \delta_0 a] \\
& + 2\eta[\bar{b}\bar{b}^{mk} + \bar{b}^m \bar{b}^k] a^{ij} D_{ik}u D_{jm}u - 2\eta \bar{b}\bar{b}_t \\
& + 2\eta[a^{ij}(\bar{b}_j + \bar{z}D_ju)(\bar{b}_i + \bar{b}_z D_iu) + a^{ij}(\bar{b}_j + \bar{b}_z D_ju)\bar{b}^k D_{ik}u] \\
& + 2\eta \bar{b} a^{ij}(\bar{b}_{ij} + 2\bar{b}_{iz} D_ju + \bar{b}_i^k D_{jk}u + \bar{b}_{zz} D_iu D_ju) \\
& + 2\eta \bar{b}\bar{b}^k [a^{ij,k} D_{ij}u + a^k] \\
& - \frac{2}{\zeta} a^{ij} D_j\zeta [\sum_{k<n} D_{ik}u D_ku + 2\eta \bar{b}(\bar{b}_i + \bar{b}_z D_iu + \bar{b}^k D_{ik}u)].
\end{aligned}$$

Taking advantage of our structure conditions, we see that

$$L(v_1^2)(X_0) \geq -C|Du|^4 + \eta\lambda |D^2u|^2.$$

Applying similar reasoning, we find that $Lu \geq -C|Du|^2 - C|D^2u|$, and $Lv \geq -C|Du|^3 - C|Du||D^2u|$. Since $1/3 \leq u^* \leq 1$ and $1/M_1R^2 \leq 8\alpha_1^2/\alpha^2$, it follows (using the form of α_2) that

$$Lw(X_0) \geq M_1 u^* \mathscr{E} \left(\frac{\alpha_1}{\alpha^2} - \frac{C_2\alpha_1}{\alpha} - \frac{C_3\alpha_1^2}{\alpha^2} - C_4 \right).$$

and hence there cannot be an interior maximum if

$$\alpha_1 \leq \frac{1}{4C_3}, \alpha \leq \min\{\frac{1}{4C_2}, \left(\frac{\alpha_1}{4C_4}\right)^{1/2}\}.$$

Finally, if $x_0^n = \rho$, then $M_1 \leq 8\zeta v^2 \leq 8C_1M^2$. With these choices for α_1 and α_2, and θ sufficiently small, it follows that w has a maximum at a point where $|Du| \leq M$ or $M_1 \leq (8\alpha/\alpha_1 R)^2$ or $M_1 \leq 8C_1M^2$. In any case, we infer (13.42). □

Next, we have a Hölder estimate for u by modifying Corollary 7.51 along the lines of Lemma 11.4.

LEMMA 13.14. *Let* a^{ij} *satisfy* (13.40a,b) *for some positive constants* λ *and* Λ, *let* a *satisfy*

$$|a| \le \Lambda_2[|p|^2 + 1], \tag{13.40d)'}$$

and let b *satisfy* (13.13a) *and*

$$|b(X,z,p')| \le \beta_0(1 + |p'|) \tag{13.43}$$

for $X \in Q^0(R)$. *If* u *solves* (13.41), *with* $|u| \le M_0$, *then there are constants* C *and* α *determined only by* b_0, n, M_0, R, β_0, λ, Λ, *and* Λ_2 *such that* $|u|_{\alpha,Q^+(R/2)} \le C$.

Since our hypotheses on the coefficients are invariant under an H_3 change of variables, we have the following global gradient bound.

THEOREM 13.15. *Let* $\mathscr{P}\Omega \in H_3$. *Suppose that* a^{ij} *and* a *satisfy conditions* (13.40a,b) *and* (13.40d)′ *for all* p *and conditions* (13.40c), (13.40e), *and* (13.40f) *for* $|p| \ge M$. *Suppose also that* b *satisfies* (13.13a) *and* (13.43) *for all* p *and*

$$|b_z| \le b_1 |p| b_p \cdot \gamma, \tag{13.13b)'}$$

$$|b_x| \le b_2 |p|^2 b_p \cdot \gamma, \tag{13.13c)'}$$

$$|b_t| \le b_3 |p|^3 b_p \cdot \gamma \tag{13.13d)'}$$

for $|p| \ge M$. *If* $u \in C^{2,1}(\overline{\Omega})$ *solves* (13.2), *then*

$$\sup_\Omega |Du| \le C(n, b_0, \ldots, b_3, \lambda, \Lambda, \Lambda_0, \ldots, \Lambda_4, \sup |u|, \sup |D\varphi|). \tag{13.44}$$

4. The Hölder gradient estimate for the conormal problem

As we have already proved interior and initial Hölder gradient estimates for equations in divergence form, we only need to estimate the Hölder norm of the gradient near $S\Omega$. When $\mathscr{P}\Omega \in H_2$, this estimate is easy to obtain: We flatten the boundary locally so that $D_k u$ $(k < n)$ solves a linear conormal problem in Q^+. Theorem 6.44 gives a Hölder estimate for $D_k u$, and then the boundary condition gives a Hölder estimate for $D_n u$ on Q^0. Finally, Theorem 6.32 provides the Hölder estimate for $D_n u$. Thus we have the following result.

THEOREM 13.16. *Let* $\Omega = \omega \times (0,T)$ *for some domain* $\omega \subset \mathbb{R}^n$ *with* $\partial\omega \in H_2$ *and let* $\varphi \in H_2$. *Let* $u \in C^{2,1}(\overline{\Omega})$ *be a solution of* (13.7) *with* A *a* C^1 *function of* (x,z,p), ψ *a* C^1 *function of* (x,z) *and* A *and* B *uniformly continuous with respect to* (X,z,p) *and suppose that there are positive constants* K, λ_K, Λ_K, *and* μ_K *such that* $|u| + |Du| \leq K$ *and*

$$\frac{\partial A^i}{\partial p_j}\xi_i\xi_j \geq \lambda_k |\xi|^2, \tag{13.45a}$$

$$|A_p| \leq \Lambda_K, \tag{13.45b}$$

$$|A_z| + |A_x| + |B| \leq \mu_K, \tag{13.45c}$$

$$|\psi_z| + |\psi_x| + |\psi| \leq \mu_K \tag{13.45d}$$

for (X,z,p) *with* $|z| + |p| \leq K$, *and*

$$|A(x,t,z,p) - A(x,s,z,p)| + |\psi(x,t,z) - \psi(x,s,z)| \leq \mu_K |t-s|^{1/2} \tag{13.46}$$

for all $(x,z,p) \in \partial\omega \times \mathbb{R} \times \mathbb{R}^n$ *with* $|z| + |p| \leq K$ *and all s and t in* $(0,T)$. *If also*

$$A(X,\varphi,D\varphi)\cdot\gamma + \psi(X,\varphi) = 0 \text{ on } C\Omega, \tag{13.47}$$

then there is a positive constant $\alpha = \alpha(K,n,\lambda_K,\Lambda_K)$ *such that*

$$|Du|_\alpha \leq C(n,\lambda_K,\Lambda_K,\mu_K,|\varphi|_2,\Omega). \tag{13.48}$$

5. Nonlinear boundary conditions with linear equations

For problems in the general form (13.2), the ideas already developed will give a Hölder gradient estimate if all coefficients are C^1 with respect to all their arguments and $\mathscr{P}\Omega \in H_3$. Since the existence theory is more complicated if we need to work with these regularity assumptions, we shall prove a Hölder gradient estimate under weaker assumptions, which are almost minimal for such an estimate.

Our first step is to study the Hölder gradient estimate when the problem is in a simple form. To this end, we recall some definitions from Chapter IV with a slight change. For $R > 0$, $\bar{\mu} \in (0,1)$ and $X_0' \in \mathbb{R}^{n+1}$ with $x_0^n = 0$, we set $x_0 = (x_0^1,\ldots,x_0^{n-1},-\bar{\mu}R,t_0)$ and we write $\Sigma^+(X_0',R)$ and $\Sigma^0(X_0',R)$ for the subsets of $Q(X_0,R)$ on which $x^n > 0$ and $x^n = 0$, respectively. We also define $\sigma(X_0,R) = \mathscr{P}\Sigma^+(X_0,'R) \setminus \Sigma^0(X_0',R)$. When the point X_0' is clear from context, we omit it from the notation.

LEMMA 13.17. *Let* (A^{ij}) *be a constant positive definite matrix with eigenvalues in the interval* $[\lambda,\Lambda]$, *let* $g \in H_{4,loc}(\mathbb{R}^{n-1})$ *with* $|Dg| \leq \mu$ *for some nonnegative constant* μ, *and suppose that* $\bar{\mu} > \mu/(1+\mu^2)^{1/2}$. *If v solves*

$$-v_t + A^{ij}D_{ij}v = 0 \text{ in } \Sigma^+(R),\ D_nv = g(D'v) \text{ on } \Sigma^0(R), \tag{13.49}$$

for some X_0' *and R, then there are constants C and* θ *determined only by n,* λ, Λ, *and* μ *such that*

$$\operatorname*{osc}_{\Sigma^+(r)} Dv \leq C\left(\frac{r}{R}\right)^\theta \operatorname*{osc}_{\Sigma^+(R)} Dv \tag{13.50}$$

for $0 < r < R$.

PROOF. For k and j in $\{1,\dots,n-1\}$, we set $w = D_k u$ and $g^j = \partial g/\partial p_j$ evaluated at $D'v$. Then w solves

$$-w_t + A^{ij} D_{ij} w = 0 \text{ in } \Sigma^+(R),\ D_n w - g^j D_j w = 0 \text{ on } \Sigma^+(0),$$

so Corollary 7.51 implies that there are constants C_1 and α determined by the same quantities as C and θ such that

$$\operatorname*{osc}_{\Sigma^+(r)} w \le C_1 \left(\frac{r}{R}\right)^\alpha \operatorname*{osc}_{\Sigma^+(R)} w.$$

Then the boundary condition implies that

$$\operatorname*{osc}_{\Sigma^0(\rho)} D_n v \le \mu \sum_{k=1}^{n-1} \operatorname*{osc}_{\Sigma^+(\rho)} D_k v \le C \left(\frac{\rho}{R}\right)^\alpha \operatorname*{osc}_{\Sigma^+(R)} Dv.$$

The proof is completed by applying Corollary 7.45 to $D_n v$ and choosing $\theta < \alpha$. □

Our next step is to consider more general linear equations.

LEMMA 13.18. *Let (a^{ij}) be a positive definite matrix-valued function with eigenvalues in the interval $[\lambda,\Lambda]$, let a be a scalar-valued function, let g be a $H_\alpha(\Sigma^+ \times \mathbb{R}^{n-1})$ scalar-valued function with $|g_p| \le \mu$ for some nonnegative constant μ, and suppose that $\bar\mu > \mu/(1+\mu^2)^{1/2}$. Suppose also that there are positive constants a_0, α, and μ_1 such that*

$$|a(X)| \le a_0 (x^n)^{\alpha-1} \tag{13.51}$$

for all X and Y in $\Sigma^+(X_0,R)$, and

$$|g(X',p') - g(Y',p')| \le \mu_1 (1+|p'|)\,|X'-Y'|^\alpha \tag{13.52}$$

for all X' and Y' in Σ^0 and $p' \in \mathbb{R}^{n-1}$. If u solves

$$-u_t + a^{ij} D_{ij} u + a = 0 \text{ in } \Sigma^+(X_0,R),\ D_n u = g(X',D'u) \text{ on } \Sigma^0(X_0,R), \tag{13.53}$$

for some X_0 and R, then there are constants C, σ, and θ determined only by A_0, n, α, λ, Λ, and μ such that $\operatorname{osc}_{\Sigma^+(R)} a^{ij} \le \sigma$ implies

$$\operatorname*{osc}_{\Sigma^+(r)} Du \le C \left(\frac{r}{R}\right)^\theta [\operatorname*{osc}_{\Sigma^+(R)} Du + a_1 R^\alpha] \tag{13.54}$$

for $0 < r < R$ and $a_1 = a_0 + \mu_1[1+\sup|Du|]$.

PROOF. We imitate the proof of Lemma 12.13. Fix $X_1' \in \Sigma^0(X_0',R/2)$. For each $r < R/2$, we want to determine a vector $V(X_1',r)$ with certain properties. In addition to an inequality like (12.19), this vector should solve the equation $V_n = g(X_1',V')$. Hence, we choose r and suppose that $V(X_1',r)$ has been determined. We then define $\bar u$ by

$$\bar u(X) = u(X) - V(X_1',r)\cdot(x-x_1') - u(X_1'),$$

we abbreviate $V(X_1',r)$ to V, and we set $b(X',p) = p_n - g(X',p')$. With β defined by

$$\beta(X) = \int_0^1 \frac{\partial b}{\partial p}(X, sDu(X) + (1-s)V)\,ds,$$

we set $Mw = \beta \cdot Dw$ and $Lw = -w_t + a^{ij}D_{ij}w$. Setting also $\psi = g(X',V') - g(X_1',V')$, we see that $L\bar{u} + a = 0$ in $\Sigma^+(r)$ and $M\bar{u} = \psi$ on $\Sigma^0(r)$. Now we set $M_0 = \sup_{\Sigma^+(r)} |\bar{u}|$, $M_1 = \mu_1(1+|V'|) + a_0$, and $H = M_0 + M_1 r^{1+\alpha}$. Since $|\psi| \le \mu_1(1+|V'|)r^\alpha$ on $\Sigma^0(r)$, Corollary 7.51 implies that

$$[\bar{u}]_{\theta,\Sigma^+(r/2)} \le Cr^{-\theta}H.$$

Hence there is a function $U \in H_{2+\alpha}^{(-\theta)}(\Sigma^+(r/2) \cup \Sigma^0(r/2))$ with $U = \bar{u}$ on $\sigma(r/2)$ and

$$|U_t| + \left|D^2U\right| \le Cr^{-\theta}Hd^{\theta-2}, |DU| \le Cr^{-\theta}Hd^{\theta-1}$$

where d denotes parabolic distance to $\sigma(r/2)$. Finally, let us set

$$w_1 = \frac{a_0}{\alpha}[r^{1+\alpha} - (x^n)^{1+\alpha}],\ h^\pm = \pm[\bar{u} - U] - w_1.$$

Then

$$\begin{aligned} L(h^\pm) &\ge Cr^{-\theta}Hd^{\theta-2} \text{ in } \Sigma^+(r/2), \\ Mh^\pm &\ge Cr^{-\theta}Hd^{\theta-1} \text{ on } \Sigma^+(r/2), \\ h^\pm &\le 0 \text{ on } \sigma(r/2), \end{aligned}$$

so the proof of Lemma 4.16 implies that $h^\pm \le CH\left(\frac{d}{r}\right)^\theta$, and therefore

$$|\bar{u} - U| \le C[\left(\frac{d}{r}\right)^\theta M_0 + M_1 r^{1+\alpha}].$$

As before, we write v for the solution of

$$\begin{gathered} -v_t + a^{ij}(X_1')D_{ij}v = 0 \text{ in } \Sigma^+(r) \\ Mv = g(X_1', Dv) \text{ on } \Sigma^0(r),\ v = u \text{ on } \mathscr{P}\Sigma^+(r) \setminus \Sigma^0(r) \end{gathered}$$

and then a similar estimate holds for $\bar{v} - U$.

We now take $\delta \in (0,1)$ to be further specified and $\delta_0 \in (0,1)$ so small that $C\delta_0^\theta \le \delta/2$. Setting $Q = \Sigma^+(r/2)$ and $Q' = \Sigma^+((1-\delta_0)r/2)$, we infer that

$$\sup_{Q \setminus Q'} |u - v| \le \frac{\delta}{2}M_0 + M_1 r^{1+\alpha},$$

with C independent of δ. On Q', we infer that there is a positive constant $C(\delta)$ (independent of M_1) such that

$$[Dv]^*_{\theta,Q'} \le C(\delta).$$

The linear theory now implies that

$$|D^2v| \le C(\delta)r^{-1-\varepsilon}(x^n)^{\varepsilon-1}M_0.$$

To estimate $|u-v|$ in Q', we define M^* by $M^*w = \beta_1 \cdot Dw$, with

$$\beta_1 = \int_0^1 \frac{\partial b}{\partial p}(X_1', sDu + (1-s)Dv)\, ds.$$

We then have that

$$|L(u-v)| \le CM_1(x^n)^{\alpha-1} + C(\delta)M_0\sigma r^{-1-\varepsilon}(x^n)^{\varepsilon-1} \text{ in } Q',$$
$$|M^*(u-v)| \le C[1+\sup|D'u|]r^\alpha \text{ on } \Sigma^0((1-\delta_0)r/2).$$

If we now set

$$w_2 = \frac{C}{\alpha}M_1[r^{1+\alpha} - (x^n)^{1+\alpha}] + \frac{C(\delta)\sigma}{\varepsilon}M_0 r^{-1-\varepsilon}[r^{1+\varepsilon} - (x^n)^{1+\varepsilon}],$$
$$w_3 = C\mu_1[1+\sup|D'u|]r^\alpha[r-x^n] + \sup_{Q\setminus Q'}|u-v|,$$

then the maximum principle gives $|u-v| \le w_2 + w_3$ in Q'. It follows that

$$|u-v| \le [C(\delta)\sigma + \frac{\delta}{2}]M_0 + C[M_1 + \sup|D'u|]r^{1+\alpha}.$$

Now we use Lemma 13.17 to infer that

$$\operatorname*{osc}_{\Sigma^+(\rho)} Dv \le C\left(\frac{\rho}{r}\right)^\theta \operatorname*{osc}_{\Sigma^+(r/2)} Dv$$

for $0 < \rho < r/4$. The desired result follows from these last two inequalities as in Lemma 12.13 after we note that our choice of V guarantees that $|V'| \le C\sup|Du|$. □

If the differential equation is of the form

$$-u_t + a^{ij}D_{ij}u + b^iD_iu + a = 0,$$

with a^{ij} continuous and $|b| \le B(x^n)^{\alpha-1}$ for some constant B, then the form of the estimate in Lemma 13.18, along with interpolation, gives

$$|Du|_{\theta;\Sigma^+(R/2)} \le C(B,n,\alpha,\lambda,\Lambda,\mu,\mu_1,\omega)[a_0 + 1 + |u|_{0;\Sigma^+(R)}].$$

It then follows via Theorem 8.11 and an approximation argument that the problem

$$-u_t + a^{ij}D_{ij}u + a = 0 \text{ in } \Omega,\ b(X,Du) = 0 \text{ on } S\Omega,\ u = \varphi \text{ on } B\Omega$$

is solvable in $H_{2+\alpha}^{(-1-\theta)}$ for $\mathscr{P} \in H_{1+\alpha}$ provided $a^{ij} \in H_\alpha^*$, a^{ij} is uniformly continuous in Ω, $a \in H_\alpha^{(-1-\theta)}$, and $b(X,D\varphi) = 0$ on $C\Omega$. We finally infer a conditional solvability result for quasilinear equations. As before, we define P and N by

$$Pu = -u_t + a^{ij}(X,u,Du)D_{ij}u + a(X,u,Du), Nu = b(X,u,Du).$$

We also define, for $\varepsilon > 0$, Ω_ε to be the subset of Ω on which $t > \varepsilon$, assuming without loss of generality that $I(\Omega) = (0,T)$.

THEOREM 13.19. *Let $\mathscr{P}\Omega \in H_{1+\alpha}$, and suppose that a^{ij} and a are in $H_\alpha(K)$ for any bounded subset K of $\Omega \times \mathbb{R} \times \mathbb{R}^n$, and that (a^{ij}) is uniformly positive definite on each K. Suppose that $b \in H_\alpha(K^*)$ for any bounded subset K^* of $S\Omega \times \mathbb{R} \times \mathbb{R}^n$, that b is Lipschitz with respect to p with $b_p \cdot \gamma$ bounded away from zero on each such K^*. If there are constants $\delta \in (0,1)$ and C (independent of ε) such that any solution of $Pu = 0$ in Ω_ε, $Nu = 0$ on $S\Omega_\varepsilon$, $u = \varphi$ on $B\Omega$ satisfies the estimate $|u|_{1+\delta} \le C$, then there is a solution of* (13.2).

PROOF. As in Section VIII.1, we need only verify local solvability. To this end, we define J by $u = Jv$ if

$$-u_t + a^{ij}(X,v,Dv)D_{ij}u + a(X,v,Dv) = 0 \text{ in } \Omega_\varepsilon,$$
$$b(X,v,Du) = 0 \text{ on } S\Omega_\varepsilon,\ u = \varphi \text{ on } B\Omega,$$

which we have just shown to be well-defined. The estimates just proved can be used as in Theorem 8.2 to infer local solvability, and then the proof of Theorem 8.3 gives the desired result. □

In deriving estimates for quasilinear equations, we shall also use the following existence result for globally smooth solutions.

PROPOSITION 13.20. *Suppose that a^{ij} and a are in $H_\alpha(\Sigma^+(R))$ with the eigenvalues of (a^{ij}) in the interval $[\lambda, \Lambda]$ and suppose that g is as in Lemma 13.18. If $\varphi \in H_{1+\delta}$, then the solution u of* (13.53) *is in $H_{2+\alpha}^{(-1-\theta)}$ for some $\theta \in (0,1)$.*

PROOF. We only need to show that $u \in H_{1+\theta}$. Near the set on which $x^n = 0$ and $|x - x_0| = R$, we use Lemma 4.33 and then combine this estimate with Lemmata 12.4, 12.12, and 13.18. □

6. The Hölder gradient estimate for quasilinear equations

The ideas just presented can be modified to give Hölder gradient estimates when the equation is quasilinear. The first step in this modification is an estimate along with an existence result when the equation and boundary have a simple, nonlinear form.

LEMMA 13.21. *Suppose (a^{ij}) is a Lipschitz, matrix-valued function defined on $\mathbb{R}^n$ and that there are positive constants K, λ, λ_0, and Λ such that*

$$a^{ij}(p)\xi_i\xi_j \ge \lambda|\xi|^2, \tag{13.55a}$$
$$|a^{ij}(p)| \le \Lambda, |a_p^{ij}(p)| \le \lambda_0 \tag{13.55b}$$

for all $p \in \mathbb{R}^n$ *and*

$$\left|a_p^{ij}(p)\right| = 0 \text{ if } |p| > K. \tag{13.55c}$$

Suppose also that g *is a Lipschitz, scalar-valued function defined on* $\mathbb{R}^{n-1}$ *such that*

$$|g(p)| \le \Lambda, \ \left|g_p(p)\right| \le \mu \tag{13.56a}$$

for all $p \in \mathbb{R}^{n-1}$ *and*

$$\left|g_p(p)\right| = 0 \text{ if } |p| > K. \tag{13.56b}$$

Then there is a positive constant θ *determined only by* K, n, λ, λ_0, *and* Λ *such that any solution* $v \in C^{2,1} \cap H_1(\Sigma^+(R))$ *with* $Dv \in C(\Sigma^+ \cup \Sigma^0)$ *of*

$$-v_t + a^{ij}(Dv)D_{ij}v = 0 \text{ in } \Sigma^+, D_n v = g(D'v) \text{ on } \Sigma^0 \tag{13.57}$$

satisfies the estimate

$$\operatorname*{osc}_{\Sigma^+(r)} Dv \le C(K, n, \lambda, \lambda_0, \Lambda, \sup_{\Sigma^+(R)} |Dv|) \left(\frac{r}{R}\right)^\theta \operatorname*{osc}_{\Sigma^+(R)} Dv. \tag{13.58}$$

In addition, for any continuous function φ, *there is a solution of* (13.57) *with* $v = \varphi$ *on* $\Sigma^+ \setminus \Sigma^0$.

PROOF. With $\bar{b}$ and η as usual, set $V = |D'v|^2 + \eta \bar{b}^2$, and, for ε a positive constant to be further specified and $k = 1, \ldots, n$, define

$$w_k^\pm = \pm D_k v + \varepsilon V.$$

A straightforward calculation gives

$$-w_{k,t}^\pm + a^{ij} D_{ij} w_k^\pm \le C\left|Dw_k^\pm\right|^2 \text{ in } \Sigma^+.$$

We also have $a^{nj} D_j w_k^\pm = 0$ on Σ^0 if $k < n$, while $w_n^\pm = \pm g(D'v) + \varepsilon V$ on Σ^0. If we now define

$$\omega_k(r) = \operatorname*{osc}_{\Sigma^+(r)} w_k^+ + \operatorname*{osc}_{\Sigma^+(r)} w_k^-,$$

the proof of Theorem 12.3 (with Proposition 7.50 replacing Theorem 6.18) shows that

$$\operatorname*{osc}_{\Sigma^+(4r)} D_k v - 2\varepsilon \operatorname*{osc}_{\Sigma^+(4r)} V \le C_1[\omega_k(4r) - \omega_k(r)]$$

for $k < n$.

Now we define $W_k^\pm = \sup_{\Sigma^+(4r)} w_k^\pm$ and infer from Theorem 7.44 (applied to $W_n^\pm - w_n^\pm$) that

$$\inf_{\Sigma^0(2r)} (W_n^\pm - w_n^\pm) \le C_2[\omega_n(4r) - \omega_n(r)],$$

and direct calculation gives

$$\inf_{\Sigma^0(2r)}(W_n^+ - w_n^+) + \inf_{\Sigma^0(2r)}(W_n^- - w_n^-) \geq \operatorname*{osc}_{\Sigma^+(4r)} D_n v - 2\varepsilon \operatorname*{osc}_{\Sigma^0(2r)} V - \operatorname*{osc}_{\Sigma^0(2r)} D_n u$$

Using the boundary condition, we find that

$$\operatorname*{osc}_{\Sigma^0(2r)} D_n v \leq C_3 \sum_{k<n} \operatorname*{osc}_{\Sigma^+(4r)} D_k v \leq C_3 C_1 \sum [\omega_k(4r) - \omega_k(r)] + 2C_3\varepsilon \operatorname*{osc}_{\Sigma^+(4r)} V.$$

Combining all these estimates and setting $\omega = \sum \omega_k$, we infer that

$$\sum \operatorname*{osc}_{\Sigma^+(4r)} D_k v - 2(n+C_3)\varepsilon \operatorname*{osc}_{\Sigma^+(4r)} V \leq C[\omega(4r) - \omega(r)].$$

If we now take $\varepsilon = 1/[10(n+C_3)\sup V^{1/2}]$, we have

$$\omega(4r) \leq C[\omega(4r) - \omega(r)],$$

and this inequality gives (13.58) in the usual way.

For the existence result, we first assume that $u \in H_{1+\delta}$ with $D_n\varphi = g(D'\varphi)$ on $\{X \in \mathscr{P}\Omega, t = -R^2, x^n = 0\}$. From Proposition 13.20 and the argument in Theorem 8.2, we infer local existence. The arguments in that proposition along with the estimate just proved give global existence via the argument in Theorem 8.3. To reduce the regularity hypothesis on φ, we approximate by smooth functions φ_m satisfying the compatibility condition and note that the corresponding u_m's converge uniformly to some limit function u. The local estimates show that $Du_m \to Du$ uniformly on compact subsets of $\Sigma^+ \cup \Sigma^0$ and that $D^2 u_m \to D^2 u$, $u_{m,t} \to u_t$ uniformly on compact subsets of Σ^+. □

From this estimate, we infer a Hölder gradient estimate in $\Sigma^+(R)$ if R is small enough and a^{ij} is continuous with respect to X.

LEMMA 13.22. *Suppose there are positive constants a_0, K, R, α, λ, λ_0, Λ, μ, μ_1 such that*

$$a^{ij}(X,p)\xi_i\xi_j \geq \lambda |\xi|^2, \tag{13.59a}$$

$$|a^{ij}(X,p)| \leq \Lambda, \ |a_p^{ij}(X,p)| \leq \lambda_0, \tag{13.59b}$$

$$|a(X)| \leq a_0 \tag{13.59c}$$

for all $(X,p) \in \Sigma^+(R) \times \mathbb{R}^n$ with $|p| \leq K$, and

$$|g_{p'}(X,p')| \leq \mu, \ |g(X,p') - g(Y,p')| \leq \mu_1 |X-Y|^\alpha \tag{13.60}$$

for all $(X,p) \in \Sigma^0(R) \times \mathbb{R}^{n-1}$. *If* $u \in C^{2,1} \cap H_1(\Sigma^+(R))$ *with* $Du \in C(\Sigma^+ \cup \Sigma^0)$ *is a solution of*

$$-u_t + a^{ij}(X,Du)D_{ij}u + a(X) = 0 \text{ in } \Sigma^+, \tag{13.61a}$$

$$D_n u - g(X,D'u) = 0 \text{ on } \Sigma^0 \tag{13.61b}$$

with $|Du| \le K$, *then there are positive constants* θ *and* σ *determined only by* K, n, α, λ, λ_0, Λ, μ, μ_1 *such that*

$$\left|a^{ij}(X,p) - a^{ij}(Y,p)\right| \le \sigma \tag{13.62}$$

for all $(X,Y,p) \in \Sigma^+ \times \Sigma^+ \times \mathbb{R}^n$ *with* $|p| \le K$ *implies*

$$\operatorname*{osc}_{\Sigma^+(r)} Du \le C(K,n,\alpha,\lambda,\lambda_0,\Lambda)\left(\frac{r}{R}\right)^{\theta}\left[\operatorname*{osc}_{\Sigma^+(R)} Du + a_0 R + \mu_1 R^{\alpha}\right]. \tag{13.63}$$

PROOF. As before, we use d to denote distance from $\mathcal{P}\Sigma^+ \setminus \Sigma^0$. Then we fix $X_0 \in \Sigma^0$ and define $R_0 = \frac{1}{2}d(X_0)$, $M = [Du]^*_{\alpha}$. For $r < R_0$, let v be the solution of

$$-v_t + a^{ij}(X_0,Dv)D_{ij}v = 0 \text{ in } \Sigma^+(r/2),$$
$$D_n v - g(X_0,D'v) = 0 \text{ on } \Sigma^0(r/2)$$
$$v = u \text{ on } \sigma(r/2)$$

given by Proposition 13.20. With V, $\bar{u}$, $\bar{v}$, and M_0 as in Lemma 13.18, and $\delta_0 \in (0,1)$ to be chosen, we set $Q = \Sigma^+(r/2)$ and $Q' = \Sigma^+((1-\delta_0)r/2)$. Then for any $\delta \in (0,1)$, there is δ_0 such that

$$\sup_{Q \setminus Q'} |u - v| \le \delta M_0 + C[a_0 r^2 + \mu_1 r^{1+\alpha}]$$

Then Lemma 13.21 gives a constant $\varepsilon = \varepsilon(\delta)$ such that

$$\left|D^2 v\right| \le C(\delta)(x^n)^{\varepsilon-1} r^{-1-\varepsilon} M_0.$$

By our choice of V, we know that $M_0 \le C \sup|Du| r$ and hence

$$\left|D^2 v\right| \le C(\delta)(x^n)^{\varepsilon-1} r^{-\varepsilon}.$$

Next, we define the vector b by

$$b = [a^{ij}(X,Du) - a^{ij}(X,Dv)]D_{ij}v \frac{Du - Dv}{|Du - Dv|^2}$$

and the operator L by

$$Lw = -w_t + a^{ij}(X,Dw)D_{ij}w + b^i D_i w.$$

Since $|b| \le C(\delta)(x^n)^{\varepsilon-1} r^{-\varepsilon}$, a simple calculation shows that

$$|L(u - v)| \le C(\delta)\sigma(x^n)^{\varepsilon-1} r^{-1-\varepsilon} M_0 + a_0.$$

Similarly, for M^* as in Lemma 13.18, we have

$$|M^*(u - v)| \le \mu_1 r^{\alpha}.$$

Now we define f by

$$f(\tau)=\int_{\tau}^{r}\exp\left(\frac{C_1}{\varepsilon}s^{\varepsilon}r^{-\varepsilon}\right)ds$$

with the constant C_1 to be determined. Then

$$\begin{aligned}L(f(x^n)) &\leq -C(\delta,C_1)(x^n)^{\varepsilon-1}r^{-\varepsilon}[C_1\lambda - C(\delta)]\\ &\leq -C(\delta)\lambda(x^n)^{\varepsilon-1}r^{-\varepsilon-1}\end{aligned}$$

provided $C_1\lambda \geq 1+C(\delta)$. It follows that

$$w=C(\delta)[a_0r+C(\delta)\sigma+\mu_1r^{\alpha}]f(x^n)$$

satisfies the inequalities $Lw \leq -|L(u-v)|$, $M^*w \leq -|M^*(u-v)|$, so

$$\begin{aligned}\sup_{Q'}|u-v| &\leq w+\sup_{Q\setminus Q'}|u-v|\\ &\leq [\delta+C(\delta)\sigma]M_0+C(\delta)[a_0r^2+\mu_1r^{1+\alpha}].\end{aligned}$$

The proof is now the same as for Lemma 12.13 with this inequality in place of (12.23). □

7. Existence theorems

With our estimates in hand, the existence theorems are easily stated. We shall only give simple results. Note that we always need a compatibility condition of the form

$$b(X,\varphi,D\varphi)=0 \text{ on } C\Omega. \tag{13.64}$$

For the conormal problem, condition (13.64) can be written

$$A(X,\varphi,D\varphi)\cdot\gamma+\psi(X,\varphi)=0 \text{ on } C\Omega,$$

and estimates can be proved under suitable combinations of the structure conditions from Section XIII.2.

Specifically, we have a result for the parabolic capillarity problem.

THEOREM 13.23. *Suppose that* $A=\nu$, *that* $\psi_t=0$, *and that* $\psi_z\leq 0$ *and suppose that* $\partial\omega\in H_2$. *Suppose also that there is a positive constant* $c_0<1$ *such that* $z\psi(X,z)\leq c_0z$, *that*

$$|B|=O(|p|),\ |p|B_z\leq O(|p|),\ |B_x|=O(1),\ |B_p|+|B_p\cdot p|=O(1/|p|) \tag{13.65}$$

as $|p|\to\infty$, *and that* (13.64) *holds. If* B *is* H_α *with respect to* X, z *and* p, *then there is a solution* $u\in H_{2+\delta}^{(-1-\alpha)}$ *(for any* $\delta\in(0,1)$*) of* (13.8).

For the uniformly parabolic conormal form, the result can be written as follows.

THEOREM 13.24. *Suppose that $\partial\omega \in H_2$, and that conditions* (13.7) *and* (13.64) *hold. Suppose also that there are positive constants $\theta_0 < \theta_1$ such that*

$$\theta_0 |\xi|^2 \le \frac{\partial A^i}{\partial p_j}\xi_i\xi_j \le \theta_1 |\xi|^2 \tag{13.66a}$$

and that

$$|p|\,|A_z| + |A_x| + |B| = O(|p|^2),\ |A_t| = O(|p|) \tag{13.66b}$$

as $|p| \to \infty$. If B is H_α with respect to X, z and p, then there is a solution $u \in H_{2+\alpha}^{(-1-\delta)}$ (for any δ in $(0,1)$) of (13.8).

Of course, with smoother data, we have smoother solutions. In both theorems, if B and the derivatives of A are in H_α, ψ is $H_{1+\alpha}$ with respect to X and z, and $\partial\omega \in H_{1+\alpha}$, then $u \in H_{2+\alpha}$.

Our existence result in general form is also quite easy.

THEOREM 13.25. *Suppose that a^{ij}, a, and b satisfy* (13.3), (13.40a,b), *and*

$$|p|^2\,|a_p^{ij}| + |p|\,|a_z^{ij}| + |a_x^{ij}| = O(|p|^2), \tag{13.67a}$$

$$|p|^2\,|a_p| + |p|\,|a_z| + |a_x| = O(|p|^4), \tag{13.67b}$$

$$|p|^3\,|b_p| + |p|^2\,|b_z| + |p|\,|b_x| + |b_t| = O(|p|^3\,b_p\cdot\gamma) \tag{13.67c}$$

as $|p| \to \infty$. If $\mathscr{P}\Omega \in H_3$, then there is a solution $u \in H_{2+\alpha}^{(-1-\delta)}$ for any α and δ in $(0,1)$.

The case of one spatial dimension is even easier, since the boundary condition can be written as $u_x = g(X,u)$. Then conditions (13.3a,b,c) give an estimate on $|u|_0$. For the gradient bound, we proceed as in Section XI.6, writing G_1 for the subset of $\mathscr{P}G$ on which neither (x,t) nor (y,t) is in $S\Omega$ and $G_2 = \mathscr{P}G \setminus G_1$. Then there are positive constants L_0 and L_1 such that the function v defined there satisfies the boundary conditions $v \le C_1|x-y|$ on G_1 and $\beta\cdot Dv \ge -L_1$ on G_2 if β is defined by

$$\beta(x,y,t) = \begin{cases} (0,1) & \text{if } (x,t) \in S\Omega \text{ but } (y,t) \notin S\Omega \\ (-1,0) & \text{if } (x,t) \notin S\Omega \text{ but } (y,t) \in S\Omega \\ (-1,1) & \text{if } (x,t) \text{ and } (y,t) \text{ are both in } S\Omega. \end{cases}$$

Hence we have a gradient bound if $|a| = O(a^{11}|p|^2)$. Similarly, the Hölder gradient estimate follows if a^{11} and a are bounded and g is Hölder in both arguments. Hence we obtain a solution of

$$-u_t + a^{11}(X,u,u_x)u_{xx} + a(X,u,u_x) = 0 \text{ in } \Omega,$$

$$u_x = g(X,u) \text{ on } S\Omega,\ u = \varphi \text{ on } B\Omega$$

for $\Omega \subset \mathbb{R}^2$ under the hypotheses just outlined.

Notes

The nonlinear oblique derivative problem for elliptic equations was first studied in detail by Fiorenza [84–86] who showed how to infer an existence theorem from suitable estimates. He also proved some of these estimates in certain special cases. For the uniformly parabolic conormal problem, a version of his existence program was carried out by Ladyzhenskaya and Ural′tseva [183, Section V.7] under more restrictive hypotheses than considered here. Their primary additional assumption is that a^{ij} depends only on X and u, and this hypothesis is made in order to study the gradient bound more simply; the Hölder and gradient Hölder estimates and a pointwise bound for the solution are easy to obtain. Subsequently, Ural′tseva [318–321] studied a generalization of the elliptic capillary problem:

$$\operatorname{div} A(x,u,Du) + B(x,u,Du) = 0 \text{ in } \Omega,$$
$$A(x,u,Du)\cdot\gamma + \psi(x,u) = 0 \text{ on } \partial\Omega,$$

when $A = \partial F/\partial p$ for some scalar function F which behaves like $(1+|p|^2)^{1/2}$. Again, the main point is to estimate the gradient of a solution of such a problem. Gradient bounds for the elliptic capillary problem were also derived independently via different methods by Spruck [298] in two space dimensions and by Simon and Spruck [293] in higher dimensions. Simon and Spruck only proved a bound on the tangential gradient (suitably extended into Ω) and then used the boundary condition to infer a bound on the full gradient. In fact, they were able to bound the tangential gradient when $|\psi| = 1$ (which corresponds to an infinite normal derivative) provided this condition holds on an open subset of $\partial\Omega$. This technique was used by Gerhardt [96, Section 4] (along with a simple pointwise bound on u_t) to study the parabolic capillary problem. A key step is to note in the special case that $\partial\Omega$ is the hyperplane $x^n = 0$ that the differential equation allows us to estimate $D_{nn}u$ in terms of the other second derivatives.

A combination of Ural′tseva's method and Simon's method for proving interior gradient bounds [292] was the basis for my work on gradient bounds for the variational elliptic conormal problem [203]. In that work $A = \partial F/\partial p$ for a large class of functions F. Particular examples are $F(p) = (1+|p|^2)^{1/2}$, which is the capillary problem, $F(p) = (1+|p|^2)^m$ with $m > 1/2$, which is uniformly elliptic, and $F(p) = \exp(\frac{1}{2}|p|^2)$, which is the false mean curvature problem. At about the same time [205], I proved a simple existence result for general quasilinear elliptic oblique derivative problems, including a maximum bound, a gradient bound and a Hölder gradient estimate. Only the gradient bound requires strong hypotheses on the coefficients. The techniques of this paper were expanded considerably by Lieberman and Trudinger [238] to obtain estimates for general uniformly elliptic oblique derivative problems. In addition, [238] is concerned with fully nonlinear equations, of which quasilinear equations are a special case. Our proof of Theorem 13.13 is modelled on the proof of [238, Theorem 2.3]. Various other gradient bounds for solutions of the elliptic oblique derivative problem were proved using the maximum principle by, for example, Korevaar [161, Sections 3 and 4] (the capillary problem) and Lieberman [216,220] (two classes of non-uniformly elliptic equations), but a key element of most maximum principle proofs is that the tangential gradient is estimated by using the equation to estimate $D_{nn}u$ in terms of the other second derivatives. An exception is Korevaar's work in which the variational nature of the boundary condition is used. It was first noted explicitly in [214, p. 116] that a key step in bounding the full gradient is the existence of a function $H(x,z,p)$ satisfying the following conditions, at least for large $|p|$:

(1) $H(x,z,p) \to \infty$ as $p \to \infty$,
(2) $H_{pp} > 0$,
(3) $H_p \cdot \gamma = 0$ on $\partial\Omega$.

The first condition implies that a bound on H gives a bound on p, while the last two imply that H solves a suitable boundary value problem and hence (under additional structure conditions) H satisfies some sort of maximum principle. In [66], Dong showed that a function H (denoted by f on [66, page 19]) with these listed properties can be constructed for any *parabolic* oblique derivative problem assuming H has second derivatives with respect to all variables and assuming some mild growth conditions of various combinations of derivatives of H. It seems, though, that some of his second derivative conditions are unrealistic (see the introduction to [323] for details). We have modified Dong's construction as in [219, Lemma 1.4] to avoid any hypotheses on the second derivatives of b. With the modification, it was shown in [219] that the gradient bounds in [203] hold even without the variational structure condition. Unfortunately, some of the conditions in [219] (such as conditions (3.8d,f,h)) are not quite correct as stated. We have made the adjustments here. In addition, we have used the weighted Sobolev inequality Lemma 13.5, following [232, Section 5], to improve the hypotheses on A and B.

An alternative to our approach, in particular our use of the function $\bar{b}$, is to consider the function $v_1 = \min\{|D'u|^2 + \varepsilon b^2, \varepsilon_1 |Du|^2\}$ for suitable positive constants ε and ε_1 provided b is written in the form $Du \cdot \gamma - g(X,u,D'u)$. Ural'tseva [323, 324] has used this function to obtain gradient bounds for various quasilinear and fully nonlinear parabolic oblique derivative problems. In particular, [324] gives a gradient bound like the one in Theorem 13.13 with two improvements: a need not be differentiable with respect to (x,z,p) and $\mathscr{P}\Omega$ need only be in H_2. On the other hand, some assumptions are made about the second derivatives of b. The method is a refinement of that presented in Section XI.7, but the behavior of b_{pp} cannot be replaced by that of $\bar{b}_{pp}$. Further refinements of this estimate appear in [267].

The Hölder gradient estimate for conormal problems is due to Ladyzhenskaya and Ural'tseva. As in Chapter XII, this estimate can be proved under even weaker hypotheses on the problem; see [217], Exercise 12.3, and Exercise 13.8.

The results in Section XIII.5 are new, but the techniques involved are not. The estimates are all proved by modifications of ideas already used in Chapter XII.

The Hölder gradient estimate for general elliptic oblique derivative problems was first proved in [205, Lemma 2.4] and this proof was based on the one in [85, Theorem 5.2] for the Hölder continuity of solutions of divergence structure equations with nonconormal, oblique boundary conditions. Another proof, under different hypotheses, was given in [238, Theorem 4.1] and this second proof is the basis for our Hölder gradient bound Lemma 13.21 in Section XIII.6. (As in [238], the proof can be extended to include the full structure on a^{ij}, a, and b provided these functions are all Lipschitz with respect to x, z, and p, and $\mathscr{P}\Omega \in H_3$.) An important difference between the elliptic and parabolic cases is that [238] first proves a Hölder estimate for $D'u$ and then one for $Du \cdot \gamma$. As previously noted for the gradient bound, such an approach is not possible for parabolic equations without a bound on u_t. This difficulty was overcome in [225, Theorem 2] and the method here is a simpler version of the argument there. Alternative proofs (under slightly different hypotheses) were given by Dong [66, Theorem 4.5] and Ural′tseva [323, Theorem 2].

Existence theorems for the parabolic capillary problem were given by Gerhardt [96, Theorems 2.5 and 4.5] and for general conormal problems by Lieberman [219, Section 7]. Existence theorems for general quasilinear (in fact, even for fully nonlinear) parabolic equations with oblique boundary conditions were given by Dong [67, Section 7], Ural′tseva [323, Theorem 6], and Nazarov and Ural′tseva [267, Section 4]. Our path (via linear equations with nonlinear boundary conditions) seems to be new.

Exercises

13.1 Prove a maximum estimate for solutions of (13.2) if the conditions

$$\operatorname{sgn} z b(X,z,-zM_1\gamma) < 0,$$
$$\operatorname{sgn} z a(X,z,p) \le k[|z|+|p|],$$
$$\Lambda(X,z,p) \le k$$

are satisfied for $|z| \ge M$, where k, M, and M_1 are positive constants.

13.2 Show that Theorem 13.1 remains valid if (13.3a) is replaced by

$$\operatorname{sgn} z b(X,z,-zM_1\gamma) < 0$$

for $|z| \ge M$ provided $I(\Omega)$ is small enough.

13.3 Let $m \in (1,2)$ and suppose that in Lemma 13.3 conditions (13.6a,c) are replaced by

$$p \cdot A(X,z,p) \ge a_0 |p|^m - a_1 |z|^2,$$
$$z\psi(X,z) \le a_0 c_0 |z|^{3-2/m}.$$

Show that

$$\sup |u| \leq C(a_0, a_1, b_0, b_1, c_0, n, m) \int_\Omega |u|^{1+(2n/m)-n} \, dX + 2M,$$

where $M = M_0 + (c_0 K_0)^{m/(2-m)}$. Show also that (13.9b) holds. What happens when $m = 2$?

13.4 Let $m > 2$ and suppose that in Lemma 13.3 conditions (13.6a,b,c) are replaced by

$$p \cdot A(X,z,p) \geq a_0 |p|^m - a_1 |z|^m,$$
$$zB(X,z,p) \leq b_0 p \cdot A(X,z,p) + b_1 |z|^m,$$
$$z\psi(X,z) \leq a_0 c_0 |z|^m.$$

Show that there are constants $M_1 = M_1(M_0, K_0, C_0, m)$ and $q = q(m,n)$ such that, if $M \geq M_1$, then

$$\sup |u| \leq C \int |u|^q \, dX + 2M.$$

If a_1, b_1, and c_0 are small enough, show also that (13.9b) holds.

13.5 Modify the arguments in Section XIII.2 to estimate the gradient of a solution of the capillary problem when ψ depends on x and t.

13.6 State and prove analogs of Lemmata 13.8, 13.9, and 13.10 which are valid locally in space and in time.

13.7 Suppose that A and B are as in Exercise 11.9. Prove a gradient bound for solutions of (13.8) if ψ depends only on x.

13.8 Suppose u is a bounded weak solution of (13.8) and that A and B satisfy the hypotheses of Exercise 12.3. If also ψ satisfy the condition

$$|\psi(X,z) - \psi(Y,w)| \leq \mu_4[|X-Y|^\alpha + |z-w|^\alpha],$$

prove that $u \in H_{1+\theta}$ for some $\theta \in (0,1)$. Show that $\alpha = \theta$ if $\partial a^i / \partial p_j$ is a continuous function of p uniformly on bounded subsets of $\Omega \times \mathbb{R} \times \mathbb{R}^n$. (See [217].)

13.9 Prove the analog of Lemma 11.14 for the conormal problem and then formulate a corresponding existence result for the conormal problem for the false mean curvature equation.

CHAPTER XIV

FULLY NONLINEAR EQUATIONS I. INTRODUCTION

Introduction

In this chapter and the next, we turn our attention to nonlinear equations which are not necessarily quasilinear. Such an equation can be written in the form

$$Pu = P(X,u,Du,D^2u,-u_t) = 0 \tag{14.1}$$

for some nonlinear function P defined on $\Gamma = \Omega \times \mathbb{R} \times \mathbb{R}^n \times \mathbb{S}^n \times \mathbb{R}$, where $\mathbb{S}^n$ denotes the set of all real symmetric $n \times n$ matrices. (The utility of taking $-u_t$ as a basic quantity will become apparent in our investigation of Monge-Ampère type equations in Chapter XV). We shall denote a typical point in Γ by (X,z,p,r,τ). When F is differentiable with respect to r and τ, we have the following extension of the definitions in Chapter VIII.

We say that the operator P is *parabolic* on some subset Γ_1 of Γ if the matrix (P_r/P_τ) is positive definite, where the subscripts denote differentiation. In particular P_τ is not zero on Γ_1. If equation (14.1) can be solved for u_t, that is, if the equation can be written as

$$u_t = F(X,u,Du,D^2u), \tag{14.2}$$

and if λ and Λ denote the maximum and minimum eigenvalues of the matrix (F_r), respectively, we say that P is *uniformly parabolic* on Γ_1 if the ratio Λ/λ is uniformly bounded on Γ_1.

When P is not differentiable with respect to r or τ, we can further extend the definitions as follows. We say that P is *parabolic* in Γ_1 if

$$P(X,z,p,r+\eta,\tau+\sigma) > P(X,z,p,r,\tau) \tag{14.3}$$

for any positive definite matrix η, any positive number σ, and any $(X,z,p,r,\tau) \in \Gamma_1$. P is *uniformly parabolic* on Γ_1 if (14.1) can be written in the form (14.2) and if there are positive functions λ and Λ such that

$$\lambda \operatorname{tr}\eta \le F(X,z,p,r+\eta) - F(X,z,p,r) \le \Lambda \operatorname{tr}\eta \tag{14.4}$$

for any (X,z,p,r) such that $(X,z,p,r,F(X,z,p,r)) \in \Gamma_1$ and any positive definite matrix η, where $\operatorname{tr}\eta$ denotes the trace of η and if the ratio Λ/λ is bounded on Γ_1.

(Note that this definition guarantees that F is Lipschitz with respect to r and that, wherever F_r exists, we can take λ and Λ to be the minimum and maximum eigenvalues of that matrix.) For any $r \in \mathbb{S}^n$, we define $|r|$ to be the sum of the absolute values of its eigenvalues, so $|r| = \operatorname{tr} r$ if and only if r is positive semidefinite.

We also use the following abbreviations for derivatives of f, which are very useful in our investigation:

$$F^{ij} = \partial F/\partial r_{ij},\ F^i = \partial F/\partial p_i,\ F_i = \partial F/\partial x^i.$$

Also F_z and F_t denote the derivatives with respect to z and t, respectively.

The following examples illustrate some important classes of fully nonlinear parabolic operators.

0.1. Monge-Ampère type operators. If the operator in (14.1) can be written as

$$Pu = (-u_t)\det(D^2u) - \psi(X,u,Du)$$

for some positive function ψ, we say that P is a Monge-Ampère type operator, in analogy with the Monge-Ampère equation $\det(D^2u) = \psi(x)$ for functions of x only. Since ψ is positive, the equation will be parabolic only if r is positive definite and τ is positive. Hence, unlike the quasilinear equations studied so far, this equation must be considered only for functions that are concave with respect to x and decreasing with respect to t. This example and a generalization are studied in Chapter XV.

0.2. Pucci extremal operators. For $\alpha \in (0, 1/n]$, let $\mathscr{L}_\alpha$ denote the set of all linear operators of the form $Lu = a^{ij}D_{ij}u$ for some constant matrix (a^{ij}) with $\mathscr{T} = 1$ and $a^{ij}\xi_i\xi_j \geq \alpha|\xi|^2$. Pucci's maximal and minimal operators are then defined by

$$M_\alpha[u](x) = \sup_{L\in\mathscr{L}_\alpha} Lu(x),\ m_\alpha[u](x) = \inf_{L\in\mathscr{L}_\alpha} Lu(x).$$

It is clear that M_α and m_α are fully nonlinear and satisfy $M_\alpha[-u] = -m_\alpha[u]$. If $\lambda_1(r)$ and $\lambda_n(r)$ denote the maximum and minimum eigenvalues of the matrix r, then an easy calculation shows that

$$M_\alpha[u] = \alpha\Delta u + (1-n\alpha)\lambda_1(D^2u),$$
$$m_\alpha[u] = \alpha\Delta u + (1-n\alpha)\lambda_n(D^2u)$$

and hence M_α and m_α satisfy (14.4) with $\lambda = \alpha$ and $\Lambda = 1$. (In fact, the condition $\mathscr{T} = 1$ is a convenient normalization for the corresponding elliptic operators. Here, it is possible to assume only uniform bounds on the upper and lower eigenvalues of the matrix (a^{ij}).) Thus the equations

$$-u_t + M_\alpha[u] = f(X),\ -u_t + m_\alpha[u] = f(X)$$

are uniformly parabolic although the coefficients are not differentiable with respect to r.

0.3. Parabolic Bellman equations. The family $\mathscr{L}_\alpha$ of example (ii) can be replaced by more general families of linear or quasilinear operators with some uniform control on their coefficients. When the operators are linear, so that $\mathscr{L} = (L_\nu)_{\nu\in\mathscr{V}}$ for some index set $\mathscr{V}$ with each L_ν linear, the equation

$$-u_t + \inf_{\nu\in\mathscr{V}} (L_\nu u - f_\nu(X)) = 0 \tag{14.5}$$

is a parabolic Bellman equation. The solution u is the minimal expected cost for an optimally controlled diffusion process. If we assume that

$$L_\nu u = a_\nu^{ij} D_{ij}u + b_\nu^i D_i u + c_\nu u,$$

then equation (14.5) will be uniformly parabolic provided there are positive constants λ_0 and Λ_0 such that

$$\lambda_0 |\xi|^2 \leq a_\nu^{ij}\xi_i\xi_j \leq \Lambda_0 |\xi|^2 \tag{14.6}$$

for all $\xi \in \mathbb{R}^n$ and $\nu \in \mathscr{V}$. We shall also consider the equation

$$-u_t + \inf_{\nu\in\mathscr{V}} L_\nu u = 0, \tag{14.5$'$}$$

when each L_ν is a quasilinear operator:

$$L_\nu u = a_\nu^{ij}(X,u,Du)D_{ij}u + a_\nu(X,u,Du)$$

and (14.6) holds for suitable positive functions $\lambda_0(X,z,p)$ and $\Lambda_0(X,z,p)$.

1. Comparison and maximum principles

The comparison principle of Chapter IX is easily extended to fully nonlinear equations.

THEOREM 14.1. *Suppose that there is a constant $K \geq 0$ such that*

$$P(X,z_1,p,r,\tau - Kz_1) < P(X,z_2,p,r,\tau - Kz_2) \tag{14.7}$$

whenever $z_1 \geq z_2$ and P is parabolic at $(X,z_i,p,r,\tau - Kz_i)$ for $i = 1$ or $i = 2$. If u and v are in $C^{2,1}(\overline{\Omega}\setminus\mathscr{P}\Omega)\cap C(\overline{\Omega})$ with P parabolic at u or at v and if $Pu \geq Pv$ in $\overline{\Omega}\setminus\mathscr{P}\Omega$ with $u \leq v$ on $\mathscr{P}\Omega$, then $u \leq v$ in Ω.

PROOF. Set $\bar{u} = \exp(-Kt)u$ and $\bar{v} = \exp(-Kt)v$. If $u > v$ somewhere in $\overline{\Omega}\setminus\mathscr{P}\Omega$, then $\bar{u} - \bar{v}$ has a positive maximum at some X_0. Now set $E = \exp(Kt_0)$ and suppress the argument X_0. If P is parabolic at u, we have that

$$\begin{aligned} Pu(X_0) &= P(X_0,E\bar{u},ED\bar{u},ED^2\bar{u},-E(\bar{u}_t + K\bar{u})) \\ &\leq P(X_0,E\bar{u},ED\bar{v},ED^2\bar{v},-E(\bar{v}_t + K\bar{u})) \end{aligned}$$

because $D\bar{v} = D\bar{u}$, $D^2\bar{u} \leq D^2\bar{v}$ and $\bar{u}_t \geq \bar{v}_t$ at X_0. Now (14.7) implies that

$$\begin{aligned} &P(X_0,E\bar{u},ED\bar{v},ED^2\bar{v},-E(\bar{v}_t + K\bar{u})) \\ &\qquad < P(X_0,E\bar{v},ED\bar{v},ED^2\bar{v},-E(\bar{v}_t + K\bar{v})), \end{aligned}$$

and this inequality gives $Pu(X_0) < Pv(X_0)$ which contradicts the assumed inequality between Pu and Pv. A similar argument applies when P is parabolic with respect to v, and hence the theorem is proved. □

Note that (14.7) is equivalent to the inequality $P_z < KP_\tau$ when P is differentiable. In many instances, this condition is easy to check directly.

Uniqueness of solutions is an easy consequence of Theorem 14.1.

COROLLARY 14.2. *Suppose that the operator P satisfies condition* (14.7) *whenever $z_1 \geq z_2$ and P is parabolic at $(X, z_i, p, r, \tau - Kz_i)$ for $i = 1$ or $i = 2$. If u and v are in $C^{2,1}(\overline{\Omega} \setminus \mathscr{P}\Omega) \cap C(\overline{\Omega})$ with P parabolic at u or at v and if $Pu = Pv$ in $\overline{\Omega} \setminus \mathscr{P}\Omega$ with $u = v$ on $\mathscr{P}\Omega$, then $u = v$ in Ω.*

For many applications, an alternative result, analogous to Lemma 2.1, is more useful because it requires no quantitative monotonicity hypotheses.

THEOREM 14.3. *Suppose that u and v are in $C^{2,1}(\overline{\Omega} \setminus \mathscr{P}\Omega) \cap C(\overline{\Omega})$. If P is parabolic with respect to u or to v and if $Pu > Pv$ in $\overline{\Omega} \setminus \mathscr{P}\Omega$ with $u < v$ on $\mathscr{P}\Omega$, then $u < v$ in Ω.*

PROOF. As in Lemma 2.1, we consider the first time (if any) that $u = v$. At such a point, we have

$$Pu = P(X, v, Dv, D^2u, -u_t) \leq Pv$$

and hence there is no such time. □

From this comparison principle, we infer the following maximum estimate.

THEOREM 14.4. *Suppose that $u \in C^{2,1}(\overline{\Omega} \setminus \mathscr{P}\Omega) \cap C(\overline{\Omega})$ and that P is parabolic at u. Suppose also that the equation $P(X, z, 0, 0, \tau) = 0$ can be solved for τ as*

$$\tau = a(X, z) \tag{14.8a}$$

for some function a and that there are positive constants k and b_1 such that

$$-za(X, z) \leq k|z|^2 + b_1. \tag{14.8b}$$

If $0 < t < T$ for $X \in \Omega$, and if $Pu = 0$ in $\overline{\Omega} \setminus \mathscr{P}\Omega$, then

$$\sup_\Omega u \leq \exp((k+1)T)(\sup_{\mathscr{P}\Omega} u^+ + b_1^{1/2}). \tag{14.9}$$

PROOF. Set $A = \sup_{\mathscr{P}\Omega} u^+ + b_1^{1/2}$ and $v = Ae^{(k+1)t}$. Then

$$Pv = P(X, v, 0, 0, -(k+1)v).$$

Since $(k+1)v > -a(X, v)$, it follows that $Pv < 0$, and $v > u$ on $\mathscr{P}\Omega$ by construction, so $Ae^{(k+1)T} \geq v > u$. □

2. Simple uniformly parabolic equations

In our analysis of fully nonlinear equations, a key role is played by the Hölder estimates for second derivatives. To derive them, we use an auxiliary existence theorem which relies on Hölder second derivative estimates for a simple class of equations. In this section, we prove the Hölder estimates along with the existence theorem.

Our first step is an algebraic lemma which is crucial to our Hölder estimate. For brevity, we define the *tensor product* $u \otimes v$ of two vectors u and v to be the matrix with entries given by $(u \otimes v)_{ij} = u_i v_j$.

LEMMA 14.5. *Let λ and Λ be positive constants with $\lambda \leq \Lambda$ and use $S[\lambda,\Lambda]$ to denote the set of all positive definite, symmetric $n \times n$ matrices A^{ij} with eigenvalues in the interval $[\lambda,\Lambda]$. Then there are positive constants λ^* and Λ^* determined only by n, λ, and Λ and a finite set of unit vectors $F = \{v_1,\dots,v_N\}$ such that any matrix $A \in S[\lambda,\Lambda]$ can be written as*

$$A = \sum_{k=1}^{N} \beta_k v_k \otimes v_k \tag{14.10}$$

with the numbers $\beta_1,\dots,\beta_n$ in the interval $[\lambda^*,\Lambda^*]$. Moreover, F can be chosen to include any finite set of unit vectors.

PROOF. First note that $A = \sum \lambda_k v_k \otimes v_k$, where $\Sigma(A) = \{v_1,\dots,v_n\}$ is an orthonormal set of eigenvectors of A with corresponding eigenvalues $\lambda_1,\dots,\lambda_n$.

Now for any set of vectors $\Sigma = \{v_k\}$, define

$$U(\Sigma) = \{\sum \beta_k v_k \otimes v_k : \beta_k > 0\},$$

so that the collection of all $U(\Sigma(A))$ with $A \in S[\lambda,\Lambda]$ forms an open cover of $S[\lambda,\Lambda]$, and hence it has a finite subcover $U(\Sigma(A_1)),\dots,U(\Sigma(A_N))$. Now let $\Gamma(\lambda,\Lambda) = \cup_i \Sigma(A_i)$ and note that

$$A = \sum_{v_k \in \Gamma} \beta_k v_k \otimes v_k$$

with each $\beta_k \geq 0$. It's easy to check that $\beta_k \leq \Lambda$. For a lower bound on β_k, we note that once we have our set of vectors $\Gamma(\lambda,\Lambda)$, we can set

$$\bar{A} = A - \lambda^* \sum_{v_k \in \Gamma(\lambda/2,\Lambda)} v_k \otimes v_k,$$

which will be in $S[\lambda/2,\Lambda]$ if $\lambda^* = \lambda/(2N)$. Then

$$A = \bar{A} + \lambda^* \sum v_k \otimes v_k = \sum (\bar{\beta}_k + \lambda^*) v_k \otimes v_k.$$

A similar consideration shows that we can include any finite set of vectors in our set $\{v_1,\dots,v_N\}$. $\square$

With this algebra in hand, we can now prove a simple Hölder estimate for second derivatives.

LEMMA 14.6. *Let $u \in C^{2,1}(Q(R)) \cap H_2(Q(R))$ satisfy $-u_t + F(D^2u) = 0$ in $Q(R)$. If F is concave and satisfies the inequality (14.4) with positive constants λ and Λ. Then there are positive constants α and C determined only by n, λ, and Λ such that*

$$\operatorname*{osc}_{Q(\rho)} u_t + \operatorname*{osc}_{Q(\rho)} D^2u \le C(n,\lambda,\Lambda)\left(\frac{\rho}{R}\right)^\alpha (\operatorname*{osc}_{Q(R)} u_t + \operatorname*{osc}_{Q(R)} D^2u) \tag{14.11}$$

for $\rho < R$.

PROOF. First note that, for any $h \in (0,R)$ the function

$$v(X) = v(X;h) = \frac{u(x,t) - u(x,t-h)}{h}$$

satisfies the equation

$$-v_t + a^{ij}(X)D_{ij}v = 0$$

for some matrix (a^{ij}) (depending on h) with eigenvalues in the interval $[\lambda,\Lambda]$. Hence, from Corollary 7.42, there is a constant $\delta_1(n,\lambda,\Lambda) \in (0,1)$ such that

$$\operatorname*{osc}_{Q((R-h)/2)} v \le \delta_1 \operatorname*{osc}_{Q(R-h)} v.$$

Sending $h \to 0$, we find that

$$\operatorname*{osc}_{Q(R/2)} u_t \le \delta_1 \operatorname*{osc}_{Q(R)} u_t.$$

Next, for $\xi \in \mathbb{R}^n$ and h as before, set

$$u_h(X,\xi) = \frac{1}{2}u(x+h\xi,t) + \frac{1}{2}u(x-h\xi,t) - u(X).$$

By the concavity of F, we have

$$F[\frac{1}{2}u(x+h\xi,t) + \frac{1}{2}u(x-h\xi,t)] \ge \frac{1}{2}F[u(x+h\xi,t)] + \frac{1}{2}F[u(x-h\xi,t)],$$

and hence

$$-(u_h)_t + F[\frac{1}{2}u(x+h\xi,t) + \frac{1}{2}u(x-h\xi,t)] - F[u] \ge 0.$$

Again, there is a matrix $(a^{ij}) \in S[\lambda,\Lambda]$ such that

$$-(u_h)_t + a^{ij}D_{ij}u_h \ge 0.$$

Now, we define $H_{s,h} = \sup_{Q(sR/4)} u_h$ for $s = 1,2$ and use Theorem 7.37 to infer that

$$(R^{-n-2}\int_{\Theta(R/4)} (H_{2,h} - u_h)^p \, dX)^{1/p} \le C(H_{2,h} - H_{1,h}).$$

Now we divide this inequality by $\frac{1}{2}h^2$ and send h to zero. Setting

$$w(X,\xi) = D_{ij}u(X)\xi_i\xi_j, M_s = \sup_{Q(sR/4)} w,$$

we find that

$$(R^{-n-2}\int_{\Theta(R/4)} (M_2 - w)^p \, dX)^{1/p} \leq C(M_2 - M_1). \tag{14.12}$$

To complete the proof, we use Lemma 14.5 to obtain a corresponding inequality for $w - m_2$. First, for any X and Y in $Q(R/2)$, there is a matrix $(a^{ij}(X,Y)) \in S[\lambda, L]$ such that

$$a^{ij}[D_{ij}u(X) - D_{ij}u(X)] = F[u(Y)] - F[u(X)] = u_t(X) - u_t(Y).$$

It follows from Lemma 14.5 that there are vectors $\xi_1, \ldots, \xi_N$ and constants λ^* and Λ^* determined only by n, λ and Λ such that

$$\sum \beta_k(X,Y)(w(Y,\xi_k) - w(X,\xi_k)) = u_t(X) - u_t(Y) \tag{14.13}$$

with $\beta_k \in [\lambda^*, \Lambda^*]$. Fixing this choice for $\xi_1, \ldots, \xi_N$, we set

$$w^{(k)} = w(\cdot, \xi_k), \; M_s^{(k)} = \sup_{Q(sR/4)} w^{(k)}, \; m_s^{(k)} = \inf_{Q(sR/4)} w^{(k)},$$

and

$$\omega(sR/4) = \sum_k M_s^{(k)} - m_s^{(k)} = \sum_k \operatorname{osc} w^{(k)}.$$

Now we fix an integer j and sum the inequalities (14.12) with $\xi = \xi_k$ over $k \neq j$ to see that

$$\begin{aligned}(R^{-n-2}\int_{\Theta(R/4)} &\sum_{k\neq j} (M_2^{(k)} - w^{(k)})^p \, dX)^{1/p} \\ &\leq n^{1/p}\sum_{k\neq j}(R^{-n-2}\int_{\Theta(R/4)} (M_2^{(k)} - w^{(k)})^p \, dX)^{1/p} \\ &\leq C\sum_{k\neq j}(M_2^{(k)} - M_1^{(k)}) \leq C(\omega(R/2) - \omega(R/4)).\end{aligned}$$

Now, for any $Y \in Q(R/2)$ and $X \in Q(R/2)$ chosen so that $w^{(j)}(X) = m_2^{(j)}$, we have

$$\beta_j(X,Y)(w^{(j)}(Y) - w^{(j)}(X)) \geq \lambda^*(w^{(j)}(Y) - m_2^{(j)}),$$

and, from (14.13),

$$\begin{aligned}\beta_j(X,Y)&(w^{(j)}(Y) - w^{(j)}(X)) \\ &\leq \operatorname*{osc}_{Q(R/2)} u_t + \sum_{k\neq j} \beta_k(X,Y)(w^{(k)}(X) - w^{(k)}(Y)) \\ &\leq \operatorname*{osc}_{Q(R/2)} u_t + \Lambda^*\sum_{k\neq j}(M_2^{(k)} - w^{(k)}(Y)).\end{aligned}$$

It follows that

$$(R^{-n-2}\int_{\Theta(R/4)}(w^{(j)}-m_2^{(j)})^p\,dX)^{1/p}\le C[\omega_t(R/2)+\omega(R/2)-\omega(R/4)],$$

where

$$\omega_t(\rho)=\operatorname*{osc}_{Q(\rho)}u_t.$$

Now we sum this inequality over all j and sum (14.12) over all k and add the resulting inequalities to infer that

$$\omega(R/2)\le C[\omega(R/2)-\omega(R/4)+\omega_t(R/2)]$$

and therefore

$$\omega(R/4)\le\delta_2\omega(R/2)+\omega_t(R/2)$$

with $\delta_2\in(0,1)$. Now we set

$$\sigma(\rho)=\omega(\rho)+\frac{1}{1-\delta_1}\omega_t(\rho)$$

and $\delta_3=\max\{\delta_2,\delta_1\}$ to infer that $\sigma(R/4)\le\delta_3\sigma(R)$. Simple iteration shows that

$$\sigma(\rho)\le C\left(\frac{\rho}{R}\right)^\alpha\sigma(R)$$

and (14.11) follows immediately from this inequality since we can choose the set of unit vectors $\{\xi_j\}$ to include all vectors of the forms e_j and $(e_j+e_k)/\sqrt{2}$ with $j\ne k$. □

Now we reduce the interior second derivative Hölder estimate for equations with lower order terms to an existence result for an auxiliary problem.

THEOREM 14.7. *Let F be defined on $\Omega\times\mathbb{S}^n$ and suppose that F is concave with respect to r and that* (14.4) *holds. Suppose further that there are positive constants b_1, b_2, and β such that*

$$|F(X,r)-F(Y,r)|\le|X-Y|^\beta\,[b_1+b_2|r|] \tag{14.14}$$

for any $r\in\mathbb{S}^n$ and any X and Y in Ω. Suppose finally that for any $X_0\in\Omega$, $\rho\le d(X_0)$, and any continuous φ, there is a unique solution v of

$$-v_t+F(X_0,D^2v)=0\ \text{ in } Q(X_0,\rho),\ v=\varphi\ \text{ on } \mathscr{P}Q(X_0,\rho). \tag{14.15}$$

If $-u_t+F(X,D^2u)=0$ in Ω, then there are positive constants α determined only by n, β, λ and Λ, and C determined also by b_2 such that

$$\langle u\rangle^*_{2+\alpha}+[u]^*_{2+\alpha}\le C(|u|^*_2+b_1). \tag{14.16}$$

PROOF. Fix $X_0\in\Omega$, set $d=d(X_0)$, let $\rho\le d/2$, and let v be the solution of (14.15) with $\varphi=u$. Then there is a matrix-valued function (a^{ij}) with eigenvalues in the interval $[\lambda,\Lambda]$ such that $w=u-v$ satisfies the equation

$$-w_t+a^{ij}D_{ij}w=F(X,D^2u)-F(X_0,D^2u).$$

Defining P by $Ph = -h_t + a^{ij}D_{ij}h$, we conclude that

$$|Pw| \le [b_1 + b_2 |D^2 u|]\rho^\beta$$

in $Q(X_0,\rho)$. It then follows from the maximum principle that

$$|u - v| \le C(n,\lambda,\Lambda)[b_1 + b_2 \sup_{Q(X_0,\rho)} |D^2 u|]\rho^{2+\beta}.$$

Now we write P_2 for the set of all polynomials f of the form

$$f(X) = F_{n+1}t + F_{ij}x^i x^j + F_i x^i + F_0.$$

Similarly to the situation in Lemma 12.13, let us assume that we have already determined a polynomial $f_\rho \in P_2$. From Lemma 14.6 and a simple interpolation argument, we infer that

$$[v - f_\rho]^*_{2+\theta;Q(\rho)} \le C |v - f_\rho|_0$$

Therefore if we define $f_{\tau\rho}$ (with τ to be determined) as the Taylor polynomial of v centered at X_0, it follows that

$$\sup_{Q(X_0,\tau\rho)} |v - f_{\tau\rho}| \le C\tau^{2+\theta} \sup_{Q(X_0,\rho)} |v - f_\rho|.$$

It follows that

$$\sup_{Q(X_0,\tau\rho)} |u - f_{\tau\rho}| \le C\tau^{2+\theta} \sup_{Q(X_0,\rho)} |u - f_\rho| + C[b_1 + b_2 \sup_{Q(X_0,\rho)} |D^2 u|]\rho^{2+\beta}.$$

Choosing τ sufficiently small and $\alpha \le \beta$, $\alpha < \theta$, we infer (14.16) as in Lemma 12.13. □

In fact, if F_r is continuous with respect to r, then we can take $\theta < 1$ to be arbitrary and hence $\alpha = \beta$ in (14.16); see Exercise 14.1.

To complete the proof of our estimates, we need to prove a suitable existence theorem. We use the exact form of Theorem 14.7 to infer existence of solutions for (14.15).

LEMMA 14.8. *If F is a concave function of r which satisfies* (14.4), *then* (14.15) *is solvable.*

PROOF. Suppose first that F_r is Lipschitz continuous with respect to X and r and that $\varphi \in H_\theta$ for some $\theta \in (0,1)$. For $s \in [0,1]$, define

$$F^{(s)}(r) = sF(r) + (1-s)\lambda \operatorname{tr} r.$$

The linear theory implies that

$$-v_t + F^{(s)}(D^2 v) = 0 \text{ in } Q(X_0,\rho), v = \varphi \text{ on } \mathscr{P}Q(X_0,\rho) \tag{14.17}$$

is solvable for $s=0$. We shall use the method of continuity to show that it is solvable for $s\in[0,1]$. So suppose that (14.17) is solvable for some s and let $|\sigma-s|(\Lambda+\lambda)\le\varepsilon$ for some constant ε to be further specified. Then

$$-u_t+F^{(\sigma)}(D^2u)=g(X)$$

if and only if

$$-u_t+F^{(s)}(D^2u)=g(X)+[\sigma-s][F(D^2u)-\lambda\Delta u].$$

Now set $U=H_{2+\alpha}^{(-\theta)}$ and $\mathscr{B}=H_\alpha^{(2-\theta)}$ in Lemma 8.5. Therefore, from the argument in Theorem 8.8 (the assumed smoothness of F_r is needed to obtain the desired properties of the Gateaux variation), we have existence of a solution to

$$-v_t+F^{(\sigma)}(D^2v)=g(X)\ \text{ in } Q,\ v=\varphi\ \text{ on } \mathscr{P}Q$$

with $Q=Q(X_0,\rho)$ and arbitrary $g\in\mathscr{B}$ provided we can prove a bound in the U norm for such a solution in terms of $|g|_\alpha^{(2-\theta)}$. First, note that we can write the differential equation as

$$-v_t+a^{ij}D_{ij}v=F(0)+g(X)$$

for some matrix (a^{ij}) with eigenvalues in $[\lambda,\Lambda]$. Extending φ to a function in U, we infer from a simple modification of Lemma 4.16 that

$$|u-\varphi|\le C[|\varphi|_\theta+|F(0)|+|g|_0^{(2-\theta)}]$$

for some C determined only by n, θ, λ, Λ, and ρ. Setting $N=[u]_{2+\alpha}^{(-\theta)}$, we infer from the interior bound of Theorem 14.7 and interpolation that

$$N\le C[|\varphi|_\theta+|F(0)|+|g|_\alpha^{(2-\theta)}+\varepsilon N].$$

If $C\varepsilon\le 1/2$, this inequality gives the appropriate estimate for N, so (14.17) has a solution for arbitrary H_θ boundary values and $s=\sigma$. By taking uniform limits of such boundary values, we infer solvability of (14.17) for arbitrary continuous φ and $s=\sigma$. It follows that the set of s for which (14.17) has a solution for arbitrary φ is open. Since Theorem 14.7 gives estimates which are uniform with respect to s, it follows that this set is also closed. As we have already shown it to be nonempty, it follows that (14.17) is solvable for any $s\in[0,1]$ and, in particular, for $s=1$. The Lipschitz smoothness of F_r can be removed by a simple approximation argument, namely replace F by its mollification since mollification preserves the properties of F needed to obtain uniform estimates. □

From Theorem 14.7 and Lemma 14.8, we infer the following second derivative Hölder estimate.

COROLLARY 14.9. *Let F be concave in r and satisfy* (14.4) *and* (14.14). *If $u\in H_{2+\alpha}^{(-2)}$ solves $-u_t+F(X,D^2u)=0$ in Ω with α the constant from Theorem 14.7, then* (14.16) *holds.*

From the proof of Lemma 14.8, we also infer the following general existence result.

THEOREM 14.10. *Let F be as in Corollary 14.9. Then there is a constant $\theta_0 \in (0,1)$ determined only by n, λ, and Λ such that if $\mathcal{P}\Omega \in H_{1+\theta}$ with $\theta \in (0,\theta_0]$, then $-u_t + F(X,D^2u) = 0$ in Ω, $u = \varphi$ on $\mathcal{P}\Omega$ has a unique solution $u \in C(\overline{\Omega}) \cap C^{2,1}(\Omega)$. In addition $u \in H^*_{2+\alpha}$ for some $\alpha \in (0,1)$. If $\varphi \in H_{1+\theta}$, then $u \in H^{(-1-\theta)}_{2+\alpha}$.*

PROOF. We only need to prove appropriate bounds if $\varphi \in H_{1+\theta}$. In this case, an easy modification of the boundary gradient estimate in Chapter X shows that $|u-\varphi| \le Cd$ and Lemma 7.47 shows that $[u-\varphi]/d$ is Hölder continuous up to $\mathcal{P}\Omega$ with exponent θ. Coupling these estimates with the interior estimate of Theorem 14.7 (and using interpolation inequalities) gives a bound for $|u|^{(-1-\theta)}_{2+\alpha}$. □

3. Higher regularity of solutions

For our further study of fully nonlinear equations, we show that $C^{2,1}$ solutions of such equations are smoother when the function F is sufficiently smooth. The basic result is quite simple.

LEMMA 14.11. *Let $u \in C^{2,1}$ be a solution of* (14.1). *If $P \in C^1(\Gamma)$, then $Du \in W^{2,1}_{q,loc}$ for any $q > 1$. If $P \in H_{1+\alpha}(K)$ for any bound subset K of Γ, then $u \in H_{3+\alpha,loc}$. If also $\partial\Omega \in H_{3+\alpha}$ and $\varphi \in H_{3+\alpha}$ and if φ satisfies the appropriate compatibility condition, then $u \in H_{3+\alpha}$.*

PROOF. For the local regularity result, we fix a coordinate vector e_m and, for $h \in \mathbb{R}$, we set

$$w(X) = w_h(X) = \frac{u(X) - u(x+he_m,t)}{h}.$$

Then, for $\theta \in (0,1)$, we define

$$u_\theta(X) = \theta u(X) + (1-\theta)u(x+he_m,t),$$

$$\gamma_\theta = (X + \theta he_m, u_\theta, Du_\theta, D^2u_\theta, -u_{\theta,t}),$$

$$a^0(X) = 1/\int_0^1 P_\tau(\gamma_\theta)\,d\theta,$$

$$a^{ij}(X) = a^0(X)\int_0^1 P^{ij}(\gamma_\theta)\,d\theta,$$

$$b^i(X) = a^0(X)\int_0^1 P^i(\gamma_\theta)\,d\theta,$$

$$c(X) = a^0(X)\int_0^1 P_z(\gamma_\theta)\,d\theta,$$

$$f(X) = a^0(X)\int_0^1 P_m(\gamma_\theta)\,d\theta,$$

where $P^{ij} = \partial P/\partial r_{ij}$, $P^i = \partial P/\partial p_i$, and $P_m = \partial P/\partial x^m$. It follows that w is a solution of the uniformly parabolic linear equation

$$-w_t + a^{ij}D_{ij}w + b^iD_iw + cw = f$$

in any cylinder $Q \subset\subset \Omega$ provided $h \le \operatorname{dist}(Q, \mathscr{P}\Omega)$. If $P \in C^1$, we can now use the regularity theory of Chapter VII to infer a bound on w_h in $W^{2,1}_q$ which is uniform in h, and sending $h \to 0$ gives the $W^{2,1}_q$ regularity for Du. If $P \in H_{1+\alpha}$, we use the regularity theory of Chapters IV and V to infer an $H_{2+\alpha}(Q)$ estimate on w_h and hence on Du.

The boundary regularity result follows by similar arguments. □

This regularity result will be useful in deriving gradient bounds in the next section and also for deriving second derivative estimates in Chapter XV.

4. The Cauchy-Dirichlet problem

Now we consider equations of the general form

$$-u_t + F(X,u,Du,D^2u) = 0 \tag{14.18}$$

when F is continuously differentiable with respect to (X,z,p,r) and (F^{ij}) satisfies the condition

$$\lambda\,|\xi|^2 \le F^{ij}(X,z,p,r)\xi_i\xi_j \le \Lambda\,|\xi|^2 \tag{14.19}$$

for some positive constants λ and Λ. We also assume that there are constants k and b_1 such that

$$zF(X,z,0,0) \le k\,|z|^2 + b_1, \tag{14.20}$$

so that the maximum of u can be estimated, via Theorem 14.4, in terms of the maximum of its boundary values. Next, we suppose that there is an increasing function μ so that

$$|F(X,z,p,0)| \le \mu(|z|)(|p|^2+1). \tag{14.21}$$

Then there is a matrix $(a^{ij}(X,z,p,r)) \in S[\lambda,\Lambda]$ such that

$$F(X,z,p,r) - F(X,z,p,0) = a^{ij}r_{ij}, \tag{14.22}$$

and hence Lemma 11.4 provides a local Hölder estimate for u. In addition, Theorem 10.4 gives a boundary gradient estimate provided $\mathscr{P}\Omega \in H_{1+\theta}$, $\varphi \in H_{1+\theta}$ for some $\theta \in (0,1]$.

To proceed in our existence program, we next prove estimates for the gradient of a solution of (14.18). Although these estimates are not immediate consequences of the corresponding results in previous sections, the method of proof requires only a slight modification from that for quasilinear equations. We recall that the operator δ is defined by $\delta F = F_z + |p|^{-2}p\cdot F_x$.

THEOREM 14.12. *Suppose that $u \in C^{2,1}(Q)$ is a bounded solution of* (14.18) *in $Q = Q(R)$. Suppose that F satisfies condition* (14.19) *and that there are non-negative constants $M \geq 1$, μ_1, and μ_2 such that*

$$|p|\,|F_p(X,z,p,r)| \leq \mu_1(|p|^2 + |r|), \tag{14.23a}$$

$$\delta F(X,z,p,r) \leq \mu_1(|p|^2 + |r|), \tag{14.23b}$$

$$|F(X,z,p,0)| \leq \mu_2 |p|^2 \tag{14.23c}$$

for $X \in Q$, $z = u(X)$, and $|p| \geq M$. If $F \in C^1(\Gamma)$, then there are positive constants C and η determined only by n, λ, Λ, μ_1, and μ_2 such that $\operatorname{osc}_{Q(R)} u \leq \eta$ implies that

$$\sup_{Q(R/2)} |Du| \leq M + C\frac{\operatorname{osc}_Q u}{R}. \tag{14.24}$$

PROOF. From Lemma 14.11, $Du \in C^{2,1}$. With ζ as in Theorem 13.13, we set

$$M_0 = \sup_Q u, M_1 = \sup_Q \zeta\,|Du|^2.$$

Now for $\alpha = \operatorname{osc}_Q u$ and $\alpha_1 \in (0, \frac{1}{2})$ to be further specified, set

$$u^* = \exp(\frac{u - M_0}{\alpha}), w = \zeta\,|Du|^2 + \alpha_1 M_1 u^*.$$

Next we define $B^i = F^i - \frac{2}{\zeta}F^{ij}D_j\zeta$. Applying the operator $D_k u D_k$ to (14.18) now yields

$$\begin{aligned} Lw = {} & |Du|^2 L\zeta + \zeta L(|Du|^2) + 2F^{ij}D_i(|Du|^2)D_j\zeta \\ & + \alpha M_1 u^* Lu + \frac{\alpha_1}{\alpha^2} M_1 u^* \mathscr{E}. \end{aligned}$$

The terms in this expression for Lw are estimated for appropriate η and α_1 as in Theorem 13.13. □

The combination of this interior estimate with our boundary gradient estimate gives a global gradient estimate.

THEOREM 14.13. *Let $u \in C^{2,1}(\Omega) \cap C(\overline{\Omega})$ be a bounded solution of* (14.18) *in Ω with $|u| \leq M_0$ and $u = \varphi$ on $\mathscr{P}\Omega$ and suppose that $\mathscr{P}\Omega \in H_{1+\theta}$ and $\varphi \in H_{1+\theta}$ for some $\theta \in (0,1]$. Suppose also that F satisfies conditions* (14.19) *and* (14.21) *and that there are nonnegative constants $M \geq 1$, μ_1,and μ_2 such that conditions* (14.23a,b,c) *hold for $X \in \Omega$, $|z| \leq M_0$, and $|p| \geq M$. Then there is a positive constant C determined only by M_0, M, n, λ, Λ, μ_1, μ_2, and* diam Ω *such that*

$$\sup_\Omega |Du| \leq C. \tag{14.25}$$

Next, we consider the Hölder gradient estimate. The pointwise boundary estimate from Lemma 7.47 follows by using the representation (14.22). The interior estimate is proved by a simple modification of the proof of Theorem 12.3.

THEOREM 14.14. *Let $u \in C^{2,1}$ solve* (14.18) *in $Q(R)$ with $|Du| \le K$ and suppose that there are constants λ, Λ, and μ_3 such that*

$$\left|F_p\right| \le \mu_3(|r|+1), \tag{14.26a}$$

$$|F_z| + |F_x| \le \mu_3(|r|+1). \tag{14.26b}$$

If $F \in C^1(\Gamma)$, then there are constants C and δ determined only by K, n, λ, Λ, and μ_3 such that

$$\operatorname*{osc}_{Q(\rho)} Du \le C\left(\frac{\rho}{R}\right)^{\delta}[1+R^2] \tag{14.27}$$

for $\rho \le R$.

PROOF. Since $F \in C^1$, Lemma 14.11 implies that $Du \in W^{2,1}_{n+1,\mathrm{loc}}$. As in Theorem 12.3, we define

$$w = w_k^{\pm} = \pm D_k u + \varepsilon |Du|^2$$

with ε a positive constant to be further specified. Then

$$\begin{aligned}-w_t + F^{ij}D_{ij}w + F^iD_iw - 2\varepsilon\mathscr{E}_2 \\ + 2\varepsilon|Du|^2\,\delta F \pm (F_zD_ku + F_k) = 0,\end{aligned}$$

and (14.26) implies that

$$-w_t + F^{ij}D_{ij}w \ge C_1(K,n,\lambda,\Lambda,\mu_3)(|Dw|^2+1).$$

We now follow the proof of Theorem 12.3 with Theorem 7.37 in place of Theorem 6.18 to infer (14.27). □

These estimates, along with our existence theorem for simple fully nonlinear equations, give an existence theorem for the more general equations considered here.

THEOREM 14.15. *Let $\mathscr{P}\Omega \in H_{1+\theta}$ and $\varphi \in H_{1+\theta}$. Suppose that F satisfies conditions* (14.19), (14.20), (14.21), *and*

$$|F(X,z,p,r) - F(Y,w,q,r)| \le \left[|X-Y| + |z-w| + |p-q|\right]^{\beta}[b_1 + b_2|r|]. \tag{14.14$'$}$$

Suppose also that F satisfies conditions (14.23a,b) *with μ_1 and μ_2 increasing functions of $|z|$ and* (14.26b) *with μ_3 an increasing function of $|z|+|p|$. If, finally F is concave with respect to r, then there is a unique solution $u \in C^{2,1}(\Omega) \cap C(\overline{\Omega})$ of the initial-boundary value problem*

$$-u_t + F(X,u,Du,D^2u) = 0 \ \text{in}\ \Omega, u = \varphi \ \text{on}\ \mathscr{P}\Omega. \tag{14.28}$$

Moreover, $u \in H_{1+\theta}$ for some $\theta > 0$.

PROOF. For θ sufficiently small and $\beta \in (0,\theta)$, define the map $J : H_{1+\beta} \to H_{1+\beta}$ by $u = Jv$ if u is the solution of

$$-u_t + F(X,v,Dv,D^2u) = 0 \text{ in } \Omega, u = \varphi \text{ on } \mathscr{P}\Omega$$

given by Theorem 14.10. Arguing as in Theorem 8.2, we infer small time existence of a smooth solution to (14.28) and then long time existence provided a uniform estimate is known for $|u|_{1+\theta}$. This estimate was just proved. Uniqueness follows from Corollary 14.2. □

A simple approximation argument allows us to replace condition (14.19) by the monotonicity condition (14.4) with constant λ and Λ. Hence this theorem is applicable to the quasilinear Bellman operator (14.5)′ provided the operators L_ν are given by

$$L_\nu = a_\nu^{ij}(X,u,Du)D_{ij}u + a_\nu(X,u,Du),$$

where the coefficients a_ν^{ij} and a_ν are differentiable with respect to X,z,p with derivatives uniformly bounded (independent of ν) on compact subsets of $\overline{\Omega} \times \mathbb{R} \times \mathbb{R}^n$ and when there are constants λ, Λ, and μ such that the inequalities

$$\lambda |\xi|^2 \le a_\nu^{ij}\xi_i\xi_j \le \Lambda |\xi|^2,$$
$$|p| \left|a_{\nu,p}^{ij}\right| + \left|\delta a_\nu^{ij}\right| \le \mu,$$
$$|a_\nu|, |p| \left|a_{\nu,p}\right|, \ \delta a_\nu \le \mu(1+|p|^2)$$

hold. When the operators L_ν are linear, we only need a uniform Hölder estimate on the coefficients since the $H_{1+\delta}$ estimates can be derived via interpolation. Further weakening of these hypotheses is discussed in the Notes. Improved boundary regularity is the topic of our next section.

5. Boundary second derivative estimates

When the boundary data of the Cauchy-Dirichlet problem for a fully nonlinear parabolic equation are in $H_{2+\alpha}$, one expects the solution to have this improved regularity. For quasilinear equations, this improved regularity is an immediate consequence of the linear theory. For general fully nonlinear equations, it requires a little more work.

Since we shall always assume that a Hölder estimate is known for u and Du, the dependence of F on the variables z and p is no longer important. In addition, our hypotheses will be invariant under an $H_{2+\beta}$ change of variable with β a known constant in $(0,1)$. For these reasons, we consider solutions of

$$-u_t = F(X,D^2u) \text{ in } Q^+(R), \ u = 0 \text{ on } Q^0(R). \tag{14.29}$$

Our first estimate is for the case that F depends only on r.

LEMMA 14.16. *Suppose that F_0 is a C^2, concave function of r and satisfies*

$$\lambda|\xi|^2 \le F_0^{ij}(r)\xi_i\xi_j \le \Lambda|\xi|^2 \tag{14.30}$$

for positive constants λ and Λ. Then there are positive constants C_1 and α determined only by n, λ and Λ such that if $u \in C^{2,1} \cap H^*_{1+\alpha}(Q^+)$ satisfies $-u_t = F_0(D^2u)$ in $Q^+(R)$ and $u = 0$ on $Q^0(R)$, then

$$\langle u\rangle^*_{2+\alpha} + [u]^*_{2+\alpha} \le C_1 |u|^*_{1+\alpha}, \tag{14.31}$$

where the norms are weighted with respect to distance from $\mathscr{P}Q^+ \setminus Q^0$.

PROOF. We first note by taking differences in the directions $x^1, \dots, x^{n-1}$ that

$$\operatorname*{osc}_{Q^+(R/2)} \frac{D_iu}{x^n} \le C\frac{\operatorname{osc}_{Q^+(R)} D_iu}{R} \tag{14.32a}$$

by elementary barrier arguments and that

$$\operatorname*{osc}_{Q^+(\rho)} \frac{D_iu}{x^n} \le C\left(\frac{\rho}{R}\right)^\alpha \operatorname*{osc}_{Q^+(R/2)} \frac{D_iu}{x^n} \tag{14.32b}$$

by Lemma 7.47. Next, we fix $h \in (0,R)$ and set

$$w(X) = \frac{u(x,t)-u(x,t-h^2)}{h^{1+\alpha}}.$$

Then $w = 0$ on $Q^0(R-h)$ and $-w_t + a^{ij}D_{ij}w = 0$ in $Q^0(R-h)$ for some matrix (a^{ij}) with eigenvalues in the interval $[\lambda,\Lambda]$, and hence

$$\sup_{Q(R/2)} \frac{w}{x^n} \le C\frac{\sup_{Q^+(3R/4)} w}{R}$$

provided $h \le R/4$. It's easy to check that $w \le CR^{-1-\alpha}|u|^*_{1+\alpha}$ in $Q^+(3R/4)$, and hence (by sending x^n to zero and using the previous estimate as well) we have

$$|D_nu|^{(1)}_{1+\alpha,Q^0} \le C|u|^*_{1+\alpha},$$

and hence (by arguing as above) $w_1 = D_nu(x',x^n) - D_nu(X',0,t)$ satisfies the estimate

$$\operatorname*{osc}_{Q^+(\rho)} \frac{w_1}{x^n} \le CR^{-2-\alpha}\rho^\alpha|u|^*_{1+\alpha}. \tag{14.33}$$

Next, fix ρ and X_1 such that $Q(X_1,\rho) \subset Q^+$, let η be the usual cutoff function in this cylinder and set $w_2 = \eta^2u_t^2 + C\rho^{-2}|Du-L|^2$ for constants C and L to be chosen. A simple calculation shows that

$$-w_{2,t} + F_0^{ij}D_{ij}w_2 \ge 0 \text{ in } Q(X_1,\rho)$$

for C chosen appropriately (here is where we use the assumption that $F_0 \in C^2$) and determined only by n, λ and Λ. Now choose $L = Du(x_1', 0, t_1)$ and $\rho = x_1^n/2$ to infer that

$$|u_t(X_1)| \leq C \sup_{\{|X'-X_1'| \leq x_1^n, x^n \leq 2x_1^n\}} \frac{|Du(X) - Du(x_1', 0, t_1)|}{x_1^n}.$$

From our previous estimates, it follows that u_t is bounded up to Q^0 with

$$|u_t|_0^{(-2)} \leq C|u|_{1+\alpha}^*.$$

Now we can use Corollary 7.45 (applied to difference quotients) to infer that

$$[u_t]_\alpha^{(2)} \leq C|u|_{1+\alpha}^*.$$

In conjunction with (14.32), (14.33), and Theorem 14.7, this inequality gives the desired regularity and estimates. □

Note that it is possible to infer the regularity of u up to Q^0 from the estimates. See Exercise 14.2 for details.

Next we use this estimate to prove an existence result.

LEMMA 14.17. *Suppose that F_0 is as in Lemma 14.16. Then, for any $\varphi \in C(\overline{Q^+})$ with $\varphi = 0$ on Q^0, there is a unique solution $v \in C^{2,1} \cap C(\overline{Q^+})$ of $-u_t = F_0(D^2u)$ in $Q^+(R)$, $u = \varphi$ on $\mathscr{P}Q^+$. Moreover, there are positive constants C_1 and α determined only by n, λ and Λ such that $u \in H_{2+\alpha}^*(Q^+)$ and, if $F_0(0) = 0$, then $|u|_{2+\alpha}^* \leq C[\varphi]_0$, where the norms are weighted with respect to distance from $\mathscr{P}Q^+ \setminus Q^0$.*

PROOF. Existence of a unique solution (even without the assumption that $\varphi = 0$ on Q^0) follows from Theorem 14.10 and a simple approximation argument. The regularity result follows from Lemma 14.16. □

Now we can state a global regularity result.

THEOREM 14.18. *Let $\mathscr{P}\Omega \in H_{2+\beta}$ and $\varphi \in H_{2+\beta}$, and suppose that F satisfies the hypotheses of Theorem 14.15 in Ω. Then there is a unique solution $u \in C^{2,1}(\Omega) \cap C(\overline{\Omega})$ of (14.28). Moreover, if*

$$-\varphi_t + F(X, \varphi, D\varphi, D^2\varphi) = 0 \text{ on } C\Omega, \tag{14.34}$$

then $u \in H_{2+\alpha}$ for some positive constant α determined only by n, λ, Λ, and β, and

$$|u|_{2+\alpha} \leq C(n, \lambda, \Lambda, \beta, b_2, \Omega)(b_1 + |\varphi|_{2+\beta}). \tag{14.35}$$

PROOF. As we have already seen, problem (14.28) is uniquely solvable and the solution is in $H_{2+\alpha}^{(-1-\theta)}$. Hence we only need to estimate the $H_{2+\alpha}$ norm, which we do locally. In a neighborhood of a point in $\overline{\Omega} \setminus \mathscr{P}\Omega$, this was done in Theorem 14.10, so we only need to consider points in $\mathscr{P}\Omega$.

By subtracting φ from u, we may assume that $\varphi = 0$ on $\mathscr{P}\Omega$. If the point is in $S\Omega$, then a change of variables allows us to restrict our attention to problem (14.29), and an easy variant of the proof of Theorem 14.7 gives an estimate of u in the $H_{2+\alpha}(Q^+(R/2))$ norm. Points in $C\Omega$ or $B\Omega$ are handled as in Theorem 5.14. □

6. The oblique derivative problem

We now consider the oblique derivative problem for fully nonlinear parabolic equations:

$$-u_t + F(X,u,Du,D^2u) = 0 \text{ in } \Omega, \tag{14.36a}$$

$$b(X,u,Du) = 0 \text{ on } S\Omega,\ u = \varphi \text{ on } B\Omega, \tag{14.36b}$$

where we always assume that $b(X,\varphi,D\varphi) = 0$ on $C\Omega$. We also assume that

$$\chi = b_p(X,z,p)\cdot\gamma > 0 \text{ on } S\Omega, \tag{14.37}$$

so that the boundary condition in (14.36) is oblique.

As before, we can write the differential equation in (14.36a) as a quasilinear equation. Hence, if we assume that there are nonnegative constants M_0 and M_1 such that

$$\operatorname{sgn} z b(X,z,-\operatorname{sgn} z M_1\gamma) < 0 \tag{14.38a}$$

for $|z| \geq M_0$ and if there are increasing functions a_0, a_1 and Λ_0 such that

$$\operatorname{sgn} z F(X,z,p,0) \leq a_0(|p|)\,|z| + a_1(|p|), \tag{14.38b}$$

$$|F_r(X,z,p,r)| \leq \Lambda_0(|p|) \tag{14.38c}$$

then Theorem 13.1 gives a pointwise estimate for u. If also F satisfies conditions (14.19) and (14.21) and if b satisfies the condition

$$|b(X,z,p')| \leq \beta_0\chi(X,z,p)(1+|p'|), \tag{14.39}$$

then Lemma 13.14 provides a Hölder estimate for u. A gradient bound then follows by modifying the proof of Theorem 13.15 along the lines indicated in Theorem 14.12.

THEOREM 14.19. *Let $u \in C^{2,1}(\overline{\Omega})$ be a solution of* (14.36) *with $|u| \leq M_0$ and $|u|_\alpha \leq M_\alpha$ for some $\alpha \in (0,1)$. Suppose that $\mathscr{P}\Omega \in H_3$ and that there are constants $M > 1$, μ_1, μ_2 such that conditions* (14.23a,c) *and*

$$F_z(X,z,p,r) \leq \mu_1(|p|^2 + |r|), \tag{14.40a}$$

$$|p|\,|F_x(X,z,p,r)| \leq \mu_1(|p|^2 + |r|), \tag{14.40b}$$

hold for $|p| \geq M$. *Suppose also that there are nonnegative constants* $b_0, \ldots, b_3$ *such that*

$$\left|b_p\right| \leq b_0 \chi, \tag{14.41a}$$

$$|p|^2 |b_z| + |p| |b_x| + |b_t| \leq b_1 |p|^3 \chi \tag{14.41b}$$

for $|p| \geq M$. *Then*

$$\sup_{\Omega} |Du| \leq C(n, b_0, \ldots, b_3, \mu_1, M, M_0, M_\alpha, |D\varphi|, \Omega). \tag{14.42}$$

In fact, Theorem 14.19 is true even if we weaken the boundary regularity to $\mathcal{P}\Omega \in H_{2+\varepsilon}$ for some $\varepsilon > 0$. We leave the details to the reader in Exercise 14.3.

From the gradient bound, we obtain a Hölder gradient estimate by a simple modification of the arguments leading to Lemma 13.21 and Theorem 14.14.

THEOREM 14.20. *Let* $u \in C^{2,1}(\overline{\Omega})$ *be a solution of* (14.36) *with* $|u| \leq M_0$ *and* $|Du| \leq M_1$. *If* $\mathcal{P}\Omega \in H_3$ *and if conditions* (14.26a,b), (14.41a), *and*

$$|b_t| + |b_x| + |b_z| \leq b_1 \chi, \tag{14.41$'$}$$

then (14.27) *holds.*

Of course, we can relax the regularity of F and b by using the method of Section XIII.6. The details are left to the reader in Exercise 14.4.

When combined with the known interior $H_{2+\alpha}$ estimates, the results of Theorems 14.19 and 14.20 are insufficient to infer existence via the theory of Chapter VIII. One alternative is to imitate the arguments of Chapter XIII by using the global $H_{1+\delta}$ estimate of Exercise 14.4. The alternative we shall use is to prove a global $H_{2+\alpha}$ estimate under suitable conditions. For this approach, we first use an idea of Dong [66] (cf. Chapter IV) to obtain the estimate for problems in a particularly simple form. We recall the following definitions from Section IV.4. Let R and $\bar{\mu}$ be positive constants with $\bar{\mu} < 1$ and let $X_0' \in \mathbb{R}^{n+1}$ with $x_0^n = 0$. We then set $X_0 = (x_0^1, \ldots, x_0^{n-1}, -\bar{\mu}R, t_0)$ and we write $\Sigma^+(X_0', R)$ and $\Sigma^0(X_0', R)$ for the subsets of $Q(X_0, R)$ on which $x^n > 0$ and $x^n = 0$, respectively. We also define $\sigma(X_0', R) = \mathcal{P}\Sigma^+(X_0', R) \setminus \Sigma^0(X_0', R)$.

LEMMA 14.21. *Let* $v \in C^{2,1}(\overline{\Sigma^+(0,R)})$ *be a solution of the problem*

$$-v_t + F(D^2 v) = 0 \text{ in } \Sigma^+(R), \tag{14.43a}$$

$$D_n v* = \beta \cdot D' v + \beta_1 \cdot x' \text{ on } \Sigma^0(R). \tag{14.43b}$$

Suppose that F *is* C^2, *concave and satisfies* (14.4) *and* β *and* β_1 *are constant vectors with* $\beta^n = 0$, $|\beta| \leq \mu$ *and* $\bar{\mu} > \mu/(1+\mu^2)^{1/2}$. *Then there are constants* C *and* θ *determined only by* n, λ, Λ, μ, $\bar{\mu}$, *and* $\sup|\beta_1|$ *such that*

$$\operatorname*{osc}_{\Sigma^+(\rho)} D^2 v + \operatorname*{osc}_{\Sigma^+(\rho)} v_t \leq C\left(\frac{\rho}{R}\right)^\theta \left[\operatorname*{osc}_{\Sigma^+(R)} D^2 v + \operatorname*{osc}_{\Sigma^+(R)} v_t\right] \tag{14.44}$$

for $\rho \in (0,R)$.

PROOF. Set

$$w = D_n v - \beta \cdot D'v - \beta_1 \cdot x' \text{ and } M = \sup_{\Sigma^+(R/2)} |D^2 v| + |v_t|.$$

Then $w = 0$ on Σ^0 and $-w_t + F^{ij} D_{ij} w = 0$ in Σ^+. From Lemma 7.47, we infer that

$$[w]^*_{1+\delta;\Sigma^0(X'_0,\rho)} \le CM$$

as long as $|X'_0| \le R/2$ and $\rho \le R/4$ for some δ determined only by n, λ and Λ. Then by direct calculation, we have that $D_{nn}v = D_n w + \beta^k D_{kn} v$, and $D_{kn} v = D_k v - \beta^j D_{jk} v - \beta_1^k = -\beta^j D_{jk} v - \beta_1^k$ on Σ^0, where j and k go only from 1 to $n-1$. Now on Σ^0, v satisfies the differential equation $-v_t + \tilde{F}(X, D_T^2 v) = 0$, where $D_T^2 v$ is the matrix of second derivatives with respect to $x^1, \ldots, x^{n-1}$ and $\tilde{F}(X,r) = F(\hat{r})$ with

$$\hat{r}_{ij} = \begin{cases} r_{ij} & \text{if } i < n, j < n \\ -\beta^k r_{kj} - \beta_1^j & \text{if } i = n, j < n \\ -\beta^k r_{ki} - \beta_1^i & \text{if } i < n, j = n \\ D_n w + \beta^k \beta^m r_{km} & \text{if } i = n, j = n \end{cases}$$

(with m and k being summed only up to $n-1$). A simple calculation shows that

$$\lambda |\xi|^2 \le \tilde{F}^{ij} \xi_i \xi_j \le \Lambda [1 + \mu^2] |\xi|^2$$

and that $\tilde{F}$ is concave with respect to r. From the previously derived Hölder gradient estimate on w and Theorem 14.7, we infer that

$$[v]^*_{2+\theta;\Sigma^0(X'_0,R/2)} \le CM.$$

The desired result now follows from this estimate via Lemma 14.16 and a standard interpolation argument. □

From this result, we easily infer global $H_{2+\theta}$ estimates for solutions of the fully nonlinear oblique derivative problem.

THEOREM 14.22. *Suppose F is as in Theorem 14.7 and $b \in H_{1+\alpha}(\mathscr{K})$ for any bounded subset $\mathscr{K}$ of $S\Omega \times \mathbb{R}^n$ with $\mathscr{P}\Omega \in H_{2+\alpha}$. Suppose also that there are positive functions G and μ such that $|b|_{1+\alpha;\mathscr{K}} \le G(K)$ and $|b_{p'}| \le \mu(K) b_p \cdot \gamma$ on $\mathscr{K} = S\Omega \times \{p \in \mathbb{R}^n : |p| \le K\}$ for any positive number K. Let $u \in C^{2,1}$ solve (14.36a,b), and suppose that $b(X, D\varphi) = 0$ on $C\Omega$. If $|u|_1 \le K$ and $[u]_{1+\varepsilon} \le K_\varepsilon$ for some $\varepsilon \in (0,1]$, then there are constants θ determined only by n, α, λ, Λ, $\mu(K)$, and C determined also by b_1, b_2, $G(K)$, K, K_ε, ε, and Ω such that $|u|_{2+\theta} \le C$.*

PROOF. Since the result is a local one, we consider first the estimate near a point of $S\Omega$. Without loss of generality, we may assume that u satisfies

$$-u_t + F(X, D^2 u) = 0 \text{ in } \Sigma^+(0,1),\ b(X, Du) = 0 \text{ on } \Sigma^0(0,1).$$

We now define $\beta_0 = b_p(0,Du(0))$, $\beta = \beta_0/\beta_0^n$, and $\beta_1 = b_x(0,Du(0))$. As before, there is a solution of (14.43) with $u = v$ on σ. Since

$$|D_n u - \beta \cdot D'u - \beta_1 \cdot x| \le G(K)R^{1+\alpha} + G(K)\,|Du - Du(0)|^{1+\alpha},$$

and

$$|Du - Du(0)|^{1+\alpha} \le |u|_2\, G(K) R^{\alpha\varepsilon}$$

on $\Sigma^0(0,R)$, we find that $|u-v| \le C[1+|u|_2]R^{\alpha\varepsilon}$ in $\Sigma^+(0,R)$. Following the proof of Theorem 14.7 gives an estimate for u in $H_{2+\theta}$ with $\theta \le \alpha\varepsilon$ and then we can repeat the argument with $|u|_2$ bounded.

Near a point of Ω, we have the desired estimate by Corollary 14.9, and near points of $B\Omega$ or $C\Omega$, similar arguments apply. □

Combining all these estimates gives the following existence result.

THEOREM 14.23. *Let $\mathscr{P}\Omega \in H_3$ and $\varphi \in H_{2+\alpha}$. Suppose that F satisfies conditions* (14.4), (14.14)′, (14.19), (14.21), *and* (14.38b,c) *and that b satisfies conditions* (14.37), (14.38a), *and* (14.39). *Suppose also that F and b satisfy conditions* (14.40) *and* (14.41) *for $|p|$ large. If also $b \in H_{1+\alpha}(\mathscr{K})$ for any bounded subset $\mathscr{K}$ of $S\Omega \times \mathbb{R} \times \mathbb{R}^n$, then there is a solution $u \in C^{2,1} \cap C(\overline{\Omega})$ of* (14.36a,b). *Moreover $u \in H_{2+\beta}$ for some $\beta \in (0,\alpha)$.*

As we have already pointed out, the smoothness hypothesis on $\mathscr{P}\Omega$ can be relaxed to $\mathscr{P}\Omega \in H_{2+\alpha}$. Existence of solutions under weaker hypotheses is discussed briefly in the Notes.

7. The case of one space dimension

When there is only one space dimension, the situation is considerably simpler as we already noted for quasilinear equations. In fact, we can reduce fully nonlinear equations to suitable quasilinear equations.

Specifically, suppose that there are positive constants a_0 and a_1 such that

$$a_0 \le F_r \le a_1 \tag{14.45}$$

and nonnegative constants b_1 and k and an increasing function μ_0 such that conditions (14.20) and (14.21) hold. By using the representation (14.22) along with Theorems 14.14, 10.17, 11.16, and 12.10, we obtain global estimates on $|u|_{1+\theta}$ for $H_{1+\alpha}$ boundary data if u is a solution of the Cauchy-Dirichlet problem $-u_t + F(X,u,u_x,u_{xx}) = 0$ in Ω, $u = \varphi$ on $\mathscr{P}\Omega$.

For the second derivative estimate, we suppose that $F_0 : \mathbb{R} \to \mathbb{R}$ is smooth and satisfies the inequality $a_0 \le F_{0,r} \le a_1$ for positive constants. If v solves $-v_t + F_0(v_{xx}) = 0$, then $w = v_x$ solves $-w_t + F_{0,r}(v_{xx})w_{xx} = 0$. Then Theorems 11.18 and 12.2 give local $H_{1+\theta}$ estimates for w and hence $H_{2+\theta}$ estimates for v. Standard approximation shows that this estimate is still valid if F_0 is only Lipschitz, and then the argument in Section XIV.3 gives a local Hölder estimate for u_{xx} and hence u_t provided F is Hölder with respect to X, z, and p. Note that, unlike the situation in higher dimensions, here F need not be concave as a function of r. Boundary estimates can be proved by imitating Lemma 14.17, assuming as usual that $-\varphi_t + F(X,\varphi,\varphi_x,\varphi_{xx}) = 0$ on $C\Omega$.

We then have the following simple existence theorem.

THEOREM 14.24. *Let $\mathscr{P}\Omega \in H_{2+\alpha}$, and let F satisfy conditions* (14.19), (14.20), *and* (14.45). *Suppose also that for any $K \geq 0$, there are constants $b_1 = b_1(K)$ and $b_2 = b_2(K)$ such that* (14.14)′ *holds for X and Y in Ω and $|z| + |w| + |p| + |q| \leq K$. Then, for any $\varphi \in H_{2+\alpha}$, there is a solution $u \in C^{2,1} \cap C(\overline{\Omega})$ of $-u_t + F(X,u,u_x,u_{xx}) = 0$ in Ω, $u = \varphi$ on $\mathscr{P}\Omega$. Moreover, $u \in H_{2+\theta}(\Omega)$ for some θ determined only by a_0 and a_1 provided $-\varphi_t + F(X,\varphi,\varphi_x,\varphi_{xx}) = 0$ on $C\Omega$.*

For the oblique derivative problem, we can write the boundary condition in the form $u_x = g(X,u)$. Therefore, in addition to the simplifications noted in Chapter XIII for the first derivative estimates, we see that the function w mentioned above satisfies a Dirichlet condition on $S\Omega$. We therefore infer the following result.

THEOREM 14.25. *Let $\Omega \subset \mathbb{R}^2$ with $\mathscr{P}\Omega \in H_{2+\alpha}$, and suppose F satisfies conditions* (14.14)′, (14.21), (14.38b,c), *and* (14.45). *If there are nonnegative constants M_0 and M_1 such that g satisfies the condition*

$$\operatorname{sgn} z g(X,z) \geq -M_1 \tag{14.46}$$

for $|z| \geq M_0$ and if $g \in H_{1+\alpha}(S\Omega \times [-K,K])$ for any positive constant K, then for any $\varphi \in H_{2+\alpha}(B\Omega)$ with $\varphi_x = g(X,\varphi)$ on $C\Omega$, there is a solution $u \in C^{2,1}(\Omega) \cap C(\overline{\Omega})$ of $-u_t + F(X,u,u_x,u_{xx}) = 0$ in Ω, $u_x = g(X,u)$ on $S\Omega$, $u = \varphi$ on $B\Omega$. Moreover, $u \in H_{2+\theta}(\Omega)$ for some θ determined only by a_0 and a_1.

Notes

A general study of fully nonlinear parabolic equations was developed by Kruzhkov [163] for the case of one space dimension, and the results of Section XIV.7 are based on that work. In the 1970's, Ivanov (see [120]) and Ivochkina [123] studied fully nonlinear elliptic equations in several dimensions. A major step towards the study of fully nonlinear parabolic equations was the work of Evans and Lenhart [79] in which the Bellman type equation, example (iii) with sufficiently smooth coefficients, was studied. The second critical step for this investigation came in the early 1980's when Evans [76] and Krylov [169] discovered a $C^{2,\alpha}$ estimate for solutions of Bellman type equations. (Evans only looked at elliptic equations while Krylov looked at parabolic equations, but their approaches, discovered independently, are very similar.) Krylov's boundary $C^{1,\alpha}$ estimate [170] for linear equations led the way for a global $C^{2,\alpha}$ estimate; in fact, this is the context in which Krylov proved his result.

Krylov's study of uniformly parabolic equations [170] required stronger conditions than those used here; he used a one-sided condition on the second derivatives of P with respect to (x,z,p,r) and the first derivative with respect to t and he used smoother boundary data. Trudinger [309] (in the context of uniformly elliptic equations) showed that Krylov's conditions could be generalized, but he still needed some one-sided control on the second derivatives of P. Safonov [282] (see [283] for a more detailed description of the method) showed that P need only be Hölder with respect to (X,z,p) and Lipschitz and concave (or convex) with respect to r. In fact, recent work [29, 36, 345] has come up with some examples of nonconvex elliptic operators for which a $C^{2,\alpha}$ estimate holds. Ivanov [119] showed that the concavity condition can be relaxed slightly (again, in the elliptic case) to obtain bounds on the first and second derivatives. The regularity of the boundary data was relaxed to that in Section XIV.4 (although with the stronger hypotheses on F) by Krylov [171] and Lieberman [211, Theorem 6]. Our proof of the interior $C^{2,\alpha}$ estimate is an easy modification of the method recently introduced by Safonov [285, Sections 4 and 5], although we handle the technicalities of the existence result differently. Moreover, Safonov assumed a linear Bellman structure for his equations; the extension to quasilinear structure is straightforward. The interior gradient bound is proved by variation of the method from Chapter XI, with details from [238, Section 3]. It can be proved under weaker hypotheses, namely, without assuming that F is differentiable with respect to x and z by invoking methods of viscosity solutions; see [315].

Further results on the regularity of solutions of fully nonlinear uniformly parabolic equations were obtained by Crandall, Kocan, and Święch [57]. They developed an L^p theory for such equations; in particular, they showed that the $W^{2,p}$ estimates of Wang [333, Section 5] apply to a broader class of equations (which include some gradient dependence) and they allow $p = n+1$. A key ingredient of the estimates in [57] is the existence result of [56, Theorem 2.1], which gives the existence of a viscosity solution to the Cauchy-Dirichlet problem in a cylinder $\Omega = \omega \times (0,T)$, with ω satisfying an exterior cone condition. Unlike the theory developed here, the operator F need not be concave with respect to r.

The oblique derivative problem for fully nonlinear elliptic equations was first studied systematically by Lieberman and Trudinger [238], who showed the key gradient estimate for such problems. Their global $C^{2,\alpha}$ estimate also assumed the continuity of F_r, but this hypothesis was shown to be extraneous by Trudinger [310, Theorem 6]. Dong extended this work to the parabolic case [67], and he made several important contributions. First, he introduced the function H, which is more fully discussed in the notes to Chapter XIII. Second, he gave a much simpler proof of the global $C^{2,\alpha}$ estimate. Ural'tseva [322, 323] has also studied the oblique derivative problem for fully nonlinear parabolic equations. By using an alternative approach, she relaxed some of Dong's hypotheses; see the Notes from Chapter XIII for more detail. The regularity of $\mathscr{P}\Omega$ can be relaxed to $H_{1+\alpha}$

in Theorem 14.22 by modifying the elliptic argument in [236, Theorem 5.4] (see Section 6 of that paper and also [286, Theorem 3.3]).

Exercises

14.1 Suppose F in Lemma 14.6 is continuously differentiable with respect to r. Show that the constant α can be chosen arbitrarily in $(0,1)$ provided C depends also on the modulus of continuity of F_r in (14.11). State and prove the corresponding result for Theorem 14.7.

14.2 Suppose that Lemma 14.16 has only been proved assuming that $u \in H^*_{2+\alpha}$. Use this weaker result to prove Lemma 14.17 and then infer the full strength of Lemma 14.16.

14.3 Prove Theorem 14.19 if $\mathscr{P}\Omega \in H_{2+\varepsilon}$ for some $\varepsilon \in (0,1)$.

14.4 Set up an existence program for the oblique derivative problem using a global $H_{1+\delta}$ estimate (rather than a global $H_{2+\alpha}$ estimate). Also state and prove the corresponding version of Theorem 14.20.

CHAPTER XV

FULLY NONLINEAR EQUATIONS II. HESSIAN EQUATIONS

Introduction

In this chapter, we finish our study of fully nonlinear equations by examining a class of nonuniformly parabolic equations based on the elliptic Monge-Ampère equation $\det D^2u = f(x)$. The theory of such equations is roughly analogous to the theory of nonuniformly parabolic quasilinear equations developed in Chapters X, XI, XII, and XIII. We note here that a complete theory would encompass various equations depending on the curvatures of the graph of the solutions, but we shall not study such equations. A more complete discussion of such equations, with relevant references, can be found in the Notes to this chapter.

As indicated in Chapter XIV, we consider the parabolic Monge-Ampère equation

$$(-u_t)\det D^2u = \psi(X,u,Du)^{n+1} \tag{15.1}$$

and the Hessian equation

$$f(\lambda(D^2u,-u_t)) = \psi(X,u,Du), \tag{15.2}$$

where we use $\lambda(r,\tau)$ to denote the vector of eigenvalues of the matrix $\bar{r} = \begin{pmatrix} r & 0 \\ 0 & \tau \end{pmatrix}$. As we shall see, our hypotheses on f include the special case $f(\lambda) = (\prod \lambda_i)^{1/(n+1)}$, so equation (15.1) can be considered as a particular example of (15.2). We separate them because our hypotheses on the inhomogeneous function ψ need to be much stronger for general Hessian equations than for the Monge-Ampère equation.

In Section XV.1, we collect some key facts about general Hessian equations, which will also be important for the Monge-Ampère equation. A key element of our approach is the assumed existence of what we call strict subsolutions. Included in Section XV.1 is a condition which guarantees the existence of these strict subsolutions. Section XV.2 is concerned with deriving appropriate estimates on solutions of Hessian equations under suitable hypotheses, and the application of these estimates to the existence theory is given in Section XV.3. Section XV.5 provides the proof of estimates for the parabolic Monge-Ampère equation under more general assumptions than those made in Sections XV.2 and XV.3.

1. General results for Hessian equations

Our interest is with initial-boundary value problems of the type

$$F(D^2u, -u_t) = \psi(X, u, Du) \text{ in } \Omega, \tag{15.3a}$$

$$u = \varphi \text{ on } \mathscr{P}\Omega \tag{15.3b}$$

where $F = f \circ \lambda$ with f satisfying suitable hypotheses, and $\mathscr{P}\Omega$ and φ are smooth enough. (We always assume, at least, that $\mathscr{P}\Omega$ and φ are in H_2.) A typical example for f is $f(\lambda) = (\prod \lambda_i)^{1/(n+1)}$, so that (15.3a) is just the parabolic Monge-Ampère equation which is easily seen not to be parabolic at an arbitrary function u but only at those functions for which each λ_i is positive. This strong restriction on the class of functions for which the equation is parabolic is an important feature of the equations we now consider. More generally, we wish to study functions f like S_k, the elementary symmetric polynomial of degree k, and ratios S_k/S_m with $n+1 \geq k > m \geq 1$. All these cases, and more, are covered by the following set of hypotheses.

We assume that there is a convex symmetric proper subcone $\mathscr{K}$ of $\mathbb{R}^{n+1}$ with vertex at the origin such that

$$\text{the positive cone } \{\lambda_i > 0 : i = 1, \ldots, n+1\} \text{ is a subset of } \mathscr{K}, \tag{15.4a}$$

$$D_i f > 0 \text{ in } \mathscr{K} \qquad \text{for } i = 1, \ldots, n+1, \tag{15.4b}$$

$$f \text{ is concave with respect to } \lambda \text{ in } \mathscr{K}, \tag{15.4c}$$

$$\limsup_{\lambda \to \lambda_0} f(\lambda) \leq 0 \text{ for all } \lambda_0 \in \partial\mathscr{K}, \tag{15.4d}$$

$$\inf_{\Omega \times \mathbb{R} \times \mathbb{R}^n} \psi \geq \psi_0 > 0. \tag{15.4e}$$

(Further structure conditions will be given below.) A function u is called *admissible* if $\lambda(D^2u(X), -u_t(X)) \in \mathscr{K}$ for all $X \in \Omega$, and an admissible function u which satisfies (15.3a,b) is called an *admissible solution* of (15.3). We show in Section XV.4 that S_k and S_k/S_m can be brought into this framework by taking $f = S_k^{1/k}$ and $f = (S_k/S_m)^{1/(k-m)}$, respectively, with

$$\mathscr{K} = \mathscr{K}_k = \{S_j(\lambda) > 0 \text{ for } j \leq k\}$$

in either case.

We collect here some consequences of conditions (15.4b,c) which show their connection to our previous hypotheses on P.

LEMMA 15.1. *Suppose that $F = f \circ \lambda$. Then*

(a) the matrices r and F_r commute.

(b) condition (15.4b) *implies that $\partial F/\partial \bar{r} > 0$.*

(c) conditions (15.4b,c) *imply that F is concave with respect to r.*

PROOF. We first note that F is invariant under orthogonal transformations. In other words, $F(UrU^T, \tau) = F(r, \tau)$ for any orthogonal matrix U because r and

UrU^T have the same eigenvalues. Hence r and F_r can be simultaneously diagonalized, and therefore they commute. Thus (a) is proved.

To prove (b), we rotate the spatial axes so that the coordinate directions are the eigenvectors of r (and hence of F_r as well). Then $r = \mathrm{diag}(\lambda_1, \dots, \lambda_n)$ and $F_r = \mathrm{diag}(\varphi_1, \dots, \varphi_n)$ for numbers λ_i and φ_i. Since $F^{ij} = (\partial f/\partial\lambda_k)(\partial\lambda_k/\partial r_{ij})$, it follows that $\varphi_i = f_k\delta_i^k > 0$. It is clear that $F_\tau > 0$.

Finally, to prove (c), we first write the eigenvalues (λ_i) of $\bar{r}$ in increasing order $\lambda_1 \leq \cdots \leq \lambda_{n+1}$, and note the variational characterization of the sum of the first k eigenvalues:

$$\sum_{m=1}^{k} \lambda_m = \inf\{\sum_{m=1}^{k} \bar{r}_{ij}\xi_m^i\xi_m^j : \xi_m \cdot \xi_j = \delta_{mj}\}.$$

It follows that $\sum_{m=1}^{k} \lambda_m$ is a concave function of $\bar{r}$. Since f is a concave function of λ, we can write

$$f(\lambda) = \inf\{\sum_{j=1}^{n+1} \mu_j\lambda_j + \mu_0 : \mu = (\mu_0, \mu_j) \in S\}$$

for some set S on which $\mu_j \geq 0$ for $j > 0$. Writing $\bar{f}(\lambda) = \bar{f}_\mu(\lambda)$ for this sum with $\mu \in S$, let us fix $\lambda \in \mathscr{K}$ and choose $\mu \in S$ so that $f(\lambda) = \bar{f}(\lambda)$. Now we fix $j < k$ such that $\lambda_j < \lambda_k$. Since $\mathscr{K}$ is symmetric, it follows that

$$\lambda^* = (\lambda_1, \dots, \lambda_{j-1}, \lambda_k, \lambda_{j+1}, \dots, \lambda_{k-1}, \lambda_j, \lambda_{k+1}, \lambda_{n+1})$$

is also in $\mathscr{K}$. Therefore, since f is symmetric ,

$$\bar{f}(\lambda^*) \geq f(\lambda^*) = f(\lambda) = \bar{f}(\lambda),$$

and hence

$$0 \leq \bar{f}(\lambda^*) - \bar{f}(\lambda) = \mu_j[\lambda_k - \lambda_j] + \mu_k[\lambda_j - \lambda_k].$$

Hence $\mu_j \geq \mu_k$ if $\lambda_j < \lambda_k$. Thus we can write

$$\bar{f}(\lambda) = \sum_{k=1}^{n} (\mu_k - \mu_{k+1}) \sum_{m=1}^{k} \lambda_k + \mu_{n+1} \sum_{m=1}^{n+1} \lambda_m + \mu_0,$$

which implies that $\bar{f}$ is a concave function of r for each $\mu \in S$. It follows that F is a concave function of r and (c) is proved. □

As with the prescribed mean curvature equation, we need to impose some condition on $\mathscr{P}\Omega$ (similar to the boundary curvature conditions of Chapter X) for (15.3) to be solvable. Here, instead of an explicit geometric condition, we use the strict subsolution hypothesis suggested by the investigations of Caffarelli, Nirenberg, and Spruck, and studied in more detail by Guan [106] and Lijuan Wang [336]. We say that $\underline{u}$ is a *strict subsolution* of (15.3) if $\underline{u} \in C^{2,1}(\overline{\Omega})$ is admissible and if there is a positive constant δ_0 such that

$$F(D^2\underline{u}, -\underline{u}_t) \geq \psi(X, \underline{u}, D\underline{u}) + \delta_0 \text{ in } \Omega,\ \underline{u} = \varphi \text{ on } S\Omega,\ \underline{u} \leq \varphi \text{ on } B\Omega. \qquad (15.5)$$

A strict subsolution exists if simple geometric and structure conditions are satisfied. To state them, we use $\kappa = (\kappa_0, \dots, \kappa_{n-1})$ to denote the space-time curvatures of $S\Omega$ as defined in Section X.3.

LEMMA 15.2. *Suppose that* $F = f \circ \lambda$ *and that conditions* (15.4a–d) *are satisfied. Suppose also that there is a positive constant* R_1 *such that* $(\kappa, R_1) \in \mathscr{K}$. *If, for any compact subset* K *of* $\overline{\Omega} \times \mathbb{R} \times \mathbb{R}^n \times \mathscr{K}$, *there is a positive constant* $R_2(K)$ *such that*

$$f(R\lambda) > \psi(X, Rz, Rp) \tag{15.6}$$

for any $R \geq R_2(K)$ *and any* $(X, z, p, \lambda) \in K$, *then there is a strict subsolution of* (15.3).

PROOF. Write d for the distance function to $S\Omega$, and, for σ a positive constant to be chosen, set $v = [\exp(-\sigma d) - 1]/\sigma$. If $X_0 \in S\Omega$, then

$$(D^2 v, -v_t)(X_0) = \operatorname{diag}(\kappa_1, \dots, \kappa_{n-1}, \sigma, \kappa_0)(X_0)$$

in a principal coordinate system and hence $(D^2 v, -v_t)(X_0) \in \mathscr{K}$ for $\sigma \geq R_1$. With σ now fixed, there is a positive constant ε such that $(D^2 v, -v_t)(X) \in \mathscr{K}$ if $v(X) \geq -\varepsilon$.

We now choose g to be a convex, $C^2((-\infty, 0])$ function such that $g \equiv -1$ on $(-\infty, -\varepsilon]$, $g' > 0$ on $(-\varepsilon, 0)$ and $g(0) = 0$ and we define $w = g(v)$. It follows that $w \in C^{2,1}(\overline{\Omega})$ and that $w = 0$ on $S\Omega$. Since g is convex, a simple calculation shows that $\lambda(D^2 w, -w_t)(X) \in \mathscr{K}$ if $v(X) \geq -\varepsilon/2$ and that $\lambda(D^2 w, -w_t)(X) \in \mathscr{K} \cup \{0\}$ if $v(X) < -\varepsilon/2$.

Next, we take ζ to be a nonnegative, $C^2(\overline{\Omega})$ function with support in $\{v < -\varepsilon/4\}$ such that $\zeta \equiv 1$ in $\{v < -\varepsilon/2\}$. For positive constants A and ε_1 to be chosen, we now define $W = \varepsilon_1 \zeta |X|^2 + w$ and $\underline{u} = AW + \varphi$. Then $\underline{u} = \varphi$ on $S\Omega$ and, if ε_1 is sufficiently small, $\underline{u} < \varphi$ on $B\Omega$.

If $v(X) \leq -\varepsilon/2$, then $\lambda(D^2 W, -W_t)(X) \in \mathscr{K}$ because $D^2 W = D^2 w + \varepsilon_1 I$, $-W_t = -w_t + \varepsilon_1$, and $\lambda(D^2 w, -w_t) \in \mathscr{K} \cup \{0\}$. Hence $\lambda(D^2 W + D^2 \varphi/A, -W_t - \varphi_t/A) \in \mathscr{K}$ provided A is sufficiently large. It follows that $\lambda(D^2 \underline{u}, -\underline{u}_t) \in \mathscr{K}$ in this case.

On the other hand, if $v(X) > -\varepsilon/2$, then $|D^2 w - D^2 W| \leq C\varepsilon_1$ and $|w_t - W_t| \leq C\varepsilon_1$. Since $\lambda(D^2 w, -w_t) \in \mathscr{K}$, it follows that $\lambda(D^2 W, -W_t) \in \mathscr{K}$ if ε_1 is small enough. As before, it follows that $\lambda(D^2 \underline{u}, -\underline{u}_t) \in \mathscr{K}$ in this case as well.

With ε_1 now fixed, we can choose A sufficiently large by virtue of (15.6) to guarantee that $\underline{u}$ is a strict subsolution. □

2. Estimates on solutions

Now we note that

$$\sum_{i=1}^{n+1} \lambda_i > 0 \qquad \text{in } \mathscr{K}$$

because $\mathscr{K}$ is a symmetric proper subcone of $\mathbb{R}^{n+1}$. Hence, if we use $\overline{u}$ to denote the solution of

$$-\overline{u}_t + \Delta\overline{u} = 0 \text{ in } \Omega,\ \overline{u} = \varphi \text{ on } \mathscr{P}\Omega, \tag{15.7}$$

then Theorem 2.4 gives $u \le \overline{u}$. On the other hand, if ψ is continuous, then there is a positive constant δ_1 such that

$$F(D^2\underline{u}, -\underline{u}_t) \ge \psi(X, \underline{u} - \theta, D\underline{u}) + \delta_0/2$$

for all $\theta \in (0, \delta_1)$, so Theorem 14.3 implies that $\underline{u} - \theta < u$ for all sufficiently small θ and hence $\underline{u} \le u$. We immediately infer a pointwise bound for u as well as a boundary gradient estimate.

LEMMA 15.3. *Suppose that f and $\mathscr{K}$ satisfy conditions* (15.4a–c). *Suppose also that $\mathscr{P}\Omega$ and φ are in H_2 and that there is a strict subsolution of* (15.3). *If u is an admissible solution of* (15.3), *then*

$$\inf_{\Omega} \underline{u} \le u \le \sup_{\mathscr{P}\Omega} \varphi, \tag{15.8a}$$

$$\sup_{\mathscr{P}\Omega} |Du| \le C(\varphi, |D\underline{u}|_0, \Omega). \tag{15.8b}$$

Further estimates can be proved under additional hypotheses. We start with a simple gradient estimate.

LEMMA 15.4. *Suppose that the hypotheses of Lemma 15.3 are satisfied along with* (15.4d,e) *and that there are nonnegative constants M, μ_1, and μ_2 such that*

$$\psi_z + \frac{p}{|p|^2} \cdot \psi_x \ge -\mu_1 |p|^2, \tag{15.9a}$$

$$p \cdot \psi_p \le \mu_2 |p|^2 \tag{15.9b}$$

for $|p| \ge M$. Suppose also that there is a positive constant μ_3 such that

$$f_i(\lambda) \ge \mu_3 \text{ if } f(\lambda) \ge \psi_0 \text{ and } \lambda_i < 0, \tag{15.9c}$$

and that

$$\lim_{R\to\infty} f(R\lambda) = \infty \tag{15.9d}$$

for any $\lambda \in \mathscr{K}$. If μ_1 is sufficiently small (determined only by μ_2, μ_3 and $\operatorname{osc} u$*), and if $Du \in C^{2,1}$, then*

$$\sup_{\Omega} |Du| \le C(\mu_1, \mu_2, \mu_3, \operatorname{osc} u, M, \sup_{\mathscr{P}\Omega} |Du|). \tag{15.10}$$

PROOF. For g a positive $C^2(\mathbb{R})$ function to be further determined, set $v = g(u)|Du|^2$ and define L by

$$Lw = -F_\tau w_t + F^{ij} D_{ij} w - \psi^i D_i w,$$

with F_τ and F_r evaluated at $(D^2u, -u_t)$ and ψ_p evaluated at (X, u, Du). A direct calculation gives

$$Lv = (g'' - \frac{(g')^2}{g})|Du|^2 F^{ij} D_i u D_j u + 2F^{ij}\zeta_{ik}\zeta_{jk}$$
$$+ g'|Du|^2 [F^{ij}D_{ij}u - F_\tau u_t - \psi^i D_i u] + g|Du|^2 [\psi_z + \frac{Du}{|Du|^2} \cdot \psi_x],$$

where $\zeta_{ij} = \sqrt{g}D_{ij}u + (g'/\sqrt{g})D_i u D_j u$. Note that $F^{ij}\zeta_{ik}\zeta_{jk} \geq 0$. Moreover, from (15.9d), for any $\lambda \in \mathscr{K}$, there is a constant $R > 1$ such that $f(R\lambda) - f(\lambda) > 0$. Concavity implies that $f_i(\lambda)(R-1)\lambda_i > 0$ and hence $F^{ij}D_{ij}u - F_\tau u_t > 0$. Therefore,

$$Lv \geq (g'' - \frac{(g')^2}{g})|Du|^2 F^{ij} D_i u D_j u - g\mu_1 |Du|^4 - g'\mu_2 |Du|^4$$

wherever $|Du| \geq M$.

Suppose now that v attains its maximum at $X_0 \in \overline{\Omega} \setminus \mathscr{P}\Omega$. Then

$$0 = D_i v = g'|Du|^2 D_i u + 2g D_{ik} u D_k u$$

at X_0, for $i = 1, \dots, n$. If also $|Du(X_0)| > M$, then we set $\xi = Du(X_0)/|Du(X_0)|$ to conclude that

$$D_{ik} u \xi_k = -\frac{g'}{2g}|Du|^2 \xi_i,$$

so ξ is an eigenvector of the matrix D^2u with eigenvalue $-g'|Du|^2/g$, which is negative if $g' > 0$. In this case, $F^{ij}D_i u D_j u \geq \mu_3 |Du|^2$ (at X_0) by (15.9c) and therefore

$$Lv(X_0) \geq |Du|^2 F^{ij} D_i u D_j u[g'' - \frac{(g')^2}{g} - g\frac{\mu_1}{\mu_3} - g'\frac{\mu_2}{\mu_3}].$$

Now we set $k = 2\mu_2/\mu_3$, $A = 9\exp(k \operatorname{osc} u)$, and

$$g(s) = \frac{k}{A - e^{k(s - \min u)}}.$$

A simple calculation shows that $g' > 0$ and that

$$g'' - \frac{2(g')^2}{g} - g\frac{\mu_1}{\mu_3} - g'\frac{\mu_2}{\mu_3} > 0$$

for this choice of g provided μ_1 is small enough. Since v attains its maximum at X_0, we must have $Lv(X_0) \leq 0$, which contradicts what we have just shown. Hence v attains its maximum either where $|Du| \leq M$ or on $\mathscr{P}\Omega$. In either case, we infer (15.10). □

For our estimates on D^2u, we use the following upper bound for $L(u - \underline{u})$.

LEMMA 15.5. *Suppose that ψ is convex with respect to p and define the operator L as in Lemma 15.4. If*

$$\psi_z(X,z,D\underline{u}) \le c_1, \tag{15.11}$$

for all $z \in R$, then there is a positive constant ε determined only by $\underline{u}$ and F such that

$$L(u-\underline{u}) \le c_1[u-\underline{u}] - \varepsilon[1+F_\tau+\sum_i F^{ii}]. \tag{15.12}$$

PROOF. For ε_1 a positive constant to be further specified, set

$$v = \underline{u} - \varepsilon_1[\frac{1}{2}|x|^2 - t].$$

Then there is a positive function σ with $\sigma(\varepsilon_1) \to 0$ as $\varepsilon_1 \to 0$ such that

$$F(D^2v,-v_t) \ge F(D^2\underline{u},-\underline{u}_t) - \sigma(\varepsilon_1) \ge \psi(X,\underline{u},D\underline{u}) + \delta_0 - \sigma(\varepsilon_1).$$

On the other hand, the concavity of F implies that

$$\begin{aligned} F(D^2v,-v_t) &\le F(D^2u,-u_t) + F^{ij}D_{ij}(v-u) + F_\tau(u-v)_t \\ &= \psi(X,u,Du) - \varepsilon_1[F_\tau + \sum_i F^{ii}] - L(u-\underline{u}) + \psi^i D_i(u-\underline{u}). \end{aligned}$$

If we now combine these two inequalities and use the convexity of ψ, we find that

$$\psi(X,\underline{u},D\underline{u}) - \sigma(\varepsilon_1) + \delta_0 \le \psi(X,u,D\underline{u}) - \varepsilon_1[F_\tau + \sum_i F^{ii}] + L(u-\underline{u}),$$

so

$$L(u-\underline{u}) \le c_1[u-\underline{u}] + \varepsilon_1[F_\tau + \sum_i F^{ii}] + \delta_0 + \sigma(\varepsilon_1).$$

The proof is completed by choosing ε_1 so small that $\sigma(\varepsilon_1) \le \delta_0/2$ and then taking $\varepsilon = \min\{\varepsilon_1, \delta_0/2\}$. □

Our next step is to prove second derivative estimates on $S\Omega$. To this end, we fix a point $X_0 \in S\Omega$ and rotate the spatial axes so that the x^n-axis is parallel to $\gamma(X_0)$ and X_0 is the origin. Then we can represent Ω locally as the supergraph of a function, that is,

$$\Omega[R] = \{X \in Q(R) : x^n > g(x',t)\} \tag{15.13}$$

for some function $g \in H_2(Q'(R))$ such that $Dg(0) = 0$, and if $\mathscr{P}\Omega$ is smoother, then so is g. We can now evaluate certain second order derivatives of functions on the boundary in terms of their boundary values and lower order derivatives. For example, suppose that $v = 0$ on $\mathscr{P}\Omega$. Then we may write $v(x',t,g(x',t)) = 0$ for

(x',t) near the origin. Hence, for $i,j<n$, we have

$$\begin{aligned}
0 &= \frac{d}{dx^i}v(x',t,g(x',t)) = D_iv+g_iD_nv, \\
0 &= \frac{d}{dx^j}\Big(\frac{d}{dx^i}v(x',t,g(x',t))\Big) \\
&= D_{ij}v+g_jD_{in}v+g_{ij}D_nv+g_iD_{nj}v+g_ig_jD_{nn}v,
\end{aligned}$$

where the derivatives of v and g are evaluated at $(x',t,g(x',t))$ and (x',t), respectively. In particular, $D_{ij}v(0) = g_{ij}(0)D_nv(0)$. In general, if $u=\varphi$ on $\mathscr{P}\Omega$ and if we assume, without loss of generality that φ has been extended into Ω to be constant along normals, then

$$D_{ij}u(X_0) = D_{ij}\varphi(X_0) - D_nu(X_0)g_{ij}(X_0)$$

for $i,j<n$ and hence we can estimate all double tangential derivatives of u on $S\Omega$.

To estimate the mixed tangential-normal derivatives, we use Lemma 15.1(a) to infer that $F^{ij}r_{jk} = F^{jk}r_{ij}$ and therefore

$$\begin{aligned}
F^{ij}D_{ij}(x^kD_mu - x^mD_ku) &= F^{ik}D_{im}u + F^{kj}D_{jm}u + x^kF^{ij}D_{ijm}u \\
&\quad - F^{im}D_{ik}u - F^{mj}D_{jk}u - x^mF^{ij}D_{ijk}u \\
&= x^kF^{ij}D_{ijm}u - x^mF^{ij}D_{ijk}u \\
&= x^k(F_\tau D_mu_t + D_m\psi) - x^m(F_\tau D_ku_t + D_k\psi).
\end{aligned}$$

It follows that

$$\begin{aligned}
&L(x^kD_mu - x^mD_ku) \\
&\qquad = \psi^kD_mu - \psi^mD_ku + \psi_z[x^kD_mu - x^mD_ku] + x^k\psi_m - x^m\psi_k,
\end{aligned}$$

and hence there is a constant c_2 determined only by $\sup|Du|$, Ω, and $\sup\big|\psi_p\big| + |\psi_z| + |\psi_x|$ such that

$$\Big|L_1(x^kD_mu - x^mD_ku)\Big| \le c_2,$$

where $L_1w = Lw - c_1w$. To continue, we assume that the axes have been rotated so that $D^2g(0)$ is diagonal. For a fixed $k<n$, we take $m=n$ and set

$$w = D_k(u-\varphi) + g_{kk}(0)[x^kD_n(u-\varphi) - x^nD_k(u-\varphi)].$$

A direct calculation gives

$$L_1(D_ku) = -F_\tau D_ku_t + F^{ij}D_{ijk}u + \psi^iD_{ik}u = \psi_zD_ku + \psi_k,$$

so the triangle inequality implies that

$$|L_1w| \le c_3[1 + F_\tau + \sum_i F^{ii}].$$

Now we fix R so that the representation (15.13) holds. If A and B are now positive constants, Lemma 15.5 implies that

$$L_1(w - A[\frac{1}{2}|x|^2 - t] - B[u - \underline{u}]) \geq 0$$

in $\Omega[R]$ provided $B\varepsilon \geq c_3 + A[1 + 2(c_1R^2 + R\sup|\psi_p|)]$.

On $\mathscr{P}\Omega[R]$, we have $D_k(u-\varphi) = -g_kD_n(u-\varphi)$ and hence

$$w = D_n(u-\varphi)[-g_k + g_{kk}(0)(x^k + x^n g_k)]$$

there. Because $D^2g(0)$ is diagonal, we have

$$|x^n| + \left|g_k - x^k g_{kk}(0)\right| \leq C(\Omega)\left|X'\right|^2$$

on $\mathscr{P}\Omega[R]$ if $\mathscr{P}\Omega \in H_3$. It then follows that $w \leq A[\frac{1}{2}|x|^2 - t] + B[u - \underline{u}]$ on $\mathscr{P}\Omega[R]$ for $B \geq 0$ and $A \geq 2C(\Omega)\sup(|Du| + |D\varphi|)$. Finally, on $\mathscr{P}Q(R) \cap \Omega$, we have $w \leq c_4(\sup(|Du| + |D\varphi|), \Omega, R)$, so

$$w - A[\frac{1}{2}|x|^2 - t] - B[u - \underline{u}] \leq 0$$

on the parabolic boundary of $\Omega[R]$ provided $A \geq 2c_4R^{-2}$. The maximum principle implies that $w \leq A[\frac{1}{2}|x|^2 - t] + B[u - \underline{u}]$ in $\Omega[R]$ and similar reasoning shows the corresponding estimate for $-w$. Hence $|w| \leq A[\frac{1}{2}|x|^2 - t] + B[u - \underline{u}]$ in $\Omega[R]$. Evaluating this inequality at $X = (0, x^n, 0)$ gives

$$|w(0,x^n,0)| \leq \frac{A}{2}(x^n)^2 + B\sup|D(u - \underline{u})|x^n \leq c_5 x^n.$$

Since $w(0,x^n,0) = D_k(u-\varphi)(0,x^n,0)[1 - g_{kk}(0)x^n]$, it follows that

$$|D_{kn}u(0)| \leq c_5 \tag{15.14}$$

and the mixed derivative estimate is proved.

To prove the double normal estimate, we need the following lemma on eigenvalues of matrices.

LEMMA 15.6. *Let $r \in \mathbb{S}^N$ with $r_{NN} = 0$, and let r' be the $(N-1)\times(N-1)$ matrix with $r'_{ij} = r_{ij}$ for $i,j < N$. For $a \in \mathbb{R}$, define*

$$r_{ij}(a) = \begin{cases} a & \text{if } i = j = N, \\ r_{ij} & \text{otherwise.} \end{cases} \tag{15.15}$$

Then there are constants a_0, c_1, and c_2 determined only by N and $|r|$ such that

$$\left|\lambda_k(r(a)) - \lambda_k(r')\right| \leq c_1/a \quad \text{if } k < N \tag{15.16a}$$

$$a \leq \lambda_N(r(a)) \leq a + c_2 \tag{15.16b}$$

for $a \geq a_0$, where the eigenvalues of r' and $r(a)$ are written in increasing order.

PROOF. We note that

$$\lambda_N(r(a)) = \max_{|\xi|=1} r_{ij}(a)\xi^i\xi^j \geq a,$$

which proves the first inequality in (15.16b). In addition, there are constants K_1 and K_2 determined only by N and $\max\{|r_{ij}|\}$ such that

$$r_{ij}(a)\xi^i\xi^j \leq K_1|\xi'|^2 + K_2|\xi'||\xi^n| + a|\xi^n|^2.$$

Since $|\xi| = 1$, it follows from this inequality and Cauchy's inequality that the second estimate in (15.16b) holds with $c_2 = K_1 + K_2^2 + 1$.

For (15.16a), we have

$$r_{ij}(a)\xi^i\xi^j \geq r'_{ij}\xi^i\xi^j - K_2|\xi'||\xi^n| + a|\xi^n|^2 \geq (r'_{ij} - \frac{K_2^2}{a}\delta'_{ij})(\xi')^i(\xi')^j + \frac{a}{2}|\xi^n|^2.$$

It follows that $\lambda_k(r(a)) \geq \lambda_k(r') - K_2^2/a$ if a is large enough, and a similar argument shows that $\lambda_k(r(a)) \leq \lambda_k(r') + K_2^2/a$ if a is large enough. □

Next, we observe that if $\lambda \in \mathscr{K}$ (with the components not necessarily written in increasing order), then there is a constant σ^* (determined by λ) such that $(\lambda_1, \ldots, \lambda_n, \sigma) \in \mathscr{K}$ for $\sigma > \sigma^*$. In particular, if we consider a point $X_0 \in S\Omega$ and work in a principal coordinate system there, we can write λ' for the eigenvalues of the matrix

$$\begin{pmatrix} [D_{ij}u]_{i,j<n} & 0 \\ 0 & -u_t \end{pmatrix}.$$

From Lemma 15.1(b), we see that $F(D^2u + \sigma\gamma\otimes\gamma, -u_t) > F(D^2u, -u_t)$ for $\sigma > 0$. Moreover, Lemma 15.6 implies that

$$\lambda_k(D^2u + \sigma\gamma\otimes\gamma, -u_t) \leq \lambda'_k + c_1/\sigma,$$
$$\lambda_{n+1}(D^2u + \sigma\gamma\otimes\gamma, -u_t) \geq \sigma + c_3$$

for some positive constants c_1 and c_3 (determined by bounds on the full second derivative matrix) provided σ is large enough, say $\sigma > \sigma^*$. For such a σ, there is a constant $\varepsilon > 0$ such that

$$f(\lambda_k(D^2u + \sigma\gamma\otimes\gamma, -u_t) - \varepsilon, \lambda_{n+1}(D^2u + \sigma\gamma\otimes\gamma, -u_t)) > F(D^2u, -u_t).$$

Hence, for $\sigma \geq \max\{c_1/\varepsilon, \sigma^*\} + c_3$, it follows that $f(\lambda', \sigma) > F(D^2u, -u_t) = \psi(X, u, Du)$ and $(\lambda', \sigma) \in \mathscr{K}$ and therefore

$$R_0 = \inf\{\sigma > \sigma_0 : f(\lambda', \sigma) > \psi(X, u, Du), (\lambda', \sigma) \in \mathscr{K} \text{ for all } X \in S\Omega\},$$

with σ_0 a constant to be further specified, is finite. We now define

$$f_\infty(\lambda) = f(\lambda', R_0), \quad G(\bar{r}) = f(\lambda'(\bar{r}), R_0),$$

so that G is concave and increasing with respect to $\bar{r}$.

Now we define the tangential derivative operator ∂ on $S\Omega$ by $\partial_i w = D_i w - \gamma_i \gamma^j D_j w$. It then follows that $D_{ij}u = D_\gamma u \partial_i \gamma_j + \partial_{ij}\varphi$ for $i,j<n$, where $D_\gamma u = \gamma \cdot Du$.

Now choose $X_0 \in S\Omega$ such that

$$g_1 = f_\infty(\lambda(D^2u, -u_t)) - \psi(X,u,Du)$$

is minimized (over $S\Omega$) at X_0. We fix a principal coordinate system centered at X_0 and write $\xi^{(1)},\dots,\xi^{(n-1)}$ for a set of $C^{2,1}$ vector fields such that $\xi_i^{(j)}(0) = \delta_i^j$ and $\xi^{(j)} \cdot \gamma = 0$ on $S\Omega$. We now define ∇_i, ∇_{ij} and $\mathscr{C}_{ij}$ by

$$\nabla_i u = \xi_k^{(i)} D_k u, \nabla_{ij} u = \xi_m^{(i)} \xi_k^{(j)} D_{km} u, \mathscr{C}_{ij} = \xi_m^{(i)} \xi_k^{(j)} D_m \gamma_k.$$

The boundary condition implies that

$$\nabla_{ij} u = (D_\gamma u)\mathscr{C}_{ij} + \nabla_{ij}\varphi - (D_\gamma \varphi)\mathscr{C}_{ij},$$

and the eigenvalues $\lambda_1',\dots,\lambda_{n-1}'$ of the matrix $[D_{ij}u]_{i,j<n}$ agree with those of the matrix $[\nabla_{ij}u]$ on $\mathscr{P}\Omega[R]$ for R so small that (15.13) is valid. Recalling that $g_1(X_0) \le g_1(X)$ for any $X \in S\Omega$, we have

$$\psi(X) - \psi(X_0) \le f_\infty(X) - f_\infty(X_0),$$

where we have suppressed the arguments u, Du, D^2u, and $-u_t$. With this same suppression in force, we infer from the concavity of G that

$$\begin{aligned}
\psi(X) - \psi(X_0) &\le G^{ij}[D_{ij}u(X) - D_{ij}u(X_0)] - G_\tau[u_t(X) - u_t(X_0)] \\
&= G^{ij}[D_\gamma u(X)\mathscr{C}_{ij}(X) + \nabla_{ij}\varphi(X) - D_\gamma\varphi(X)\mathscr{C}_{ij}(X) \\
&\qquad - D_\gamma u(X_0)\mathscr{C}_{ij}(X_0) - \nabla_{ij}\varphi(X_0) - D_\gamma\varphi(X_0)\mathscr{C}_{ij}(X_0)] \\
&\quad - G_\tau[(\varphi_t(X) - \varphi_t(X_0)) + D_\gamma(u-\varphi)(X)\bar{g}(X) \\
&\qquad - D_\gamma(u-\varphi)(X_0)\bar{g}(X_0)],
\end{aligned}$$

where $\bar{g} = -g_t/(1+|Dg|^2)^{1/2}$, and G_τ and G^{ij} are evaluated at X_0. Rearranging this inequality and setting $G_0 = G^{ij}\mathscr{C}_{ij}(X_0) + G_\tau \bar{g}(X_0) - \psi_p \cdot \gamma(X_0)$ then gives

$$\begin{aligned}
\psi(X) - \psi(X_0) \le {} & [G_0 + \psi_p \cdot \gamma(X_0)][D_\gamma u(X) - D_\gamma u(X_0)] \\
& + G^{ij}[\mathscr{C}_{ij}(X) - \mathscr{C}_{ij}(X_0)][D_\gamma u(X) - D_\gamma u(X_0)] \\
& + G_\tau[\bar{g}(X) - \bar{g}(X_0)][D_\gamma u(X) - D_\gamma u(X_0)] \\
& + G^{ij} D_\gamma\varphi(X_0)[\mathscr{C}_{ij}(X) - \mathscr{C}_{ij}(X_0)] \\
& + G_\tau D_\gamma \varphi(X_0)[\bar{g}(X) - \bar{g}(X_0)] \\
& + G^{ij}[(\nabla_{ij}\varphi + \mathscr{C}_{ij} D_\gamma\varphi)(X) - (\nabla\varphi + \mathscr{C}_{ij} D_\gamma \varphi)(X_0)] \\
& + G_\tau[(\varphi_t - \bar{g} D_\gamma\varphi)(X) - (\varphi_t - \bar{g} D_\gamma \varphi)(X_0)].
\end{aligned}$$

Since ψ is twice differentiable, we can find a vector Ψ_1 and a constant k_1 such that

$$\psi(X)-\psi(X_0)\geq \psi_p(X_0)\cdot\gamma(X_0)[D_\gamma u(X)-D_\gamma u(X_0)] \\ +\Psi_1\cdot(x-x_0)-k_1|X-X_0|^2$$

on $\mathscr{P}\Omega[R]$. Hence, if we set

$$H(X,X_0)=G^{ij}(\mathscr{C}_{ij}(X)-\mathscr{C}_{ij}(X_0))-G_\tau(\bar{g}(X)-\bar{g}(X_0)),$$

these two inequalities give a vector g_0 and a constant k_2 such that

$$0\leq [G_0+H(X,X_0)][D_\gamma u(X)-D_\gamma u(X_0)] \\ +g_0\cdot(x-x_0)+k_2|X-X_0|^2.$$

Now we estimate G_0. For this estimate, we note that $[\mathscr{C}_{ij}(X_0)]$ is the diagonal matrix with diagonal entries $\kappa_1(X_0),\dots\kappa_{n-1}(X_0)$, and that $\bar{g}(X_0)=-g_t(X_0)$ and hence

$$G_0=\sum_i\kappa_i(X_0)G^{ii}-G_\tau g_t(X_0)-\psi_p\cdot\gamma(X_0).$$

If we set $v=\underline{u}-u$, then the concavity of G implies that

$$\begin{aligned} G(\nabla^2\underline{u},-\underline{u}_t)-G(\nabla^2u,-u_t)&\leq G^{ij}\nabla_{ij}v-G_\tau v_t\\ &=[\sum_i\kappa_iG^{ii}+\bar{g}G_\tau]D_\gamma v\\ &=[G_0+\psi_p\cdot\gamma(X_0)]D_\gamma v,\end{aligned}$$

where G^{ij} and G_τ are evaluated at $(\nabla^2u,-u_t)$. If we also assume that

$$G(X_0)\leq\psi(X_0)+\delta_0/2 \tag{15.17}$$

and take σ_0 so large (determined by F and $\underline{u}$) that

$$G(\nabla^2\underline{u},-\underline{u}_t)\geq F(D^2\underline{u},-\underline{u}_t),$$

then

$$G(\nabla^2\underline{u},-\underline{u}_t)-G(\nabla^2u,-u_t)\geq\psi(X_0,\underline{u},D\underline{u})-\psi(X_0,u,Du)+\delta_0/2.$$

In addition, because $\underline{u}(X_0)=u(X_0)$ and ψ is convex in p, we have

$$\psi(X_0,\underline{u},D\underline{u})-\psi(X_0,u,Du)\geq\psi^iD_iv(X_0)=\psi_p\cdot\gamma(X_0)D_\gamma v(X_0).$$

It follows that

$$G_0D_\gamma v\geq\frac{\delta_0}{2},$$

so $D_\gamma v<0$, $G_0<0$ and

$$G_0\leq\frac{\delta_0}{2D_\gamma v(X_0)}.$$

Now we note that $|H(X,X_0)| \leq k_3 |X - X_0|$ for some constant k_3, so

$$D_\gamma u(X) - D_\gamma u(X_0) \leq \frac{g_0}{G_0} \cdot (x - x_0) + k_4 |X - X_0|^2$$

for some constant k_4. Using the upper bound for $|Dv|$, we infer that

$$D_n u(X) \leq D_n u(X_0) + v_1 \cdot (x - x_0) + k_5 |X - X_0|^2$$

on $\mathscr{P}\Omega[R]$ for some constant vector v_1 and constant scalar k_5. Since $D_n u$ satisfies a linear inequality $|L(D_n u)| \leq k_5$, an easy barrier argument shows that $D_{nn} u(X_0) \leq C$. Recalling that $-u_t + \Delta u \geq 0$ and that u_t is bounded on $S\Omega$ (because $u_t = \varphi_t + g_t D_n(u - \varphi)$), we see that

$$|D_{nn} u(X_0)| \leq C$$

with the constant C determined independent of R_0 and σ_0. If we take σ_0 large enough (determined also by C) that $f(\lambda_0', \sigma_0) > \psi(X_0)$ and $(\lambda_0', \sigma_0) \in \mathscr{K}$ for $\lambda_0' = \lambda'(D^2 u, -u_t)(X_0)$, it follows that $R_0 = \sigma_0$. Since $f_i > 0$, we conclude that there is a positive constant δ_1 (determined by our second derivative bound only at X_0) such that

$$f_\infty(X_0) \geq \psi(X_0) + \delta_1.$$

If (15.17) is false, this inequality is clear with $\delta_1 = \delta_0/2$. By the choice of X_0, we then have $f_\infty \geq \psi + \delta_1$ on $S\Omega$. It follows that $F(D^2 u + (R_1 - D_{\gamma\gamma} u)\gamma \otimes \gamma, -u_t) \geq F(D^2 u, -u_t) + \delta_1$ for some known constant R_1, and this inequality (along with our previous second derivative bounds) implies that

$$\left|D^2 u\right| \leq C \text{ on } \mathscr{P}\Omega.$$

Let us collect the results of our calculations.

LEMMA 15.7. *Suppose that the hypotheses of Lemma 15.3 are satisfied and that ψ is convex with respect to p. Suppose also that $\mathscr{P}\Omega \in H_4$ and $\varphi \in H_4$. If $Du \in C^{2,1}(\overline{\Omega})$, then*

$$\sup_{\mathscr{P}\Omega} \left|D^2 u\right| \leq C(\sup |Du|, |\underline{u}|_2, \Omega, |\varphi|_4, f, \psi). \tag{15.18}$$

To derive global bounds for the second derivatives, we consider two slightly different situations. In the first, we suppose that there is an admissible function $\underline{v} \in C^{2,1}(\Omega) \cap C(\overline{\Omega})$ such that

$$F(D^2 \underline{v}, -\underline{v}_t) \geq \psi(X, u, D\underline{v}) + \delta_0, \ \underline{v} \leq u \text{ in } \Omega. \tag{15.5$'$}$$

If $\psi_z = 0$, then we can take $\underline{v} = \underline{u}$. More generally, if there are constants $R \in \mathbb{R}$ and $p \in \mathbb{R}^n$ such that $F(R\lambda) > \psi(X, z, R(p + x))$ for $\lambda = (1, \ldots, 1)$ and $|z| \leq \sup |u|$, we can take $\underline{v} = R[\frac{1}{2}|x|^2 - t] + v_0$ for a suitable constant v_0. We note from the proof of Lemma 15.5 that there is a positive constant ε such that $L(u - \underline{v}) \leq -\varepsilon$.

Let us now suppose that

$$\psi_z \geq 0, \ \psi_t \leq \mu_4 \tag{15.19}$$

for some nonnegative constant μ_4. Since $w = -u_t$ satisfies the equation $Lw = \psi_z w - \psi_t$, it follows that

$$L_1(w + \frac{\mu_4}{\varepsilon}(u - \underline{v})) \leq 0$$

wherever $w < 0$ and therefore

$$-u_t \geq -\frac{\mu_4}{\varepsilon}\sup(u - \underline{v}) - \sup_{\mathscr{P}\Omega}(-u_t)^-.$$

Since $-u_t + \Delta u > 0$ on $\mathscr{P}\Omega$, it follows that

$$-u_t \geq C \text{ on } \mathscr{P}\Omega, \tag{15.20}$$

and this estimate will be needed to give a global bound on the second spatial derivatives.

To estimate the second derivatives, we fix a unit vector $\xi \in \mathbb{R}^n$ to be further specified and define

$$w(X) = \xi^i \xi^j D_{ij} u(X).$$

Since F is jointly concave with respect to r and τ, we have

$$-F_\tau w_t + F^{ij} D_{ij} w \geq \xi^i \xi^j D_{ij} \psi.$$

By direct calculation, we have

$$\begin{aligned}\xi^i \xi^j D_{ij}\psi = \xi^i \xi^j \psi_{ij} &+ 2\psi_{iz}\xi^i Du\cdot\xi + \psi_i^k \xi^i \zeta_k \\ &+ \psi_{zz}(Du\cdot\xi)^2 + \psi_z^k(Du\cdot\xi)\zeta_k + \psi^{ij}\zeta_i\zeta_j \\ &+ \psi_z w + \psi^k D_{ijk} u \xi^i \xi^j,\end{aligned}$$

where $\zeta_k = D_{km} u \xi^m$. Recalling that ψ is convex in p and setting

$$c_3 = \sup\left[|\psi_{xx}| + |Du|\,|\psi_{xz}| + |Du|^2\,|\psi_{zz}|\right],$$
$$c_4 = \sup\left[|\psi_{xp}| + |Du|\,|\psi_{zp}|\right],$$

we infer that

$$L_1 w \geq \psi_z w - c_3 - c_4|\zeta|.$$

Now we write M_2 for the maximum of w on $\Omega^* = \{(X,\xi) \in \Omega \times \mathbb{R}^n : |\xi| = 1\}$, and we set $W = w(X) - A[u - \underline{v}]$ with A a positive constant to be chosen. If the maximum of W over Ω^* is attained at (X_0, ξ), then a simple Lagrange multiplier argument shows that ξ is an eigenvector of $D^2u(X_0)$ with corresponding eigenvalue no larger than M_2, so $|\zeta| \leq M_2$. It follows that

$$L_1(w - A[u - \underline{v}])(X_0) \geq -c_3 - c_4 M_2 - c_4 C + A\varepsilon \geq 0$$

for $A \geq C(c_3, c_4, \varepsilon) + c_4 M_2/\varepsilon$. If we assume that

$$c_4 \leq \varepsilon/(2\sup(u - \underline{v})), \tag{15.21}$$

then we obtain an upper bound for w and hence for $|D^2u|$ because $-u_t + \Delta u \geq 0$ and we have a lower bound for $-u_t$ from (15.20).

Note that (15.21), in general, requires $|\psi_{xp}|$ and $|\psi_{zp}|$ to be quite small. One way to achieve this condition is to assume that $\psi(X,z,p) = \psi_1(X,z) + \psi_2(p)$ so that $c_4 = 0$. Alternatively, we may suppose that ψ is strictly convex with respect to p, that is, the minimum eigenvalue of the matrix ψ_{pp} is at least some positive constant, which we may take to be ψ_0. In this case, we have

$$\xi^i\xi^j D_{ij}\psi \geq -c_3 - c_4|\zeta| + \psi_0|\zeta|^2 \geq -C$$

for the vector ζ as before. Thus we obtain a global second derivative bound in this case without the smallness assumption for c_4.

Instead of assuming that there is a function $\underline{v}$ satisfying (15.5)′, we now assume that (15.19) is satisfied and that there is a positive constant k_0 such that

$$f_i(\lambda) \geq \frac{k_0 f(\lambda)}{|\lambda|}. \tag{15.22}$$

For A and k positive constants to be further specified, we define

$$w_1(X) = e^{-kt}[-u_t + A|Du|^2]$$

and for $\xi \in \mathbb{R}^n$ with $|\xi| = 1$, we define

$$\overline{w}(X,\xi) = D_{ij}u\xi^i\xi^j,\ w(X,\xi) = e^{-kt}[\overline{w} + A|Du|^2].$$

Suppose first that $\sup_\Omega w_1 \geq \sup_{\Omega\times\{|\xi|=1\}} w$ and choose X_0 so that $w_1(X_0) = \sup w_1$. If $u_t(X_0) < 0$, it follows from (15.22) that

$$F_\tau(D^2u, -u_t)(X_0) \geq C(n)k_0\psi_0/(-u_t(X_0)),$$

and hence, at X_0, we have

$$e^{kt}Lw_1 = kF_\tau w_1 + \psi_z w_1 + A\psi_x \cdot Du + AF^{ij}D_{ik}uD_{jk}u.$$

Then

$$e^{kt}Lw_1 \geq kF_\tau(-u_t) - c_5(A, \sup|Du|, \sup|\psi_x|, |\psi_t|) \geq 0$$

if $k \geq c_5/[C(n)k_0\psi_0]$ and hence the maximum principle implies that

$$\sup\overline{w},\ \sup(-u_t) \leq C \tag{15.23}$$

in this case. Of course, if $u_t(X_0) \geq 0$, this inequality is obvious. Since $-u_t + \Delta u \geq 0$, we then infer that

$$|D^2u| + |u_t| \leq C. \tag{15.24}$$

On the other hand, if $\sup w_1 < \sup w$ and the maximum of w occurs at a point (X_0,ξ) such that $\overline{w}(X_0,\xi) \geq 0$, we have

$$F^{ij}\zeta_i\zeta_j \geq \frac{C(n)k_0\psi_0}{w}|\zeta|^2$$

for any $\zeta \in R^n$. It follows that

$$F^{ij}D_{ik}uD_{jk}u \geq c_6(n,k_0,\psi_0)|D^2u|$$

at X_0 and hence

$$e^{kt}Lw = kF_\tau w + L\overline{w} + L(A|Du|^2) \geq -c_7 - c_8|D^2u| + Ac_6|D^2u|$$

with c_8 independent of A. We now choose A so large that $Ac_6 - c_8 \geq 1$ to infer that $|D^2u(X_0)| \leq c_7$ which again implies (15.23) and hence (15.24).

The global second derivative bound can also be proved under slightly weaker hypotheses. We indicate these results in Exercise 15.1.

In addition, we also have a Hölder estimate for the second derivatives of u, which we state in the following form combined with our global second derivative bounds.

COROLLARY 15.8. *Suppose the hypotheses of Lemma 15.7 are satisfied along with condition* (15.19). *Suppose also that ψ is strictly convex with respect to p,* (15.21) *holds or* (15.22) *holds. If D^2u and u_t are in $C^{2,1}$ and if*

$$F(D^2\varphi, -\varphi_t) = \psi(X,\varphi,D\varphi) \text{ on } C\Omega, \tag{15.25}$$

then there are positive constants β and C_1 determined only by F, the constant C from (15.18), *c_2, c_3 and c_4 (and k_0 if* (15.22) *holds) such that $|u|_{2+\beta} \leq C_1$.*

PROOF. Lemma 15.7 along with the preceding remarks gives a bound for $|D^2u|$ and $|u_t|$ in terms of the indicated quantities. For the Hölder estimate, we use the functions $\bar{f}_\mu$ from the proof of Lemma 15.1. Let Σ denote the image of $\overline{\Omega}$ under the map $(D^2u, -u_t)$. Since Σ is a compact subset of $\mathscr{K}$, there is a compact set S^* such that $f(\lambda)$ agrees with

$$f^*(\lambda) = \inf\{\bar{f}_\mu(\lambda) : \mu \in S^*\}$$

if $\lambda = \lambda(\bar{r})$ for some $\bar{r} \in \Sigma$. Hence μ_j is bounded above and below by positive constants as long as $\mu \in S^*$ and $j \geq 0$. Hence, if we set $F^* = f^* \circ \lambda$, it follows that u is a solution of $F^*(D^2u, -u_t) = \psi(X,u,Du)$ in Ω. Because of our bounds on μ_j, this differential equation can be written in the form from Chapter XIV, and hence a Hölder estimate follows from Theorem 14.18. □

3. Existence of solutions

To see that our estimates imply the existence of solutions for (15.3), we use our previous existence theorem for uniformly parabolic equations.

THEOREM 15.9. *Let $F = f \circ \lambda$, where f is defined on a convex symmetric, proper subcone $\mathscr{K}$ of $\mathbb{R}^{n+1}$ with vertex at the origin with conditions* (15.4a–e), (15.19), *and* (15.25) *satisfied. Suppose that ψ is convex with respect to p, that there is a strict subsolution $\underline{u}$, and that there is a function $\Phi \in H_2(B\Omega)$ such that $\Phi = -\varphi_t$ on $C\Omega$ and $\lambda(D^2\varphi, \Phi) \in \mathscr{K}$. Suppose also that conditions* (15.9) *hold with μ_1 sufficiently small. Suppose finally that either* (15.21) *holds,* (15.22) *holds, or else ψ is strictly convex with respect to p. If $F \in H_2$, $\psi \in H_2$, $\mathscr{P}\Omega \in H_4$,* *and $\varphi \in H_4$, then there is a unique admissible solution u of* (15.3). *Moreover, $u \in H_{3+\alpha}$ for any $\alpha \in (0,1)$.*

PROOF. Let φ be an arbitrary H_4 extension into Ω such that $\varphi_t = -\Phi$ on $B\Omega$. Now choose $\delta_0 > 0$ so small that φ is admissible in $\Omega(\delta_0)$ and then $\varepsilon > 0$ so small that v is admissible on $\Omega(\delta_0)$ if $v \in C^{2,1}$ and $|v - \varphi|_2 < \varepsilon$. Now take $\tilde{F}$ to be uniformly parabolic and $\tilde{\psi}$ to be uniformly bounded so that

$$\tilde{F}(r,\tau) = F(r,\tau), \tilde{\psi}(X,z,p) = \psi(X,z,p)$$

for

$$\left|r - D^2\varphi\right| + |\tau + \varphi_t| + |p - D\varphi| + |z - \varphi| < \varepsilon.$$

We also assume that $\tilde{\psi}_x$, $\tilde{\psi}_z$, and $\tilde{\psi}_p$ are uniformly bounded. From Theorem 14.15, there is a unique solution $\tilde{u} \in H_{1+\theta}$ of

$$\tilde{F}(D^2\tilde{u}, -\tilde{u}_t) = \tilde{\psi}(X,\tilde{u},D\tilde{u}) \text{ in } \Omega(\delta_0), \tilde{u} = \varphi \text{ on } \mathscr{P}\Omega(\delta_0),$$

and Theorem 14.18 implies that $\tilde{u} \in H_{2+\alpha}$ for some positive α.

It follows that there is a positive constant δ such that

$$|\tilde{u} - \varphi|_{2,\Omega(\delta)} < \varepsilon,$$

and hence $\tilde{u}$ solves (15.3) in $\Omega(\delta)$, $\tilde{u} = \varphi$ on $\mathscr{P}\Omega(\delta)$ and $\tilde{u}$ is admissible there. By uniqueness, $\tilde{u}$ is the desired solution in $\Omega(\delta)$. The uniform estimates already proved in Corollary 15.8 and Lemma 14.11 then imply global solvability because at each step, we can take $\Phi = -u_t$. □

4. Properties of symmetric polynomials

In this section, we study the symmetric polynomials

$$S_k(\lambda) = \sum(\prod_{j=1}^{k} \lambda_{i(j)}),$$

where the sum is over all multiindices i with $i(j+1) > i(j)$ and λ is a vector in $\mathbb{R}^N$. We also study the ratios S_k/S_j and other related functions. We start by defining $\sigma_k = S_k/\binom{n}{k}$ if $k > 0$ and $\sigma_0 \equiv 1$. To facilitate matters, we recall the following calculus lemma.

LEMMA 15.10. *If f is defined by*

$$f(x,y) = \sum_{j=0}^{m} c_j x^{m-j} y^j, \tag{15.26}$$

if c_0 and c_m are nonzero, and if all roots x/y of the equation $f(x,y) = 0$ are real, then x/y is real for any roots of $f_x(x,y) = 0$ and $f_y(x,y) = 0$.

PROOF. For each fixed y, f is a polynomial of degree m in x and therefore it has m real roots. By Rolle's theorem, the equation $f_x(x,y) = 0$ has $m-1$ real roots for each fixed y and hence all the roots are real. A similar argument shows that, for each fixed x, $f_y(x,y) = 0$ has $m-1$ real roots. □

From this lemma, we can infer the Newton inequalities.

LEMMA 15.11. *For any real numbers $\lambda_1, \ldots, \lambda_N$ and any $k \geq 2$,*

$$\sigma_k(\lambda)\sigma_{k-2}(\lambda) \leq \sigma_{k-1}(\lambda)^2. \tag{15.27}$$

PROOF. Suppose first that $\sigma_j(\lambda)$ is nonzero for all $j \leq n$. Then we can use the function f defined by

$$f(x,y) = \prod(x + \lambda_i y) = \sum \binom{n}{j} \sigma_j(\lambda) x^{n-j} y^j$$

in Lemma 15.10. If we differentiate this function $k-2$ times with respect to y and $n-k$ times with respect to x, we obtain the function

$$F(x,y) = \sigma_{k-2}(\lambda)x^2 + 2\sigma_{k-1}(\lambda)xy + \sigma_k(\lambda)y^2,$$

and Lemma 15.10 implies (after a simple induction argument) that the roots of $F(x,y) = 0$ are real. It follows that $\sigma_{k-2}\sigma_k \leq \sigma_{k-1}^2$ in this case. For the general case, we note that the set on which $\sigma_j \neq 0$ for $j = 1, \ldots, n$ is dense in $\mathbb{R}^N$. □

These inequalities lead to the Maclaurin inequalities.

LEMMA 15.12. *Fix integers $k > j \geq 1$. If $\sigma_m(\lambda) > 0$ for $m \leq k$, then*

$$\sigma_k(\lambda)^{1/k} \leq \sigma_j(\lambda)^{1/j}. \tag{15.28}$$

PROOF. A simple induction argument shows that we only need to prove (15.28) for $k = j+1$. We prove (15.28) in this case by induction on j. If $j = 1$, then

$$\sigma_2 = \sigma_0\sigma_2 \leq \sigma_1^2$$

by Lemma 15.11, so (15.28) holds. For $j > 1$, we have by the induction hypothesis and Lemma 15.11 that

$$\sigma_{j+1}\sigma_j^{(j-1)/j} \leq \sigma_{j+1}\sigma_{j-1} \leq \sigma_j^2.$$

It follows that

$$\sigma_{j+1} \leq \sigma_j^{2-(j-1)/j} = \sigma_j^{(j+1)/j},$$

and (15.28) is obtained by simple algebra from this inequality. □

Another set of inequalities is known as the Newton-Maclaurin inequalities. We prove only two special cases which will be used below.

LEMMA 15.13. *If $k > j$ are positive integers and if $\sigma_m(\lambda) > 0$ for $m \leq k$, then*

$$\frac{\sigma_k(\lambda)}{\sigma_{k-1}(\lambda)} \leq \frac{\sigma_j(\lambda)}{\sigma_{j-1}(\lambda)}. \tag{15.29}$$

If also $k > j+1$, then

$$\left(\frac{\sigma_{k-1}(\lambda)}{\sigma_j(\lambda)}\right)^{1/(k-j-1)} \geq \left(\frac{\sigma_k(\lambda)}{\sigma_j(\lambda)}\right)^{1/(k-j)}. \tag{15.30}$$

PROOF. Note that it suffices to prove (15.29) when $k-j=1$, and in this case

$$\frac{\sigma_k}{\sigma_{k-1}} = \frac{\sigma_k\sigma_{k-2}}{\sigma_{k-1}\sigma_{k-2}} \leq \frac{\sigma_{k-1}^2}{\sigma_{k-1}\sigma_{k-2}} = \frac{\sigma_j}{\sigma_{j-1}}$$

by Lemma 15.11.

To prove (15.30), we rewrite the inequality in the form

$$\sigma_{k-1}^{k-j} \geq \sigma_k^{k-j-1}\sigma_j.$$

When $k-j=2$, this inequality is just Lemma 15.11. Once we have proved it for $j=k-m$, we set $j=k-(m+1)$ and write

$$\sigma_{k-1}^{k-j} = \sigma_{k-1}^m\sigma_{k-1} \geq \sigma_k^{m-1}\sigma_{j+1}\sigma_{k-1} \geq \sigma_k^{m-1}\sigma_k\sigma_j$$

by the induction hypothesis and (15.29). But this inequality is what we wanted to prove. □

We shall also consider the quantities $S_{j,i}$ defined by $S_{j,i}(\lambda) = S_j(\lambda^{(i)})$, where $\lambda_k^{(i)} = \lambda_k$ if $i \neq j$ and $\lambda_i^{(i)} = 0$. These quantities are connected to the symmetric polynomials by the equation

$$S_j(\lambda) = S_{j,i}(\lambda) + \lambda_i S_{j-1,i}(\lambda) \tag{15.31}$$

for each i. These quantities (with suitable i and j) are positive on $\mathscr{K}_k$.

LEMMA 15.14. *If $\lambda \in \mathscr{K}_k$, then $S_{j,i}(\lambda) > 0$ for $1 \leq j < k$.*

PROOF. We argue by induction on j. First, $S_{1,i}(\lambda) \equiv 1 > 0$, so the lemma is true for $j=1$. Now suppose that $S_{j,i} > 0$ for $j=m-1$ with $m<k$. Then Newton's inequality and (15.31) give

$$S_{m,i}^2 \geq S_{m+1,i}S_{m-1,i} = (S_{m+1} - \lambda_i S_{m,i})S_{m-1,i}.$$

Since $S_{m+1} > 0$ and $S_{m-1,i} > 0$, it follows that

$$S_{m,i}^2 > -\lambda_i S_{m,i}S_{m-1,i},$$

and hence

$$S_{m,i}S_m = S_{m,i}(S_{m,i} + \lambda_i S_{m-1,i}) > 0.$$

The desired result follows from this inequality because $S_m > 0$. □

Our next step is an algebraic inequality, which is equivalent to the concavity of the function $f(\lambda) = (\prod_{i=1}^m \lambda_i)^{1/m}$.

LEMMA 15.15. *Let $f_1, \ldots, f_m$ be superadditive functions on some convex set Σ and suppose that $f_i > 0$ on Σ. Then $(\prod f_i)^{1/m}$ is superadditive.*

PROOF. First we define P by $P(\lambda) = \prod_{j=1}^m \lambda_j$. Then if a and b are m-tuples with positive components, we have

$$\sum_{j=1}^m a_j D_j P(b) \le m \left(\prod_{j=1}^m a_j D_j P(b) \right)^{1/m} = mP(a)^{1/m} P(b)^{(m-1)/m}$$

by the arithmetic-geometric mean inequality. It follows that

$$\prod_{j=1}^m a_j D_j P^{1/m}(b) \le P^{1/m}(a),$$

so $P^{1/m}$ is concave. By homogeneity, we also have that $P^{1/m}$ is superadditive, that is

$$P^{1/m}(a+b) \ge P^{1/m}(a) + P^{1/m}(b).$$

Now let λ and μ be elements of Σ and write f for $(f_1, \cdots, f_m)$. The superadditivity of $(\prod f_i)^{1/m} = P^{1/m} \circ f$ now follows by taking $a = f(\lambda)$ and $b = f(\mu)$ in this inequality. $\square$

We are now ready to prove our concavity result for $f = (S_k/S_j)^{1/(k-j)}$. Since f is homogeneous of degree one, it suffices to show that f is superadditive.

THEOREM 15.16. *Let m and k be positive integers with $m \le k \le N$. If λ and μ are in $\mathscr{K}_k$, then*

$$\left(\frac{S_k(\lambda+\mu)}{S_{k-m}(\lambda+\mu)} \right)^{1/m} \ge \left(\frac{S_k(\lambda)}{S_{k-m}(\lambda)} \right)^{1/m} + \left(\frac{S_k(\mu)}{S_{k-m}(\mu)} \right)^{1/m}. \tag{15.32}$$

PROOF. First, we prove (15.32) when $m = 1$ by induction on k. If $k = 1$, then inequality (15.32) is an obvious equality. For $k > 1$, we use the homogeneity of S_k to infer that

$$kS_k(\lambda) = \sum_{i=1}^N \lambda_i S_{k-1,i}(\lambda),$$

so (15.31) implies that

$$(n-k)S_k(\lambda) = \sum_{i=1}^N S_{j,i}(\lambda). \tag{15.33}$$

Now we use (15.31) twice to write

$$S_k(\lambda) = S_{k,i}(\lambda) + \lambda_i S_{k-1}(\lambda) - \lambda_i^2 S_{k-2,i}(\lambda).$$

Summing on i and using (15.33) now gives

$$kS_k(\lambda) = S_1(\lambda)S_{k-1}(\lambda) - \sum_{i=1}^{N} \lambda_i^2 S_{k-2,i}(\lambda). \tag{15.34}$$

Now we define $g_k = S_k/S_{k-1}$ and $g_{k,i} = S_{k,i}/S_{k-1,i}$ to see from (15.31) and (15.34) that

$$kg_k(\lambda) = S_1(\lambda) - \sum_{i=1}^{N} \frac{\lambda_i^2}{\lambda_i + g_{k-1,i}(\lambda)}.$$

Next we define

$$\varphi_k(\lambda,\mu) = g_k(\lambda+\mu) - g_k(\lambda) - g_k(\mu),$$

and infer that

$$\begin{aligned}\varphi_k(\lambda,\mu)&\\ &= \frac{1}{k}\sum_{i=1}^{N}\left(\frac{\lambda_i^2}{\lambda_i + g_{k-1,i}(\lambda)} + \frac{\mu_i^2}{\mu_i + g_{k-1,i}(\mu)} - \frac{(\lambda_i+\mu_i)^2}{\lambda_i+\mu_i+g_{k-1,i}(\lambda+\mu)}\right).\end{aligned}$$

By the induction hypothesis and Lemma 15.14, we have that $g_{k-1,i}$ is superadditive, and therefore for each i (abbreviating $g_{k-1,i}$ to g)

$$\begin{aligned}\frac{\lambda_i^2}{\lambda_i+g(\lambda)} &+ \frac{\mu_i^2}{\mu_i+g(\mu)} - \frac{(\lambda_i+\mu_i)^2}{\lambda_i+\mu_i+g(\lambda+\mu)}\\ &\geq \frac{\lambda_i^2}{\lambda_i+g(\lambda)} + \frac{\mu_i^2}{\mu_i+g(\mu)} - \frac{(\lambda_i+\mu_i)^2}{\lambda_i+\mu_i+g(\lambda)+g(\mu)}\\ &= \frac{(\lambda_i g(\mu) - \mu_i g(\lambda))^2}{(\lambda_i+g(\lambda))(\mu_i+g(\mu))(\lambda_i+g(\lambda_i)+\mu_i+g(\mu_i))} \geq 0.\end{aligned}$$

It follows that $\varphi_k(\lambda,\mu) \geq 0$ and hence (15.32) is proved for $m = 1$ and any $k \geq 1$.

If $m > 1$, we write

$$\left(\frac{S_k(\lambda)}{S_{k-m}(\lambda)}\right)^{1/m} = \left(\prod_{j=1}^{m} g_{k-j+1}(\lambda)\right)^{1/m}$$

By the case $m = 1$, each g_{k-j+1} is superadditive, and Lemma 15.15 completes the proof of (15.32). □

We are now ready to show that our structure conditions are satisfied if $f = S_k^{1/k}$.

THEOREM 15.17. *Conditions* (15.4), (15.9c,d), *and* (15.22) *hold for* $f = S_k^{1/k}$ *and* $\mathscr{K} = \mathscr{K}_k$. *Moreover,* $\mathscr{K}_k$ *is a convex, proper subcone of* $\mathbb{R}^{n+1}$.

PROOF. It's easy to check (15.4a,d) hold while (15.4b) follows from Lemma 15.14. Lemma 15.15 and the homogeneity of f give (15.4c).

Now suppose that $\lambda_i < 0$. Then $f_i = \frac{1}{k} S_k^{(1-k)/k} S_{k-1,i}$, and

$$S_{k-1,i} = S_{k-1} - S_{k-2,i}\lambda_i \geq S_{k-1} \tag{15.35}$$

by (15.31). Therefore

$$f_i \geq \frac{1}{k} S_k^{(1-k)/k} S_{k-1} \geq \frac{1}{k},$$

by Maclaurin's inequality and (15.9c) is proved. The homogeneity of S_k implies (15.9d), and $\mathscr{K}_k$ is a proper subset of $\mathbb{R}^{n+1}$ because the Maclaurin inequalities imply that $\sum \lambda_i > 0$ on $\mathscr{K}_k$. It is a cone because f is homogeneous, and it is convex because it is a component of the set on which f, a concave function, is positive.

Finally, we write

$$S_k = S_{k,i} + \lambda_i S_{k-1,i} \leq S_{k-1,i}^{(k-1)/k} + \lambda_i S_{k-1,i} \leq C(n,k)\,|\lambda|\,S_{k-1,i}$$

to infer (15.22). □

For the Hessian quotient problems, we need to work only a little harder.

THEOREM 15.18. *Suppose* $f = (S_k/S_j)^{1/(k-j)}$ *with* $k > j$. *Then conditions* (15.4) *and* (15.9c,d) *are satisfied for* $\mathscr{K} = \mathscr{K}_k$.

PROOF. Conditions (15.4a,c,d) are proved just as in Theorem 15.17.

To prove (15.4b), we have

$$f_i = \frac{1}{k-j}\left(\frac{S_k}{S_j}\right)^{(1-k+j)/(k-j)} \frac{A}{S_j^2},$$

where

$$\begin{aligned} A &= S_{k-1,i}S_j - S_k S_{j-1,i} \\ &= S_{k-1,i}S_{j,i} + \lambda_i S_{k-1,i}S_{j-1,i} - S_{k,i}S_{j-1,i} - \lambda_i S_{k-1,i}S_{j-1,i} \\ &= S_{k-1,i}S_{j,i} - S_{k,i}S_{j-1,i} \end{aligned}$$

by (15.31). From the Newton-Maclaurin inequality (15.29)inequalities, we infer that

$$\sigma_{k,i}\sigma_{j-1,i} \leq \sigma_{k-1,i}\sigma_{j,1},$$

so that

$$\begin{aligned} A &= \binom{n}{k-1}\binom{n}{j}[\sigma_{k-1,i}\sigma_{j,i} - \frac{n-k+1}{k}\frac{j}{n-j+1}\sigma_{k,i}\sigma_{j-1,i}] \\ &\geq \binom{n}{k-1}\binom{n}{j}\sigma_{k-1,i}\sigma_{j,i}(1 - \frac{n-k+1}{k}\frac{j}{n-j+1}). \end{aligned}$$

It follows that $f_i \geq c(n,j,k) f^{1-k+j} S_{k-1,i}S_{j,i}/S_j^2 > 0$ in $\mathscr{K}_k$.

If $\lambda_i < 0$, we use (15.35) to infer that

$$f_i \geq c(n,j,k)\left(\frac{S_k}{S_j}\right)^{(1-k+j)/(k-j)}\frac{S_{k-1}}{S_j}.$$

It then follows from (15.30) that $f_i \geq C(n,j,k)$, so (15.9c) is proved. Again (15.9d) is an immediate consequence of the homogeneity of f. □

It is a simple matter to write the corresponding existence theorems for these operators.

COROLLARY 15.19. *Let* $f = S_k^{1/k}$ *with* $1 < k \leq n+1$ *and let* $\mathscr{P}\Omega \in H_4$ *and* $\varphi \in H_4$. *Suppose* $\psi \in H_{2,loc}$ *is convex with respect to* p *with* $\psi_z \geq 0$. *Suppose* φ *satisfies the compatibility condition* (15.25) *and* $S_m(\lambda(D^2\varphi)) > 0$ *on* $B\Omega$ *for* $m < k$. *If there is a strict subsolution of* (15.3) *and if conditions* (15.9a,b) *are satisfied with* μ_1 *sufficiently small, then there is a unique admissible solution of* (15.3).

COROLLARY 15.20. *Let* $f = (S_k/S_j)^{1/(k-j)}$ *with* $1 \leq j < k \leq n+1$ *and let* $\mathscr{P}\Omega \in H_4$ *and* $\varphi \in H_4$. *Suppose* $\psi \in H_{2,loc}$ *is convex with respect to* p *with* $\psi_z \geq 0$. *Suppose* φ *satisfies the compatibility condition* (15.25) *and* $S_m(\lambda(D^2\varphi)) > 0$ *on* $B\Omega$ *for* $m < k$. *Suppose also that* (15.9a,b) *hold with* μ_1 *sufficiently small and that either* (15.21) *holds or else* ψ *is strictly convex with respect to* p. *If there is a strict subsolution of* (15.3), *and if there is a function* $\underline{v}$ *satisfying* (15.5)$'$, *then there is a unique admissible solution of* (15.3).

In general, it is not known if these results are true when the hypotheses, especially the convexity of ψ, are relaxed. (Although if the smallest eigenvalue of the matrix ψ_{pp} is not too negative, some existence theorems are known; see Exercise 15.1 and the Notes for details. In addition, condition (15.9b) is actually superfluous for these operators; see Exercise 15.2.) On the other hand, for $k = n+1$ in Corollary 15.19, we show in the next section that our hypotheses can be weakened considerably.

5. The parabolic analog of the Monge-Ampère equation

In this section, we look at a special fully nonlinear parabolic equation of the type considered in Sections XV.2 and XV.3 for which the estimates hold under weaker hypotheses on the inhomogeneous term ψ. Specifically, we consider the boundary value problem

$$-u_t \det D^2u = \psi(X,u,Du)^{n+1} \text{ in } \Omega, \tag{15.36a}$$

$$u = \varphi \text{ on } \mathscr{P}\Omega. \tag{15.36b}$$

This problem is the special case of (15.3) in which $f(\lambda) = (\prod \lambda_i)^{1/(n+1)}$ and $\mathscr{K}$ is just the positive cone $\{\lambda_i > 0, i = 1,\ldots,n+1\}$. As in Section XV.2, we can estimate $|u|_0$ and $|Du|_{0,\mathscr{P}\Omega}$ if a strict subsolution exists. In addition, since $u(\cdot,t)$ is convex for any $t \in I(\Omega)$, it follows that $|Du(\cdot,t)|$ attains its maximum value on $\partial\Omega(t)$ and hence $|Du|_0$ is estimated in terms of a strict subsolution without imposing any additional structure conditions on ψ.

A key step in our second derivative estimates is the following modification of the barrier construction in Lemma 15.5.

LEMMA 15.21. *Suppose that $\psi \in C^1$ and $|Du| \le K$. Then there are positive constants M, δ_0, δ_1, and θ such that*

$$Lv \ge \delta_1(1 + \sum_i F^{ii} + F_\tau) \tag{15.37}$$

in $\Omega[\delta_0]$, where $v = \underline{u} - u + M|x^n|^2 - \theta[|x'|^2 - 2t]$. If also $0 \in \mathscr{P}\Omega$ and if there are nonnegative constants c_1 and c_2 such that

$$x^n \ge -c_1 \text{ in } \Omega[\delta_0],\ x^n \le c_2|X'|^2 \text{ on } \mathscr{P}\Omega[\delta_0], \tag{15.38}$$

then there are positive constants $\delta < \delta_0$ and δ_2 such that

$$v \le -\delta_2|X'|^2 \text{ on } \mathscr{P}\Omega[\delta], \tag{15.39a}$$

$$v \le -\delta_2 \text{ on } \Omega \cap \mathscr{P}Q(\delta). \tag{15.39b}$$

PROOF. We compute

$$\begin{aligned} Lv &= L(\underline{u} - u) + MF^{nn} - 2\theta(F_\tau + \sum_{i<n} F^{ii}) + 2M\psi_p \cdot x - \theta\psi_p \cdot x' \\ &\ge (\varepsilon - 2\theta)(F_\tau + \sum_i F^{ii}) + MF^{nn} + \varepsilon \\ &\quad - \psi_1\delta_0[\theta + M] - \psi_1 \sup|Du - D\underline{u}| + \inf[\psi(X,u,Du) - \psi(X,\underline{u},D\underline{u})] \end{aligned}$$

(as in Lemma 15.6), where $\psi_1 = \sup|\psi_p|$. The choice $\theta = \varepsilon/4$ yields

$$Lv \ge MF^{nn} + \frac{1}{2}\varepsilon(F_\tau + \sum_i F^{ii}) - c_3,$$

where c_3 depends only on ψ, c_1, and $\sup|D(u - \underline{u})|$, provided $M\delta_0 \le 1$.

Next, we observe that the minimum eigenvalue of the matrix

$$\begin{pmatrix} F_\tau & 0 \\ 0 & F_r \end{pmatrix}$$

is $f(\lambda)/((n+1)\lambda_{n+1})$, where λ_{n+1} is the largest λ_i. It follows that

$$Lv \ge \psi_0\left[\frac{M}{(n+1)\lambda_{n+1}} + \frac{\varepsilon}{4}\sum_{i\le n}\frac{1}{\lambda_i}\right] + \frac{\varepsilon}{4}(F_\tau + \sum_i F^{ii}) - c_3.$$

Now we use the arithmetic-geometric mean inequality to see that

$$\frac{M}{(n+1)\lambda_{n+1}} + \frac{\varepsilon}{4}\sum_{i\le n}\frac{1}{\lambda_i} \ge c_4\frac{M^{1/(n+1)}}{\Psi}$$

for $c_4 = ((n+1)\varepsilon/4)^{n/(n+1)}$ and $\Psi = \sup \psi(X,u,Du)$. It follows that

$$Lv \geq c_4 \frac{\psi_0}{\Psi} M^{1/(n+1)} - c_3 + \frac{\varepsilon}{4}(F_\tau + \sum_i F^{ii}).$$

Now we take $M \geq 1$ so large that $c_4 \psi_0 M^{1/(n+1)}/\Psi - c_3 \geq \varepsilon/4$ and then $\delta_0 \leq 1/M$ to infer (15.37) with $\delta_1 = \varepsilon/4$.

If (15.38) holds, then we have

$$v = M|x^n|^2 - \theta[|x'|^2 - 2t] \leq [M\delta(c_1+c_2) - \theta]\,|X'|^2$$

on $P_1 = \mathscr{P}\Omega[\delta]$. Hence (15.39a) holds with $\delta_2 = -\theta/2$ if $\delta \leq \delta_0$ and $M\delta(c_1 + c_2) \leq \theta/2$. Moreover,

$$v \leq [2M\delta c_2 - \theta]\,|X'|^2 \leq -\frac{\theta}{2}|X'|^2 \leq -\frac{\theta}{2}\delta^2$$

on

$$P_2 = \Omega \cap \mathscr{P}Q(\delta) \cap \{x^n < 2c_2|X'|^2,\ t = -\delta^2\}$$

provided $4c_2 M\delta \leq \theta$. Next, on

$$P_3 = \Omega \cap \mathscr{P}Q(\delta) \cap \{x^n < 2c_2|X'|^2,\ t > -\delta^2\}$$

we have $|x'|^2 + (x^n)^2 = \delta^2$, and therefore

$$\delta^2 > |x'|^2(1 - 2c_2\delta^2) \geq \frac{1}{2}|x'|^2 \geq |X'|^2/4$$

provided $4c_2\delta^2 \leq 1/2$. It follows that $v \leq -\frac{\theta}{2}|X'|^2 \leq -\frac{\theta}{8}\delta^2$ on P_3.

Now note that $\underline{u} + c_5 d$ is a subsolution for c_5 a sufficiently small positive constant, so $D_\gamma(u - \underline{u}) \geq c_5$. Hence there is a positive constant c_6 such that $\underline{u} - u \leq -c_5\delta^2$ on

$$P_4 = \Omega \cap \mathscr{P}Q(\delta) \cap \{x^n \geq 2c_2|X'|^2\}.$$

It follows that

$$v \leq -c_6\delta^2 + M|x^n|^2 \leq -\frac{c_6}{2}\delta^2$$

on P_4 if $M\delta \leq c_6/2$. Combining all these possibilities gives (15.39b) because $\Omega \cap \mathscr{P}Q(\delta) = P_1 \cup P_2 \cup P_3 \cup P_4$. □

Since the differential equation was not used in bounding the double tangential derivatives, we obtain the estimate $|u_{ij}| \leq C$ on $\mathscr{P}\Omega$ for $i,j < n$ just as before. For the mixed derivatives, we infer (15.14) by using $-Av$ in place of $A[\frac{1}{2}|x|^2 - t] + B[u - \underline{u}]$.

The double normal derivative is estimated by modifying the proof of (15.18) appropriately. First, for σ_0 a positive constant at our disposal, we define R_0 by

$$R_0 = \inf\{\sigma > \sigma_0 : f(\lambda', \sigma) > \psi_0, (\lambda', \sigma) \in \mathscr{K} \text{ for all } X \in S\Omega\}.$$

With this choice for R_0, we define f_∞ and G as before and set

$$g_1 = f_\infty(\lambda(D^2u, -u_t)).$$

If X_0 is now the point in $S\Omega$ at which g_1 attains its minimum, we have that

$$0 \leq [G_0 + H(X, X_0)][D_\gamma u(X) - D_\gamma u(X_0)] + g_0 \cdot (x - x_0) + k_1 |X - X_0|^2$$

with G_0, H, g_0 and k_1 as before. If

$$G(D^2 u, -u_t) \leq \psi_0 + \delta_0/2 \tag{15.17$'$}$$

and σ_0 is sufficiently large, then

$$G(\nabla^2 \underline{u}, -\underline{u}_t) - G(\nabla^2 u, -u_t) \geq \psi(X_0, \underline{u}, D\underline{u}) - \psi_0 + \delta_0/3 \geq \psi_0/3$$

and hence $G_0 \geq \delta_0/(2D_\gamma v(X_0))$. It follows that

$$D_\gamma u(X) - D_\gamma u(X_0) \leq \frac{g_0}{G_0} \cdot (x - x_0) + k_3 |X - X_0|^2$$

on $\mathscr{P}\Omega[R]$, and hence that $|D_{nn} u(X_0)| \leq C$. Whether or not (15.17)$'$ holds, by choosing σ_0 sufficiently large, we conclude that there is a positive constant δ_1 such that

$$f_\infty(X_0) \geq \psi_0 + \delta_1$$

and also that $R_0 = \sigma_0$. Since $f_\infty(\lambda(r,\tau)) = [R_0 \prod \lambda_i'(r,\tau)]^{1/(n+1)}$, we infer a positive lower bound for $\prod \lambda_i(D^2 u, -u_t)$ and hence an upper bound for $D_{nn} u$. Therefore, we obtain a boundary estimate on the second derivatives of u assuming only that $\mathscr{P}\Omega \in H_4$, $\varphi \in H_4$ and that there is a strict subsolution.

For a global second derivative estimate, we first define $\Phi(\bar{r}) = \sum \ln(\lambda_i(\bar{r}))$ and $\Psi = (n+1) \ln \psi$. Thus, we can write the differential equation as

$$\Phi(D^2 u, -u_t) = \Psi(X, u, Du), \tag{15.40}$$

and by direct calculation, we have

$$\Phi^{ij} = u^{ij}, \Phi^{ij,km} = -u^{ik} u^{jm}, \Phi^{ij}_\tau = 0, \Phi_{\tau\tau} = -\tau^{-2},$$

where u^{ij} denote the entries of the inverse matrix of r. Differentiating (15.40) once with respect to t shows that $v = -u_t$ solves the equation

$$-\frac{v_t}{v} + u^{ij} D_{ij} v - \Psi^i D_i v - \Psi_z v + \Psi_t = 0.$$

If (15.19) holds, it follows that

$$-v_t + v u^{ij} D_{ij} v - v \Psi^i D_i v - \mu_4 v \leq 0$$

wherever $v \leq 0$ and then Corollary 2.5 gives a lower bound for v:

$$v \geq \exp(\mu_4 T) \inf_{\mathscr{P}\Omega} v.$$

Now we let $\xi \in \mathbb{R}^n$ and $h \in C^2(\mathbb{R}^n)$, so that $w = \exp(h(Du))D_{\xi\xi}u$ satisfies

$$\frac{-w_t}{w} = h^k D_{tk}u + \frac{D_{t\xi\xi}u}{D_{\xi\xi}u},$$

$$\frac{D_i w}{w} = h^k D_{ik}u + \frac{D_{i\xi\xi}u}{D_{\xi\xi}u},$$

$$\frac{D_{ij}w}{w} = \frac{D_i w D_j w}{w^2} + h^{km} D_{ik}uD_{jm}u + h^k D_{ijk}u + \frac{D_{ij\xi\xi}u}{D_{\xi\xi}u} - \frac{D_{i\xi\xi}uD_{j\xi\xi}u}{(D_{\xi\xi}u)^2}.$$

In addition, differentiating the equation (15.40) twice in the direction ξ gives

$$-\Phi_\tau D_{t\xi\xi}u + \Phi^{ij}D_{ij\xi\xi}u = \Phi^{ik}\Phi^{jm}D_{ij\xi}uD_{km\xi}u + \tau^{-2}D_{\xi t}u^2 + D_{\xi\xi}\Psi,$$

and hence, after defining the operator L_0 by

$$L_0 v = -\Phi_\tau v_t + \Phi^{ij}D_{ij}v,$$

we infer that

$$\exp(-h)L_0 w \geq D_{\xi\xi}u\left[h^{km}\Phi^{ij}D_{ik}uD_{jm}u + h^k L_0(D_k u)\right] + \Phi^{ik}\Phi^{jm}D_{ij\xi}uD_{km\xi}u - \frac{1}{D_{\xi\xi}u}\Phi^{ij}D_{i\xi\xi}uD_{j\xi\xi}u + D_{\xi\xi}\Psi. \tag{15.41}$$

If we choose $h(p) = A|p|^2/2$ with A a positive constant to be further specified, we find that

$$h^{km}\Phi^{ij}D_{ik}uD_{jm}u = A\Delta u,$$

and the differential equation gives us $L_0(D_k u) = D_k\Psi$.

Next, we suppose that the constant μ_2 has been chosen so that

$$|D\Psi| + |D^2\Psi| \leq \mu_2 \tag{15.42}$$

for $|z| + |p| \leq |u|_0 + |Du|_0$, where here the symbol D represents differentiation with respect to (x,z,p). It follows that

$$D_{\xi\xi}uh^k L_0(D_k u) + D_{\xi\xi}\Psi \geq \frac{D_{\xi\xi}u}{w}\Psi^i D_i w - C\left[1 + |D^2u|^2 + A(1 + |D^2u|)\right]$$

for some constant C determined only by n, μ_2, and $|Du|_0$.

The remaining third derivative terms in (15.41) are handled by considering w as a function of (X,ξ) on $\Omega \times \{|\xi| = 1\}$. If (Y,ξ) is a point at which w attains its maximum on this set, we see that $D_{\xi\xi}u(Y)$ will be the maximum eigenvalue of

$D^2u(Y)$, and hence $1/D_{\xi\xi}u$ is the minimum eigenvalue of Φ_r. Since Φ_r is positive definite, it follows that

$$\frac{1}{D_{\xi\xi}u}\Phi^{ij}D_{i\xi\xi}uD_{j\xi\xi}u \leq \Phi^{ik}\Phi^{jm}D_{ij\xi}uD_{km\xi}u$$

at (Y,ξ). We also have $Dw(Y)=0$ and $L_0w(Y)\leq 0$, and hence

$$0 \geq A\Delta uD_{\xi\xi}u - C(1+|D^2u|^2+A(1+|D^2u|))$$

at Y. This inequality, with A chosen sufficiently large, gives the desired estimate on $|D^2u|$. We therefore obtain the following global bounds for solutions of the parabolic Monge-Ampère equation.

THEOREM 15.22. *Let $u \in C^{2,1}(\overline{\Omega})$ be a convex solution of* (15.36) *in Ω. If $u=\varphi$ on $\mathscr{P}\Omega$ with $\mathscr{P}\Omega \in H_4$ and $\varphi \in H_4$, if there is a strict subsolution $\underline{u}$ of* (15.36), *and if ψ is a positive, C^2 function satisfying* (15.19) *and* (15.42), *then*

$$|D^2u| \leq C(\mu_2,\mu_4,|\underline{u}|_2,\Omega,|\varphi|_4). \tag{15.43}$$

This estimate immediately leads to the following existence theorem.

THEOREM 15.23. *Let $\mathscr{P}\Omega \in H_4$ and $\varphi \in H_4$. Suppose that ψ is a positive function defined on $\overline{\Omega}\times\mathbb{R}\times\mathbb{R}^n$ with $\psi_z \leq 0$, and suppose that $\psi \in H_2(K)$ for any compact subset K of $\overline{\Omega}\times\mathbb{R}\times\mathbb{R}^n$. If the compatibility condition*

$$-\varphi_t \det D^2\varphi = \psi(X,\varphi,D\varphi)^{n+1} \tag{15.25$'$}$$

is satisfied, if φ is convex on $B\Omega$, and if there is a strict subsolution of (15.36), *then* (15.36) *has a unique solution.*

In fact, the existence of a strict subsolution can be relaxed considerably if we strengthen our geometric assumptions on Ω. According to Lemma 15.2, we have a strict subsolution if Ω is convex in x and strictly increasing in t provided ψ grows linearly with sufficiently slow linear growth in z and p. In other words, $\psi \leq C_1 + C_2(|z|+|p|)$ for some positive constant C_1 and a sufficiently small positive constant C_2. By modifying the argument in Lemma 15.2 (see [33]), we can relax this condition so that C_2 is an arbitrary constant, but more is true.

THEOREM 15.24. *Suppose that there are positive constants M_0 and C_2 and an increasing function g such that*

$$(-u_t)\det(D^2u) \leq g(|Du|) + C_2|u| \tag{15.44}$$

wherever $X \in \Omega$, $u \leq -M_0$ and $-u_t > 0$ and $D^2u > 0$. Then

$$\sup_{\Omega}(-u) \leq C(C_2, g(1), \sup_{\mathscr{P}\Omega} u^- + M_0, \Omega). \tag{15.45}$$

PROOF. Assume without loss of generality that $t > 0$ in Ω. Then set $R = \operatorname{diam}\Omega$ and $c_0 = \sup_{\mathscr{P}\Omega} u^- + M_0$, and define

$$v = -Ke^{Kt} + \frac{|x|^2}{2R} - c_0,$$

with K a positive constant to be determined. Since $|Dv| \le 1$, it follows that

$$-v_t \det(D^2 v) = K^2 e^{Kt} R^{-n} > g(1) + 2C_2 K e^{Kt} \ge g(|Dv|) + C_2 |v|.$$

provided $K = 1 + c_0 + 2(C_2 + g(1))R^n$. Theorem 14.3 implies that $v < u$ in Ω and the explicit form of c_0 gives (15.45). □

Since $-u_t + \Delta u \ge 0$, it follows that $u \le \sup \varphi$, and hence an explicit subsolution is not needed for an L^∞ bound on u.

To obtain a boundary gradient estimate, we need only construct a lower barrier. For this construction, we assume the existence of constants $\beta \ge 0$ and $a_0 > 0$ and an increasing function μ such that

$$0 \le \psi(X, z, p)^{n+1} \le \mu(|z|) d(X)^\beta |p|^{n+2+\beta} \tag{15.46}$$

for $d(X) \le a_0$, $z \in \mathbb{R}$, and $|p| \ge \mu(|z|)$. We also suppose that for any $X_0 \in S\Omega$, there are constants k and R and a point $Y = (y, t_0)$ such that $|x_0 - y| = R$ and $|x - y|^2 + k(s - t) < R^2$ for any $X \in \Omega$ with $t \le t_0$. (In other words, Ω is strictly convex-increasing.) We then extend φ to Ω so that $D^2\varphi \ge 0$ and $-\varphi_t \ge 0$ in Ω, which is possible by Exercise 15.3, and set $\rho(X) = R^2 - |x - y|^2 - k(s - t)$ and $w = \varphi - f \circ \rho$ for f a function to be determined. We suppose first that $f' > 0$ and $f'' < 0$ and note that

$$(-w_t) \det D^2 w \ge (-f_t) \det D^2 f = -2^n k [f'' |x - y|^2 + f'](f')^n,$$

and

$$\psi(X, u, Dw)^{n+1} \le 2^{n+2+\beta} \mu(M) d^\beta (f')^{n+2+\beta},$$

where $M = \sup |u|$. If we take $f(\rho) = \frac{1}{\nu} \ln(1 + A\rho)$ with A and ν positive constants, it follows that $f' > 0$ and $f'' < 0$. In addition, $\rho f' \le 1$ if $\nu \ge 1$. Let us choose a so that $f(a) = M$ and suppose that A and ν have been chosen so that $a \le \min\{a_0 R, R^2/4\}$ and $A \ge \mu(M)\nu[1 + Aa]$. (Note that $1 + Aa = \exp(\nu M)$ and $d \le \rho/R$.) Then

$$(-w_t) \det D^2 w \ge kW[AR^2/4 - \exp(\nu M)],$$

$$\psi(X, u, Dw)^{n+1} \le W[2^{2+\beta} A \mu(M) R^{n+2}/\nu],$$

where $W = (f')^n 2^n A/(\nu(1 + A\rho)^2)$. If we now take $\nu = 1 + 2^{3+\beta}\mu(M)R^n/k$ and then A sufficiently large, we infer that

$$(-w_t) \det D^2 w \ge \psi(X, u, Dw)^{n+1},$$

so the maximum principle implies that $w \le u$ in a neighborhood of X_0. As before, this inequality gives a lower bound on $D_\gamma u(X_0)$ and the fact that u is subharmonic gives an upper bound. Therefore $|Du|$ is bounded on $S\Omega$ and hence in Ω because u is convex with respect to x.

We obtain our second derivative bounds in this case by taking $\underline{u}$ to satisfy

$$(-\underline{u}_t)\det D^2\underline{u} \geq \psi(X,u,Du)+1.$$

The construction of such a function follows from the proof of Lemma 15.2 since ψ is bounded. Combining these results, we infer the following estimates and existence results.

THEOREM 15.25. *Suppose there are positive constants ψ_0 and d_0 and an increasing function μ such that*

$$\psi_0 \leq \psi(X,z,p) \leq \mu(|z|)\,|p|^{(n+2)/(n+1)} \tag{15.47}$$

for $d < d_0$. Suppose also that conditions (15.19) *and* (15.42) *hold with c_2 and μ_2 increasing functions of $|z|+|p|$. If $\mathscr{P}\Omega \in H_4$ and $\varphi \in H_4$ and if Ω is strictly convex-increasing, then there is a constant C determined only by Ω, μ, ψ_0, β, c_2, $|\varphi|_4$, and such that $|D^2u| \leq C$ in Ω for any admissible solution of* (15.36).

THEOREM 15.26. *Let Ω, ψ, and φ satisfy the hypotheses of Theorem 15.25. Suppose also that $\psi \in H_2(K)$ for any compact subset K of $\overline{\Omega}\times\mathbb{R}\times\mathbb{R}^n$. If the compatibility condition* (15.25)′ *is satisfied and if φ is convex on $B\Omega$, then* (15.36) *has a unique admissible solution.*

If the domain Ω is a cylinder of the form $\omega\times(0,T)$, and if ω is a strictly convex domain in $\mathbb{R}^n$ with sufficiently smooth boundary, then Theorems 15.25 and 15.26 are true provided φ_t is bounded from above by a negative constant and (15.47) is replaced by the stronger condition

$$\psi_0 \leq \psi(X,z,p) \leq \mu(|z|)\,|p|\,. \tag{15.47$'$}$$

The proof of this fact is left to the reader as Exercise 15.4.

Notes

The approach to parabolic Hessian equations in this chapter is a combination of ideas from several sources, most of which study elliptic Hessian equations. Until the 1970's the only elliptic Hessian equation to receive much attention was the elliptic Monge-Ampère equation in two space dimensions. In the late 1970's and early 1980's a number of authors considered the Monge-Ampère equation in higher dimensions, and shortly thereafter more general Hessian equations were examined. We refer the interested reader to the Notes in Chapter 17 of [101] (and to [107]) for a more detailed history of the elliptic Monge-Ampère equation.

Our structure conditions (15.4a–e) come from [34] (see, in particular, conditions (2), (3), (5), and (6) of that work) as do Lemmata 15.1 (Section 3 of [34]) and 15.2 (Section 2 of [34]). The maximum bound and boundary gradient estimate in Lemma 15.3 are immediate; Lijuan Wang [336] and Guan [106] first used the existence of a strict subsolution to infer these estimates for solutions of elliptic equations. The proof of Lemma 15.4 is a simple modification of the corresponding elliptic results in [336, Section 3.1] while Lemma 15.5 is the parabolic analog of the corresponding result in [106, Lemma 2.1]. Our proof of double tangential and mixed tangential-normal second derivative estimates at the boundary is, like the same results in [281], based on the elliptic results in [33,34]. The original proof of the double normal derivative estimate (see [33, 34, 123]) used a barrier argument to obtain a careful upper bound on the inner normal derivative and then the equation provides the estimate on the double normal derivative. A key assumption in these works is that

$$\lim_{R\to\infty} f(\lambda_1,\ldots,\lambda_n,R) = \infty. \tag{15.48}$$

Our proof of double normal derivative estimates at the boundary is based on the results of Trudinger [313, Sections 2 and 3], who showed that such estimates are possible in the elliptic case without assuming (15.48). Our proof of global second derivative estimates in the first case (in terms of $\underline{v}$) is based on the first derivative estimates in [106, Section 2]; the observation on strict convexity seems to be new. Global bounds when (15.22) is valid follow [126, Lemma 2.3], in which the convexity of ψ with respect to p is relaxed to the condition

$$\psi^{ij}\zeta_i\zeta_j \geq -\mu_5\psi|\zeta|^2 \tag{15.49}$$

for a suitably small, positive constant μ_5; see Exercise 15.1. In addition, Caffarelli, Nirenberg, and Spruck prove a global $C^{2,\alpha}$ estimate by first using the one-sided estimates (of Calabi and Nirenberg) on some of the third derivatives of u at the boundary to infer a two-sided boundary logarithmic modulus of continuity on all of the second derivatives via a barrier argument due to Caffarelli [33, Lemma 5.1]. There does not seem to be a simple parabolic analog of this result. Caffarelli, Nirenberg, and Spruck also introduced a division of elliptic Hessian equations into two types: type 1 equations, for which the positive λ_i axes are in $\partial\mathscr{K}$, and type 2 equations, for which the positive λ_i axes are in $\mathscr{K}$. The refined analysis in this chapter is essential for type 1 equations, but, for type 2 equations, the curvature condition of Lemma 15.2 is always satisfied, and estimates can be obtained more easily. Note that the Hessian and Hessian quotient equations studied in this chapter are all of type 1.

Theorem 15.9 is a variation on an old idea, which does not seem to appear in recent literature. The point of using the local existence theory is to avoid compatibility conditions as much as possible. In [329] Wang studies the parabolic Monge-Ampère equation in cylindrical domains but with a fourth order compatibility condition (to insure that the two different ways of computing u_{tt} on $C\Omega$ give the same answer). This compatibility condition was removed in [337] but the solutions obtained there are not necessarily smooth up to $\mathscr{P}\Omega$.

The results in Section XV.4 come from a variety of sources; we found [109] and [258] very useful. Although they are mostly well-known by experts in the field, there does not seem to be any other one place where all are collected. As the names suggest, the Newton inequalities (15.27) were first proved by Newton [269] and the Maclaurin inequalities are due to Maclaurin [247].The Newton-Maclaurin inequalities, which are straightforward consequences of these other inequalities, are certainly quite old, but the first reference to them known to this author is the paper of Trudinger [312]. Lemma 15.14 is a special case of Lemma 8 from [126]; we have modified the proof from [239, Lemma 2.4] to make it direct rather than *reductio ad absurdum.* Lemma 15.15 is due to Minkowski [257] (see also Theorem 10 in [109]); our proof is based on Gårding's proof [92] of the more general inequality

$$\sum_{j=1}^{N} a_j D_j P(b) \leq mP(a)^{1/m} P(b)^{(m-1)/m},$$

which holds whenever P is a polynomial in N variables of degree m with is hyperbolic with respect to a and b, and a and b lie in the same component of the set on which $P \neq 0$. (In particular this inequality holds if $P = S_m$ for any $m \leq N$ and any a and b in $\mathscr{K}_k$.) Theorem 15.16 was proved for $m = 1$ by Marcus and Lopes [250] in 1957, and then for general m (without using the result of Marcus and Lopes) by McLeod [254] in 1959. We have followed the proof of Marcus and Lopes for the case $m = 1$ and then that of Bullen and Marcus [26] for the case $m > 1$. That $f = S_k^{1/k}$ and $\mathscr{K} = \mathscr{K}_k$ satisfy (15.4a–d) was proved, via slightly different methods (i.e., Gårding's inequality for concavity of f and using the analyticity of f^k to infer (15.4b)) by Caffarelli, Nirenberg, and Spruck. They first noted condition (15.9c) in [35], and condition (15.22) comes from Ivochkina's work [126]. Theorem 15.18 was proved in [312]. It is useful to notice that (15.22) does not hold for general Hessian quotient equations; see Exercise 15.5.

The parabolic Monge-Ampère equation was first shown to have classical solutions by Reye [281], who proved a slightly different version of Theorem 15.26; he considered noncylindrical domains for which $B\Omega$ consists of only one point. Except for the boundary double normal second derivative estimate, our proof of the estimates follows his, which in turn follows [33]. Similar results (with ψ depending only on X) were also proved by G. Wang [329] and G. Wang and W. Wang [332]. R. H. Wang and G. L. Wang [337] also proved a parabolic version of Caffarelli's result on interior regularity [31] for the elliptic Monge-Ampère equation when ψ is Lipschitz with respect to x and t. Chen, Wang, and Lian [48] further developed the techniques from [337] to study generalized solutions of this equation.

In 1976, Krylov [167] had proposed three alternatives for the parabolic analog of the Monge-Ampère equation:

$$-u_t \det D^2 u = \psi^{n+1}, \tag{15.50a}$$

$$\det D^2 u = [(u_t + \psi)_+]^n, \tag{15.50b}$$

$$\det(D^2u - u_tI) = \psi^n, \tag{15.50c}$$

in suitable classes of functions. We have followed Reye [281] in using the first equation as a model. The reason for our choice was elucidated by Krylov in [167]: the quantity $(-u_t)\det D^2u$ appears naturally in the Krylov-Tso parabolic version of the Bakel′man-Aleksandrov-Pucci maximum principle (see Theorem 7.1 and the notes to that chapter). In the past ten years, considerable work has been done on these alternatives and their generalizations to Hessian equations.

First, Wang and Liu [241] looked at a different Hessian version of (15.50a):

$$-u_tf(\lambda(D^2u)) = \psi(X)$$

with $\psi > 0$ in $\overline{\Omega}$ and Ω a cylindrical domain. They showed existence of solutions (under appropriate curvature conditions) in the obvious class of admissible functions (namely, $-u_t > 0$ and $\lambda(D^2u)$ in a suitable convex symmetric proper subcone of $\mathbb{R}^n$).

Equation (15.50b), when rewritten as

$$-u_t + \left(\det(D^2u)\right)^{1/n} = \psi,$$

has attracted the most attention. Ivochkina and Ladyzhenskaya [131] showed that the Cauchy-Dirichlet problem for this equation has an admissible solution (that is, a convex solution) in a cylindrical domain $\Omega = \omega \times (0,T)$ if ω and $\varphi(\cdot,0)$ are strictly convex, and if either

$$\inf_{\Omega}\psi + \inf_{\mathscr{P}\Omega}\varphi_t > \frac{1}{2}ad^2 \tag{15.51a}$$

or

$$\inf_{\mathscr{P}\Omega}(\psi+\varphi_t) > 0, \quad \psi \text{ and } \left(\det(D^2\varphi(\cdot,0))\right)^{1/n} \text{ are concave,} \tag{15.51b}$$

where $a = \max\{\sup_\Omega g_t, 0\}$ and $d = \operatorname{diam}\Omega$. A straightforward maximum principle argument [131, Theorems 3.1 and 3.2] shows that either of the conditions in (15.51) implies that $u_t + \psi$ has a positive lower bound and hence $\det D^2u$ must be strictly positive. Wang and Wang [338] improved this result by considering generalized solutions. In this way, they were able to relax the compatibility conditions but not (15.51). Another variant of this result was given in [47] by relaxing (15.51).

Ivochkina and Ladyzhenskaya [130] expanded the approach in [131] to examine the equation

$$-u_t + \left[S_k(\lambda(D^2u)\right]^{1/k} = \psi.$$

Now, admissible solutions must have $\lambda(D^2u) \in \mathscr{K}_k$, and conditions (15.51) are only modified in that the determinant of $D^2\varphi$ is replaced by $\left[S_k(\lambda(D^2u)\right]^{1/k}$. The operator $[S_k(\lambda)]^{1/k}$ was replaced in [134] by a general $f(\lambda)$ with f satisfying (15.4a–e) and (15.48). Similar results appear in [242].

A more general class of fully nonlinear operators was introduced by Ivochkina in [128]. They have the form

$$G(a_0(Du)u_t + w(X),\ A(Du)D^2uA(Du) - a_1(Du)u_tI + W(X)) = \psi(X)$$

for suitable scalar functions G, a_0, a_1, and w and symmetric matrix-valued functions A and W. The key assumptions are that a_0 and a_1 are nonnegative with at least one everywhere positive, and A is positive definite. The special case $G(\tau,r) = -\tau + [S_k(\lambda(r))]^{1/k}$ leads to the previously discussed Hessian form of (15.50b), and this general form includes many other equations of interest. As with the Hessian equations in this chapter, a critical assumption is that this equation is only parabolic for certain values of u_t and D^2u; for this reason, Ivochkina calls the equation "nontotally parabolic". In [128], the main interest is with showing that appropriate *a priori* estimates imply the existence of admissible solutions. These estimates for uniformly parabolic equations (which means that the eigenvalues of the matrix

$$\begin{pmatrix} G_r & 0 \\ 0 & -G_\tau \end{pmatrix}$$

lie between some fixed positive multiples of some positive function $\lambda(\tau,r)$ and that the derivatives of the functions involved can be suitably estimated in terms of λ) are derived in [135, 140]. The ideas (but not the estimates) in these papers are further developed in [129].

In a different direction, G. Wang and W. Wang [332] study a different analog of the Monge-Ampère equation,

$$-u_t + \log\det D^2u = \psi(X,u,Du);$$

when ψ is independent of p, this equation arises in the study of certain eigenvalue problems for the Monge-Ampère operator; see [317], in which the restricted parabolic problem is also solved. The corresponding equation for symmetric functions

$$-u_t + \log S_k(\lambda((D^2u)) = \psi(X,u)$$

was discussed by X.-J. Wang [339, Appendix A]. Chou and Wang [53] looked at a variation on this last equation

$$-u_t + f(S_k(\lambda(D^2u))) = \psi(X,u)$$

with f a strictly increasing, strictly concave function which satisfies the additional conditions

$$f(t) = \begin{cases} t^{1/p} & \text{if } t \geq 1, \\ \log t & \text{if } t < t_0, \end{cases}$$

where p and t_0 are positive constants with $p > k$. Trudinger and Wang [314, Section 2] look at

$$-u_t + log\left(\frac{S_k(\lambda(D^2u))}{S_m(\lambda(D^2u))}\right) = \psi(X).$$

Another variation on the class of Hessian equations is the class of curvature equations in which the eigenvalues of the Hessian matrix are replaced by the curvatures of the solution surface. At one extreme is the parabolic mean curvature equation $-u_t/v+\operatorname{div}(Du/v)=H(X,u)$, where $v=(1+|Du|^2)^{1/2}$, which has already been studied in Chapter XII. At the other extreme is the parabolic Gauss curvature equation which can be written as $(-u_t)\det D^2u=\psi(X,u)v^{n+2}$, which was studied in this chapter. We refer the interested reader to [138] for a thorough discussion of the elliptic version of such equations. The underlying method of proof for the estimates is similar to the contents of this chapter, but often a much more detailed analysis is needed. Moreover, many of the ideas used here in studying parabolic Hessian equations are based on a consideration of curvature equations. For example, condition (15.9c) was first considered by Caffarelli, Nirenberg, and Spruck in their study [35] of elliptic curvature equations, and the interest in Hessian quotient equations comes from Trudinger's work on elliptic curvature quotient equations [312, 313]. Although it seems natural from the point of view of this chapter to study the equation $f(\kappa,-u_t/v)=\psi(X,u,Du)$, such research has not yet been carried out. On the other hand, Ivochkina and Ladyzhenskaya [132, 133, 136, 137] have studied the equation $-u_t+f(\kappa)=\psi(X)$, which is the natural generalization of their parabolic Hessian equations.

A quite different approach to global first and second derivative bounds was introduced by Krylov. In the series of papers [175, 176], he was able to prove the elliptic version of many of the estimates in this chapter by considering the equation as a Bellman equation of the form

$$\inf_{\alpha\in I}\{L_\alpha u+f_\alpha\}=0$$

for a suitable family of linear operators (L_α). In particular, this approach works for equations such as

$$S_k(\lambda(D^2u))=\sum_{j=0}^{k-1}c_j(x)S_j(\lambda(D^2u))$$

for appropriate coefficient functions c_k.

Other classes of fully nonlinear equations have also been studied. Ivanov [120] looked at a class of nonuniformly elliptic equations which are not necessarily concave with respect to the r variable. Kutev [181] considered fully nonlinear elliptic equations from the point of view of rewriting the Serrin curvature conditions in terms of their nonlinear analogs. Finally, Laptev [194] studied some nonuniformly parabolic equations in one space dimension.

Finally, we remark that the theory of oblique derivative problems for fully nonlinear parabolic equations is in its infancy. Urbas [325, 326] has shown that only in two space dimensions is there a possibility of obtaining second derivative estimates for solutions of oblique derivative problems for (elliptic) Monge-Ampère and Hessian equations with nonlinear boundary conditions. In this situation, the second derivative bounds are still not simple to prove.

Exercises

15.1 Show that Lemmata 15.5 and 15.7 and Corollary 15.8 (if (15.22) holds) are still valid if we replace the hypothesis of convexity of ψ with respect to p by inequality (15.49) with μ_5 sufficiently small.

15.2 Show that condition (15.9b) in Lemma 15.4 can be removed if f is homogeneous of order 1 and ψ is convex with respect to p. Conclude that Corollaries 15.19 and 15.20 are valid even if we remove the restriction (15.9b). (Compare with the gradient bounds in [106] and [336].)

15.3 Suppose that Ω is a strictly convex-increasing domain with $\mathscr{P}\Omega \in H_a$ for some $a > 2$. If $\varphi \in H_a(\mathscr{P}\Omega)$ and if φ is strictly convex on $B\Omega$, show that φ can be extended to all of Ω so that $D^2\varphi > 0$ and $\varphi_t < 0$.

15.4 Verify the remarks following Theorem 15.26.

15.5 Show that (15.22) fails in case $n = 1$ and $f = S_2/S_1$. Show also that the conditions

$$\lim_{|\lambda|\to\infty} \sum f_i(1+\lambda_i^2) = \infty \text{ in } \mathscr{K}(\delta_0) = \{\lambda \in \mathscr{K} : f(\lambda) \geq \delta_0\}$$

from [106] and

$$\sum f_i\lambda_i^2 \geq c\sum \lambda_i \text{ in } \mathscr{K}(\delta_0)$$

from [244] fail to hold for this example.

Bibliography

[1] D. R. Adams, *Traces of potentials arising from translation invariant operators*, Ann. Scuola Norm. Sup. Pisa **25** (1971), 203–217.

[2] R. A. Adams, *Sobolev Spaces*, Academic Press, New York, 1975.

[3] A. D. Aleksandrov, *Certain estimates for the Dirichlet problem*, Dokl. Akad. Nauk SSSR **134** (1960), 1001–1004 (Russian). English transl. in Soviet Math. Dokl. **1** (1960), 1151–1154.

[4] H. Amann, *Linear and Quasilinear Parabolic Problems*, Birkhäuser, Basel, 1995.

[5] S. Angenent and M. Fila, *Interior blow-up in a semilinear parabolic equation*, Differential Integral Equations **9** (1996), 865–877.

[6] D. E. Apushkinskaya and A. I. Nazarov, *An initial-boundary value problem with a Venstell′ boundary condition for parabolic equations not in divergence form*, Alg. i Anal. **6** (1994), no. 6, 1–29 (Russian). English transl. in St. Petersburg Math. J. **6** (1995), 1127–1149.

[7] ______, *Boundary estimates for the first-order derivatives of a solution to a nondivergent parabolic equation with composite right-hand side and coefficients of lower-order derivatives*, Prob. Mat. Anal. **14** (1995), 3–27 (Russian). English transl. in J. Math. Sci. **77** (1995), 3257–3276.

[8] O. Arena, *Sopra una classe di equazioni paraboliche*, Boll. Unione Mat. Ital. (4) **1** (1969), 9–24.

[9] A. Arkhipova, *On the smoothness of the solution of nonuniformly elliptic equations*, Prob. Mat. Anal. **7** (1979), 14–25 (Russian). English transl. in Selecta Math. Sov. **5** (1986), 127–136.

[10] D. Aronson and J. Serrin, *A maximum principle for nonlinear parabolic equations*, Ann. Scuola Norm. Sup. Pisa **21** (1967), 291–305.

[11] A. Azzam and E. Kreyszig, *Smoothness of solutions of parabolic equations in regions with edges*, Nagoya Math. J. **84** (1981), 159–168.

[12] I. Ya. Bakel′man, *Theory of quasilinear elliptic equations*, Sibirsk. Mat. Zh. **2** (1961), 179–186 (Russian).

[13] R. B. Barrar, *Some estimates for solutions of linear parabolic equations*, Ph.D. Thesis, University of Michigan, Ann Arbor, MI, 1952.

[14] ———, *Some estimates for solutions of parabolic equations*, J. Math. Anal. Appl. **3** (1961), 373–397.

[15] J. Bell, A. Friedman, and A. A. Lacey, *On solutions to a quasi-linear diffusion problem from the study of soft tissue*, SIAM J. Appl. Math. **51** (1991), 484–493.

[16] S. Bernstein, *Sur la généralisation du problème de Dirichlet II*, Math. Ann. **69** (1910), 82–136.

[17] ———, *Sur les équations du calcul des variations*, Ann. Sci. École Norm. Sup. **29** (1912), 431–485.

[18] M. Bertsch, R. Dal Passo, and B. Franchi, *A degenerate parabolic equation in noncylindrical domains*, Math. Ann. **294** (1992), 551–578.

[19] T. Bhattacharya and F. Leonetti, *A new Poincaré inequality and its application to the regularity of minimizers of integral functionals with nonstandard growth*, Nonlinear Anal. **17** (1991), 833–839.

[20] E. Bombieri, E. DeGiorgi, and M. Miranda, *Una maggiorazione a priori relativa alle ipersuperfici minimali non parametriche*, Arch. Rational Mech. Anal. **32** (1969), 255–267.

[21] E. Bombieri and E. Giusti, *Local estimates for the gradient of non-parametric surfaces of prescribed mean curvature*, Comm. Pure Appl. Math. **26** (1973), 381–394.

[22] M. Bourgoing, *$C^{1,\beta}$ regularity of viscosity solutions via a continuous-dependence result*, Adv. Differential Equations **9** (2004), 447–480.

[23] M. Bramanti and M. C. Cerutti, *$w_p^{1,2}$ solvability for the cauchy-dirichlet problem for parabolic equations with VMO coefficients*, Comm. Partial Differential Equations **18** (1993), 1735–1763.

[24] A. Brandt, *Interior Schauder estimates for parabolic differential- (or difference-) equations via the maximum principle*, Israel J. Math. **7** (1969), 254–262.

[25] R. M. Brown, W. Hu, and G. M. Lieberman, *Weak solutions of parabolic equations in noncylindrical domains*, Proc. Amer. Math. Soc. **125** (1997), 1785–1792.

[26] P. S. Bullen and M. Marcus, *Symmetric means and matrix inequalities*, Proc. Amer. Math. Soc. **12** (1961), 285–290.

[27] S.-S. Byun and Lihe Wang, *Parabolic equations in Reifenberg domains*, Arch. Rational Mech. Anal. **176** (2005), 271–301.

[28] X. Cabrè, *On the Alexandroff-Bakelman-Pucci estimate and the reversed Hölder inequality for solutions of elliptic and parabolic equations*, Comm. Pure Appl. Math. **48** (1995), 539–570.

[29] X. Cabré and L. A. Caffarelli, *Interior $C^{2,\alpha}$ regularity theory for a class of nonconvex fully nonlinear elliptic equations*, J. Math. Pures Appl. (9) **82** (2003), 573–612.

[30] L. A. Caffarelli, *Interior estimates for fully nonlinear equations*, Ann. Math. **130** (1989), 189–213.

[31] ______, *Interior $W^{2,p}$ estimates for solutions of the Monge-Ampère equation*, Ann. Math. **131** (1990), 135–150.

[32] L. A. Caffarelli and C. E. Kenig, *Gradient estimates for variable coefficient parabolic equations and singular perturbation problems*, Amer. J. Math. **120** (1998), 391–439.

[33] L. A. Caffarelli, L. Nirenberg, and J. Spruck, *The Dirichlet problem for nonlinear second order elliptic equations I. Monge-Ampère equations*, Comm. Pure Appl. Math. **37** (1984), 369–402.

[34] ______, *The Dirichlet problem for nonlinear second order elliptic equations III. functions of the Hessian*, Acta Math. **155** (1985), 261–301.

[35] ______, *Nonlinear second order elliptic equations V. the Dirichlet problem for Weingarten hypersurfaces*, Comm. Pure Appl. Math. **41** (1988), 47–70.

[36] L. A. Caffarelli and Y. Yuan, *A priori estimates for solutions of fully nonlinear equations with convex level set*, Indiana Univ. Math. J. **49** (2000), 681–695.

[37] A. P. Calderón and A. Zygmund, *On the existence of certain singular integrals*, Acta Math. **88** (1952), 85–139.

[38] ______, *Local properties of solutions of elliptic partial differential equations*, Studia Math. **20** (1961), 171–225.

[39] S. Campanato, *Proprietà di hölderianità di alcune classi di funzioni*, Ann. Scuola Norm. Sup. Pisa **(3) 18** (1964), 175–188.

[40] ______, *Equazioni ellittiche del II° ordine e spazi $\mathscr{L}^{2,\lambda}$*, Ann. Mat. Pura Appl. **(4) 69** (1965), 321–381.

[41] ______, *Equazioni paraboliche del secondo ordine e spazi $\mathscr{L}^{2,\theta}(\Omega,\delta)$*, Ann. Mat. Pura Appl. **(4) 73** (1966), 55–102.

[42] P. Cannarsa, G. DaPrato, and J.-P. Zolèsio, *Evolution equations in noncylindrical domains*, Atti. Naz. Accad. Lincei Rend. Cl. Sci. Fis.Mat. Natur. (8) **83** (1990), 73–77.

[43] J. R. Cannon, *Regularity at the boundary for solutions of linear parabolic differential equations*, Ann. Scuola Norm. Sup. Pisa **(3) 19** (1965), 415–427.

[44] J. Caristi, *Fixed point theorems for mappings satisfying inwardness conditions*, Trans. Amer. Math. Soc. **215** (1976), 241–251.

[45] C. Chen, M. Nakao, and Y. Ohara, *Global existence and gradient estimates for quasilinear parabolic equations of the m-Laplacian type with a strong perturbation*, Adv. Math. Sci. Appl. **10** (2000), 225–237.

[46] ______, *Global existence and gradient estimates for a quasilinear parabolic equation of the mean curvature type with a strong perturbation*, Differential Integral Equations **14** (2001), 59–74.

[47] L. Chen and G. Wang, *Some remarks on the solution of one type of parabolic Monge-Ampère equation*, Chinese Ann. Math. Ser. A **24** (2003), 33–40 (Chinese). English transl. in Chinese J. Contemp. Math. **24** (2003), 35–44.

[48] L. Chen, G. Wang, and S. Lian, *Generalized solution of the first boundary value problem for parabolic Monge-Ampère equation*, J. Partial Differential Equations **14** (2001), 149–162.

[49] Y. Z. Chen, *$C^{1,\alpha}$ regularity of viscosity solutions of fully nonlinear elliptic PDE under natural structure conditions*, J. Partial Differential Equations **6** (1993), no. 3, 193–216.

[50] Y. Z. Chen and E. DiBenedetto, *On the local behavior of solutions of singular parabolic equations*, Arch. Rational Mech. Anal. **103** (1988), 319–345.

[51] ______, *Hölder estimates of solutions of singular parabolic equations with measurable coefficients*, Arch. Rational Mech. Anal. **118** (1992), 257–271.

[52] Y. Z. Chen and S. Zhou, *A note on three classes of nonlinear parabolic equations with measurable coefficients*, J. Partial Differential Equations **8** (1995), 310–320.

[53] K. Chou and X.-J. Wang, *A variational theory of the Hessian equation*, Comm. Pure Appl. Math. **54** (2001), 1029–1064.

[54] C. Ciliberto, *Formule di maggiorazione e teoremi di esistenza per le soluzioni delle equazioni paraboliche in due variabili*, Ricerche Mat. **3** (1954), 40–75.

[55] A. Constantin and J. Escher, *Global solutions for quasilinear parabolic problems*, J. Evol. Equations **2** (2002), 97–111.

[56] M. G. Crandall, P. Fok, M. Kocan, and A. Święch, *Remarks on nonlinear uniformly parabolic equations*, Indiana Math. J. **47** (1998), 1293–1326.

[57] M. G. Crandall, M. Kocan, and A. Święch, *L^p-theory for fully nonlinear uniformly parabolic equations*, Comm. Partial Differential Equations **25** (2000), 1997–2053.

[58] M. G. Crandall, P.-L. Lions, and P. E. Souganidis, *Maximal solutions and universal bounds for some partial differential equations of evolution*, Arch. Rational Mech. Anal. **105** (1989), 163–190.

[59] G. DaPrato, *Spazi $\mathscr{L}^{(p,\theta)}(\omega, \delta)$ e loro proprietà*, Ann. Mat. Pura Appl. **(4) 69** (1965), 383–392.

[60] E. DeGiorgi, *Sulla differenziabilitá e l'analiticita delle estremali degli integrali multipli regolari*, Mem. Accad. Sci. Fis. Mat. Natur. **3** (1957), 25–43.

[61] E. DiBenedetto, *On the local behaviour of solutions of degenerate parabolic equations with measurable coefficients*, Ann. Scuola Norm. Sup. Pisa **13** (1986), 487–535.

[62] ———, *Intrinsic Harnack type inequalities for solutions of certain degenerate parabolic equations*, Arch. Rational Mech Anal. **100** (1988), 129–147.

[63] ———, *Degenerate Parabolic Equations*, Springer-Verlag, Berlin, 1993.

[64] E. DiBenedetto and A. Friedman, *Hölder estimates for nonlinear parabolic degenerate equations*, J. Reine Angew. Math. **357** (1985), 1–22.

[65] E. DiBenedetto and N. S. Trudinger, *Harnack inequalities for quasi-minima of variational integrals*, Ann. Inst. H. Poincarè Anal. Non Linèaire **1** (1984), 295–308.

[66] G. C. Dong, *Initial and nonlinear oblique boundary value problems for fully nonlinear parabolic equations*, J. Partial Differential Equations **1** (1988), 12–42.

[67] ———, *Hölder estimate of quasilinear parabolic equations with nonlinear oblique derivative boundary condition*, J. Partial Differential Equations **3** (1990), 49–53.

[68] ———, *Nonlinear Partial Differential Equations of Second Order*, American Mathematical Society, Providence, RI, 1991.

[69] J. L. Doob, *Classical Potential Theory and its Probabilistic Counterpart*, Springer-Verlag, Berlin, 1984.

[70] K. Ecker, *Estimates for evolutionary surfaces of prescribed mean curvature*, Math. Z. **180** (1982), 179–192.

[71] D. Edmunds and L. A. Peletier, *Quasilinear parabolic equations*, Ann. Scuola Norm. Sup. Pisa **(3) 25** (1971), 397–421.

[72] E. G. Effros and J. L. Kazdan, *On the Dirichlet problem for the heat equation*, Indiana Univ. Math. J. **20** (1971), 683–693.

[73] S. D. Èidel′man, *On fundamental solutions of parabolic systems II*, Mat. Sb. **53 (95)** (1961), 73–136 (Russian). English transl. in Amer. Math. Soc. Transl. (2) **41** (1964), 49–120.

[74] I. Ekeland, *Nonconvex minimization problems*, Bull. Amer. Math. Soc. (N. S.) **1** (1979), 443–474.

[75] L. Escauriaza, *A note on Krylov-Tso's inequality*, Proc. Amer. Math. Soc. **115** (1992), 1053–1056.

[76] L. C. Evans, *Classical solutions of fully nonlinear, convex, second order elliptic equations*, Comm. Pure Appl. Math. **25** (1982), 333–363.

[77] L. C. Evans and R. P. Gariepy, *Wiener's criterion for the heat equation*, Arch. Rational Mech. Anal. **78** (1982), 293–314.

[78] ———, *Measure Theory and Fine Properties of Functions*, CRC Press, Boca Raton, 1992.

[79] L. C. Evans and S. M. Lenhart, *The parabolic Bellman equation*, Nonlinear Anal. **5** (1982), 765–773.

[80] H. Federer and W. Fleming, *Normal and integral currents*, Ann. Math. **72** (1960), 458–520.

[81] M. Fila and G. M. Lieberman, *Derivative blow-up and beyond for quasilinear parabolic equations*, Differential Integral Equations **7** (1994), 811–821.

[82] A. F. Filippov, *Conditions for the existence of a solution of a quasi-linear parabolic equation*, Dokl. Akad. Nauk SSSR **141** (1961), 568–570. English transl. in Soviet Math. Dokl. **2** (1961), 1517–1519.

[83] R. Finn, *On equations of minimal surface type*, Ann. Math. **60** (1954), 397–416.

[84] R. Fiorenza, *Sui problemi di derivata obliqua per le equazioni ellittiche*, Ricerche Mat. **8** (1959), 83–110.

[85] ———, *Sulla hölderianità della soluzioni dei problemi de derivata obliqua regolare del secondo ordine*, Ricerche Mat. **14** (1965), 102–123.

[86] ———, *Sui problemi di derivata obliqua per le equazioni ellittiche quasi lineare*, Ricerche Mat. **15** (1966), 74–108.

[87] A. Friedman, *Boundary estimates for second order parabolic equations and their applications*, J. Math. Mech. **7** (1958), 771–792.

[88] ———, *Interior estimates for parabolic systems of partial differential equations*, J. Math. Mech. **7** (1958), 393–418.

[89] ———, *Partial Differential Equations of Parabolic Type*, Prentice-Hall, Englewood Cliffs, N. J., 1964. Reprinted by Krieger, Malabar, FL, 1983.

[90] K. Friedrichs, *On the differentiability of the solutions of linear elliptic differential equations*, Comm. Pure Appl. Math. **6** (1953), 299–326.

[91] L. Gårding, *Dirichlet's problem for linear elliptic partial differential equations*, Math. Scand. **1** (1953), 55–72.

[92] ———, *An inequality for hyperbolic polynomials*, J. Math. Mech. **8** (1959), 957–965.

[93] E. Gagliardo, *Problema al contorno per equazioni differentiali lineari di tipo parabolico in n variabili*, Richerche Mat. **5** (1956), 239–257.

[94] R. Gariepy and W. P. Ziemer, *Thermal capacity and boundary regularity*, J. Differential Equations **45** (1982), 374–388.

[95] M. G. Garroni and J. L. Menaldi, *Green Functions for Second Order Parabolic Integro-differential Problems*, Longman Scientific and Technical, Harlow, Essex, 1992.

[96] C. Gerhardt, *Evolutionary surfaces of prescribed mean curvature*, J. Differential Equations **36** (1980), 139–172.

[97] M. Giaquinta and E. Giusti, *Global $C^{1,\alpha}$-regularity for second order quasilinear elliptic equations in divergence form*, J. Reine Angew. Math. **351** (1984), 55–65.

[98] Y. Giga, *Interior derivative blow-up for quasilinear parabolic equations*, Discrete and Continuous Dynamical Systems **1** (1995), 449–461.

[99] D. Gilbarg, *Boundary value problems for nonlinear elliptic equations in n variables*, Nonlinear Problems, 1963, pp. 151–159.

[100] D. Gilbarg and L. Hörmander, *Intermediate Schauder estimates*, Arch. Rational Mech. Anal. **74** (1980), 297–318.

[101] D. Gilbarg and N. S. Trudinger, *Elliptic Differential Equations of Second Order*, Springer-Verlag, Berlin, 2001. Reprint of the third edition (1998).

[102] B. H. Gilding, *Hölder continuity of solutions of parabolic equations*, J. London Math. Soc. **13** (1976), 103–106.

[103] E. Giusti, *Boundary behavior of nonparametric minimal surfaces*, Indiana U. Math. J. **22** (1975), 435–444.

[104] R. Ya. Glagoleva, *A priori estimate of the Hölder norm and the Harnack inequality for the solution of a second order linear parabolic equation with discontinuous coefficients*, Mat. Sb. (N.S.) **76 (118)** (1968), 167–185 (Russian). English transl. in Math. USSR-Sb. **5** (1968), 159–176.

[105] M. Gruber, *Harnack inequalities for solutions of general second order parabolic equations and estimates of their Hölder constants*, Math. Z. **185** (1984), 23–43.

[106] B. Guan, *The Dirichlet problem for a class of fully nonlinear elliptic equations*, Comm. Partial Differential Equations **19** (1994), 399–416.

[107] C. E. Gutíerrez, *The Monge-Ampère Equation*, Progress in Nonlinear Differential Equations and their Applications, vol. 44, Birkhäuser, Boston, 2001.

[108] G. Han, $C^{1,\alpha}$ *regularity of viscosity solutions of fully nonlinear parabolic PDE under natural structure conditions*, J. Partial Differential Equations **12** (1999), 1–10.

[109] G. H. Hardy, J. E. Littlewood, and G. Pòlya, *Inequalities*, Cambridge University Press, London, 1934.

[110] K. Hayasida and Y. Ikeda, *Dirichlet problem for evolutionary surfaces of prescribed mean curvature in a non-convex domain*, J. Math. Soc. Japan **56** (2004), 1169–1202.

[111] D. Hilbert, *Über das Dirichletsche Prinzip*, Jber. Deutsch. Math.-Verein. **8** (1900), 184–188.

[112] J. van der Hoek, *Existence via interior estimates for second order parabolic equations*, Proc. Centre Math. Anal. **8** (1984), 221–232.

[113] G. Huisken, *Non-parametric mean curvature evolution with boundary conditions*, J. Differential Equations **77** (1989), 369–375.

[114] M. Iannelli and G. Vergara Caffarelli, *On the boundary value problem for surfaces of prescribed mean curvature*, Sympos. Math. **14** (1974), 473–480.

[115] A. M. Il′in, A. S. Kalashnikov, and O. A. Oleinik, *Second order linear equations of parabolic type*, Uspekhi Mat. Nauk **17** (1962), 3–146 (Russian). English transl. in Russian Math. Surveys **17** (1962), 1–143.

[116] A. V. Ivanov, *Interior estimates of the first derivatives of the solutions of quasilinear non-uniformly elliptic and non-uniformly parabolic equations*, Zap. Nauchn. Sem. Leningrad. Otdel. Math. Inst. Steklov. (LOMI)

14 (1969), 24–74 (Russian). English transl. in Sem. V. A. Sem. Steklov Math. Inst. Leningrad **14** (1971), 9–21.

[117] ———, *The first boundary value problem for second order quasi-linear parabolic equations*, Zap. Nauchn. Sem. Leningrad. Otdel Mat. Inst. Steklov. (LOMI) **38** (1973), 10–32 (Russian). English transl. in J. Soviet Math. **8** (1977), 354–372.

[118] ———, *Quasilinear degenerate and non-uniformly elliptic and parabolic equations of second order*, Trudy Mat. Inst. Steklov. **160** (1982), 1–285 (Russian). English transl. in Proc. Steklov Inst. Math. **160** (1982), 1–287.

[119] ———, *Second-order nonlinear nonuniformly elliptic equations*, Zap. Nauchn. Sem. Leningrad. Otdel. Mat. Inst. Steklov. (LOMI) **138** (1984), 35–64 (Russian). English transl. in J. Soviet Math.**32** (1986), 448–469.

[120] ———, *On classical solvability of the Dirichlet problem for nonlinear, nonuniformly elliptic equations*, Trudy Mat. Inst. Steklov. **179** (1989), 54–79 (Russian). English transl. in Proc. Steklov Inst. Math. **179** (1989), 55–83.

[121] A. V. Ivanov, N. M. Ivochkina, and O. A. Ladyzhenskaya, *On a strengthening of results on the solvability of the Dirichlet problem for uniformly elliptic equations of second order*, Vestnik Leningrad. Univ **1975** (1975), no. 1, 61–72 (Russian). English transl. in Vestnik Leningrad Univ. **8** (1980), 51–64.

[122] N. M. Ivochkina, *The integral method of barrier functions and the Dirichlet problem for equations with operators of Monge-Ampère type*, Mat. Sb. **112 (154)** (1980), 193–206 (Russian). English transl. in Math. USSR- Sb. **40** (1981), 179–192.

[123] ———, *A priori estimates of* $\|u\|_{C^2(\overline{\Omega})}$ *of convex solutions of the Dirichlet problem for the Monge-Ampère equation*, Zap. Nauchn. Sem. Leningrad. Otdel. Mat. Inst. Steklov. (LOMI) **96** (1980), 69–79 (Russian). English transl. in J. Soviet Math. **21** (1983), 689–697.

[124] ———, *Classical solvability of the Dirichlet problem for the Monge-Ampère equation*, Zap. Nauchn. Sem. Leningrad. Otdel. Mat. Inst. Steklov. (LOMI) **131** (1983), 72–79 (Russian). English transl. in J. Soviet Math. **30** (1985), 2287–2292.

[125] ———, *A description of the stability cones generated by differential operators of Monge-Ampère type*, Mat. Sb. **122 (164)** (1983), 265–275 (Russian). English transl. in Math. USSR- Sb. **50** (1985), 259–268.

[126] ———, *Solution of the Dirichlet problem for some equations of Monge-Ampère type*, Mat. Sb. **128** (1985), 403–415 (Russian). English transl. in Math. USSR-Sb. **56** (1987), 403–415.

[127] ———, *The Dirichlet problem for the curvature equation of order m*, Algebra i Anal. **2** (1991), 192–217 (Russian). English transl. in Lenigrad Math. J. **2** (1991), 631–654.

[128] ______, *On the Dirichlet problem for fully nonlinear parabolic equations*, Zap. Nauchn. Sem. S.-Peterburg. Otdel. Mat. Inst. Steklov. (POMI) **233** (1996), 101–111, 256 (Russian). English transl. in J. Math. Sci. **93** (1999), 698–696.

[129] ______, *On some directions in the development of the theory of second-order fully nonlinear evolution equations*, Proceedings of the St. Petersburg Mathematical Society, Vol. 8, 2000, pp. 121–151 (Russian). English transl. in American Mathematical Society Translations, Series 2, Vol. 205 (2002), 105–131.

[130] N. M. Ivochkina and O. A. Ladyzhenskaya, *On parabolic problems generated by some symmetric functions of the Hessian*, Top. Meth. Nonlinear Anal. **4** (1994), 19–29.

[131] ______, *Parabolic equations generated by symmetric functions of the eigenvalues of the Hessian or by the principal curvatures of a surface. I. Parabolic Monge-Ampère equations*, Algebra i Analiz **6** (1994), no. 3, 141–160 (Russian). English transl. in St. Petersburg Math. J. **6** (1995), 575–594.

[132] ______, *Estimation of the second derivatives for surfaces evolving under the action of their principal curvatures*, Topol. Methods Nonlinear Anal. **6** (1995), no. 2, 265–282.

[133] ______, *The first initial-boundary value problem for evolution equations generated by symmetric functions of the principal curvatures*, Dokl. Akad. Nauk **340** (1995), 155–157 (Russian). English transl. in Russian Akad Sci. Dokl. Math. **51** (1995), 33–35.

[134] ______, *Flows generated by symmetric functions of the eigenvalues of the Hessian*, Zap. Nauchn. Sem. S.-Peterburg. Otdel. Mat. Inst. Steklov. (POMI) **221** (1995), 127–144, 258 (Russian). English transl. in J. Math. Sci. **87** (1997), 3353–3365.

[135] ______, *Estimates for first-order derivatives for solutions of some classes of non-totally parabolic equations*, Alg. i Anal. **9** (1997), no. 6, 109–131 (Russian). English transl. in St. Petersburg Math. J. **9** (1998), 1111-1128.

[136] ______, *Estimation of the second derivatives on the boundary for surfaces evolving under the action of their principal curvatures*, Alg. i Anal. **9** (1997), no. 2, 30–50 (Russian). English transl. in St. Petersburg Math. J. **9** (1998), 199-217.

[137] ______, *On classical solvability of the first initial-boundary value problem for equations generated by curvatures*, Topol. Methods Nonlinear Anal. **11** (1998), no. 2, 375–395.

[138] N. M. Ivochkina, M. Lin, and N. S. Trudinger, *The Dirichlet problem for the prescribed curvature quotient equation with general boundary values*, Geometric Analysis and the Calculus of Variations, 1996, pp. 125–141.

[139] N. M. Ivochkina and A. P. Oskolkov, *Nonlocal estimates for the first derivatives of solutions of the first boundary problem for certain classes of nonuniformly elliptic and nonuniformly parabolic equations and systems*, Trudy Mat. Inst. Steklov. **110** (1970), 65–101 (Russian). English transl. in Proc. Steklov Inst. Math. **110** (1970), 72–115.

[140] N. M. Ivochkina and S. I. Prokofieva, *A priori estimate for the second-order derivatives of solutions to the Dirichlet problem for fully nonlinear parabolic equations*, Problem. Mat. Anal. **16** (1997), 112–133. English transl. in J. Math. Sci. (New York) **92** (1998), 4302–4315.

[141] H. Jenkins and J. Serrin, *The Dirichlet problem for the minimal surface equation in higher dimensions*, J. Reine Angew. Math. **229** (1968), 170–187.

[142] L. I. Kamynin, *A boundary value problem in the theory of heat conduction with a nonclassical boundary condition*, Zh. Vychisl. Mat. i Mat. Fiz. **4** (1962), 1006–1024 (Russian). English transl. in USSR Comput. Math. Math. Phys. **4** (1962), no.6, 33–59.

[143] ———, *On the smoothness of heat potentials I. Heat potentials on a surface of type* $\Lambda^{0,1,(1+\alpha)/2}_{1,\alpha,\alpha/2}$, Differentsial'nye Uravneniya **1** (1965), 799–839 (Russian). English transl. in Differential Equations **1** (1965), 613–647.

[144] ———, *On the smoothness of heat potentials III. The special heat potential* $P(x,t)$ *of the simple layer on surfaces of type* $\Lambda^{0,2,(1+\alpha)/2}_{1,\alpha,\alpha/2}$ *and* $\Lambda^{1,\alpha,\alpha/2}_{1,1,(1+\alpha)/2}$, Differentsial'nye Uravneniya **2** (1966), 1333–1357,1484–1501 (Russian). English transl. in Differential Equations **2**, (1966), 690–701, 767–775.

[145] ———, *On the smoothness of heat potentials II. Heat potentials on a surface of type* $\Lambda^{1,\alpha,\alpha/2}_{1,1,(1+\alpha)/2}$, Differentsial'nye Uravneniya **2** (1966), 647–687 (Russian). English transl. in Differential Equations **2** (1966), 337–356.

[146] ———, *The maximum principle and boundary* α*-estimates of the solution of the first boundary value problem for a parabolic equation in a non-cylindrical domain*, Zh. Vychisl. Mat. i Mat. Fiz. **7** (1967), 551–567 (Russian). English transl. in USSR Comput. Math. Math. Phys. **7** (1967), no. 3, 104–127.

[147] ———, *Smoothness of heat potentials VI. The special heat potentials* P *and* Q *on surfaces of type* $\Lambda^{m+1,\alpha,\alpha/2}_{2m+1,1,(1+\alpha)/2}$ *and* $\Lambda^{m+1,1,(1+\alpha)/2}_{2m+3,\alpha,\alpha/2}$, Differentsial'nye Uravneniya **4** (1968), 1867–1891, 2034–2055 (Russian). English transl. in Differential Equations **4** (1968), 962–974, 1051–1061.

[148] ———, *The smoothness of heat potentials V. The heat potentials* U, V, *and* W *on surfaces of the type* $\Lambda^{m+1,\alpha,\alpha/2}_{2m+1,1,(1+\alpha)/2}$ *and* $\Lambda^{m+1,1,(1+\alpha)/2}_{2m+3,\alpha,\alpha/2}$, Differentsial'nye Uravneniya **4** (1968), 347–365, 881–895 (Russian). English transl. in Differential Equations **4** (1968), 185–195, 459–465.

[149] ______, *Solution of the boundary-value problem of the fifth kind for second-order parabolic equations in noncylindrical domains*, Sibirsk. Mat. Zh. **9** (1968), 1153–1166 (Russian). English transl. in Siberian Math. J. **9** (1968), 859-870.

[150] L. I. Kamynin and B. N. Khimchenko, *The principle of the maximum and boundary Lipschitz bounds for the solution of a second order elliptic-parabolic equation*, Sibirsk. Mat. Zh. **15** (1974), 343–367 (Russian). English transl. in Siberian Math. J. **15** (1974), 242–260.

[151] ______, *The maximum principle and local Lipschitz estimates near the lateral boundary for the solutions of a second order parabolic equation*, Sibirsk. Mat. Zh. **16** (1975), 1172–1187 (Russian). English transl. in Siberian Math. J. **16** (1975), 897–909.

[152] ______, *A Nadirashvili-type theorem for a second order parabolic equation with nonnegative characteristic form*, Differentsial′nye Uravneiya **27** (1986), 52–66 (Russian). English transl. in Differential Equations **27** (1986), 511–523.

[153] L. I. Kamynin and V. N. Maslennikova, *The solution of the first boundary value problem in the large for a quasilinear parabolic equation*, Dokl. Akad. Nauk SSSR **137** (1961), 1049–1052 (Russian). English transl. in Soviet Math. Dokl. **2** (1962), 405–408.

[154] ______, *The solution of the first boundary value problem for a quasilinear parabolic equation in a noncylindrical region*, Mat. Sb. **57 (99)** (1962), 241–264 (Russian).

[155] ______, *Boundary estimates of the Schauder type for the solution of the problem with oblique derivative for a parabolic equation in a noncylindrical domain*, Sibirsk. Mat. Zh. **7** (1966), 83–128 (Russian). English transl. in Siberian Math. J. **7** (1966), 65–103.

[156] B. Kawohl and N. Kutev, *Global behavior of solutions to a parabolic mean curvature equation*, Differential Integral Equations **8** (1995), 1923–1946.

[157] J. L. Kazdan, *Prescribing the Curvature of a Riemannian Manifold*, CBMS Regional Conference Series in Mathematics, vol. 57, American Mathematical Society, 1985.

[158] W. A. Kirk and J. Caristi, *Mapping theorems in metric and Banach spaces*, Bull. Acad. Polon. Sci. Sèr. Math. Astronom. Phys. **23** (1979), 891–894.

[159] B. Knerr, *Parabolic interior Schauder estimates by the maximum principle*, Arch. Rational Mech. Anal. **75** (1980), 51–58.

[160] H. Koch, *On fully nonlinear mixed parabolic problem with oblique boundary conditions*, Universität Heidelberg, 1995.

[161] N. J. Korevaar, *Maximum principle gradient estimates for the capillary problem*, Comm. Partial Differential Equations **13** (1988), 1–31.

[162] H. Krüger, *Lineare parabolische Differentialgleichungen auf nichtzylindrischen Gebieten*, Ph.D. Thesis, Universität Augsburg, Augsburg, Germany, 1984.

[163] S. N. Kruzhkov, *Nonlinear parabolic equations in two independent variables*, Trudy Moskov. Mat. Obshch. **16** (1967), 329–346 (Russian). English transl. in Trans. Moscow Math. Soc. **16** (1967), 355-373.

[164] ______, *Remarks concerning the nature of the continuity of solutions of parabolic equations and some of their applications*, Mat. Zametki **6** (1969), 97–108 (Russian). English transl. in Math. Notes **6** (1969), 517–523.

[165] S. Kruzhkov, A. Castro, and M. Lopez, *Mayoraciones de Schauder y teorema de existencia de las soluciones del problema de Cauchy para ecuaciones parabolicas lineali y no lineali (I)*, Cienc. Mat. (Havana) **1** (1980), 55–76.

[166] ______, *Mayoraciones de Schauder y teorema de existencia de las soluciones del problema de Cauchy para ecuaciones parabolicas lineali y no lineali (II)*, Cienc. Mat. (Havana) **3** (1982), 37–56.

[167] N. V. Krylov, *Sequences of convex functions and estimates of the maximum of the solution of a parabolic equation*, Sibirsk. Math. Zh. **17** (1976), 290–303 (Russian). English transl. in Siberian Math. J. **17** (1976), 226–236.

[168] ______, *On the maximum principle for nonlinear parabolic and elliptic equations*, Izv. Akad. Nauk SSSR **42** (1978), 1050–1062 (Russian). English transl. in Math. USSR-Izv. **13** (1979), 335–347.

[169] ______, *Boundedly inhomogeneous elliptic and parabolic equations*, Izv. Akad. Nauk SSSR **46** (1982), 487–523 (Russian). English transl. in Math. USSR-Izv. **20** (1983), 459–493.

[170] ______, *Boundedly inhomogeneous elliptic and parabolic equations in a domain*, Izv. Akad. Nauk SSSR **47** (1983), 75–108 (Russian). English transl. in Math. USSR-Izv. **22** (1984), 67–98.

[171] ______, *On estimates for the derivatives of solutions of nonlinear parabolic equations*, Dokl. Akad. Nauk. SSSR **274** (1984), 23–26 (Russian). English transl. in Soviet Math. Dokl. **29**, (1984), 14–17.

[172] ______, *Nonlinear Elliptic and Parabolic Equations*, "Nauka", Moscow, 1985. English transl. D. Reidel, Dordrecht, Holland, 1987.

[173] ______, *Barriers for derivatives of solutions of nonlinear elliptic equations on a surface in Euclidean space*, Comm. Partial Differential Equations **19** (1994), 1909–1944.

[174] ______, *Weak interior second order derivative estimates for degenerate nonlinear elliptic equations*, Differential Integral Equations **7** (1994), 133–156.

[175] ______, *On the general notion of fully nonlinear elliptic equations*, Trans. Amer. Math. Soc. **347** (1995), 857–895.

[176] ———, *A theorem about degenerate elliptic Bellman equations in bounded domains*, Differential Integral Equations **8** (1995), 961–980.

[177] ———, *On the Calderón-Zygmund theorem with applications to parabolic equations*, Algebra i Analiz **13** (2001), no. 4, 1–25 (Russian). English transl. in St. Petersburg Math. J. **13** (2002), 509–526.

[178] ———, *The Calderón-Zygmund theorem and parabolic equations in $L_p(\mathbb{R}, C^{2+\alpha})$-spaces*, Ann. Scuola Norm. Sup. Pisa **(5)1** (2002), 799–820.

[179] ———, *Parabolic equations in L_p spaces with mixed norms*, Algebra i Analiz **14** (2002), no. 4, 91–106 (Russian). English transl. in St. Petersburg Math. J. **14** (2003), 603–614.

[180] N. V. Krylov and M. V. Safonov, *Certain properties of solutions of parabolic equations with measurable coefficients*, Izv. Akad. Nauk **44** (1980), 161–175 (Russian). English transl. in Math. USSR-Izv. **16** (1981), 155–164.

[181] N. Kutev, *Existence and nonexistence of classical solutions of the Dirichlet problem for a class of fully nonlinear nonuniformly elliptic equations*, Banach Center Publ. **19** (1987), 169–178.

[182] O. A. Ladyzhenskaya, *Solution of the first boundary value problem in the large for quasilinear parabolic equations*, Trudy Moskov. Math. Obšč. **7** (1958), 149–177 (Russian).

[183] O. A. Ladyzhenskaya, V. A. Solonnikov, and N. N. Ural′tseva, *Linear and Quasilinear Equations of Parabolic Type*, Izdat. "Nauka", Moscow, 1967 (Russian). English transl. : American Mathematical Society, Providence R. I., 1968.

[184] O. A. Ladyzhenskaya and N. N. Ural′tseva, *A boundary value problem for linear and quasilinear parabolic equations*, Izv. Akad. Nauk SSSR Ser. Mat. **26** (1962), 5–52 (Russian). English transl. in Amer. Math. Soc. Transl. (2) **47** (1965), 217–267.

[185] ———, *A boundary value problem for linear and quasilinear parabolic equations II*, Izv. Akad. Nauk SSSR Ser. Mat. **26** (1962), 753–780 (Russian). English transl. in Amer. Math. Soc. Transl. (2) **47** (1965), 268–299.

[186] ———, *A boundary value problem for linear and quasilinear parabolic equations III*, Izv. Akad. Nauk SSSR Ser. Mat. **27** (1963), 161–240 (Russian). English transl. in Amer. Math. Soc. Transl. (2) **56** (1966), 103–192.

[187] ———, *Global estimates of first derivatives of the solutions of quasilinear elliptic and parabolic equations*, Zap. Nauchn. Sem. Leningrad. Otdel. Mat. Inst. Steklov (LOMI) **14** (1969), 127–155 (Russian). English transl. in Sem. V. A. Steklov Math. Inst. Leningrad **14** (1971), 63–77.

[188] ———, *Local estimates for gradients of solutions of non-uniformly elliptic and parabolic equations*, Comm. Pure Appl. Math. **23** (1970), 677–703.

[189] ———, *Estimates of* $\max|u_x|$ *for the solutions of quasilinear elliptic and parabolic equations of the general form and existence theorems*, Zap. Nauchn. Sem. Leningrad. Otdel. Mat. Inst. Steklov. (LOMI) **138** (1984), 90–107 (Russian). English transl. in J. Soviet Math. **32** (1986), 486–499.

[190] ———, *Estimates for the Hölder constant for functions satisfying a uniformly elliptic or a uniformly parabolic quasilinear inequality with unbounded coefficients*, Zap. Nauchn. Sem. Leningrad. Otdel Mat. Inst. Steklov. (LOMI) **147** (1985), 72–94 (Russian). English transl. in J. Soviet Math. **37** (1987), 837–851.

[191] ———, *Estimates on the boundary of a domain for the first derivatives of functions satisfying an elliptic or parabolic inequality*, Trudy Mat. Inst. Steklov. **179** (1988), 102–125 (Russian). English transl. in Proc. Steklov Inst. Math. **179** (1989), 109–135.

[192] E. M. Landis, *s-capacity and its applications to the study of solutions of a second-order elliptic equation with discontinuous coefficients*, Mat. Sbornik **76 (118)** (1968), 186–213 (Russian). English transl. in Math. USSR Sb. **5** (1968), 177–204.

[193] ———, *Second Order Equations of Elliptic and Parabolic Type*, Nauka, Moscow, 1971 (Russian). English transl. American Mathematical Society, Providence, R. I., 1998.

[194] G. I. Laptev, *The first boundary value problem for a second order nonlinear parabolic equation with one space dimension*, Izv. Vyssh. Uchebn. Zaved. Mat. **32** (1988), no. 5, 84–86 (Russian). English transl. in Soviet Math. (Iz. VUZ) **6** (1988), no. 5, 109–112.

[195] H. Lebesgue, *Sur le problème de Dirichlet*, Rend. Circ. Mat. Palermo **24** (1907), 371–402.

[196] J. Leray and J. Schauder, *Topologie et équations fonctionelles*, Ann. Sci. École Norm. Sup. **51** (1934), 45–78.

[197] H. A. Levine, *The role of critical exponents in blowup theorems*, SIAM Rev. **32** (1990), 262–288.

[198] J. L. Lewis and M. A. Murray, *The method of layer potentials for the heat equation in time-varying domains*, Memoirs Amer. Math. Soc. **114** (1995), 1–157.

[199] A. Lichnewsky and R. Temam, *Pseudosolutions of the time-dependent minimal surface problem*, J. Differential Equations **30** (1978), 340–364.

[200] G. M. Lieberman, *The quasilinear Dirichlet problem with decreased regularity at the boundary*, Comm. Partial Differential Equations **6** (1981), 437–497.

[201] ———, *The Dirichlet problem for quasilinear elliptic equations with Hölder continuous boundary values*, Arch. Rational Mech. Anal. **79** (1982), 305–323.

[202] ———, *Solvability of quasilinear elliptic equations with nonlinear boundary conditions*, Trans. Amer. Math. Soc. **273** (1982), 753–765.

[203] ———, *The conormal derivative problem for elliptic equations of variational type*, J. Differential Equations **49** (1983), 281–257.

[204] ———, *Interior gradient bounds for non-uniformly parabolic equations*, Indiana Univ. Math. J. **32** (1983), 579–601.

[205] ———, *The nonlinear oblique derivative problem for quasilinear elliptic equations*, Nonlinear Anal. **8** (1984), 49–65.

[206] ———, *Solvability of quasilinear elliptic equations with nonlinear boundary conditions. II*, J. Funct. Anal. **56** (1984), 210–219.

[207] ———, *The Perron process applied to oblique derivative problems*, Advances in Math. **55** (1985), 161–172.

[208] ———, *Regularized distance and its applications*, Pac. J. Math. **117** (1985), 329–352.

[209] ———, *The Dirichlet problem for quasilinear elliptic equations with continuously differentiable boundary data*, Comm. Partial Differential Equations **11** (1986), 167–229.

[210] ———, *The first initial-boundary value problem for quasilinear second order parabolic equations*, Ann. Scuola Norm. Sup. Pisa **(4) 13** (1986), 347–387.

[211] ———, *Global regularity of solutions of nonlinear second order elliptic and parabolic differential equations*, Math. Z. **193** (1986), 331–346.

[212] ———, *Intermediate Schauder theory for second order parabolic equations II. Existence, uniqueness, and regularity*, J. Differential Equations **63** (1986), 32–57.

[213] ———, *Intermediate Schauder theory for second order parabolic equations I. Estimates*, J. Differential Equations **63** (1986), 1–31.

[214] ———, *Quasilinear elliptic equations with nonlinear boundary conditions*, Proc. Symp. Pure Math. **45** (1986), 113–117.

[215] ———, *Regularity of solutions of nonlinear elliptic boundary value problems*, J. Reine Angew. Math. **369** (1986), 1–13.

[216] ———, *Gradient bounds for solutions of nonuniformly elliptic oblique derivative problems*, Nonlinear Anal. **11** (1987), 49–61.

[217] ———, *Hölder continuity of the gradient of solutions of uniformly parabolic equations with conormal boundary conditions*, Ann. Mat. Pura Appl. **148** (1987), 77–99, 397–398.

[218] ———, *Local estimates for subsolutions and supersolutions of oblique derivative problems for general second order elliptic equations*, Trans. Amer. Math. Soc. **304** (1987), 343–353.

[219] ———, *The conormal derivative problem for non-uniformly parabolic equations*, Indiana Univ. Math. J. **37** (1988), 23–72.

[220] ______, *Gradient estimates for capillary-type problems via the maximum principle*, Comm. Partial Differential Equations **13** (1988), 33–59.

[221] ______, *Intermediate Schauder theory for second order parabolic equations III. The tusk condition*, Appl. Anal. **33** (1989), 25–43.

[222] ______, *Optimal Hölder regularity for mixed boundary value problems*, J. Math. Anal. Appl. **143** (1989), 572–586.

[223] ______, *Boundary regularity for solutions of degenerate parabolic equations*, Nonlinear Anal. **14** (1990), 501–524.

[224] ______, *Initial regularity for solutions of degenerate parabolic equations*, Nonlinear Anal. **14** (1990), 525–536.

[225] ______, *On the Hölder gradient estimate for solutions of nonlinear elliptic and parabolic oblique boundary value problems*, Comm. Partial Differential Equations **15** (1990), 515–523.

[226] ______, *The natural generalization of the natural conditions of Ladyzhenskaya and Ural′tseva for elliptic equations*, Comm. Partial Differential Equations **16** (1991), 311–361.

[227] ______, *The conormal derivative problem for equations of variational type in nonsmooth domains*, Trans. Amer. Math. Soc. **330** (1992), 41–67.

[228] ______, *Existence of solutions of the first initial-boundary value problem for parabolic equations via elliptic regularization*, Progress in Partial Differential Equations: Elliptic and Parabolic Problems, 1992, pp. 193–206.

[229] ______, *Intermediate Schauder theory for second order parabolic equations IV. Time irregularity and regularity*, Differential Integral Equations **5** (1992), 1219–1236.

[230] ______, *Boundary and initial regularity for solutions of degenerate parabolic equations*, Nonlinear Anal. **20** (1993), 557–569.

[231] ______, *Sharp form of estimates for subsolutions and supersolutions of quasilinear elliptic equations involving measures*, Comm. Partial Differential Equations **18** (1993), 1191–1212.

[232] ______, *Maximum estimates for solutions of degenerate parabolic equations in divergence form*, J. Differential Equations **113** (1994), 543–571.

[233] ______, *Time-periodic solutions of linear parabolic differential equations*, Comm. Partial Differential Equations **24** (1999), 631–663.

[234] ______, *The maximum principle for equations with composite coefficients*, Electronic J. Differential Equations **2000** (2000), 1–17.

[235] ______, *Pointwise estimates for oblique derivative problems in nonsmooth domains*, J. Differential Equations **173** (2001), 178–211.

[236] ______, *Higher regularity for nonlinear oblique derivative problems in nonsmooth domains*, Ann. Scuola Norm Sup. Pisa **1** (2002), 111–151.

[237] ______, *A mostly elementary proof of Morrey space estimates for elliptic and parabolic equations with VMO coefficients*, J. Funct. Anal. **201** (2003), 457–479.

[238] G. M. Lieberman and N. S. Trudinger, *Nonlinear oblique boundary value problems for fully nonlinear elliptic equations*, Trans. Amer. Math. Soc. **295** (1986), 509–546.

[239] Mi Lin, *On the solvability of some curvature quotient equations*, Ph.D. Thesis, Northwestern University, Evanston, Illinois, 1994.

[240] P.-L. Lions, N. S. Trudinger, and J. I. E. Urbas, *The Neumann problem for equations of Monge-Ampère type*, Comm. Pure Appl. Math. **39** (1986), 539–563.

[241] H. Liu and G. Wang, *Some results on evolutionary equations involving functions of the eigenvalues of the Hessian*, Northeast. Math. J. **13** (1997), 433–448.

[242] ______, *The first initial-boundary value problem for complete nonlinear parabolic equations generated by the eigenvalues of a Hessian matrix*, Acta Sci. Natur. Univ. Jilin. (1998), no. 1, 27–36.

[243] S. Li and X. Wang, *An upper estimate of solution for a general class of parabolic equations*, J. Partial Differential Equations **13** (2000), 289–300.

[244] Y. Y. Li, *Some existence results for fully nonlinear elliptic equations of Monge-Ampère type*, Comm. Pure Appl. Math. **43** (1990), 233–271.

[245] Y. Y. Li and L. Nirenberg, *The distance function to the boundary, Finsler geometry, and the singular set of viscosity solutions of some Hamilton-Jacobi equations*, Comm. Pure Appl. Math. **58** (2005), 85–146.

[246] A. Lunardi, *Analytic Semigroups and Optimal Regularity in Parabolic Equations*, Birkhäuser, Basel, 1995.

[247] C. Maclaurin, *A second letter to Martin Folkes, Esq.; concerning the roots of equations, with the demonstration of other rules in algebra*, Phil. Transactions **36** (1729), 59–96.

[248] P. Marcellini and K. Miller, *Asymptotic growth for the parabolic equations of prescribed mean curvature*, J. Differential Equations **51** (1984), 326–358.

[249] J. Marcinkiewicz, *Sur l'interpolation d'opérations*, C. R. Acad. Sci. Paris **208** (1939), 1272–1273.

[250] M. Marcus and L. Lopes, *Symmetric functions and Hermitian matrices*, Canad. J. Math. **9** (1957), 305–312.

[251] M. I. Matiĭčuk and S. D. Èidel'man, *On parabolic systems with coefficients satisfying a Dini condition*, Dokl. Akad. Nauk SSSR **165** (1965), 482–485 (Russian). English transl. in Soviet Math. Dokl. **6** (1965), 1451–1464.

[252] ______, *The correctness of the problem of Dirichlet or Neumann for second-order parabolic equations with coefficients in Dini classes*, Ukr. Mat. Zh. **26** (1974), 328–337 (Russian). English transl. in Ukr. Math. J. **26** (1974), 269–276.

[253] A. Maugeri, D. K. Palagachev, and L. G. Softova, *Elliptic and parabolic equations with discontinuous coefficients*, Wiley-VCH, Berlin, 2000.

[254] J. B. McLeod, *On four inequalities in symmetric functions*, Proc. Edinburgh Math. Soc. **11** (1959), 211–219.

[255] J. H. Michael, *A general theory for linear elliptic partial differential equations*, J. Differential Equations **23** (1977), 1–29.

[256] J. H. Michael and L. M. Simon, *Sobolev and mean-value inequalities on generalized submanifolds of* $\mathbb{R}^n$, Comm. Pure Appl. Math. **26** (1973), 361–379.

[257] H. Minkowski, *Geometrie der Zahlen, I*, Leipzig, 1896.

[258] D. S. Mitrinovic, *Analytic Inequalities*, Springer-Verlag, Berlin, 1970.

[259] C. B. Morrey Jr, *On the solutions of quasi-linear elliptic partial differential equations*, Trans. Amer. Math. Soc. **43** (1938), 126–166.

[260] ———, *Multiple Integrals in the Calculus of Variations*, Springer-Verlag, Berlin, 1966.

[261] J. Moser, *A new proof of de Giorgi's theorem concerning the regularity problem for elliptic differential equations*, Comm. Pure Appl. Math. **13** (1960), 457–468.

[262] ———, *A Harnack inequality for parabolic differential equations*, Comm. Pure Appl. Math. **17** (1964), 101–134.

[263] N. Nadirashvili, *A lemma on the inner derivative, and the uniqueness of solutions of the solutions of the second boundary value problem for second order elliptic equations*, Dokl. Akad. Nauk SSSR **261** (1981), 804–808 (Russian). English transl. in Soviet Math. Dokl. **24** (1981), 598–601.

[264] J. Nash, *Continuity of solutions of parabolic and elliptic equations*, Amer. J. Math. **80** (1958), 931–954.

[265] A. I. Nazarov, *Hölder estimates for bounded solutions of problems with an oblique derivative for parabolic equations of nondivergence structure*, Probl. Math. Anal. **11** (1990), 37–46 (Russian). English transl. in J. Soviet Math. **64** (1993), 1247–1252.

[266] A. I. Nazarov and N. N. Ural′tseva, *Convex-monotone hulls and an estimate of the maximum of the solutions of a parabolic equation*, Zap. Nauchn. Sem. Leningrad. Otdel. Mat. Inst. Steklov. (LOMI) **147** (1985), 95–109 (Russian). English transl. in J. Soviet Math. **37** (1987), 851–859.

[267] ———, *A problem with an oblique derivative for a quasilinear parabolic equation*, Zap. Nauchn. Sem. S.-Petersburg. Otdel. Mat. Inst. Steklov. **200** (1992), 118–131 (Russian). English transl. in J. Math. Sci. **77** (1995), 1587-1593.

[268] J. Nečas, *Sur une mèthode pour résoudre les équations aux dérivées partielles du type elliptiques, voisine de la variationelle*, Ann. Scuola Norm. Sup. Pisa **(3) 16** (1962), 305–326.

[269] I. Newton, *Arithmetica universalis: sive de compositione et resolutione arithmetica liber. [Opera, I.]*

[270] L. Nirenberg, *A strong maximum principle for parabolic equations*, Comm. Pure Appl. Math. **6** (1953), 167–177.

[271] O. A. Oleinik and S. N. Kruzhkov, *Quasilinear second-order parabolic equations with many independent variables*, Usp. Mat. Nauk **16** (1961), no. 5, 115–156 (Russian). English transl. in Russian Math. Surveys **16** (1961), no. 5, 105–146.

[272] O. A. Oleinik and S. N. Venstell′, *The first boundary value problem and cauchy's problem for quasi-linear equations of parabolic type*, Mat. Sb. (N. S.) **41 (83)** (1957), 105–128 (Russian).

[273] V. Oliker and N. N. Ural′tseva, *Evolution of nonparametric surfaces with speed depending on curvature. II. The mean curvature case*, Comm. Pure Appl. Math. **46** (1993), 97–135.

[274] L. Orsina and M. M. Porzio, *$L^\infty(Q)$ estimates and existence of solutions for some nonlinear parabolic equations*, Boll. Un. Mat. Ital. **B(7)-6** (1992), 631–647.

[275] D. K. Palagachev, M. A. Ragusa, and L. G. Softova, *Cauchy-Dirichlet problem in Morrey spaces for parabolic equations with discontinuous coefficients*, Boll. Unione Mat. Ital. Sez. B Artic. Ric. Mat. (8) **6** (2003), 667–683.

[276] C. V. Pao, *Nonlinear Parabolic and Elliptic Equations*, Plenum, New York, 1992.

[277] I. M. Petrushko, *On boundary and initial conditions in $\mathscr{L}_p$, $p > 1$, of solutions of parabolic equations*, Mat. Sb. **125 (167)** (1984), 489–521 (Russian). English transl. in Math. USSR-Sb. **53** (1986), 489–522.

[278] M. H. Protter and H. F. Weinberger, *Maximum Principles in Differential Equations*, Prentice-Hall, Englewood Cliffs, N. J., 1967. Reprinted by Springer-Verlag, Berlin, 1984.

[279] C. Pucci, *Limitazioni per soluzioni di equazioni ellitiche*, Ann. Math. Pura Appl. **(4) 74** (1966), 15–30.

[280] J. M. Rakotoson and W. P. Ziemer, *Local behavior of solutions of quasilinear elliptic equations with general structure*, Trans. Amer. Math. Soc. **319** (1990), 747–764.

[281] S. J. Reye, *Fully Nonlinear Parabolic Differential Equations of Second Order*, Ph.D. Thesis, Australian National University, Canberra, ACT, Australia, 1986.

[282] M. V. Safonov, *On the classical solvability of Bellman's elliptic equations*, Dokl. Akad. Nauk SSSR **278** (1984), 810–813 (Russian). English transl. in Soviet Math. Dokl. **30** (1984), 482–485.

[283] ———, *On the classical solution of nonlinear elliptic equations of second order*, Izv. Akad. Nauk SSSR **52** (1988), 1272–1287 (Russian). English transl. in Math. USSR Izv. **33** (1989), 597–612.

[284] ______, *Nonlinear Elliptic Equations of Second Order*, Dipartimento di Matematica Applicata "G. Sansome", Università degli Studie di Firenza, 1991.

[285] ______, *On the boundary value problems for fully nonlinear elliptic equations of second order*, Technical Report MRR 049-94, Australian National University, 1994.

[286] ______, *On the oblique derivative problem for second order elliptic equations*, Comm. Partial Differential Equations **20** (1995), 1349–1367.

[287] J. Schauder, *Der Fixpunktsatz in Funktionalräumen*, Studia Math. **2** (1930), 171–180.

[288] ______, *Über das Dirichletsche Problem im Grossen für nicht-lineare elliptische Differentialgliechungen*, Math. Z. **37** (1933), 623–634, 768.

[289] F. Schulz and G. Williams, *Barriers and existence results for a class of equations of mean curvature type*, Anal. **7** (1987), 359–374.

[290] J. Serrin, *The problem of Dirichlet for quasilinear elliptic differential equations with many independent variables*, Philos. Trans. Roy. Soc. London Ser. A **264** (1969), 413–496.

[291] ______, *Gradient estimates for solutions of nonlinear elliptic and parabolic equations*, Contributions to Nonlinear Functional Analysis, 1971, pp. 565–601.

[292] L. M. Simon, *Interior gradient bounds for non-uniformly elliptic equations*, Indiana Univ. Math. J. **25** (1976), 821–855.

[293] L. M. Simon and J. Spruck, *Existence and regularity of a capillary surface with prescribed contact angle*, Arch. Rational Mech. Anal. **61** (1976), 19–34.

[294] L. G. Softova, *Oblique derivative problem for parabolic operators with VMO coefficients*, Manuscripta Math. **103** (2000), 203–220.

[295] ______, *Parabolic equations with VMO coefficients in Morrey spaces*, Electron. J. Differential Equations **2001** (2001), no. 51, 1–25.

[296] V. A. Solonnikov, *On boundary value problems for linear parabolic systems of differential equations of general form*, Trudy Mat. Inst. Steklov. **83** (1965), 3–163 (Russian). English transl. in Proc. Math. Inst. Steklov **83** (1966), 1-184.

[297] E. Sperner Jr., *Schauder's existence theorem for α-Dini continuous data*, Ark. Mat. **19** (1981), 193–216.

[298] J. Spruck, *On the existence of a capillary surface with prescribed contact angle*, Comm. Pure Appl. Math. **28** (1975), 189–200.

[299] E. M. Stein, *Singular Integrals and Differentiability Properties of Functions*, Princeton University Press, Princeton, N. J., 1970.

[300] K. Sturm, *Harnack's inequality for parabolic operators with singular low order terms*, Math. Z. **216** (1994), 593–611.

[301] A. Tersenov and A. Tersenov, *Global solvability for a class of quasilinear parabolic problems*, Indiana U. Math. J. **50** (2004), 1899–1912.

[302] F. Treves, *Basic Linear Partial Differential Equations*, Academic Press, New York, 1975.

[303] N. S. Trudinger, *Some existence theorems for quasi-linear, non-uniformly elliptic equations in divergence form*, J. Math. Mech. **19** (1968/69), 909–919.

[304] ———, *Pointwise estimates and quasilinear parabolic equations*, Comm. Pure Appl. Math. **21** (1968), 205–226.

[305] ———, *Lipschitz continuous solutions of elliptic equations of the form* $\mathscr{A}(Du)D^2u = 0$, Math. Z. **109** (1969), 211–216.

[306] ———, *On the regularity of generalized solutions of linear, non-uniformly elliptic equations*, Arch. Rational Mech. Anal. **42** (1971), 50–62.

[307] ———, *The boundary gradient estimate for quasilinear elliptic and parabolic differential equations*, Indiana Univ. Math. J. **21** (1972), 657–670.

[308] ———, *Local estimates for subsolutions and supersolutions of general second order elliptic quasilinear equations*, Invent. Math **61** (1980), 67–79.

[309] ———, *Fully nonlinear, uniformly elliptic equations under natural structure conditions*, Trans. Amer. Math. Soc. **278** (1983), 217–231.

[310] ———, *Boundary value problems for fully nonlinear elliptic equations*, Proc. Centre for Math. Anal. **8** (1984), 65–83.

[311] ———, *Lectures on Nonlinear Second Order Elliptic Equations*, Nankai Institute of Mathematics, Tianjin, China, 1985.

[312] ———, *The Dirichlet problem for the prescribed curvature equations*, Arch. Rational Mech. Anal. **111** (1990), 153–170.

[313] ———, *On the Dirichlet problem for Hessian equations*, Acta Math. **175** (1995), 151–164.

[314] N. S. Trudinger and X.-J. Wang, *A Poincaré inequality for Hessian integrals*, Calc. Var. Partial Differential Equations **6** (1998), 313–328.

[315] N. S. Trudinger, *Lipschitz lecture notes*. per oral communication.

[316] K. Tso, *On an Aleksandrov-Bakel′man type maximum principle for second-order parabolic equations*, Comm. Partial Differential Equations **10** (1985), 543–553.

[317] ———, *On a real Monge-Ampère functional*, Inv. Math. **101** (1990), 425–448.

[318] N. N. Ural′tseva, *Nonlinear boundary value problems for equations of minimal-surface type*, Trudy Mat. Inst. Steklov. **116** (1971), 217–226 (Russian). English transl. in Proc. Steklov Inst. Math. **116** (1971), 227–237.

[319] ———, *Solvability of the capillary problem*, Vestnik Leningrad. Univ. **1973** (1973), no. 19, 53–64 (Russian). English transl. in Vestnik Leningrad Univ. **6** (1979), 363–375.

[320] ______, *Solvability of the capillary problem II*, Vestnik Leningrad. Univ. **1975** (1975), no. 1, 363–375 (Russian). English transl. in Vestnik Leningrad Univ. **8** (1980), 151–158.

[321] ______, *Estimates for the maxima of the moduli of the gradients for solutions of capillarity problems*, Zap. Nauchn. Sem. Leningrad. Otdel. Mat. Inst. Steklov. (LOMI) **115** (1982), 274–284 (Russian). English transl. in J. Soviet Math. **28** (1985), 806–813.

[322] ______, *Gradient estimates for solutions of nonlinear parabolic oblique derivative problems*, Technical Report 64-89, Australian National University, 1989.

[323] ______, *A nonlinear problem with an oblique derivative for parabolic equations*, Zap. Nauchn. Sem. Leningrad. Otdel. Mat. Inst. Steklov. (LOMI) **188** (1991), 143–158 (Russian). English transl. in J. Math. Sci. **70** (1994), 1817–1827.

[324] ______, *Gradient estimates for solutions of nonlinear parabolic oblique boundary problem*, Geometry and Nonlinear Partial Differential Equations, 1992, pp. 119–130.

[325] J. Urbas, *The oblique derivative problem for equations of Monge-Ampère type*, Proc. Centre for Math. Anal. **12** (1987), 171–195.

[326] ______, *Nonlinear oblique boundary value problems for Hessian equations in two dimensions*, Ann. Inst. H. Poincaré Anal. Non Linéaire **12** (1995), 507–575.

[327] V. Vespri, *L^∞ estimates for nonlinear parabolic equations with natural growth conditions*, Rend. Sem. Mat. Univ. Padova **90** (1993), 1–8.

[328] G. Wang, *Harnack inequalities for functions in De Giorgi parabolic class*, Lecture Notes in Math. **1306** (1986), 182–201.

[329] ______, *The first boundary value problem for parabolic Monge-Ampère equation*, Northeastern Math. J. **3** (1987), 463–478.

[330] G. Wang and F. Liu, *The Hölder estimate near the boundary for solutions of general second order parabolic quasilinear equations with measurable coefficients*, Northeast. Math. J. **4** (1988), 417–431.

[331] G. Wang and A. X. Sun, *Weak Harnack inequality for functions from class $\mathscr{B}_2$ and application to parabolic Q-minima*, Chinese J. Contemp. Math. **14** (1993), no. 1, 75–84.

[332] G. Wang and W. Wang, *The first boundary value problem for general parabolic Monge-Ampère equation*, J. Partial Differential Equations **3** (1990), no. 2, 1–15.

[333] Lihe Wang, *On the regularity of fully nonlinear parabolic equations I*, Comm. Pure Appl. Math. **45** (1992), 27–76.

[334] ______, *On the regularity theory of fully nonlinear parabolic equations II*, Comm. Pure Appl. Math. **45** (1992), 141–178.

[335] ______, *A geometric approach to the Calderón-Zygmund estimates*, Acta Math. Sin, English Ed. **19** (2003), 381–396.

[336] Lijuan Wang, *The Dirichlet problems for a class of fully nonlinear elliptic equations relative to the eigenvalues of the Hessian*, J. Partial Differential Equations **5** (1992), 53–68.

[337] R. H. Wang and G. Wang, *On existence, uniqueness, and regularity of viscosity solutions for the first initial-boundary value problems to parabolic Monge-Ampère equation*, Northeast. Math. J. **8** (1992), 417–446.

[338] ______, *An improvement of a result of Ivochkina and Ladyzhenskaya on a type of parabolic Monge-Ampère equation*, Chinese Ann. Math. Ser. B **18** (1997), 405–422.

[339] X.-J. Wang, *A class of fully nonlinear elliptic equations and related functionals*, Indiana Univ. Math. J. **43** (1994), 25–53.

[340] P. Weidemaier, *On the sharp initial trace of functions with derivatives in* $L^q(0,T;L^p(\omega))$, Boll. Un. Mat. Ital. **9-B** (1995), 321–338.

[341] ______, *On L^p-estimates of optimal type for the parabolic oblique derivative problem with VMO coefficients*, Differential Integral Equations **18** (2005), 935–946.

[342] W. Wieser, *Parabolic Q-minima and minimal solutions to variational flow*, Manuscripta Math. **59** (1987), 63–107.

[343] H.-M. Yin, *$\mathscr{L}^{2,\mu}$-estimates for parabolic equations and applications*, J. Partial Differential Equations **10** (1997), 31–44.

[344] J. Yong, *Weak solutions of second order parabolic equations in noncylindrical domains*, J. Partial Differential Equations **2** (1989), no. 2, 76–86.

[345] Y. Yuan, *A priori estimates for solutions of fully nonlinear special Lagrangian equations*, Ann. Inst. H. Poincaré Anal. Non Linéaire **18** (2001), 261–270.

[346] Q. Zhang, *On a parabolic equation with a singular lower order term*, Trans. Amer. Math. Soc. **348** (1996), 2811–2844.

[347] N. Zhu, *$C^{1+\alpha}$ regularity of viscosity solutions for fully nonlinear parabolic equations*, Nonlinear Anal. **38** (1999), 977–994.

[348] W. P. Ziemer, *Weakly Differentiable Functions*, Springer-Verlag, Berlin, 1989.

Index